# ELECTROLYTES

# ELECTROLYTES

## Supramolecular Interactions and Non-Equilibrium Phenomena in Concentrated Solutions

Georgii Georgievich Aseyev

CRC Press is an imprint of the
Taylor & Francis Group, an **informa** business

CRC Press
Taylor & Francis Group
6000 Broken Sound Parkway NW, Suite 300
Boca Raton, FL 33487-2742

First issued in paperback 2021



Version Date: 20141014

ISBN 13: 978-1-03-223603-2 (pbk)
ISBN 13: 978-1-4822-4938-5 (hbk)



**Publisher's Note**
The publisher has gone to great lengths to ensure the quality of this reprint but points out that some imperfections in the original copies may be apparent.

**Library of Congress Cataloging-in-Publication Data**

Aseev, G. G. (Georgii Georgievich)
Electrolytes : supramolecular interactions and non-equilibrium phenomena in concentrated solutions / Georgii Georgievich Aseyev.
pages cm
"A CRC title."
Includes bibliographical references and index.
ISBN 978-1-4822-4938-5 (hardcover : alk. paper) 1. Electrolyte solutions. 2. Electrolytes. 3. Supramolecular electrochemistry. I. Title.

QD565.A834 2015
541'.372--dc23 2014039479

**Visit the Taylor & Francis Web site at**
**http://www.taylorandfrancis.com**

**and the CRC Press Web site at**
**http://www.crcpress.com**

# Contents



## SECTION I Supramolecular Interactions

# *SECTION II Non-Equilibrium Phenomena*

# List of Tables

# Preface

Though existing empirical and semiempirical calculation methods allow high-precision determination of physical and chemical properties of concentrated multicomponent aqueous solutions of electrolytes, they require mathematical treatment of enormous experimental material, and do not give any idea about structural conditions or the nature of interactions in a host–guest (water–ion) solution [1–10].

Various early *physical* concepts (also involving the methods of statistical theory [11]) considered a certain regular order of distribution of substrates in a host, and the concept of ion atmosphere and its thickness were the fundamental points [12,13]. Ion atmosphere meant that every substrate happened to be surrounded by a certain density of electric charges. It was believed that the properties of electrolyte solutions depended only on the ratio of electrostatic forces of ionic fields, conditioned by their charges. Such concepts correctly depict quite dilute electrolyte solutions.

Further, the theory of intermolecular interactions [14–21] began to develop intensively. Proceeding from the ideas of classical and statistical mechanics and quantum chemistry, various thermodynamic ratios of intermolecular interactions were obtained, which take into account mainly pair-wise ion–ion, ion–dipole, and dipole–dipole interactions. Even if electrostatic interactions were taken into account, and Coulomb's law was used, it was not quite correct, since it does not describe the properties of stationary electrostatic fields. It only determined the force of interaction between ions, and force is not field (see Chapter 2). The Debye–Hückel theory was used in many cases, which describes the electrostatic properties of quite dilute electrolyte solutions. Much more complicated interactions occur in an electrolyte solution: the electrolyte molecule may split into a cation and an anion, which then interact with each other and with water dipoles, solvation occurs, the water structure is disrupted, solvated water dipoles form complexes already in a dilute solution, which also interact with each other with rising intensity as the electrolyte concentration increases, and finally leading to the formation of aquatic complexes and ligands The dielectric constant of a solution decreases with increased electrolyte concentration, and this significantly influences the structure formation processes in a solution. The structural state of aqueous electrolyte solutions cannot be described by the theory of intermolecular interactions, and that is not the purpose of this branch.

## SECTION I: SUPRAMOLECULAR INTERACTIONS

Chapter 1 briefly considers the views of a number of scientists from the early nineteenth century on the processes of ions solvation in electrolyte solutions, and further the evolution of the concepts on these processes and the structural states of solutions. Early approaches allowed inconsistency in the application of electrostatics formulas. Most of these formulas are valid only for distances larger than the size of the interacting particles. This condition is not met in concentrated electrolyte solutions.

In all considered and similar variants of theories, thermodynamic functions of solvation depend on a large number of parameters that could be chosen at random. For parameters such as dipole moment and polarizability, numerical values were used, which were obtained from experiments on pure gases and liquids, though it is known that these parameters have other values in a strong field of ions. The fact that all authors used different numerical values for these parameters seems interesting, but oddly enough, they obtained satisfactory results consistent with experimental data. Insufficient accuracy of the experimental data did not allow drawing a conclusion about the correctness of a particular calculation method. That dielectric permittivity $\varepsilon_S$ in a solution dramatically decreases compared with pure water was completely disregarded.

Since the second half of the twentieth century, the problem of structure formation in electrolyte solutions began to be intensively studied not only by experimental methods, but with the involvement of quantum-chemical and quantum-statistical methods as well. This chapter considers the main approaches to address this problem, and conclusions are drawn, that for now, a priori they cannot give positive results when compared with the experimental data for concentrated electrolyte solutions.

Chapter 2 analyzes the steps of forming approaches to the idea of long-range supermolecular forces. To date, the main electrostatic doctrine is the Debye–Hückel theory, which, though having been modernized by a number of researchers (Gronwall, La-Mer, Sandved, Onsager, Guggenheim, Falkenhagen, Mikulin, Fuoss, etc.), still contains some approximations with the original task settings. Ion atmosphere is not able to describe the whole concentration area of an electrolyte solution. Structure formation from dilute solutions to concentrated ones shall be considered from the perspective of three mutually intersecting concentration zones, and it is necessary to take into account the decrease in a solution's dielectric permittivity with an increase in electrolyte concentration.

The chapter under consideration sets out the principles of the supramolecular-thermodynamic approach, on the basis of which further theoretical presentations are made. For the correct description of the whole concentration area of an electrolyte solution, various fluctuations of ion concentrations are taken into account, near the one which is considered by Maxwell–Boltzmann, Fermi–Dirac, and Bose–Einstein quantum statistics, on the basis of which a new quantum statistics for concentrated solutions is obtained. There is proof of inconsistency of applying only one Maxwell–Boltzmann statistics for concentrated electrolyte solutions, which is actively applied today in various quantum-chemical or quantum-statistical theories. The function of distribution of supramolecular host–guest structure is derived, which is the basis for further presentation.

Chapter 2 also deals with the derivation of interparticle interactions potential. Taking into account the fact that the unit volume of electrolyte solution contains electric charges of ions that are immobile, the ratio between the electric field intensity created by these charges and the charge density is described by Maxwell's equation for electrostatic fields. There is proof of inconsistency of applying Coulomb's law to describe electrostatic interactions of equilibrium systems between ions in a solution. This law describes the force of interaction between charges, but not the electrostatic field properties, as it is postulated by Maxwell's equations.

In addition to electrostatic interactions, the interparticle interactions potential takes into account the supramolecular interactions (various ion–dipole and dipole–dipole interactions), van der Waals' forces (orientation interaction [Keesom's effect], induction interaction [Debye's effect], and dispersion interaction [London's effect]), and short-range forces (supramolecular weak hydrogen bonds associated with the Pauli principle manifestation, and short-range repulsion forces). All these interactions allow describing the whole concentration area of an electrolyte solution from the perspective of three mutually overlapping areas.

Chapter 3 is devoted to the development of thermodynamics of super- and supramolecular interactions. Taking into account that activity coefficients are a real physical quantity, they are defined by various methods, and they give results coinciding well with each other, which indicates the self-consistency of these ideas, the thermodynamic formula for activity coefficients calculation is derived. The activity theory provides an extremely convenient tool for processing of experimental data and for finding new regularities. The formula for activity coefficients determination is based on the interparticle interactions potential. Further, a number of necessary formulas for the practical calculation of potential are derived: minimum distance of interaction of electrostatic forces between ions in an electrolyte solution, distance between ions in a solution, and dielectric permittivity of a solution.

Chapter 4 presents statistical averaging and derivation of various formulas for the calculation of ion–dipole and dipole–dipole interactions. To date, the Boltzmann statistics have been commonly used during statistical averaging. As mentioned earlier, in Chapter 2, we conclude that the Boltzmann statistics are not suitable for electrolyte solutions and that wrong theoretical results can be obtained while using them. Quantum statistic obtained by us in Chapter 2 adequately describes the distribution of particles in an electrolyte solution in the first concentration area, which then dramatically decreases, which is a correct interpretation of experimental results. In the second concentration area, ionic or ion-aqueous clusters (supermolecules) are formed and no statistical distribution already exists; the third concentration area is not discussed here. Proceeding from such a complicated distribution of interparticle interactions, only energy values were used during statistical averaging. Proceeding from these statements, statistical formulas were derived to calculate dipole–dipole, quadrupole–quadrupole, stable dipole–induced dipole, and induced dipole–induced dipole interactions. Multipole moments of higher orders were not considered. The ion–dipole interactions in the first approximation of perturbation theory were considered: ion–dipole, ion–quadrupole, orientation and deformation of water dipoles in ion field, ion–dipole dispersion (London's), and ion–dipole polarization (induction). Numerical evaluation of ion–dipole, multipole, induction and dispersion forces and their contribution to solution properties is performed.

Further, attention is paid to weak chemical interactions in an electrolyte solution: weak hydrogen bond and repulsive interactions. Statistical averaging of weak hydrogen bond formula is performed.

Basic information about repulsion potential is obtained by quantum-chemical calculations, as well as during experimental research of scattered molecular beams. Short-range repulsion forces, which appear at close distances between chemically saturated molecules (in the simplest case between atoms with filled electron shells), have as a prototype the interaction between two hydrogen atoms in the ground state with parallel electron spins, and are associated with the Pauli principle manifestation. As the calculations show, the potential of repulsion between atoms with filled electron shells is the sum of terms of polynomial type series. On this basis, many kinds of repulsion potentials arose, and some of them have been investigated. In the author's judgment, the potential of the point center of repulsion, which has only one configurable constant, has been the most appropriate for the whole concentration area of an electrolyte and meets all the set requirements. Using this repulsion potential, adjustment of interparticle interactions potential was performed for a number of electrolytes according to activity coefficients data.

Next, the thermodynamics of water activity interaction is developed. Based on the Gibbs–Duhem equation, a thermodynamic formula for water activity calculation is obtained, in which the interparticle interactions potential has been used.

## SECTION II: NON-EQUILIBRIUM PHENOMENA

Chapter 5 presents a historical review of electricity concepts from the first half of the nineteenth century until now. It is shown that the foundation of the modern theory of electrolytic dissociation is defined by two statements: (1) Arrhenius' hypothesis of electrolytic dissociation and (2) the relation between dissociation degree and electrolyte concentration in a solution through the law of mass action (Ostwald's dilution law). Modern theories of electrical conductivity are analyzed, the majority of which include ideas of the classical Debye–Hückel theory of ion atmosphere and electrochemical mobilities of ions, and also the theories of semiconductors, ion plasma, methods of nuclear magnetic resonance, electron paramagnetic resonance, relaxation radiospectroscopy, analytical ab initio researches, and a number of other theories. But the problem of electrical conductivity in solutions was never fully resolved. It is currently important to define the concentration dependence of electrical conductivity in a wide range of concentrations from the perspective of supermolecular interactions outlined in Section I.

For this purpose, the differential equation of continuity for electrolyte solutions in a perturbed state is derived and solved, which takes into account all kinds of supermolecular interactions and structure formation processes. The obtained dependences allow drawing a conclusion about the mechanism of electrical conductivity, which differs from the generally accepted views.

Chapter 6 is devoted to the theoretical consideration of viscosity movement in concentrated solutions, for which an analysis of semiempirical and theoretical approaches to this phenomenon has been performed. Problems of the theory of viscosity, its mechanism, dependence on external factors, and internal characteristics of a system are still controversial.

The author analyzed and theoretically investigated a new model of solution shear flow. The differential equation of continuity of solution flow, which takes into account all kinds of supermolecular formations, is obtained and solved. The new model of solution shear flow allowed performing a comparatively simple theoretical solving of all necessary equations.

Chapter 7 considers isothermal isobaric molecular diffusion without gradients conditioned by external force fields. An analysis of theoretical approaches of diffusion processes shows that there are no phenomenological models to date, which would realize continuous transition in terms of any of their parameters from one model to another. Predictive capabilities of existing models are limited by their narrow field of applicability.

A theoretical consideration of diffusion parameters is made, which takes into account the role of effective super- and supramolecular interactions, the nonlocal nature of interactions, and the influence of geometric dimensions of particles and dielectric permittivity of a solution on the formation of thermodynamic and kinetic properties of electrolytic solutions. Concepts of the structure formation processes in a solution are discussed.

The equations obtained in Sections I and II are illustrated with numerous diagrams of various kinds of super- and supramolecular interactions, and tables comparing the obtained dependences with experimental data.

All comments and suggestions will be greatly appreciated by the author.

## REFERENCES

1. Aseyev G.G., Ryshchenko I.M., Savenkov A.S. 2007. Equations and determination of physical and chemical properties of solutions of sulfate-nitrate of an ammonium. *Zhurn. Prikl. Khimii.* 80 (2): 213–220.
2. Aseyev G.G. 2001. *Electrolytes: Methods for Calculation of the Physicochemical Parameters of Multicomponent Systems.* New York: Begell House, Inc. Publishers, p. 368.
3. Aseyev G.G., Ryshchenko I.M., Savenkov A.S. 2005. *Electrolytes. Physical and Chemical Parameters of Concentrated Multicomponent Systems.* Kharkov: NTU "HPI," p. 448.
4. Aseyev G.G., I.D. Zaytsev. 1992. *Properties of Aqueous Solutions of Elektrolites.* Boca Raton, FL: CRC Press, p. 1774.
5. Aseyev G.G., Zaytsev I.D. 1996. *Volumetric Properties of Elektrolite Solutions: Estimation Methods and Experimental Data.* New York: Begell House, Inc. Publishers, p. 1572.
6. Aseyev G.G. 1996. *Thermal Properties of Elektrolite Solutions. Methods for Calculation of Multicomponent Systems and Experimental Data.* New York: Begell House, Inc. Publishers, p. 498.
7. Aseyev G.G. 1999. *Electrolytes: Properties of Solutions Methods for Calculation of Multicomponent Systems and Experimental Data on Thermal Conductivity and Surface Tension.* New York: Begell House, Inc. Publishers, p. 522.
8. Aseyev G.G. 1998. *Electrolytes: Transport Phenomena. Calculation of Multicomponent Systems and Experimental Data on Electrical Conductivity.* New York: Begell House, Inc. Publishers, p. 612.
9. Aseyev G.G. 1998. *Electrolytes: Equilibria in Solutions and Phase Equilibria: Calculation of Multicomponent Systems and Experimental Data on the Activities of Water, Pressure Vapor and Osmotic Coefficients.* New York: Begell House, Inc. Publishers, p. 758.
10. Aseyev G.G. 1998. *Electrolytes: Transport Phenomena. Methods for Calculation of Multicomponent Solutions, and Experimental data on Viscosities and Diffusion Coefficients.* New York: Begell House, Inc. Publishers, p. 548.

11. Yukhnovsky I.R., Golovko M.F. 1980. *Head Statistical Theory of Classical Equilibrium Systems*. Kiev, Ukraine: Naukova Dymka, p. 372.
12. Harned G., Owen B. 1952. *Physical Chemistry of Solutions of Electrolytes*. Moscow, Russia: Publishing House of the Foreign Lit., p. 628.
13. Robinson R., Stokes R. 1963. *Solutions of Electrolytes*. Moscow, Russia: Publishing House of the Foreign Lit., pp. 1–2.
14. Prezhdo V.V., Kraynov N. 1994. *Molecular Interactions and Electric Properties of Molecules*. Kharkov, Ukraine: Publishing house "Osnova", p. 240.
15. Simkin B.Yu., Sheykhet I.I. 1989. *Quantum Chemical and Statistical Theory of Solutions. Computing Methods and Their Application*. Moscow, Russia: Chemistry, p. 256.
16. Shakhparonov M.I. 1976. *Introduction in the Modern Theory of Solutions (Intermolecular Interactions. Structure. Prime Liquids)*. Moscow, Russia: The Highest School, p. 296.
17. Eremin V.V., Kargov S.I., Uspenskaya I.A. et al. 2005. *Fundamentals of Physical Chemistry. Theory and Tasks*. Moscow, Russia: Examination, p. 480.
18. Girshfelder Dzh. Curtice Ch., Byrd R. 1961. *Molecular Theory of Gases and Liquids*. Moscow, Russia: Publishing House of the Foreign Lit., p. 930.
19. Kaplan I.G. 1982. *Introduction to the Theory of Intermolecular Interactions*. Moscow, Russia: Nauka, p. 312.
20. Barash Yu. S. 1988. *van der Waals Forces*. Moscow, Russia: Nauka, p. 344.
21. Smirnova N. A. 1987. *Molecular Theories of Solutions*. Leningrad, Russia: Chemistry, p. 336.

# Author

**Georgii Georgievich Aseyev** is head of the Department of Information Technology in the Kharkov State Academy of Culture, Kharkov, Ukraine. He earned his master's degree in the A.M. Gorkiy Kharkov State University. He earned candidate of engineering sciences and doctor of engineering sciences degrees from the Kharkov Scientific-Research and Design Institute of Inorganic Chemistry. He is a professor and an academician at the International Informatization Academy under UN jurisdiction. He teaches in the graduate department of the National Polytechnic Institute in the Department of Technology of Inorganic Substances. He conducts courses related to physical and theoretical chemistry of concentrated electrolyte solutions and mathematical modeling of technological inorganic processes. Dr. Aseyev is an author of *Properties of Aqueous Solutions of Electrolytes* published by CRC Press. Professor Aseyev's scientific interests are in the areas of theoretical chemistry, nanotechnologies, supramolecular interactions in concentrated electrolyte solutions, and artificial intelligence. He has written hundreds of magazine articles and dozens of monographs and textbooks. He is a member of the editorial staff of various magazines, and a member of the Scientific Council that awards the Doctor of Sciences degree.

# *Section I*

## *Supramolecular Interactions*

# 1 Historic Introduction

## 1.1 EARLY VIEWS ON THE PROCESS OF ION SOLVATION IN ELECTROLYTE SOLUTIONS

The structure of an electrolyte solution, to a great extent, is formed by solvation processes that are described by various thermodynamic relations engaging proofs obtained from all kinds of radiophysical measuring methods. Theoretical calculation methods of thermodynamic functions of ion solvation can be broken down into two groups. The first group involves calculations where a solvent is considered as a continuous dielectric. Historically, these calculations were the first attempts to understand and to calculate the ion solvation process. These works belong to Born [1], Webb [2,3], Grahame [4], Hasted et al. [5], Frank [6], Hush [7], and others. In this chapter, we do not take into account works in which division methods of thermodynamic functions of electrolyte solvation into ionic components are analyzed.

We will analyze one typical work of Born—the first quantitative calculation of ion solvation energy. He considered the solvated ion as a noncollapsing ball of radius $r$ with charge $Z_e$ located in a continuous medium with relative dielectric permittivity. The energy, released during solvation, was calculated as the remainder of electrostatic ion potential energies both in the solvent and the vacuum. Further development of the ion solvation theory in this direction [2–7] demonstrated inapplicability of the approach of Born to calculate solvation energy. The main drawback of Born theory is that it does not take into account solvent structure, the molecular picture of the ion solvation process is not described, and the decrease of dielectric permittivity in the electrolyte solution is not taken into account. Calculations of this species [1–7] should be regarded as particularly empirical since they are based on Born's equation that describes incorrectly the energetics of the ion solvation process. More profound understanding of the ion solvation process was achieved by deciphering solvent molecular structure.

The second group of theoretical calculation methods take into account the solvent molecular structure. The idea of ion solvation that takes into account peculiarities of water structure in electrolyte solutions was developed further in the works of Bernal and Fowler [8], Eley and Evans [9], Moelwin-Hughes [10], Melvin-Hughes [11], Attree [12], Milner [13], Buckingham [14], and others.

As an epigraph to the considered group of calculations, we will quote Van Arkel who qualitatively described hydration process, taking into account the year it was written [15]:

> Ions are bound with water molecules into complexes, i.e. several dipoles of water molecules are shaped around each ion. This complex from ions and immediately adjacent water molecules will further on draw neighboring water molecules and, therefore, a second layer of water molecules will be shaped connected with an ion, in any case,

> rather weaker, than the first layer. Since this process continues further on, we can imagine the state of the diluted solution as follows: ions are directly surrounded by a layer of water molecules, tightly connected with them; then there is the second layer which is already not that tightly bound with an ion; water molecules possess large freedom of movements and not all of them are quite accurately directed by their dipoles to the ion center. In every successive layer the orientation effect as well as packing density of water molecules decrease more and more.

Calculation of the ion solvation energy that takes into account peculiarities of the water structure was made by Bernal and Fowler. According to their idea, ions acting upon water molecules break their own structure. The effect of such a change is proportional to the ion polarizing power. It was believed that the effect on water molecules by small multicharged ions would be the highest, and that of the large single-charged ions would be the lowest. Their approach takes into account the interaction energy of the ion with the surrounding water molecules, considered as point dipoles, as well as interaction of the hydrated ion with the rest of the volume of water (Born interaction). The concepts of Bernal and Fowler did not include the repulsion energy between an ion and water molecules at small distances, energy contributions of the quadrupole, polarization, and others. Later, Bernal and Fowler added ion hydration into their calculations.

The calculation of Eley and Evans was conducted in the same way as that of Van Arkel [15], but paid more attention to details. The three-field model is assumed for a water molecule, in which the observed dipole moment of the water molecule was used for selecting values of hydrogen and oxygen atomic charges. The oxygen atom is assumed as the water molecule center. Using crystallographic ion radius and the coordination number equal to 4, Eley and Evans calculated hydration heats of a large number of ions and demonstrated that the results are in rather good concordance with experimental hydration heats. Eley and Evans commented that disregarding polarization and other interactions mentioned in the calculation of Bernal and Fowler may significantly affect the final results, which can be merely incidentally compensated by other mistakes. The required hydration heat is obtained by summing up terms, the majority of which are energies, not heats.

Melvin-Hughes calculated ion hydration heats in the framework of the general theory of intermolecular forces, in which the electric-charging method and the ionic radius concepts were not used. This calculation takes into account the interactions between ion and point permanent dipole, and ion and induced dipole, which was located closer to the ion than the permanent dipole as well as between permanent dipoles themselves. Interaction between the primary hydrate complex and the solvent was taken into consideration rather roughly through polarizability of the medium. Despite the more correct formulation of estimation compared to the previous works, the calculations of Melvin-Hughes, however, have many disadvantages. The final interaction between induced dipoles is not taken into account, and the interaction energy of permanent dipoles in tetrahedral and octahedral configurations differs only by values of the coordination number, while they actually depend on the configuration geometry too. The work of Melvin-Hughes is catching, inasmuch as its author, unlike his predecessors, pays major attention to calculation of the primary hydration energy, which introduces predominant contribution to the full hydration energy.

Attree and later his learner Milner performed the most accurate calculation of the hydration energy of a number of ions. The following interactions are taken into account in their calculation: ion–dipole; dipole–dipole; ion–induced dipole, the position of which does not coincide with the position of the noninduced dipole; and induced dipole–induced dipole. Repulsion forces and interaction of the primary hydrate complex with the solvent weight were taken into account. Energy decrease of the main ion state in the field of surrounding water molecules was considered rather roughly. The obtained results are reliable enough and can serve as good approximation to hydration energies. The main difference of this calculation from the previous ones is as follows. It is known that the principal state of some ions is degenerated. This degeneration can be withdrawn by means of an electrostatic field of the closest water molecules and it is the reason for additional stabilization of ions in the solution. In the work of Milner, the numerical value of this stabilization is calculated by means of the perturbation theory. Energies of excited states of hydrated ions are assessed too. All these allowed the interpretation of the observed absorption spectra of hydrated ions, from which numerical values of several parameters, required for calculation of hydration energy, were calculated. The calculation of Milner serves as an example of combination of electrostatic states with quantum–chemical representations. The work under question is mostly interesting by the fact that it states dependence of hydration energy from the structure of ion outer shells. It differs qualitatively from all the other works.

In the work of Buckingham, the ion and solvent molecule interaction energy is expanded in series in powers of $r^{-1}$ with coefficients being multipole moments of the corresponding order. Before, one was usually restricted only to two terms of this expansion taking into account only the dipole moment of the solvent molecule in the calculations. Buckingham once again confirmed that the ion–dipole interaction energy makes a basic contribution to the full interaction energy, but at the same time he paid attention to the importance of estimating ion–quadrupole interactions. He, in particular, proved that this interaction is the main reason for a known difference in heats of ion hydration $K^+$ and $F^-$. When changing the sign of the ion charge (presuming that such will not cause the ion radius to change), the water molecule is forced to change its orientation relating to the ion, so that the ion–dipole interaction energy stayed negative (attraction). Buckingham demonstrated that the sign of the ion–quadrupole term changes during such ration of the dipole. Hence, the term of the ion–dipole interaction energy of anions and cations is the same, and that of ion–quadrupole is different. Considering it to be a reason for difference in heats of ion hydration $K^+$ and $F^-$, they calculated quadrupole moments of water and ammonia molecules. The necessity of estimating induction and dispersion energy, which at such consideration occur naturally as the higher-order terms under $r^{-1}$, is discussed in this work too. Interaction of the primary hydrate complex with the medium taking into account dielectric saturation is considered by Born. It is also demonstrated that saturation is inappreciable beyond the first hydrate layer. As we all know, Born's equation describes incorrectly the energetics of the ion solvation process because it is empirical. Dielectric saturation beyond the first hydrate layer is estimated incorrectly while dielectric permittivity $\varepsilon_S$ in the solution decreases sharply compared to clear water.

We summarize the conclusions under the group of calculations, in which the solvent molecular structure is taken into account. Inconsistency of using electrostatic formulas is made in all calculations. Most of these formulas are valid only for distances, larger than the sizes of interacting particles. This condition is not met in electrolyte solutions. In all considered and similar variants of the theory, thermodynamic functions of solvation depend on a large number of parameters, in which the value can be chosen at random. For parameters such as dipole moment and polarizability, numerical values are used, which were found from experiments on pure gases and liquids, though it is known that these parameters have other values in a strong field of ion. It is interesting to mention the fact that all authors use different numerical values for these parameters, and oddly enough, they obtain satisfactory results consistent with the experimental data. Insufficient accuracy of the experimental data does not allow drawing a conclusion about correctness of a particular calculation method. It was highlighted by Krugliak et al. [16], and for general statistics we would like to quote the physicochemist Mikulin [17, p. 37]:

> Untutored researcher cannot but get astonished that multiple theories of electrolyte solutions, offered by various scientists, are associated nearly with the same experimental data and in the result every of the theories is declared fair by its author, though all of them differ radically one from another. It is explained most frequently by the fact that when building a theory, it is introduced indefinite and unobserved constants in an amount sufficient, as in this connection L.D. Landau expressed himself, "to place any not too much wild formula into experimental results". Sometimes in case of a successful approach one manages to "substantiate" such formula, which allows obtaining more or less satisfactory results even with one constant. All such formulas, as a rule, are of clearly interpolational nature and are of no substantial interest.

## 1.2 EVOLUTION OF SOLVATION CONCEPTS AND STRUCTURAL STATE OF ELECTROLYTE SOLUTIONS

The period of the 1960s is characterized by a sharp increase of a number of works in the field of physicochemistry of solutions, attraction of statistical mechanics, and quantum chemistry methods, in which the issues of studying solvent structures and their influence on solvation in solutions began to be considered more completely. These are the works of Mishchenko et al. [18,19], Samoylov [20], Yatsimirsky [21], Izmaylov [22–24], Krugliak et al. [16,25,26], Krestov et al. [27–29], Yukhnovsky et al. [30–32], Lyashchenko [33–38], and others.

Mishchenko and Sukhotin, studying thermodynamic properties of solutions, introduced a notion about the complete-solvation limit corresponding to the concentration at which all solvent molecules belong to cation and anion solvation shells. Solvated ions and vacant solvent, which structure changes with concentration, are located in the solution below this limit. There is no vacant solvent in the solution above this limit; all its molecules make the immediate surroundings of the ions. At such concentration, the solution can be compared with a system consisting of hydrated ions adjoined by their hydration shells. They performed calculation of the energy of ion hydration, claiming, to their point of view, possibly to be a complete estimation of constituent effects using the Born cycle. During calculations, besides ion–dipole and Born interaction, others were taken into account as well: conditioned by orientation and deformation water polarization in the ion field; dispersion interaction between the ion and water molecules; mutual repulsion of water molecules in the ion hydration shell; and repulsion of ion and water molecules of the hydrate layer at interpenetration of their electron shells. Izmaylov together with Ivanova tried to calculate enthalpy of solvation in spirits using the method of Mishchenko. In correlation with experimental data, about 30% was gained only when as a spirit molecule ($r_{sp}$) the value $r_{sp} = r_{H_2O}(M_{sp}/M_i)$ was taken, where $M_{sp}$ and $M_i$ correspond to molecular weights of spirit and water respectively. As it is seen from the mentioned formula, the spirit molecule radius is a purely empirical, adjustable value. At the present time, the thermodynamic cycle of the calculation of chemical heats of ion hydration by Mishchenko and Sukhotin, taking into account their composing effects, possesses a few substantial drawbacks: we have spoken already about drawbacks of the used Born cycle; there are no components of energy effects in the cycle of second approximation of the perturbation method; statistical averaging of cycle components over all orientations is not given; the Clausius–Mossotti equation is used incorrectly because it is not applicable for electrolyte solutions; the water molecule radius (0.193, but not 0.138 nm) is used incorrectly; and others.

Samoylov was one of the first who substantiated molecular-kinetic notion about the structure of solvents and their influence on solvation in solutions. He interpreted the hydration phenomenon not like tight binding of a certain number of water molecules but like effect of ions on heat motion of solution molecules closest to them. He differentiates two types of close ion hydration—hydrophilic and hydrophobe. Ions, which, interacting with water molecules by their electric fields, produce on their heat motion either inhibiting or accelerating effect, are hydrated hydrophylically. The inhibiting effect is increased with boost of ion charge or, at that charge, with decrease of their sizes. Large single-charged ions produce the accelerating effect. Hydrophobe hydration is found in the case of complex organic ions and molecules in a number of nonelectrolytes. It is determined by the inhibiting effect of diluted particles on translational motion of solution water molecules. In contrast to hydrophilic hydration, hydrophobe hydration does not result from strong interaction of water molecules and a solute, but most likely occurs in the result of interaction intensification among molecules $H_2O$, contributing thereby to structuring of free water. According to the ideas of Samoylov, hydrophobe hydration lies in stabilization of the water structure by solute particles.

Comparing the values of self-diffusion activation energies in clear water with potential barrier height, dividing the solvent molecule position near an ion and in volume, Samoylov introduced notions about *positive* and *negative* ion hydration. In the first case, effective binding of the closest water molecules by ions takes place, and in the second one increase of mobility of the closest molecules is observed compared to their mobility in clear water. The approach offered by Samoylov for studying the near solvation has enough in common. It can describe a continuous transition from a comparatively weak (even negative) hydration to a strong one. Near ion hydration in aqueous solutions is connected tightly with the structural state of water. It lies in the fact that increase of water order leads to weakening of ion hydration. For example, destruction of the water structure intensifies hydration. The role of the structural state of water in the phenomena of ion hydration in solutions lays emphasis upon great significance of short-range forces for solution properties. During ion hydration, the proper structure of water is changed and a new structure, typical for a solution, occurs. With that, high stability of the water structure becomes apparent. This is due to the fact that, firstly, every molecule in water participates approximately in four hydrogen bonds and, secondly, that translational motion of molecules $H_2O$ takes place generally in structure voids. With the increase of temperature and pressure, the proper structure of water becomes less ordered and near ion hydration is intensified and makes association of cations and anions, as well as contact ion paring, more difficult.

Yatsimirsky conducted theoretical investigations in the area of thermodynamics of ion hydration and lattice energy of complex salts. The stated values were embedded into the world chemistry literature and textbooks as *thermochemical radius of Yatsimirsky* and, together with ionic radius of Goldschmidt, Pauling, and Shannon, are widely used in the physical and inorganic chemistry. Major cycle of works, dedicated to the thermochemistry of complex compounds, was performed by Yatsimirsky in the 1940s. In order to solve a number of thermodynamic problems in the chemistry of coordination compounds, he determined the values of variables of energy in their crystal lattices unknown at that time. This value can be calculated by means of radius of the ions composing the lattice based on the equation of A.F. Kapustinsky. Yatsimirsky was the first to demonstrate that lattice energy can be calculated by the equation using ion solvation heat and salt dissolution heat. Applicability of the offered approach is confirmed by multiple data. The obtained data allowed Yatsimirsky to define formation energy of gaseous complex ions. The analysis of calculated values of lattice energy and formation energy of gaseous complex ions demonstrated that both cations and anions can be ranged in series by a decrease of disposition toward covalent bonding. Later, development of this approach allowed Yatsimirsky to create his own classification of metal and ligand ions in 1950, which, actually, forestalls a popular concept of Pearson's *soft* and *hard* acids and bases that appeared in 1963. He developed (together with Samoylov, 1952) a thermochemical method for defining coordination numbers of ions in aqueous solutions. Based on his multiple own and literature data, Yatsimirsky conducted a thorough analysis of the nature of chelate, polychelate, and macrocyclic effects. He was the first to demonstrate the influence of the energy of spatial organization of donor atoms and solvation on thermodynamic parameters of reactions for ion complex formation in transition metals with polydentate ligands of cyclic and noncyclic structures.

Izmaylov—the founder of the *solution** school—made an important contribution to the scientific developments of the physical chemistry of solutions, together with his successors, among which was Aleksandrov [39] who developed theoretical notions about ion solvation, acidity, as well as practical recommendations for

* It is not very clear why Izmaylov is named in many publications as the founder of the *solution* school. He was dealing mainly with experimental physical methods of researches, using methods of quantum chemistry in theoretical concepts. His scientific school could be classified as adherents of *physical* understanding of the nature of solutions. Unfortunately, at that time, the dispute between the *physical* and *chemical* areas of researches of solution structures still went on [39]. Only then in 1998 that Krestov reconciled these two areas by his structural-thermodynamic approach [29].

determining pH. Izmaylov was the first one to demonstrate that the relative strength of acids and bases depends on the nature of a solvent and to develop the quantitative theory of solvent effect on electrolyte strength. It was established that the solvent produces differentiating effect on the strength of acids and bases. Based on the vast experimental materials, Izmaylov substantiated the unified acidity scale and developed the electrolyte dissociation theory. The offered scheme of electrolyte solution forming and the introduced notion of the single zero activity coefficient $\gamma_0$ (primary medium effect, transition activity coefficient) are now generally accepted. The contribution of Izmaylov into the development of the solvation theory is considerable. He used a wide arsenal of physical–chemical research methods (electromotive force method, conductometry, polarography, electronic and oscillation spectroscopy, cryoscopy, radioactive-tracer method). Based on the data about circuit e.m.f. without transition and solubility, it was established that changes in the Gibbs energies during ion solvation to a number of solvents similar in nature are changed linearly depending on the main quantum number of the outer electron shell of an ionized atom. Based on that, the extrapolation partition method of the total Gibbs energy into ionic components was offered. Izmaylov (together with Krugliak) for the first time made an attempt to perform the theoretical calculation of the Gibbs energies of solvation of separate ions using notions about the donor–acceptor mechanism of interaction between an ion and a solvent molecule. Then, Izmaylov and Krugliak, Mikhaylov, and Drakin formulated a notion that a prerequisite of cation solvation is the availability of unoccupied orbitals and lone-pair electrons of oxygen atoms (nitrogen, fluorine) and solvent molecules with them. Formation of solvated complexes takes place by means of electron delocalization of donor orbitals in electronegative atoms of solvent molecules into acceptor orbitals of cations. Solvation is the same for anions, however the anion is a donor of electrons and the proton of the solvent molecule is an acceptor. The anion-molecule bond is implemented by the hydrogen type bond. Energetic equivalence of bonds in solvated complexes envisages hybridization of cation and anion orbitals, which in turn defines the geometrical structure of solvated complexes and coordination numbers of ions.

Krugliak was one of the first to use methods of quantum chemistry for derivation of the quantitative theory of condensed media (pure liquids, aqueous, and nonaqueous electrolyte solutions, nonelectrolyte solutions, etc.). It required accurate quantitative data about potential surfaces of interaction of ions with solvent molecules as well as ions and molecules among themselves. As he believed, model electrostatic notions about elementary ion-molecular and intermolecular interactions are obviously insufficient in such systems. One should estimate purely quantum–mechanical exchangeable components of the interaction energy.

In their works, Krugliak and Izmaylov *scolded* the electrostatic theory and swept it entirely away in their researches. Despite its imperfection at that time, it was already possessing, but shortsightedly, such a powerful instrument as quantum mechanics could lead to its improvement. Taking into account its provisions in the Hamiltonians of the Schrodinger equation, the final theory of electrolyte solutions could have been formulated then.

The sphere of duties of Krugliak included the following areas. A wide range of mathematical problems appearing in theoretical chemistry (especially in quantum chemistry) was considered: calculation of overlap integrals of orbitals *s*-, *p*-, and *d*-type, numerical integration of multicentric molecular integrals, calculation σ—of molecule structures, creation of semiempirical models for calculation of the electronic structure of molecules taking into account interaction of electrons (closed- and open-shell approximation self-consistent field restricted method, electronic configuration overlap method), solution of the complete eigenvalue problem in reference to the molecule small oscillation theory, analysis of electron paramagnetic resonance (EPR) spectra, calculation of complex formation constants and thermodynamic functions of complexes with hydrogen bond from nuclear magnetic resonance (NMR) data, and many others.

Hence, in the beginning of the 1970s a situation was created in the physical chemistry of solutions provoked by confrontation of the *solution* schools with thermodynamic and structural trend. The tight connection between these areas, seemingly distanced far away from one another in the theory about solutions, was successfully formulated by Krestov. He was one of the few who understood that the interaction of solution particles (molecules, atoms, ions), solvation, chemical processes in solutions, and their structure are so tightly connected that their study in principle makes one problem. Few of the main areas of scientific developments, performed by Krestov as well as by his multiple learners and successors were qualitative determination of energy and structural contributions into thermodynamic characteristics of solvation and complex formation, and determination of their interconnection with the type of intermolecular interactions. It should be pointed out that an immensely wide range of solvents and solutes, which differ in chemical nature and structure, was used in conducted researches. The method of structural-thermodynamic solvation characteristics, developed by Krestov, gained multifaceted momentum and at present obtained a strong presence in the science dealing with solutions as a conceptual framework and a main constructive idea of the structural-thermodynamic solvation theory. The dissolution of molecules of different nature in the solvent results in disorder of translational and rotational motions, and change in frequency and amplitude of their oscillations specific for clear liquid, which, in turn, impacts on the condition of the submolecular structure of the solvent. Macroscopically, these changes appear in so-called structural effects of solvation—changes of macrofeatures related directly or indirectly to the structural state of the system, its orderliness.

In order to describe the electrolyte solution structure, Yukhnovskiy used the collective variables method imitating the solution by the ion–dipole system, where long- and short-range forces are present. The Coulomb interaction is considered as a long-range interaction. The short-range type energy is represented by the Lennard-Jones potential.

Yukhnovskiy discovered analytic expressions for binary functions of cation–anion, ion–dipole, and dipole–dipole distribution for solutions of low and high concentrations. The binary function of cation–dipole distribution, calculated for two dipole–ion concentrations, demonstrates that positioning of the ion on the dipole axis continuation from the dipole opposite pole side is more likely than its positioning on the straight line perpendicular to the dipole axis. Availability in the solution of orientational motion of molecules in the electric field of ions closest to them is confirmed by this. From the type of binary functions of dipole–dipole distribution, discovered for five dipole orientations, it appears that dipoles are located in the solution preferably on one straight line and turned to one another with opposite poles. The probability of similar positioning of dipoles increases with decrease of ion concentration. Turn of dipoles to one another with analogous poles is least probable.

Usage of long-range interaction according to the Coulomb's law in this theory raises doubts, since the Coulomb interaction for ions in the solution only gives rise to unstable configuration of ions in the solution and provokes the necessity to use the screened Coulomb potential; usage of the Lennard-Jones potential for concentrated electrolyte solutions; absence of estimation of dielectric permittivity decrease of the concentrated solution (even increase of dielectric constant of the medium is in place); reasons for formation of complexes, various clusters in the offered idealized solution structure; absence of a possibility to simulate an ensemble of particles $>10^6$ (interaction of several particles was described).

The stated demonstrates that the problem of solutions is tightly connected with studying of their structure. Thermodynamics defines the general energy of bonds; however, it is impossible to prove the quantity of these bonds and tightness of each of them without structural data. By analyzing the works [18–21,27–29], it can be seen that with attraction of studying of the structure of concentrated electrolyte solutions and solvation problems, mainly, of newer radiophysical and other methods of research into experimental investigations, comprehension of this problem grew more completely and deeply. It was demonstrated qualitatively that interaction of ions with solvent molecules is the most complex. It is to a large extent ion–dipole and quantum–mechanical depending on the electronic structure of an atom. If talking about the quantitative theory, then, to our point of view, many researchers may have been stopped by certain drawbacks of the electrostatic theory and by the Debye–Huckel doctrine firmed up strongly up to the present time.

Proceeding from the quantum–chemical theory of the solvation donor–acceptor process of ion interaction with water, spirit, ammonia, and carboxylic acid molecules, the authors of works [16,22–26] made a conclusion that it is partially covalent but not Coulomb's and ion–dipole. Contribution of covalent interactions is considerable if dealing with small-sized ions as well as ions with incomplete electron shells. The unified mechanism of solvation of the same ions in different solvents is confirmed with them by permanence of solvation energy. All conclusions are not fully correct for setting up a finished theory of concentrated electrolyte solutions.

Attraction of the quantum–statistical theory to the problems of structural state and solvation in electrolyte solutions did not solve these problems either [30–32,40]. It follows from this theory that electrolyte solutions are characterized only by the short-range coordination and nearest orientational order in formation, of which the big role is played by short-range forces. These conclusions are not enough, for example, for explaining the reasons of appearing ion or ion–aqueous clusters or supercomplexes. It should be added that the potential was studied for several particles, which does not provide a possibility for correctly extrapolating the obtained results on a larger number of interactions.

Besides quantum–chemical and quantum–statistical methods of studying properties of electrolyte solutions, tight statistical methods of Bogolyubov and Mayer were used. Review of these researches can be found in the works of Glauberman [41,42] and Martynov [43]. The works of Levich and Kiryanov [44], Kelbg [45], Martynov [46], Martynov and Kessler [47], Vorontsov-Velyaminov et al. [48], and some other authors were not included into the reviews. Further multiple editions on the solution statistical theory and thermodynamics contributed nothing special to the science of structural properties and solvation of concentrated electrolyte solutions [49–81].

Not going into the detailed evaluation of thermodynamic and statistical theories, one should state their common weak points:

- Certain idealization of primary physical and chemical prerequisites underlying these theories.
- Underestimation of weak chemical interaction forces among ions and water molecules causing ion solvation and formation of complex compounds in a solution as well as weak hydrogen bonds caused by electron shell overlapping and leading to formation of ion or ion–aqueous complexes (supermolecules).
- In deriving initial equations, the change of dielectric permittivity of the solvent depending on the change of ions concentration is neglected and is taken equal to the macroscopic value in usual conditions.

The listed drawbacks lead to the fact that all known statistical and electrostatistical theories of solutions a priori cannot provide positive results in case of coincidence with experimental data for concentrated electrolyte solutions, which does not allow taking them as a basis for building a quantitative theory of such solutions.

Further on, we are going to consider approaches and possibilities of quantum mechanics and quantum chemistry developing rather intensively [82–122]. Creation of the relativistic section of quantum mechanics by P.A.M. Dirac dates back to 1929. It was a period of rapid growth of a new theory and its methods, used for calculation of all new physical and chemical objects. Success of the first calculations of isolated atoms and molecules of hydrogen produced euphoria among physicians. From the memories of W. Heisenberg, who developed matrix formulation of quantum mechanics in 1925–1927, it was amazing time, when it seemed that solution to all mysteries of nature had been found and there had been nothing left but to use it consistently and carefully. However, as it happened not once in the history of physics, the nature turned to be immeasurably much complicated. It turned out very soon that the exact solution of the Schrödinger molecular equation is impossible even for the elemental molecular system—ion $H_2^+$. Consistent usage of the approximate method—the Hartree–Fock–Roothaan self-consistent field (SCF) method—for molecules consisting of several atoms faced insuperable computational problems. Development of the calculation method $\pi$ of electronic systems with the same name by Hückel in 1931 was the first successful practical implementation of the program formulated by Dirac, but at the same time turned to be just a first step on a long way of development of quantum chemistry calculation methods. The difficulty of the process can be proved by the fact that a slightly modified simple Hückel method was used until the mid-1970s, and its variant with extended basis, created by Hoffmann in 1963, is still being used.

The search for optimal ways of solving the Schrödinger molecular equation in the framework of the SCF scheme initially followed two main directions. The first one includes methods *ab initio*, which even at present are used seldom for calculation of molecules and molecular fragments containing more than 10 atoms. Due to the rapid development of organic and bio-organic chemistry, the second group of calculation methods, so-called semiempirical methods, has been gaining in popularity recently. It is the creation and improvement of these methods that allowed quantum chemistry to become a tool of the chemist and to invade laboratory practice. At the same time, despite the gained progress, semiempirical methods are not panacea for solving real problems in chemistry, possessing few drawbacks and limitations. Correct usage of methods from this group envisages clear understanding by a chemist of their possibilities and a circle of tasks, solving of which they are meant for. Other existing methods for molecule calculation (electron pair method and molecular orbit method) are approximate too. Even though these methods can turn useful further on when solving separate specific problems, they, at least, in their modern look, cannot bring considerable success in solving large key problems of the modern theory of chemical composition yet.

Following are the types of problems solved by quantum chemistry at the present time:

1. One point problems
   a. Electron transfer spectroscopy (energy of excited states and transition intensity)
   b. Reactivity
   c. Analysis of chemical bond nature in given molecules
   d. Nature of intermolecular interactions
2. Problems of geometry optimization type
   a. Search of equilibrium geometry, that is, local energy minimums
   b. Oscillation problem for equilibrium geometry (resulting in oscillation spectroscopy)
   c. Search for saddle points, potential energy profiling, and other methods of determining potential barrier heights, and transient and intermediate states as a means for clarifying mechanisms of chemical reactions
   d. Nature of intermolecular interactions
3. Molecular dynamics based on quantum chemistry
   a. Determination of global energy minimums
   b. Determination of chemical transformation mechanisms by means of their modeling

Another component of the quantum theory was developing together with wave and quantum mechanics—quantum statistics or statistical physics of quantum systems consisting of a large number of particles [123–131]. Based on classical laws of motion of separate particles, the theory of behavior of their aggregate was created—classical statistics. Similarly, based on the quantum laws of particle motion, quantum statistics was created to describe behavior of macro-objects in cases when the laws of classical mechanics cannot be applied in order to describe motions of microparticles making them—in this case random quantum properties appear in properties of macro-objects. It should be noted that the system here means only particles interacting with one another. At that, the quantum system cannot be considered as the aggregate of particles preserving their individuality. In other words, quantum statistics requires abandoning the notion of distinguishability of particles—this was named as the identity principle. Two particles of one nature were considered as identical in atomic physics. However, this identity was not acknowledged as absolute. Thus, two particles of one nature could be distinguished at least conceptually. In quantum statistics, the possibility of distinguishing two particles of the same nature is completely absent. Quantum statistics assumes that two states of a system, which differ from one another only by rearrangement of two particles of the same nature, are identical and indistinguishable. Hence, the main provision of quantum statistics is the identity principle of identical particles included into the quantum system (e.g., ion–ion and dipole interactions in electrolyte solutions). Quantum systems are distinguished by this from classical systems. In the interaction of microparticles, an important role belongs to the spin—the intrinsic moment of a microparticle. The spin of electrons, protons, neutrons, neutrinos, and other particles is expressed by a half-integral value, while that of photons and $\pi$-mesons by an integral value (1 or 0). Depending on the spin, the microparticle is subordinated to one of two different types of statistics. The systems of identical particles with the whole spin (bosons) are subordinated to Bose–Einstein quantum statistics, the distinctive feature of which is that a random number of particles can be in every quantum state. There is another type of Fermi–Dirac quantum statistics for particles with the half-integer spin (fermions). The distinctive feature of this type of statistics is that a random number of particles can be in every quantum state. This requirement is called the Pauli exclusion principle. The Pauli principle allowed explaining regularities of filling-in of shells with electrons in multielectron atoms (e.g., weak hydrogen bonds in electrolyte solutions). This principle expresses specific property of particles that are subordinated to it. And now it is difficult to understand why two identical particles exclude one another to take the same state. Such type of interaction does not exist in classical mechanics. What is its physical nature? What are the physical sources of exclusion? These are the problems that need to be solved. Nowadays one thing is clear: physical interpretation of the exclusion principle in the framework of classical physics is impossible. An important conclusion of quantum statistics is a provision that a particle, included into any system, is not identical to the same particle but included into the system of another type, or a free one. This implies importance of the task of revealing the specific nature of a material carrier of a definite property of systems.

We will characterize in brief the aforementioned scientific areas: physical chemistry, chemical thermodynamics, electrostatic theory, intermolecular interactions, statistical theory, quantum chemistry, and quantum statistics possess their own definite limitations within which considerable successes were achieved. None of the theories taken independently is able to describe all, extremely complicated supramolecular interactions in concentrated electrolyte solutions because of its limitations. The quantitative theory of concentrated electrolyte solutions can be created only with usage of all best achievements of the stated scientific areas.

Nowadays successful work is underway over a contemporary theory of concentrated electrolyte solutions by Lyashchenko and his school [33–38]. Researches of his school are connected with fundamental problems of physical and structural chemistry of solutions, solvation theory, formation interactions, and interparticle interactions of solutions with homogeneous and heterogeneous equilibriums and phase diagrams of water–salt and water–nonelectrolyte systems in equilibrium and non-equilibrium conditions, and contemporary methods of solution research are developed. It is dielectric, electronic, infrared spectroscopy, electrical conductivity, physical–chemical analysis of solutions and phase diagrams, thermodynamic calculations, computer modeling (Monte Carlo methods and molecular dynamic methods), structural–geometry analysis, etc.

At the present time, a new distinctive scientific area is developed in the scientific school of Lyashchenko, which includes one of such orientations as development of theoretical bases of streamlined synthesis of intermediate products and charge in order to obtain materials at predefined structural properties of initial media and water–salt compositions. Creation of a new class of materials being absorbents and transducers of millimeter radiation based on microheterogeneous water–salt and water–polymer–salt systems [131–138]. Substantial results have been obtained in the development of a new distinctive method for numeric evaluation of relaxation time distribution functions in electrolyte and nonelectrolyte solutions from the dielectric spectra of the microwave area calculated by the computer modeling method (molecular dynamics) based on the known potentials of interparticle interaction. Using the molecular dynamics method, the investigation of the dielectric properties of methanol and lithium sulfate solutions in the frequency range from 1 GHz up to 100 GHz was conducted and molecular processes determining the dielectric relaxation at ultrahigh frequencies (time intervals $10^{-11}$ to $10^{-10}$ s) [139–146] were defined.

A new approach in the physical–chemical analysis of water–salt systems has been developed where the composition–solubility and composition–property diagram is introduced with additional coordinates determined by the dielectric properties of concentrated and saturated salt solutions as well as by the structural parameters of complexes and by the solution as a whole. The scheme of concentration structural transition to high concentrations can be graphically depicted in Figure 1.1 [132,146].

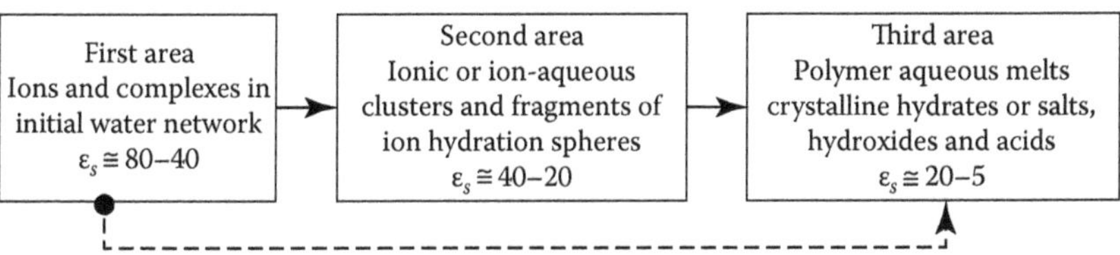


**FIGURE 1.1** Structural transition from dilute to concentrated solutions.

## REFERENCES


1. Born M. 1920. Volume and heat of hydration of ions. *J. Phys.* 1: 45–52.
2. Webb J. 1926. The free energy of hydration of ions and the electrostriction of the solvent. *J. Am. Chem. Soc.* 48: 2589–2603.
3. Webb J. 1927. *Trans. Am. Electrochem. Soc.* 51: 559.
4. Grahame D.C. 1950. Effects of dielectric saturation upon the diffuse double layer and the free energy of hydration of ions. *J. Chem. Phys.* 18: 903.
5. Hasted J.B., Ritson D.M., Collie C.H. 1948. Dielectric properties of aqueous ionic solutions: Part I and II. *J. Chem. Phys.* 16: 11.
6. Frank H.S. 1941. Local dielectric constant and solute activity. A hydration association model for strong electrolytes. *J. Am. Chem. Soc.* 63: 1789.
7. Hush N.S. 1948. The free energies of hydration of gaseous ions. *Aust. J. Sci. Ser. A* 1: 480.
8. Bernal J.P., Fowler R.H. 1933. A theory of water and ionic solution, with particular reference to hydrogen and hydroxyl ions. *J. Chem. Phys.* 1: 515.
9. Eley D.D., Evans M.G. 1938. Heats and entropy changes accompanying the solution of ions in water. *Trans. Faraday Soc.* 34: 1093.
10. Moelwin-Hughes E.A. 1949. Ionic hydration. *Proc. Cambr. Philos. Soc.* 45: 477.
11. Melvin-Hughes E.A. 1962. *Physical Chemistry*, Vol. 2. Publishing House of the Foreign Literature, Moscow, Russia, p. 769.
12. Attree R.W. 1950. Ionic solvation energies. Thesis, Princeton University, Princeton, NJ, p. 42.
13. Milner P.C. 1955. The stabilities of ions in aqueous solution. Thesis, Princeton University, Princeton, NJ, p. 23.
14. Buckingham A.D. 1957. Theory of ion–solvent interaction. *Disc. Faraday Soc.* 24: 151.
15. Van Arkel A. 1934. *Chemical Bond from the Electrostatic Point of View.* The Lane with Fr. ONTI Goskhimtekhizdat, Leningrad. otd-niye, Leningrad, Russia, p. 248.
16. Krugliak Yu.A., Kvakush V.S., Dyadyusha G.G. et al. 1967. *Methods of Calculations in a Quantum Chemistry*. Naukova Dumka Publ., Kiev, Ukraine, p. 258.
17. Mikulin G.I. (ed.). 1968. *Questions of Physical Chemistry of Solutions of Electrolytes.* Chemistry, Leningrad, Russia, p. 420.
18. Mishchenko K.P., Sukhotin A.M. 1953. Solvation of ions in solutions of electrolytes. II. Calculation of chemical energy of a solvation taking into account separate components it effects. *Zhurn. Phys. Chem.* 27 (1): 26–40.
19. Vorobyov A.F. 1972. Concerning definition in thermodynamics of reference states of individual substances and solutions. *Theor. Exp. Chem.* 3 (5): 705–708.
20. Samoylov O.Ya. 1957. *Structure of Aqueous Solutions of Electrolytes and Hydration of Ions*. Publishing House of Academy of Sciences of the USSR, Moscow, Russia, p. 182.
21. Yatsimirsky K.B. 1951. *Thermochemistry of Complex Compounds*. Publishing House of Academy of Sciences of the USSR, Moscow, Russia, p. 252.
22. Izmaylov N.A. 1976. *Electrochemistry of Solutions*, 3rd edn. Khimiya, Moscow, Russia, p. 488.
23. Izmaylov N.A. 1959. Chemical energies of a solvation of ions. *Dokl. Acad. Sci. USSR* 126 (5): 1033–1036.
24. Izmaylov N.A., Krugliak Yu.A. 1960. Round timber to a question of a solvation of ions. *Dokl. Acad. Sci. USSR* 134 (6): 1390–1393.
25. Krugliak Yu.A., Uitmen D.R. 1962. Methods of calculations in a quantum chemistry. I. Calculation of integrals of $A_n$ and $B_n$ on the electronic digital computer. *Zhurn. Struct. Chem.* 3 (5): 569–583.
26. Krugliak Yu.A., Dyadyush G.G., Kupriyevich V.A. et al. 1969. *Computational Methods of Electronic Structure and Ranges of Molecules*. Naukova Dumka, Kiev, Ukraine, p. 308.
27. Krestov G.A. 1987. *Thermodynamics of Ionic Processes in Solutions*, 2nd prod. reslave. Chemistry, Leningrad, Russia, p. 272.

28. Krestov G.A., Kobenin V.A. 1977. *From a Crystal to Solution.* Chemistry, Leningrad, Russia, p. 112.
29. Abrosimov V.K., Krestov G.A., Alper G.A. et al. 1998. *Achievements and Problems of the Theory of a Solvation: Structural and Thermodynamic Aspects.* A Series of Problems of Chemistry of Solutions. Science, Moscow, Russia, p. 247.
30. Yukhnovsky I.R., Head M.F., Vysochansky V.S. 1977. Research of binary cumulative distribution functions of the mixed ion–dipole systems which have been written down in the form of an exponential curve from potential of average force. *Ukr. Phys. Zhurn.* 22 (8): 1330–1335.
31. Yukhnovsky I.R. 1988. *Electrolytes.* Naukova Dumka, Kiev, Ukraine, p. 168.
32. Yukhnovsky I.R., Golovko M. F. 1980. *Head Statistical Theory of Classical Equilibrium Systems.* Kiev, Ukraine: Naukova dymka, p. 372.
33. Lyashchenko A.K. 1992. Structural effects of a solvation and structure of aqueous solutions of electrolytes. *Zhurn. Phys. Chem.* 66 (1): 167–184.
34. Lyashchenko A.K. (Ed.) 1994. Structure and structure-sensitive properties of aqueous solutions of electrolytes and nonelectrolytes. In: *Relaxation Phenomena in Condensed Matter.* Advances in Chemical Physics Series. LXXXVII. John Wiley & Sons Inc., New York, Chapter 4, pp. 379–426.
35. Lyashchenko A.K. 2001. Concentration transition from water-electrolyte to electrolyte-water solvent and ionic clusters in solutions. *J. Mol. Liquids* 91: 21–31.
36. Lyashchenko A.K. 2002. *Structural and Molecular-Kinetic Properties of Concentrated Solutions and Phase Equilibria in Water-Salt Systems: Concentrated and Saturated Solutions.* A Series of Problem of Chemistry of Solutions. Nauka, Moscow, Russia, pp. 93–118.
37. Balankina E.S., Lyashchenko A.K. 2002. The acoustical and structural changes of aqueous solutions due to transition to molten supercooled salts. *J. Mol. Liquids* 101 (1–3): 273–283.
38. Loginova D.V., Lileev A.S., Lyashchenko A.K., Harkin V.S. 2003. Hydrophobic hydration of anions on the example of solutions of propionates of alkali metals. *Achievements in Chemistry and Engineering Chemistry*, Vol. 17 (7 (32)). RHTU of D.I. Mendeleyev, Moscow, Russia, pp. 26–29.
39. Aleksandrov V.V. 1981. *Acidity of Nonaqueous Solutions.* Vyscha School (HGU), Kharkov, Ukraine, p. 256.
40. Solovyev Yu.I. 1959. *History of the Doctrines of Solutions.* Publishing House of Academy of Sciences of the USSR, Moscow, Russia, p. 320.
41. Glauberman A.E. 1959. *Thermodynamics and Structure of Solutions.* Publishing House of Academy of Sciences of the USSR, Moscow, Russia, pp. 5–34.
42. Glauberman A.E. 1962. Methods of the modern statistical theory of solutions of electrolytes. *Vesn. Lvovsk. Univ. Ser. Phys.* 1 (8): 10–45.
43. Martynov G.A. 1967. Statistical theory of solutions of electrolytes average concentration. *Uspekhi Fiz. Nauk* 91 (3): 455–483.
44. Levich V.G., Kiryanov V.A. 1962. Statistical theory of solutions of the strong electrolytes. *Zhurn. Phys. Chem.* 36: 1646–1654.
45. Kelbg G.U. 1960. Ntersuchungen zur statistisch-mechanischen. Theorie starker Elektrolyte. III. Coulomb-Potential mit uberlagertem Herzfeld-Potential. *Z. Phys. Chem. (DDR)* 214 (3/4): 141–152.
46. Martynov G.A. 1965. Statistical theory of solutions of electrolytes. II. Model of the charged balls. *Electrochemistry* 1 (5): 557–565.
47. Martynov G.A., Kessler Yu.M. 1967. Calculation of thermodynamic properties of solutions of the strong electrolytes taking into account amendments to model of the charged solid balls. *Electrochemistry* 3 (1): 76–84.
48. Vorontsov-Velyaminov N., Elyashevich A.M., Crown A.K. 1966. Theoretical research of thermodynamic properties of solutions of the strong electrolytes by a Monte-Carlo method. *Electrochemistry* 2 (6): 708–716.

49. Chuyev G.N. 1999. Statistical physics of a solvated electron. *Ukr. Fiz. Zhurn.* 169 (2): 155–170.
50. Kvasnikov I.A. 2011. *Thermodynamics and Statistical Physics: Theory of Equilibrium Systems*, Vol. 2. Publishing House of Editorial URSS, Moscow, Russia, p. 430.
51. Zubarev D.N., Morozov V.G. 2002. *Statistical of the Mechanic of Nonequilibrium Processes*. FIZMATLIT, Moscow, Russia, p. 432.
52. Kvasnikov I.A. 2005. *Thermodynamics and Statistical Physics: Quantum Statistics*, Vol. 4. KomKniga, Moscow, Russia, p. 352.
53. Leontovich M.A. 1983. *Introduction in Thermodynamics: Statistical Physics*. Nauka, Moscow, Russia, p. 324.
54. Yulmetyev R.M. 1972. *Introduction in a Statistical Physics of Liquids*. Publishing House of the Pedagogical Institute, Kazan, Russia, p. 218.
55. Gelfer Ya.M. 1981. *History and Methodology of Thermodynamics and Statistical Physics*, 2nd edn., Rev. and Supplementary. Higher School, Moscow, Russia, p. 536.
56. Zakharov A.Yu. 2006. *The Lattice Models of a Statistical Physics*. The Novgorod State University, Velikiy Novgorod, Russia, p. 262.
57. Bikkin H.M., Lyapilin I.I. 2009. *Nonequilibrium Thermodynamics and Physical Kinetics*. OURO RAHN, Yekaterinburg, Russia, p. 500.
58. Kirichenko N.A. 2005. *Thermodynamics, Statistical and Molecular Physics*, 3rd edn. Fizmatkniga, Moscow, Russia, p. 176.
59. Bazarov I.P., Gevorgyan E.V., Nikolaev B.P. 1989. *Nonequilibrium Thermodynamics and Physical Kinetics*. MGU Publishing House, Moscow, Russia, p. 240.
60. Vasiliev A.M. 1980. *Introduction to Statistical Physics*. Higher School, Moscow, Russia, p. 272.
61. Vasilevsky A.S., Multanovsky V.V. 1985. *Statistical Physics and Thermodynamics*. Education, Moscow, Russia, p. 256.
62. Gibbs J.V. 1982. *Termodynamiks. Statistical Mechanics*. Science, Moscow, Russia, p. 584.
63. Berezin F.A. 2008. *Lectures on a Statistical Physics*, 2nd prod. MTsNMO, Moscow, Russia, p. 198.
64. Bazarov I.P. 1979. *Methodological Problems of Statistical Physics and Thermodynamics*. Publishing House of Moscow University, Moscow, Russia, p. 87.
65. Mayer Dzh., Geppert-Mayer M. 1980. *Statistical Mechanics*. Mir, Moscow, Russia, p. 546.
66. Schrödinger E. 1999. *Statistical Thermodynamics*. Udmurt. un-t., Izhevsk, Russia, p. 96.
67. Rumer Yu.B., Ryvkin M.Sh. 2000. *Thermodynamics, Statistical Physics and Kinetics*, 2nd prod. Publishing House of Novosib. un-t., Novosibirsk, Russia, p. 608.
68. Terletsky Ya. 1973. *Statistical Physics*, 2nd prod. The Highest School, Moscow, Russia, p. 280.
69. Sadovsky M.V. 1999. *Lectures on a Statistical Physics*. Publishing House of Electrophysics Institute UrO RAHN, Yekaterinburg, Russia, p. 264.
70. Baxter R. 1985. *Exactly Solvable Models in Statistical Mechanics*. The Lane with English. Mir, Moscow, Russia, p. 488.
71. Temperley G., Rawlinson J., Rashbruk Dzh. 1971. *Physics of Prime Liquids. Statistical Theory*. The Lane with English. Mir, Moscow, Russia, p. 309.
72. Rezibua P., Lener M. 1980. *Classical Kinetic Theory of Liquids and Gases*. The Lane with English. Mir, Moscow, Russia, p. 424.
73. Bogolyubov N.N. 1979. *Selected Papers on Statistical Physics*. MGU, Moscow, Russia, p. 343.
74. Klimontovitch Y.L. 1982. *Statistical Physics*. Nauka, Moscow, Russia, p. 608.
75. Greiner W., Neise L., Stöcker H. 2001. *Thermodynamics and Statistical Mechanics*. Springer, New York, p. 480.
76. Levanov A.V., Antipenko E.Ye. 2006. *Determination of Thermodynamic Properties of Statistical Methods. Real Gases. The Liquid. Solid Body*. MGU, Moscow, Russia, p. 437.

77. Kryglyak I.I., Yukhnovsky I.R. 1982. Method of collective variables in the equilibrium statistical theory of restricted systems of the charged particles. I. Continual model of the aquasystem occupying a half-space. *Theor. Math. Phys.* 52 (1): 114–126.
78. Bazarov I.P., Gevorgyan E.V., Nikolaev B.P. 1986. *Thermodynamics and Statistical Physics: Theory of Equilibrium Systems*. MGU, Moscow, Russia, p. 312.
79. Zaitsev R.O. 2004. *Statistical Physics*. MFTI, Moscow, Russia, p. 300.
80. Kirov I.A. 2009. *Chemical Thermodynamics. Solutions*. Publishing House of OmGTU, Omsk, Russia, p. 236.
81. Bazhin N.M., Ivanchenko V.A., Parmon V.I. 2004. *Thermodynamics for Chemists*, 2nd edn. Chemistry, Kolos, Moscow, Russia, p. 416.
82. Johnson D. 1985. *Thermodynamic Aspects of Inorganic Chemistry*. Mir, Moscow, Russia, p. 328.
83. Abarenkov I.V., Brothers V.F., Tulub A.V. 1989. *Beginning of the Beginning of a Quantum Chemistry*. The Highest School, Moscow, Russia, p. 303.
84. Stepanov N.F. 2001. *Of Quantum Mechanics and Quantum Chemistry*. Mir, Moscow, Russia, p. 519.
85. Zagradnik R., Polak R. 1979. *Bases of a Quantum Chemistry*. Mir, Moscow, Russia, p. 504.
86. Mushrooms L.A., Mushtakov S. 1999. *Quantum Chemistry*. Gardarika, Moscow, Russia, p. 390.
87. Zhidomirov G.M., Bagaturyants A.A., Abronin I.A. 1979. *Applied Quantum Chemistry*. Chemistry, Moscow, Russia, p. 296.
88. Bader R. 2001. *Atoms in Molecules. Quantum Theory*. Mir, Moscow, Russia, p. 532.
89. Mayer I. 2006. *Selected Chapters of Quantum Chemistry: Proofs of Theorems and Formulas*. BINOM: Lab. Knowledge, Moscow, Russia, p. 384.
90. Dmitriyev I.S., Semenov S.G. 1980. *Quantum Chemistry—Its Past and the Present. Development of Electronic Ideas of the Nature of a Chemical Bond*. Atomizdat, Moscow, Russia, p. 160.
91. Tatevsky V.M. 1973. *Classical Theory of Molecular Structure and Quantum Mechanics*. Chemistry, Moscow, Russia, p. 320.
92. Kozman U. 1960. *Introduction to Quantum Chemistry*. Publishing House of the Aliens. Lit., Moscow, Russia, p. 558.
93. Davtyan O.K. 1962. *Quantum Chemistry*. The Highest School, Moscow, Russia, p. 784.
94. Nagakura S., Nakajima T. (eds.). 1982. *Introduction to Quantum Chemistry*. The Lane with the Jap. Mir, Moscow, Russia, p. 264.
95. Sinanoglu O. (ed.). 1968. *Modern Quantum Chemistry: In 2t.* The Lane from English Mir, Moscow, Russia.
96. Mueller M.R. 2001. *Fundamentals of Quantum Chemistry: Molecular Spectroscopy and Modern Electronic Structure Computations*. Kluwer, New York, p. 265.
97. Baranovsky V.I. 2008. *Quantum Mechanics and Quantum Chemistry*. Academy, Moscow, Russia, p. 384.
98. Brodsky A.M. 1968. *Modern Quantum Chemistry. 2*. Mir, Moscow, Russia, p. 319.
99. Gankin V.Y., Gankin Y.V. 1998. *How Chemical Bonds Form and Chemical Reactions Proceed*. ITC, Boston, MA, p. 315.
100. Gubanov V.A., Zhukov V.P., Litinskii A.O. 1976. *Semi-Empirical Molecular Orbital Methods in Quantum Chemistry*. Nauka, Moscow, Russia, p. 219.
101. Zelentsov S.V. 2006. *Introduction to Modern Quantum Chemistry*. Novosibirsk, Nizhniy Novgorod, Russia, p. 126.
102. Tsirelson V.G. 2010. *Quantum Chemistry. Molecules, Molecular Systems and Solid Bodies*. BINOM Lab. Knowledge, Moscow, Russia, p. 496.
103. Novakovskaya Yu.V. 2004. Molecular Systems. *The Structure and Interaction Theory with a Radiation: Common Bases of a Quantum Mechanics and Symmetry Theory*. Editorial of URSS, Moscow, Russia, p. 104.

104. Segal J. 1980. *Semi-Empirical Methods of Calculation of the Electronic Structure*. Mir, Moscow, Russia, p. 327.
105. Bartók-Pártay A. 2010. *The Gaussian Approximation Potential: An Interatomic Potential Derived from First Principles Quantum Mechanics*. Springer, Heidelberg, Germany, p. 96.
106. Sinanoglu O. 1966. *Multielectronic Theory of Atoms, Molecules and Their Interactions*. The Lane with English. Mir, Moscow, Russia, p. 152.
107. Gelman G. 2012. *Quantum Chemistry*. BINOM Lab. Knowledge, Moscow, Russia, p. 520.
108. Stepanov N.F., Pupyshev V.I. 1991. *Quantum Mechanics of Molecules and Quantum Chemistry*. izd-vo MGU, Moscow, Russia, p. 384.
109. Unger F.G. 2007. *Quantum Mechanics and Quantum Chemistry*. TGU, Tomsk, Russia, p. 240.
110. Flarri R. 1985. *Quantum Chemistry*. Mir, Moscow, Russia, p. 472.
111. Fudzinaga C. 1983. *Method of Molecular Orbitals*. Mir, Moscow, Russia, p. 461.
112. Hedwig P. 1977. *Applied Quantum Chemistry*. Mir, Moscow, Russia, p. 596.
113. Shalva O., Dodel R., Dean S., Malryyo Zh.-P. 1978. *Localization and Delocalization in a Quantum Chemistry. Atoms and Molecules in a Ground State*. Mir, Moscow, Russia, p. 416.
114. Fitts D.D. 2002. *Principles of Quantum Mechanics: As Applied to Chemistry and Chemical Physics*. Cambridge University Press, New York, p. 352.
115. Hehre W.J.A. 2003. *Guide to Molecular Mechanics and Quantum Chemical Calculations*. Wavefunction, Inc., Irvine, CA, p. 796.
116. Inagaki S. 2009. *Orbitals in Chemistry*. Springer, Berlin, Germany, p. 327.
117. Koch W.A., Holthausen M.C. 2001. *Chemist's Guide to Density Functional Theory*, 2nd edn. Wiley-VCH Verlag GmbH, Weinheim, Germany, 293.
118. Levine I.N. 1999. *Quantum Chemistry*, 5th edn. Prentice Hall, Brooklyn, NY, p. 739.
119. Lowe J.P., Peterson K.A. 2006. *Quantum Chemistry*, 3rd edn. Elsevier Academic Press, p. 703.
120. Maruani J., Minot C. (eds.). 2002. *New Trends in Quantum Systems in Chemistry and Physics. 2. Advanced Problems and Complex Systems*. Kluwer Academic Publishers, Dordrecht, the Netherlands, p. 313.
121. Piela L. 2007. *Ideas of Quantum Chemistry*. Elsevier Science, p. 1086.
122. Reiher M., Wolf A. 2009. *Relativistic Quantum Chemistry. The Fundamental Theory of Molecular Science*. Wiley-VCH, Weinheim, Germany, p. 691.
123. Kashurnikov V.A., Krasavin A.V. 2010. *Numerical Methods of a Quantum Statistician*. FIZMATLIT, Moscow, Russia, p. 270.
124. Brillouin L. 2004. *Quantum Statistician*. The Lane about Fr., 2nd prod. Editorial of URSS, Moscow, Russia, p. 514.
125. Kadanov L., Beym G. 1964. *Quantum Statistical Mechanics*. Mir, Moscow, Russia, p. 256.
126. Kvasnikov I.A. 2011. *Quantum Statistician*. Krasadar, Moscow, Russia, p. 576.
127. Kreft V.D., Kremp D., Ebeling V. 1988. *Turnip of Quantum Statistician of Systems of the Charged Particles*. Mir, Moscow, Russia, p. 408.
128. Perina Ya. 1987. *Quantum Statistics of Linear and Nonlinear Optical Phenomena*. Mir, Moscow, Russia, p. 368.
129. Fujita C. 1969. *Introduction to Nonequilibrium Quantum Statistical Mechanics*. Mir, Moscow, Russia, p. 208.
130. Holevo A.S. 2003. *Statistical Structure of Quantum Theory*. ISSLED. Institute of Computer, Moscow–Izhevsk, Russia, p. 192.
131. Holevo A.S. 1991. *Quantum Probability and Quantum Statistics. Appl. Modern Problems of Mathematics. Fundamental Directions*, Vol. 83. VINITI, Moscow, Russia , pp. 3–132.
132. Lyashchenko A.K., Dunyashev V.S. Ed. by edition A.M. Kutepov. 2003. *Spatial Structure of Water. Water: Structure, State, Solvation. Achievements of the Last Years: Collective Monograph*. A Series of Problems of Chemistry of Solutions. Nauka, Moscow, Russia, pp. 107–145.

133. Lyashchenko A.K., Zasetsky A.Yu. 1998. Change of a structural state, dynamics of molecules of water and properties of solutions upon transition to electrolytic water solvent. *Zhurn. Struct. Chem.* 39 (5): 851–863.
134. Lyashchenko A.K., Novskova A. 2005. Structural dynamics of water and its ranges in all area of the focused polarization. *Biomed. Technol. Radiotr.* 1–2: 40–50.
135. Lyashchenko A.K., Dunyashev L.V., Dunyashev V.S. 2006. Spatial structure of water in all area of a near order. *Zhurn. Struct. Chem.* 47 (7): 36–53.
136. Lyashchenko A.K. 2007. Struktura of water, millimetric waves and their primary target in biological objects. *Biomed. Radiotr.* (8–9): 62–76.
137. Lyashchenko A.K., Novskov A. Under the editorship of Tsivadze, A.Yu. 2008. *Structural Dynamics and Ranges of Orientation Polarization of Water and Other Liquids Structural Self-Organization in Solutions and on the Phase Boundary: Collective Monograph.* LKI Publishing House, Moscow, Russia, pp. 417–500.
138. Osokina M.D., Loginova D.V., Lileev A.S., Lyashchenko A.K. 2008. Stabilization of the structure of water solutions of ammonium fluoride. *Adv. Chem. Chem. Technol.* XXII (3): 60–63.
139. Vasilyeva A.V., Loginov A.I., Lileev A.S., Lyashchenko A.K. 2008. Rotational mobility and stabilization of water molecules in solutions phosphate and ammonium dihydrophosphate. *Success. Chem. Chem. Technol.* XXII (3): 93–96.
140. Lyashchenko A.K., Karataeva I.M. 2010. The activity of the water and the dielectric constant of water solutions of electrolytes. *J. Phys. Chem.* 84 (2): 376–384.
141. Lyashchenko A.K., Zasetsky A.Yu. 1998. Complex dielectric permittivity and relaxation parameters of concentrated aqueous electrolyte solutions in millimeter and centimeter wavelength ranges. *J. Mol. Liquid* 77: 61–65.
142. Lileev A.S., Balakaeva I.V., Lyashchenko A.K. 2001. Dielectric properties, hydration and ionic association in binary and multicomponent formate water–salt systems. *J. Mol. Liquid* 87: 11–20.
143. Lileev A.S., Lyaschenko A.K., Harkin V.S. 1992. Dielectric properties of aqueous solutions of yttrium and copper. *J. Inorg. Chem.* 37 (10): 2287–2291.
144. Lyashchenko A.K. 1994. Structure and structure-sensitive properties of aqueous solutions of electrolytes and nonelectrolytes. In: Coffey W. (ed). *Relaxation Phenomena in Condensed Matter.* Adv. Chem. Phys., Vol. LXXXVII, pp. 379–426.
145. Lyashchenko A.K., Karatayeva I.M. 2007. Communication of activity of water with a statistical dielectric constant of strong solutions of electrolytes. *Dokl. Akad. Sci.* 414 (3): 357–359.
146. Lyashchenko A.K. 2010. Communication of activity of water with a statistical dielectric constant of strong solutions of electrolytes. *Zhurn. Inorg. Chem.* 55 (11): 1930–1936.

# 2 Supramolecular Designing of a Host

## 2.1 STAGES OF FORMATION OF APPROACHES TO THE CONCEPT OF LONG-RANGE SUPERMOLECULAR FORCES

Currently, the most widespread classical electrostatic theory of dilute electrolyte solutions is the Debye–Hückel theory [1,2], which considers a certain regular order of ion distribution in a solution, with the notion of ion atmosphere and its thickness as fundamental points. The concept of ion atmosphere by Debye and Hückel explains that every ion happens to be surrounded by a certain electric charge density. The properties of strong electrolyte solutions depend on the ratio of electrostatic forces of ionic fields conditioned by their charges, and properties of a medium are expressed by the dielectric permittivity of a solvent.

Classical first and second approximations of the Debye–Hückel theory correctly describe quite dilute electrolyte solutions. One reason for this is that the initial physical interpretations of the theory about *ion atmosphere*, *ionic diameter*, etc., are valid only for dilute solutions, as Fowler, Kirkwood, and Onsager showed in their time [3]. Frank and Thompson came to a similar conclusion [4].

Ordinary solving of the fundamental equation of the Debye–Hückel theory is based on mathematical simplification and expansion of a hyperbolic sine into a series [3]:

$$\sinh\frac{ze\Psi}{kT}\approx\frac{ze\Psi}{kT}+\frac{\left(ze\Psi/kT\right)^3}{3!}+\frac{\left(ze\Psi/kT\right)^5}{5!}+\cdots,$$

taking into consideration the first term of expansion $ze\Psi/kT$. Here, $z$ is the ion charge; $e$ is the electron charge, equal to $1.6021892\cdot10^{-19}$ C; $\Psi$ is the electrostatic field potential; $k$ is the Boltzmann's constant, equal to $1.380662\cdot10^{-23}$ J/K; and $T$ is the temperature, K.

Naturally, it is allowable under the condition of $ze\Psi/kT \ll 1$, which corresponds to the situation of dilute solutions.

Further development of this electrostatic theory was associated with advancement into the area of more concentrated solutions. One course was solving Debye–Hückel theory equations more accurately. Hückel tried to extend the theory into the area of concentrated solutions, believing that the dielectric permittivity of water decreases in a solution depending on the kind of ions [5]. This attempt failed, as science has almost no data about the influence of ion charge on dielectric permittivity and its dependence on concentration. The Debye–Hückel theory uses macroscopic dielectric permittivity of a solvent and does not take into account its decrease with the increase in electrolyte concentration. Hückel modified the formula for free energy of a solution, having assumed that dielectric permittivity of a solvent decreases linearly with the increase in solution concentration. However, Gronwall and La Mer [6] proved that Hückel's application of the charging process was incorrect and that all his further conclusions were also wrong. In this connection, Hückel's formula is essentially purely empirical. In fact, the initial position taken by Hückel leads to a formula, from which it can be concluded, during comparison with experimental data, that dielectric permittivity of a solvent does not decrease but rather increases with solution concentration growth.

Further, the works of Gronwall, La Mer, and Sandved took into account the higher-order terms while expanding a hyperbolic sine into a series. Though quite a powerful mathematical apparatus was applied, it was not possible to advance into a high concentration area significantly [3]. Somewhat later, Nakajima [7], Mikulin [8,9], and Guggenheim [10] developed improved numerical methods to solve the Debye–Hückel equation and compiled special diagrams and tables to calculate the solution of thermodynamic functions, which allowed excluding the influence of the first and second approximation of the theory. Because it was not quite convenient to work with diagrams and tables, these works were not widely used.

A number of researches were devoted to the development of electrostatic theory, taking into account the occupied volume of ions, which constitute the ion atmosphere, in the form of a corresponding correction to ion distribution function. And it was found that the density of electric charges in a solution cannot exceed a certain limiting value. Though the introduction of such a correction to Debye–Hückel theory is quite reasonable, the obtained results fall short of expectations and did not explain the characteristic features of change in thermodynamic properties of concentrated electrolyte solutions depending on concentration. Falkenhagen and Kelbg [11] gave a good review of these theories at that time.

For decades, researchers tried to modernize the Debye–Hückel theory in one manner or another. Obviously, it appears necessary to change the task setting. At the end of the last century, the author was also engaged in quantitative theoretical problems of concentrated electrolyte solutions [12–16]. Based on Maxwell's equation for electrostatic fields, and Maxwell–Boltzmann, Fermi–Dirac, and Bose–Einstein quantum statistics, a quantum statistics for concentrated electrolyte solutions was obtained, and interparticle interactions potential was introduced, which took into account long- and short-range interactions and short-range repulsion forces. For every electrolyte, potential was corrected according to experimental values of activity coefficients. Structural features of a concentrated electrolyte solution were considered from the perspective of three mutually overlapping zones (see the following section), as shown in Figure 1.1.

Various applications of the theoretical part to water activity calculation [17,18] and transfer properties in concentrated electrolyte solutions (electrical conductivity, viscosity, and diffusion) were demonstrated [19,20]. All the accumulated material was summarized in [21] and, partially, later in [22]. Then, works in this branch were stopped for some time.

Unfortunately, in the said works, two ratios of the electrostatic part of interaction were demonstrated in the interparticle interactions potential. One, being quite cumbersome, described the whole concentration area and was difficult to work with, while the other one is quite compact, but the lower limit began with the electrolyte concentrations in a solution from values >0.1$m$ (mol/kg $H_2O$). It was extremely inconvenient. There were also no systematic data on the reduction of dielectric permittivity in an electrolyte solution, though the works were published [23–27], in which only disembodied data measured at low frequencies were given for aqueous electrolytic solutions.

At the end of the twentieth century, Lyaschenko and his school proceeded to do extensive researches of measuring dielectric permittivity ($\varepsilon'$) and losses ($\varepsilon''$) of electrolyte solutions at ultrahigh frequencies. From these data, one can find dielectric permittivity, $\varepsilon_S$, in an electrolyte solution. These researches emerged rather late, due to the fact that the high electrical conductivity of strong electrolytes made it impossible to directly measure $\varepsilon_S$ at low frequencies, and the latest advances of UHF radio electronics with the use of modern radio engineering as a means of measurement changed the current situation. As Lyaschenko points out in his publications, two or three laboratories are dealing with such measurements throughout the world. To date, a systematic research has been carried out for dielectric permittivity and losses of solutions of fluorides, chlorides, nitrates, sulfates of alkaline metals, and a number of other salts in the temperature range of 288–308 K and in the frequency interval of 7–120 or 7–25 GHz, which covers the area of maximum dispersion of water and solutions. In general, more than 40 systems have been investigated [28–39].

It has become possible to investigate real solutions theoretically. It is also necessary to correct the situation of applying two potentials of interparticle interactions, which will be demonstrated in the following section.

## 2.2 PRINCIPLES OF SUPRAMOLECULAR-THERMODYNAMIC APPROACH

We will present supermolecular and supramolecular features of aqueous electrolyte solution, applying the supramolecular-thermodynamic approach. By supramolecular-thermodynamic approach we mean the contribution to the integral thermodynamic property, which contains (often in a hidden form) structural information and which is a macroscopic response of the system of solvent matter particles to structural changes induced by the introduction of solute particles into it and by the formation of a system of microscopic solution fragments.

Supramolecular-thermodynamic characteristics can be used to describe structures being formed, by taking into account the change in solution dielectric permittivity. This will enable to identify with a high degree of credibility the nature of change in solvent structure during solution formation, and will serve as a reliable criterion while building theoretical models that describe structure and properties of solutions at the molecular level.

The variety of supramolecular-thermodynamic characteristics is associated with the variety of parameters of thermodynamic state: if a state is set by entropy and volume, the corresponding thermodynamic function is energy $E$; if by energy and volume, the thermodynamic function is entropy $S$, etc. Values of some thermodynamic functions may turn out to be observable (e.g., energy).

Supramolecular-thermodynamic characteristics are defined by four main features:

1. All of them are functions of state.
2. All of them are additive.
3. All of them, each in its own conditions, reach an extremum in equilibrium.
4. Through any supermolecular-thermodynamic function of this chemical system, all the rest can be restored.

Supramolecular structure and peculiarities of electrolyte solutions depend on interactions between receptors of a solvent and substrates of ions [40,41]. Initial dissolving of a guest substrate has the greatest influence on host receptor interaction—a host preorganization is formed around ion substrates in accordance with complementarity based on chelate and macrocyclic effects, which is defined by the solvated guest and the host H-bonds network. Complexes are formed in the initial water network. Further addition of the guest leads to a serial disruption of host H-bonds, which begin to be conditioned by the formation of supermolecules (aggregates of complexes). Host structure loss activation occurs in cases where guest concentrations correspond to the formation of supermolecules or supramolecular assemblies (supercomplexes), that is, in the general case, these are supermolecules of ionic or ion-aqueous, or supercomplexes of polymer aqueous melts of crystalline hydrates or salts, hydroxides, and acids. Formation of supermolecules and supramolecular assemblies in the second and third zones (Figure 1.1) leads to such a host preorganization, when, while binding, no unfavorable rearrangements, both in terms of entropy and enthalpy, occur, which would lower the total free energy of complex formation. In these zones, the primary structure of water is completely disrupted and steric, and many other factors play a defining role in the processes of new preorganization of the host structure.

In the stationary state, the host structure is presented by spherical symmetry, conditioned by the intensity of electrostatic field and its interaction between guest ion potentials. Long-range forces of electrostatic interaction appear between cations and anions. From Maxwell's equation for electrostatic fields, an equation of the electrostatic part of supermolecular interactions potential between guest ions, $\Psi^0$, is derived. To describe the mass content of salts from dilute to concentrated ones, while deriving the equation of supramolecular interactions potential, various fluctuations of ion concentrations near the matter under consideration are accounted by Maxwell–Boltzmann, Fermi–Dirac, and Bose–Einstein quantum statistics, on the basis of which the quantum statistics for concentrated solutions is obtained.

## 2.3 QUANTUM STATISTICS OF SUPERMOLECULAR INTERACTIONS

Quantum statistics is the section of statistical physics that investigates systems consisting of an abundance of particles subject to the laws of quantum mechanics. In contrast to the background of classical statistical physics, in which identical particles are distinguishable (a particle can be distinguished from all particles of the same kind), quantum statistics is based on the principle of indistinguishability of identical particles.

The state of a system does not change with rearrangement of similar particles. And it turns out, as it will be shown later, that groups of particles with integer and half-integer spins are subject to different statistics.

Maxwell–Boltzmann statistics is a statistical method for describing physical systems that contain a large number of noninteracting particles moving under the laws of classical mechanics. Boltzmann distribution is a special case of Gibbs canonical distribution for an ideal gas in an external potential field since, in the absence of interaction between particles, Gibbs distribution falls into the product of Boltzmann distributions for separate particles [42].

Independence of probabilities gives an important result: probability of a given impulse value is absolutely independent of molecule position, and, vice versa, probability of molecule position does not depend on its impulse. This means that distribution of particles according to impulses (velocities) does not depend on field. In other words, it stays the same from point to point in the space in which the particles are confined. The only change is the probability of detecting the particle, or, what is the same, the number of particles.

We write down Maxwell–Boltzmann quantum statistics for electrolyte solutions in the following form:

$$n_{i\,sr}^{0} = n_i \exp\left(\frac{-U_{ij}}{kT}\right), \quad n_{j\,sr}^{0} = n_j \exp\left(\frac{-U_{ji}}{kT}\right), \tag{2.1}$$

where

$n_{i\,sr}$ and $n_{j\,sr}$ are the average concentrations of ions of $i$ and $j$ kind
$n_i$ and $n_j$ are concentrations of ions of $i$ and $j$ kind as well
$U_{ij}$ and $U_{ji}$ is the potential energy of $i$-ion near $j$-ion
$kT$ is its kinetic energy
$k$ is the Boltzmann constant, equal to $1.380662 \cdot 10^{-23}$ J/K
$T$ is the temperature (in Kelvin)
Sign "0" designates the equilibrium (steady) state of the medium

The normalization condition is

$$\sum_i n_{i\,sr}^{0} + \sum_j n_{j\,sr}^{0} = N, \tag{2.2}$$

where $N$ is the number of guest particles.

According to modern quantum theory, all elementary and complex particles are divided into two classes. The first class includes electrons, protons, neutrons, and all particles with the so-called *half-integer spin*. These particles are subject to *Fermi–Dirac statistics*. They are called *fermions*. The second class includes photons, $\pi$- and $K$-mesons, and all particles with *integer spin*. These particles are called *bosons*. Quantum theory does not allow any other possibilities. Boltzmann statistics is the approximate limiting case, into which Fermi–Dirac and Bose–Einstein statistics evolve, under certain conditions.

Fermi–Dirac quantum statistics:

$$n_{isr}^{0} = \frac{n_i}{\left[\exp\left(U_{ij}/kT\right)+1\right]_i}, \quad n_{jsr}^{0} = \frac{n_j}{\left[\exp\left(U_{ji}/kT\right)+1\right]}. \tag{2.3}$$

The normalization condition is

$$\sum_{ij} \frac{1}{\left[\exp\left(U_{ij}/kT\right)+1\right]} + \sum_{ji} \frac{1}{\left[\exp\left(U_{ji}/kT\right)+1\right]} = N. \tag{2.4}$$

Bose–Einstein quantum statistics:

$$n_{isr}^{0} = \frac{n_i}{\left[\exp\left(U_{ij}/kT\right)-1\right]}, \quad n_{jsr}^{0} = \frac{n_j}{\left[\exp\left(U_{ji}/kT\right)-1\right]}. \tag{2.5}$$

The normalization condition is

$$\sum_{ij} \frac{1}{\left[\exp\left(U_{ij}/kT\right)-1\right]} + \sum_{ji} \frac{1}{\left[\exp\left(U_{ji}/kT\right)-1\right]} = N. \tag{2.6}$$

In all three quantum statistics, the allowable microstates are assumed as equally possible. But the statistics differ from each other in the way they define microstates and statistical weights of macrostates. Boltzmann statistics stands for the viewpoint of *fundamental distinguishability of particles*, even when particles are absolutely identical. If *A*-particle is in quantum state I and *B*-particle is in quantum state II, then a new microstate will result when these particles will interchange places, that is, *A*-particle will enter into state II and *B*-particle into state I. On the contrary, Fermi–Dirac and Bose–Einstein quantum statistics assume that *no changes will occur* with such a rearrangement—exactly the same microstate will result. Both of these statistics stand for the viewpoint of *fundamental indistinguishability of identical particles*. The difference between Fermi–Dirac and Bose–Einstein statistics is as follows. *Fermi–Dirac statistics assumes that not more than one particle can be in every quantum state. Bose–Einstein statistics does not impose such limitations. It assumes that any number of particles can be in every quantum state.* Such a different behavior of bosons and fermions is justified in quantum mechanics by the field theory [42,43]. It will not be discussed here.

For illustration, we consider two identical particles *A* and *B*, which require distribution in terms of three quantum states [44]. States will be depicted by squares. All equally possible cases allowed by Boltzmann statistics are presented in Figure 2.1.

Altogether, there are nine microstates; the mathematical probability of each of them is 1/9. According to Bose–Einstein and Fermi–Dirac statistics (Figure 2.1), states 1 and 2, 3 and 4, and 5 and 6 are fundamentally indistinguishable, and each

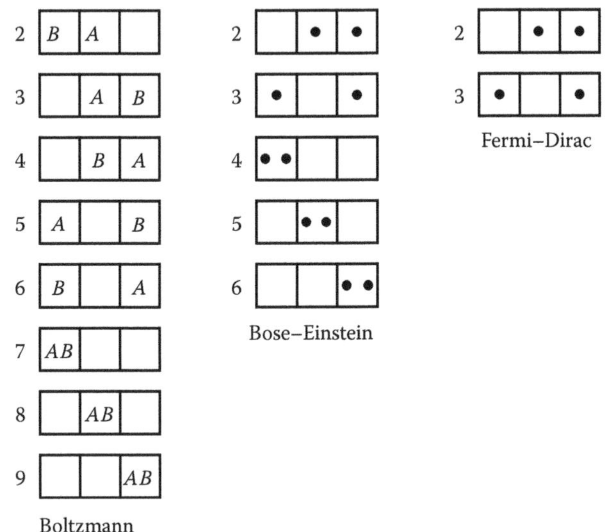


**FIGURE 2.1** Quantum states of particles *A* and *B* according to the three statistics. (From Sivukhin, D.V., *General Course of Physics. II: Thermodynamics and Molecular Physics*, FIZMATLIT, Moscow, Russia, 2005, p. 544.)

pair of such states must be regarded as one state. Particles are *depersonalized*; they already cannot be designated by different letters *A* and *B*, they are indicated by dots in our scheme. If the particles are bosons, then the number of all possible microstates will be 6, and the probability of each of them is 1/6. If probability would be determined according to Boltzmann, then each of the first three states should be attributed with the mathematical probability of 2/9, and each of the last three with 1/9. It is, therefore, evident that Boltzmann and Bose–Einstein, as well as Fermi–Dirac statistics, are fundamentally different. For fermions, the last three distributions, presented in the middle column, cannot be realized. Only three microstates remain, which are depicted on the right. The probability of each of them is equal to 1/3 (according to Boltzmann it is equal to 2/9). Transition from quantum particles to elementary ones is shown, for example, in [42–50]. It will not be demonstrated here.

Analysis of this qualitative example shows how wrong it would be to take into account macrostates of particles in an electrolyte solution, if only Boltzmann statistics would be applied.

Figure 2.2 shows the course of change in the number of particles with certain energy according to Bose (2.5), Maxwell (2.1), and Fermi (2.3) [49].

Bose distribution is most sensitive to decrease of energy, $\varepsilon$; with energy decrease, the number of particles increases rapidly. Fermi distribution weakly depends on the energy value of the particles. At low energies, it leads to a virtually constant number of particles. The number of particles depends on $\varepsilon$ for it to begin manifesting itself only at energy values close to the value of chemical potential $\mu$. Energy interval, within which this dependence is manifested, equals only to $2kT$.

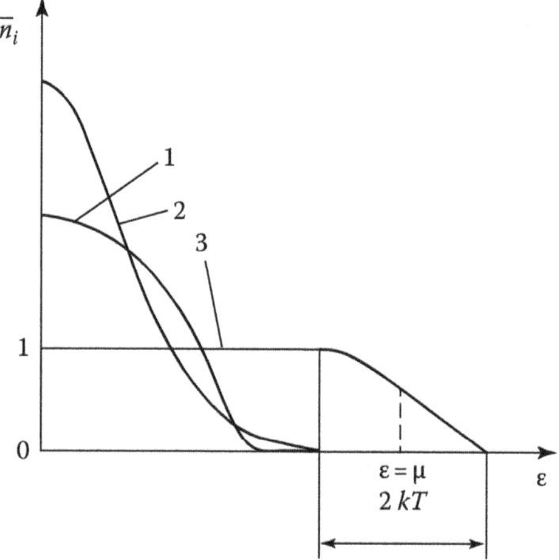


**FIGURE 2.2** Comparison of the general trend of change in the number of particles according to the three statistics: 1—Maxwell–Boltzmann statistics; 2—Bose–Einstein statistics; and 3—Fermi–Dirac statistics. (From Semenchenko, V.K., *Selected Heads of Theoretical Physics*, Education, Moscow, Russia, 1966, p. 396.)

Particles with low energies are more common in Bose distribution than in Maxwell distribution and are rarer in Fermi distribution. With energy increase, Bose–Einstein and Fermi–Dirac distributions tend to zero, and all three distributions give the same result.

We write down the quantum statistics for particles in an electrolyte solution as follows. We join the left parts of quantum characteristics (2.3) and (2.5), and then add (2.1), and, after some transformations, for particles $\overset{0}{n}_{i\,sr}$ we have

$$\overset{0}{n}_{i\,sr} = n_i \exp\left(\frac{-U_{ij}}{kT}\right) + \frac{n_i}{\left[\exp\left(U_{ij}/kT\right)+1\right]_i} + \frac{n_i}{\left[\exp\left(U_{ij}/kT\right)-1\right]}$$
$$= n_i \frac{3\exp\left(U_{ij}/kT\right)-\exp\left(-U_{ij}/kT\right)}{\exp\left(2U_{ij}/kT\right)-1}. \quad (2.7)$$

For particles $\overset{0}{n}_{j\,sr}$ of the right part of (2.1), (2.3), and (2.5), we similarly obtain

$$\overset{0}{n}_{j\,sr} = n_j \exp\left(\frac{-U_{ji}}{kT}\right) + \frac{n_j}{\left[\exp\left(U_{ji}/kT\right)+1\right]_i} + \frac{n_j}{\left[\exp\left(U_{ji}/kT\right)-1\right]}$$
$$= n_j \frac{3\exp\left(U_{ji}/kT\right)-\exp\left(-U_{ji}/kT\right)}{\exp\left(2U_{ji}/kT\right)-1}. \quad (2.8)$$

In a binary electrolyte solution, there are two kinds of ions: $n_i$ and $n_j$. In this case, we add together (2.7) and (2.8) for the two kinds of ions, and after some transformations, we obtain quantum statistics for cations and anions in the following form ($n_i \neq n_j$):

$$\overset{0}{n}_{i\,sr} + \overset{0}{n}_{j\,sr} = n_i \frac{3\exp\left(U_{ij}/kT\right) - \exp\left(-U_{ij}/kT\right)}{\exp\left(2U_{ij}/kT\right) - 1} + n_j \frac{3\exp\left(U_{ji}/kT\right) - \exp\left(-U_{ji}/kT\right)}{\exp\left(2U_{ji}/kT\right) - 1}. \tag{2.9}$$

The normalization condition for (2.9) is

$$\sum_{ij} \frac{3\exp\left(U_{ij}/kT\right) - \exp\left(-U_{ij}/kT\right)}{\exp\left(2U_{ij}/kT\right) - 1} + \sum_{ji} \frac{3\exp\left(U_{ji}/kT\right) - \exp\left(-U_{ji}/kT\right)}{\exp\left(2U_{ji}/kT\right) - 1} = N. \tag{2.10}$$

In case of symmetric electrolytes in a solution ($n_i = n_j$), after some transformations, (2.9) has the form of

$$\overset{0}{n}_{ij\,sr} = n_{ij} \frac{2\left[3\exp\left(U_{ij}/kT\right) - \exp\left(-U_{ij}/kT\right)\right]}{\exp\left(2U_{ij}/kT\right) - 1}. \tag{2.11}$$

The normalization condition for (2.11) is

$$\sum_{ij} \frac{2\left[3\exp\left(U_{ij}/kT\right) - \exp\left(-U_{ij}/kT\right)\right]}{\exp\left(2U_{ij}/kT\right) - 1} = N.$$

Figure 2.3 reflects the statistical distribution of potential energy of particles of an electrolyte solution of symmetric type from average concentration of particles in a unit of volume and radius between ions $R$ (nm) according to statistics (2.11), curve 1; Maxwell–Boltzmann statistics, curve 3, formula (2.1); and electrostatic interactions, curve 2.

The left vertical axis shows the scale of values for the potential of electrostatic part of guest particle interactions, $\left|\Psi^0_{el}\right|$, and the ratio $n_{ij\,sr}/n_{ij}$ for Maxwell–Boltzmann statistics. The right vertical axis shows the scale of values for ratio $n_{ij\,sr}/n_{ij}$ for the statistics according to formula (2.11). The experimental data to calculate points of the curves were taken from [21].

Curve 1 in Figure 2.3 demonstrates the statistical significance in the first area of concentration structural transition to average concentrations in accordance with Figure 1.1 to the conventional border of the second area. It is defined by a point

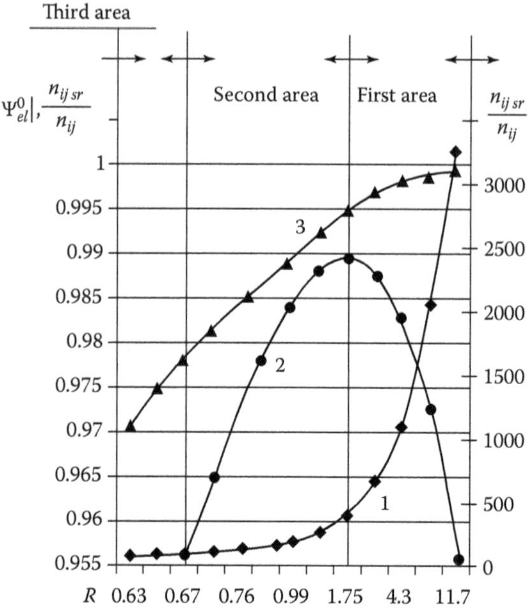


**FIGURE 2.3** The course of change of statistical distributions and electrostatic interactions of guest particles from average concentration of ions for electrolytes of symmetric type: 1—statistics according to formula (2.11); 2—electrostatic interactions; and 3—Maxwell–Boltzmann statistics, formula (2.1).

of sudden deceleration in the increase of electrostatic interactions, as it is seen in curve 2. Area 1 is a zone of formation of various complexes in the initial network of water.

Curve 2 falls down steeply in the second area of further increase of guest ion concentrations. Intensive formation of supermolecules of ion-aqueous clusters and fragments of ion hydration spheres occurs in accordance with Figure 1.1, and electrostatic interactions between ions are substantially shielded. Curve 1 has almost no statistical significance any more. Initial water network is destroyed, and further increase of ion concentrations in a solution leads to full shielding of electrostatic forces.

The third area is a zone of formation and interaction of various supramolecular assemblies in a destroyed water network. No statistics can reflect such processes, and this will be shown in the next section.

Figure 2.3 presents curve 3 of Maxwell–Boltzmann statistics. As it can be noticed, it permeates the whole concentration area, without any peculiarities. It does not carry any information on its own. It should be noted that all statistical sums and many Hamiltonians of quantum chemistry rest on this statistics. Simple exponential form of Maxwell–Boltzmann statistics allows ease of work: to integrate, to differentiate, to expand into series, etc. In its pure form, it is not applicable for electrolyte solutions; such complicated interactions, which take place with increase in ion concentration between all particles being formed, cannot be described by such a straight inclined line, as curve 3 in Figure 2.3. Certainly, Fermi–Dirac and Bose–Einstein statistics are very hard to use, they are difficult to integrate, do not expand into series, and are difficult to differentiate, but, in aggregate, suite very well for statistical description of the first concentration area in Figure 2.3. This is what we use in Section 2.6.1. Regarding the effects of applying only Maxwell–Boltzmann statistics, the reader can refer to the comments in Figure 2.1.

## 2.4 SUPRAMOLECULAR HOST–GUEST STRUCTURE DISTRIBUTION FUNCTION

By the first step, proceeding from the conditions of Section 2.2, it will be necessary to determine distribution functions $f_{ij}$ and $f_{ji}$, which shall be applied to guest solutions in the general case, irrespective of whether these solutions are in a state of equilibrium or in a disturbed one. These factors are not mutually independent. We obtain a distribution function for nonstationary fields and then, in passage to the limit, for stationary ones, and we use the obtained expressions in our case under consideration.

We consider a unit volume of an electrolyte solution containing $n_i$ of ions of various kinds with charges $|z_i|e$, where $z_i$ is the ion charge and $e$ is the electron charge, equal to $1.6021892 \cdot 10^{-19}$ C.

We derive a function that expresses the probability of simultaneous occurrence of two ions in two considered unit volumes of the solution. We consider two elementary volumes, $dV_i$ and $dV_j$, determined by vectors $\vec{R}_i$ and $\vec{R}_j$ drawn from some arbitrary point. Distance between the elementary volumes will be determined as follows:

$$\vec{R}_{ji} = \vec{R}_j - \vec{R}_i = -\vec{R}_{ij}. \tag{2.12}$$

The number of complexes and ion-aqueous clusters and fragments of ion hydration spheres (supermolecular formations or associations) of ions in the unit volume of $i$ and $j$ kind is denoted by $n_i$ and $n_j$. The average concentration of supermolecular formations for a certain time interval of $i$-ions in $dV_i$ with occurrence of $j$-ion in $dV_j$ is denoted by $n_{i\,sr}$, and the average concentration of the same formations for a certain time interval of $j$-ions in $dV_j$ with occurrence of $i$-ion in $dV_j$ is denoted by $n_{j\,sr}$.

Concentrations $n_{i\,sr}$ and $n_{j\,sr}$ depend on the following variables, which shall be taken into account in general form in distribution function $f_{ij}$ and then in passage to the limit for the stationary case, which we are now considering.

1. Concentrations of supermolecular formations of ions $n_i$ and $n_j$, and average concentrations $n_{i\,sr}$ and $n_{j\,sr}$ depend on distance $\vec{R}_{ij}$ of interaction aggregate: electrostatic forces, supermolecular interactions $U_{sp}$, van der Waals forces $U_{vv}$, and weak chemical interactions $U_x$.
2. When supermolecular associations of ions are exposed to external disturbing force (e.g., external electric field action), the dependence of $n_{i\,sr}$ and $n_{j\,sr}$ on the direction of effect of this disturbing force appears.

3. With movement of the entire solution (turbulent one), $n_{isr}$ depends on location of supermolecular formations of $i$-ions, and, respectively, $n_{jsr}$ depends on location of various associations of $j$-ions. All these conditions are expressed as follows:

$$n_{isr} = f_{ij}\left(\vec{R}_i, \vec{R}_{ij}, U_{sp}, U_{vv}, U_x\right), \quad n_{jsr} = f_{ji}\left(\vec{R}_j, \vec{R}_{ji}, U_{sp}, U_{vv}, U_x\right). \qquad (2.13)$$

To find the distribution function $f_{ij}$ during a time interval $t$, we record elements of volume $\mathrm{d}V_i$ and $\mathrm{d}V_j$ in space with supermolecular formations of ions of $i$ and $j$ kind. Charge density of various supermolecular formations is denoted by ρ. Charge density of supermolecular formations of $i$-ions is denoted by $\rho_i$, and, respectively, density of the similar formations of charges of $j$-ions through $\rho_j$. Then, the probability of occurrence of ion complex of $i$-ion in $\mathrm{d}V_i$ is equal to $\rho_i/\rho$, and the probability of occurrence of the similar complex of $j$-ion in $\mathrm{d}V_j$ is $\rho_j/\rho$. Each probability is associated with concentrations of various associations of $n_i$ and $n_j$ of ions through the following expressions:

$$\frac{\rho_i}{\rho} = K_i n_i \mathrm{d}V_i; \quad \frac{\rho_j}{\rho} = K_j n_j \mathrm{d}V_j, \qquad (2.14)$$

where $K_i$ and $K_j$ are proportionality coefficients.

If $\mathrm{d}V_i$ or $\mathrm{d}V_j$ contains simultaneously $i$-ion and $j$-ion with their complexes, then this charge density is denoted by $\rho_{ij}$ or $\rho_{ji}$. In this case, the following ratios can be written down:

$$\frac{\rho_{ij}}{\rho_j} = \rho_{isr} \mathrm{d}V_i; \quad \frac{\rho_{ji}}{\rho_i} = \rho_{jsr} \mathrm{d}V_j, \qquad (2.15)$$

where $\rho_{isr}$ and $\rho_{jsr}$ are the average charge density of ionic complexes or ion-aqueous clusters and fragments of hydration spheres $n_{isr}$ and $n_{jsr}$, respectively. Eliminating $\rho_i$ and $\rho_j$ from (2.14) and (2.15), we obtain

$$\frac{\rho_{ij}}{\rho} = K_i K_j n_i \rho_{isr} \mathrm{d}V_i \mathrm{d}V_j, \quad \frac{\rho_{ji}}{\rho} = K_j K_i n_j \rho_{jsr} \mathrm{d}V_j \mathrm{d}V_i. \qquad (2.16)$$

To determine the required distribution functions, $f_{ij}$ and $f_{ji}$, we use the fact that $\rho_{ij}/\rho$ is the probability of simultaneous occurrence of $i$-ion with its complex in $\mathrm{d}V_i$ and, similarly, $j$-ion in $\mathrm{d}V_j$. We also take into account that function $f_{ij}$ is symmetric, that is,

$$f_{ij} = f_{ji}, \qquad (2.17)$$

because the distribution functions depend on distance $R = \left|\vec{R}_{ij}\right| = \left|\vec{R}_{ji}\right|$ depending on the aggregate of interactions of electrostatic forces, supermolecular interactions $U_{sp}$, van der Waals forces $U_{vv}$, and weak chemical interactions $U_x$. This case for the distribution function will be denoted by superscript "0".

Then, based on (2.13) and (2.16), the distribution function will be equal to

$$|z_j| e f_{ij}\left(\vec{R}_i, \vec{R}_{ij}, U_{sp}, U_{vv}, U_x\right) = n_j \rho_{jsr}\left(\vec{R}_i, \vec{R}_{ij}, U_{sp}, U_{vv}, U_x\right)$$

$$= n_i \rho_{isr}\left(\vec{R}_j, \vec{R}_{ji}, U_{sp}, U_{vv}, U_x\right) = |z_j| e f_{ji}\left(\vec{R}_j, \vec{R}_{ji}, U_{sp}, U_{vv}, U_x\right). \tag{2.18}$$

Without changing similarity, we can also show that

$$|z_j| e f_{ij}\left(\vec{R}_i, \vec{R}_{ij}, U_{sp}, U_{vv}, U_x\right) = n_i n_{jsr} = n_j n_{isr} = |z_i| e f_{ji}\left(\vec{R}_j, \vec{R}_{ji}, U_{sp}, U_{vv}, U_x\right). \tag{2.19}$$

Thus, $f_{ij}$ is the concentration of ion complexes or ion-aqueous clusters and fragments of hydration spheres of $j$-ions in the unit volume, at distance $\vec{R}_{ij}$ from the similar supermolecular formations of $i$-ion, which interact with supermolecular formations $n_i$ of ions being in the unit volume at distance $\vec{R}_{ji}$, and vice versa for $f_{ji}$. In short, $f_{ij}$ is equal to the concentration of ion complexes or ion-aqueous clusters and fragments of hydration spheres of $j$-ions and the similar supermolecular formations $n_i$ of ions of $i$ kind. We also note an important condition of symmetry (2.17). Usually, values $n_i$ and $n_j$ are known, and

$$f_{ij}\left(\vec{R}_i, \vec{R}_{ij}, U_{sp}, U_{vv}, U_x\right), \quad f_{ji}\left(\vec{R}_j, \vec{R}_{ji}, U_{sp}, U_{vv}, U_x\right),$$

$$\rho_{isr}\left(\vec{R}_j, \vec{R}_{ji}, U_{sp}, U_{vv}, U_x\right) \quad \text{and} \quad \rho_{jsr}\left(\vec{R}_i, \vec{R}_{ij}, U_{sp}, U_{vv}, U_x\right),$$

shall be determined.

## 2.5 RATE OF DISTRIBUTION FUNCTION CHANGE IN TIME

When ion associations are exposed to an external disturbing force, spherical electrostatic symmetry of the interaction of ion complexes is disrupted, and supermolecule begins to move in the direction of disturbing force action. Let speed $\vec{v}$ of $i$-ion complex near $j$-ion complex be equal to

$$\vec{v}_{ij} = \varphi_{ij}\left(\vec{R}_i, \vec{R}_{ij}, U_{sp}, U_{vv}, U_x\right), \tag{2.20}$$

where $\varphi$ is some function.

A similar expression is for the rate of supermolecular formation of $j$-ion near the supermolecular formation of $i$-ion:

$$\vec{v}_{ji} = \varphi_{ji}\left(\vec{R}_j, \vec{R}_{ji}, U_{sp}, U_{vv}, U_x\right). \tag{2.21}$$

Taking into account (2.18), we write down the change in number of ion complexes in volume element $dV_j$ in a unit of time:

$$-\left[\frac{\partial}{\partial x_j}\left(\vec{v}_{ji}\rho_{i\,sr}\right) + \frac{\partial}{\partial y_j}\left(\vec{v}_{ji}\rho_{i\,sr}\right) + \frac{\partial}{\partial z_j}\left(\vec{v}_{ji}\rho_{i\,sr}\right)\right] = -\nabla_j\left(\vec{v}_{ji}\rho_{i\,sr}\right)dV_j, \tag{2.22}$$

where

$x_j$, $y_j$, and $z_j$ are coordinates of position of supermolecular formation of $j$-ion in the solution

$\nabla_j$ is the Hamiltonian operator for $j$-ion

A similar expression will be for the change in number of ion complexes in volume element $dV_i$ in a unit of time:

$$-\left[\frac{\partial}{\partial x_i}\left(\vec{v}_{ij}\rho_{j\,sr}\right) + \frac{\partial}{\partial y_i}\left(\vec{v}_{ij}\rho_{j\,sr}\right) + \frac{\partial}{\partial z_i}\left(\vec{v}_{ij}\rho_{j\,sr}\right)\right] = -\nabla_i\cdot\left(\vec{v}_{ij}\rho_{j\,sr}\right)dV_i. \tag{2.23}$$

Since distribution function (2.18) is the average concentration of supermolecules of $j$-ions in element $dV_j$ with simultaneous presence of supermolecules of $i$-ions in $dV_i$ or vice versa, then it is necessary to take into account the change in number of ion complexes in $dV_j$ over time $t_i$, during which $i$-ion complexes are in $dV_i$. Change in the number of supermolecular formations in $dV_j$ over time $t_i$ is determined as follows:

$$-t_i\nabla_j\cdot\left(\vec{v}_{ij}\rho_{j\,sr}\right)dV_j. \tag{2.24}$$

We also write down a similar expression tfor change in number of supermolecular formations of $j$-ions in $dV_i$:

$$-t_j\nabla_i\cdot\left(\vec{v}_{ji}\rho_{i\,sr}\right)dV_i. \tag{2.25}$$

We take into account that the probability $t_i/t$ of occurrence of $i$-ion complex in volume $dV_i$ and the probability $t_j/t$ of occurrence of $j$-ion complex in volume $dV_j$ are associated with concentrations of supermolecular formations $n_i$ and $n_j$ through the following ratios:

$$\frac{t_i}{t} = K_i n_i dV_i; \quad \frac{t_j}{t} = K_j n_j dV_j. \tag{2.26}$$

Proceeding from (2.22), (2.23), and (2.26), we obtain an overall change in the number of ion complexes in volume elements $dV_i$ and $dV_j$ over time $t$:

$$-t\left[n_j\nabla_j\cdot\left(\vec{v}_{ij}\rho_{jsr}\right)+n_i\nabla_i\cdot\left(\vec{v}_{ji}\rho_{isr}\right)\right]dV_idV_j. \tag{2.27}$$

Change of the distribution function in time will be determined by the following expression:

$$\frac{|z_j|e\partial f_{ij}}{\partial t}=-\left[n_i\nabla_i\cdot\left(\vec{v}_{ij}\rho_{jsr}\right)+n_j\nabla_j\cdot\left(\vec{v}_{ji}\rho_{isr}\right)\right]=\frac{|z_j|e\partial f_{ji}}{\partial t}. \tag{2.28}$$

Taking into account (2.18) and (2.19), we obtain (2.28) in the form of

$$-\frac{\partial f_{ij}\left(\vec{R}_i,\vec{R}_{ij},U_{sp},U_{vv},U_x\right)}{\partial t}=\nabla_i\cdot\left(f_{ij}\vec{v}_{ij}\right)+\nabla_j\cdot\left(f_{ji}\vec{v}_{ji}\right)=-\frac{\partial f_{ji}\left(\vec{R}_j,\vec{R}_{ji},U_{sp},U_{vv},U_x\right)}{\partial t}. \tag{2.29}$$

This expression is nothing else but continuity equation, which has application in the theory of ion motion. In steady state, $\partial f_{ji}/\partial t=0$. Then

$$\nabla_i\cdot\left(f_{ij}\vec{v}_{ij}\right)+\nabla_j\cdot\left(f_{ji}\vec{v}_{ji}\right)=0. \tag{2.30}$$

To apply Equations 2.29 and 2.30, it is necessary to be able to determine $f_{ij}$ and $\vec{v}_{ij}$.

### 2.5.1 Equations of Motion

Ions can move under the influence of, firstly, forces acting on ions and, secondly, heat (random) motion, and, thirdly, the entire motion of a solution. Forces acting on ions can be external (external electric field) and internal (concentration gradients, long-range and short-range forces, etc.), conditioned by presence of ion supermolecules themselves and of host particles.

Speed of $i$-ion complex with mobility $\omega_i$ and force $\vec{F}_i$ acting on it will be $\vec{F}_i\omega_i$. Friction coefficient $\rho_{itr}$ is determined as $\rho_{itr}=1/\omega_i$. Flow of ion supermolecules $kT\omega\nabla f$ emerges under the action of concentration gradient $\nabla f$ and diffusion coefficient according to Nernst—$kT\omega$. On the other hand, this flow of ion supermolecules is equal to $\vec{v}f$ and diffusion rate is $kT\nabla\ln f$. If we denote speed of solution motion by $\vec{V}(\vec{R}_i)$ and $\vec{V}(\vec{R}_j)$ in the points determined by vectors $\vec{R}_i$ and $\vec{R}_j$, then the total speed in these points will be determined by the following formulas:

$$\vec{v}_{ij}=\vec{V}\left(\vec{R}_i\right)+\omega_i\left(\vec{F}_{ij}-kT\nabla_i\ln f_{ij}\right),\quad \vec{v}_{ji}=\vec{V}\left(\vec{R}_j\right)+\omega_j\left(\vec{F}_{ji}-kT\nabla_j\ln f_{ji}\right). \tag{2.31}$$

From (2.30) and (2.31), we obtain

$$\nabla_i f_{ij}\left[\vec{V}\left(\vec{R}_i\right)+\omega_i\left(\vec{F}_{ij}-kT\nabla_i \ln f_{ij}\right)\right]+\nabla_j f_{ji}\left[\vec{V}\left(\vec{R}_j\right)+\omega_j\left(\vec{F}_{ji}kT\nabla_j-\ln f_{ji}\right)\right]=0. \tag{2.32}$$

To apply this equation, for example, to the theory of irreversible processes, it is necessary to be able to determine $\vec{v}_{ij}$ and $\vec{v}_{ji}$.

For stationary fields being considered in our case, equilibrium case is characterized by the fact that speed values $\vec{V}\left(\vec{R}_i\right)$ and $\vec{V}\left(\vec{R}_j\right)$ $\omega_i=\omega_j=1$ disappear in Equation 2.32 and we also use (2.17). Then we can write down

$$\vec{F}_{ij} = kT\nabla\cdot\ln f_{ij} = kT\nabla\cdot\ln f_{ji} = \vec{F}_{ji}. \tag{2.33}$$

## 2.6 SUPRAMOLECULAR INTERACTIONS POTENTIAL

Potential of supramolecular host–guest structure includes not only electrostatic forces but also various ion–dipole interactions, van der Waals forces, supramolecular weak hydrogen bonds, and forces of repulsive nature. This subsection considers electrostatic interactions in detail.

Section 2.5 determines distribution functions $f_{ij}$ and $f_{ji}$, which may be applied to electrolyte solutions in the general case, when these solutions are in equilibrium state (see (2.18), (2.30), and (2.33)). We will use the obtained expressions in our case under consideration.

Host–guest aggregate contains electric charges of ions that stay immobile; the ratio between electric field intensity generated by these charges and charge density is described by Maxwell's equation for electrostatic fields:

$$\text{div}E^0 = \frac{\overset{0}{\rho}_{sr}}{\varepsilon_S\varepsilon_0}, \tag{2.34}$$

where

$\varepsilon_S$ is the dielectric constant of a solvent
$\varepsilon_0$ is the electric constant, equal to $8.85\cdot10^{-12}$ F/m
$E^0$ is the electric field intensity, V/m
$\overset{0}{\rho}_{sr}$ is the volumetric density of electric charge, C/m

Here, as before, sign "0" further denotes the steady state of the medium.

### 2.6.1 MAXWELL'S EQUATIONS

Maxwell's equations are a theoretical generalization of experimental laws: Coulomb's law, Ampere's law, laws of electromagnetic induction, and others.

In Maxwell's equations, the following are postulated:

- Coulomb's law
- Invariance of charge in various inertial reference systems
- Superposition principle

*Superposition principle* is one of the most general laws in many subdisciplines of physics. In its simplest formulation, superposition principle states: Result of action of several external forces on a particle is the vector sum of the action these forces.

Superposition principle is best known in electrostatics, in which it claims that the energy of interaction of all particles in a many-particle system is simply the sum of energies of pair-wise interactions between all possible pairs of particles. In some publications devoted to intermolecular interactions, the question is discussed whether these interactions are additive or not. All ion, ion–dipole, dipole–dipole, including weak hydrogen and chemical bonds have common electrostatic nature and are subject to superposition principle.

Maxwell's equations are fully compatible with principles of the special relativity theory. They are also applicable with microscopic description of a substance, when charged particles are subject to the principles of quantum mechanics and electromagnetic field remains classical (not quantum). In this case, quantum objects (e.g., electrons) are described by Schrödinger's equation or Dirac's equation; however, electromagnetic interaction potentials in these equations are determined by classical Maxwell's equations.

We would like to warn future researchers about the indiscriminate use of Coulomb's law in quantum chemistry and statistical physics, when trying to describe equilibrium systems associated with the presence of one kind of elementary charges or another. A long time ago, in his review devoted to works on atom–atom potentials,* Kitaigorodsky stated [51]:

> But how one takes into account electrostatic interactions while using atom-atom potentials scheme for conformational calculations? There is only one option to attribute the so-called residual charge to each of the atoms. These charges can be determined by approximate quantum-chemical methods. However, from the author's viewpoint, such calculations are quite conditional because of uncertainty of atomic charge concept itself. Anyway, there is a large number of works, in which authors add *coulomb term* into atom-atom-potentials formula. Charge values are either tried (then the number of parameters dramatically increases, and *value of the theory becomes next to nothing*), or taken from quantum-mechanical calculations. Unfortunately, because of the mentioned uncertainty of atomic charge concept itself, their values will vary substantially (*up to difference in sign*), depending on calculation method.

And nowadays, Coulomb's law, as one of the main kinds of ion interactions in an electrolytic solution, is applied in statistical physics as well, for example [52].

* At the same time, we take into account that much more complicated interactions are present in aqueous electrolyte solutions than between neutral atoms or molecules.

The objectives of this chapter do not include reviewing the application of Hamiltonians and Schrödinger equations or statistical sums. But we make one more remark. In the electrostatic theory, Earnshaw theorem is applied, which is outlined in many textbooks on physics, for example [53]: *there is no such configuration of immobile charges, which would be stable if there are no other forces except Coulomb ones.* Or in another version: *closed system of immobile charges cannot be in stable equilibrium state (under the condition that only Coulomb forces are acting between charges).* We do not provide evidence, because, again, this is outside the scope of this chapter. In electrolytic solutions, it is necessary to describe properties of electrostatic fields occurring between ions. Otherwise, when repeatedly used by various researchers law of Coulomb forces of ions interaction in a solution it shall lead to joining of all ions into some sphere, but it does not happen in reality.

Many researchers write down Coulomb's law as $F_{1,2} = q_1q_2/\varepsilon R_{1,2}$. But this formula is not Coulomb's law. Note that this law is written down as follows: $F_{1,2} = q_1q_2/\varepsilon R_{1,2}^2$. Radius between particles is measured or calculated in nm. Squared radius in nm, being in denominator, leads to big troubles in obtaining final results. It was easier to omit the square in radius and call these forces as Coulomb ones.

We return to Equation 2.34. According to this equation, divergence of electric field intensity at any point of the medium is proportional to charge density at this point. In an electrolyte solution, charge density at distance $\vec{R}_{ij}$ from $i$-ions is equal to $\overset{0}{n}_{jsr}\left|z_{je}\right|$ and from $j$-ions $\overset{0}{n}_{isr}\left|z_{ie}\right|$.

To determine $\overset{0}{\rho}_{sr}$ in (2.34), we take into account that electrical neutrality condition is observed in a host solution:

$$n_i\left|z_i\right|e = n_j\left|z_i\right|e, \tag{2.35}$$

or, which is the same thing:

$$\overset{0}{n}_{isr}\left|z_i\right|e = \overset{0}{n}_{jsr}\left|z_i\right|e, \tag{2.36}$$

where $e$ is the electron charge, equal to $1.6021892 \cdot 10^{-19}$ C.

Then, total charge density is expressed as follows:

$$\overset{0}{\rho}_{sr} = \overset{0}{n}_{isr}\left|z_i\right|e - \overset{0}{n}_{jsr}\left|z_i\right|e. \tag{2.37}$$

We substitute Maxwell–Boltzmann (2.1), Fermi–Dirac (2.3), and Bose–Einstein (2.5) quantum statistics into (2.37), taking into account (2.12) in the following form:

$$U_{ji}\left(-R_{ij}\right) = -U_{ij}\left(R_{ij}\right), \tag{2.38}$$

$$\rho_{sr}^{0} = n_i |z_i| e \left\{ \exp\left(\frac{-U_{ij}}{kT}\right) + \frac{1}{\left[\exp\left(U_{ij}/kT\right)+1\right]} + \frac{1}{\left[\exp\left(U_{ij}/kT\right)-1\right]} \right\}$$

$$-n_j |z_i| e \left\{ \exp\left(\frac{U_{ij}}{kT}\right) + \frac{1}{\left[\exp\left(-U_{ij}/kT\right)+1\right]} + \frac{1}{\left[\exp\left(-U_{ij}/kT\right)-1\right]} \right\}. \quad (2.39)$$

From (2.39), we have

$$\rho_{isr}^{0} = n_i |z_i| e \left\{ \exp\left(\frac{-U_{ij}}{kT}\right) + \frac{1}{\left[\exp\left(U_{ij}/kT\right)+1\right]} + \frac{1}{\left[\exp\left(U_{ij}/kT\right)-1\right]} \right\},$$

$$\rho_{jsr}^{0} = n_j |z_i| e \left\{ \exp\left(\frac{U_{ij}}{kT}\right) + \frac{1}{\left[\exp\left(-U_{ij}/kT\right)+1\right]} + \frac{1}{\left[\exp\left(-U_{ij}/kT\right)-1\right]} \right\}. \quad (2.40)$$

With regard to (2.40), relation (2.18) will look like the following:

$$\begin{aligned} |z_i| e f_{ij}^{0} &= n_i n_j |z_i| e \left\{ \exp\left(\frac{-U_{ij}}{kT}\right) + \frac{1}{\left[\exp\left(U_{ij}/kT\right)+1\right]} + \frac{1}{\left[\exp\left(U_{ij}/kT\right)-1\right]} \right\} \\ &= n_j n_i |z_i| e \left\{ \exp\left(\frac{U_{ij}}{kT}\right) + \frac{1}{\left[\exp\left(-U_{ij}/kT\right)+1\right]} + \frac{1}{\left[\exp\left(-U_{ij}/kT\right)-1\right]} \right\} \\ &= |z_i| e f_{ji}^{0}. \end{aligned} \quad (2.41)$$

We substitute (2.41) into (2.33) and, after some transformations, we obtain

$$\begin{aligned} kT\nabla \cdot \ln f_{ij}^{0} &= n_j |z_i| e \left\{ \exp\left(\frac{-U_{ij}}{kT}\right) + \frac{1}{\left[\exp\left(U_{ij}/kT\right)+1\right]} + \frac{1}{\left[\exp\left(U_{ij}/kT\right)-1\right]} \right\} \\ &= n_j n_i |z_i| e \left\{ \exp\left(\frac{U_{ij}}{kT}\right) + \frac{1}{\left[\exp\left(-U_{ij}/kT\right)+1\right]} + \frac{1}{\left[\exp\left(-U_{ij}/kT\right)-1\right]} \right\} \\ &= kT\nabla \cdot \ln f_{ji}^{0}. \end{aligned} \quad (2.42)$$

Equation 2.42 will be kept only when potential energy of interaction between ions, $U$, will have spherical symmetry. In case of electrostatic field, generated by ions, potential serves as a measure of interaction between charges and, in accordance with Gauss theorem, potential field has spherical symmetry in the system of immobile continuously distributed charges. We proceed to the discovery of $U$, as applied to electrolyte solutions.

It is known from electrostatics that potential energy of interaction of two point charges $|z_i|e$ and $|z_j|e$, being in a solution at distance $R_{ij}$ from each other, can be calculated by the following formula:

$$U_{ij} = \frac{1}{4\pi\varepsilon_0}\cdot\frac{|z_i z_j|e^2}{\varepsilon_S R_{ij}} = \frac{1}{4\pi\varepsilon_0}\cdot\frac{|z_j z_i|e^2}{\varepsilon_S R_{ji}} = U_{ji}. \tag{2.43}$$

We consider a system consisting of $N$ point charges. Interaction energy of such system is equal to a sum of energies of interaction of charges taken in pairs:

$$U = \sum_{i=1}^{N} |z_i|e \sum_{j=1}^{N} \frac{1}{4\pi\varepsilon_0}\cdot\frac{|z_i|e}{\varepsilon_S R_{ij}}. \tag{2.44}$$

Formula (2.44) can be represented in the following form*:

$$U = \sum_{i=1}^{N} |z_i|e\varphi_j, \tag{2.45}$$

where $\varphi_j$ is the potential in $i$th charge location point, generated by all the other charges:

$$\varphi_j = \frac{1}{4\pi\varepsilon_0}\sum_{j=1}^{N}\cdot\frac{|z_i|e}{\varepsilon_S R_{ij}}. \tag{2.46}$$

Formula (2.45) determines not the total electrostatic energy of point charges system but only their mutual potential energy. Each charge $|z_i|e$, taken individually, has an electrical energy. It is called the intrinsic energy of charge and it is the energy of mutual repulsion of infinitely small parts into which it can be imaginatively split. Certainly, this energy is not taken into account in formula (2.45). Only work is taken into account, which is spent on interaction of charges $|z_i|e$ and $|z_j|e$ but not on their generation.

* In fact, $U = (1/2)\sum_{i=1}^{N}(q_i)\sum_{j=1}^{N}(1/4\pi\varepsilon_0)\cdot(q_j/R_{ij})$ formula is used in electrostatics. It is explained by the fact that both indices run from 0 to $N$, independently of each other. Summands, for which $i$-index value coincides with $j$-index value, are not taken into account. Coefficient 1/2 is set because, while summing-up, potential energy of each pair of charges is counted twice, that is, similar charges are considered. But we consider electrolyte solutions that contain cations and anions, besides there are electrolytes of symmetric type 1–1 (2–2) and nonsymmetric 1–2 (2–1), etc. We will take into account the similar coefficient in Section 2.3.

We return to (2.39) and consider index of power in the exponential with regard to (2.43), and write it down as follows:

$$\frac{U_{ij}}{kT} = \frac{1}{4\pi\varepsilon_0} \cdot \frac{|z_i z_j| e^2}{\varepsilon_S kTR_{ij}} = \frac{1}{4\pi\varepsilon_0} \cdot \frac{|z_j z_i| e^2}{\varepsilon_S kTR_{ji}} = \frac{U_{ji}}{kT}. \tag{2.47}$$

In (2.47), we denote

$$\beta = \frac{1}{4\pi\varepsilon_0} \cdot \frac{|z_i z_j| e^2}{\varepsilon_S kT} = \frac{1}{4\pi\varepsilon_0} \cdot \frac{|z_j z_i| e^2}{\varepsilon_S kT}. \tag{2.48}$$

With regard to (2.48) and (2.12), we rewrite (2.39) in the form of

$$\rho_{sr}^0 = n_i |z_i| e \left\{ \exp\left(\frac{-\beta}{R_{ij}}\right) + \frac{1}{\left[\exp(\beta/R_{ij})+1\right]} + \frac{1}{\left[\exp(\beta/R_{ij})-1\right]} \right\}$$

$$- n_j |z_i| e \left\{ \exp\left(\frac{\beta}{R_{ij}}\right) + \frac{1}{\left[\exp(-\beta/R_{ij})+1\right]} + \frac{1}{\left[\exp(-\beta/R_{ij})-1\right]} \right\}. \tag{2.49}$$

We simplify expression (2.49). We take the second term in the first parentheses, the second summand in the second parentheses, sum them up, and, after some transformations, we obtain

$$\frac{1}{\left[\exp(\beta/R_{ij})+1\right]} - \frac{1}{\left[\exp(-\beta/R_{ij})+1\right]} = -\frac{\exp(\beta/R_{ij})-1}{\exp(\beta/R_{ij})+1}. \tag{2.50}$$

Further, we take the third term in the first parentheses, the third summand in the second parentheses, sum them up as well, and, after some transformations, we obtain

$$\frac{1}{\left[\exp(\beta/R_{ij})-1\right]} - \frac{1}{\left[\exp(-\beta/R_{ij})-1\right]} = \frac{\exp(\beta/R_{ij})+1}{\exp(\beta/R_{ij})-1}. \tag{2.51}$$

We sum (2.50) and (2.51) up, and, after some transformations, we have

$$\frac{\exp(\beta/R_{ij})+1}{\exp(\beta/R_{ij})-1} - \frac{\exp(\beta/R_{ij})-1}{\exp(\beta/R_{ij})+1} = \frac{4\exp(\beta/R_{ij})}{\exp(2\beta/R_{ij})-1} \equiv \frac{2}{\sinh(\beta/R_{ij})}. \tag{2.52}$$

We return to (2.49). We take the first summand in the first parentheses, and also the first one in the second parentheses, and, after some transformations, we obtain

$$\exp\left(\frac{-\beta}{R_{ij}}\right)-\exp\left(\frac{\beta}{R_{ij}}\right)\equiv -2\sinh\left(\frac{\beta}{R_{ij}}\right). \tag{2.53}$$

We sum (2.52) and (2.53) up and perform some transformations:

$$\frac{2}{\sinh(\beta/R_{ij})}-2\sinh\left(\frac{\beta}{R_{ij}}\right)=2\left[\operatorname{csch}\left(\frac{\beta}{R_{ij}}\right)\cosh\left(\frac{\beta}{R_{ij}}\right)-2\sinh\left(\frac{\beta}{R_{ij}}\right)\right]. \tag{2.54}$$

With regard to (2.54), we write down Maxwell's equation (2.34) for electrostatic fields in an electrolyte solution in the following form:

$$\operatorname{div} E^0=2\frac{n_i|z_i|e}{\varepsilon_S\varepsilon_0}\left[\operatorname{csch}\left(\frac{\beta}{R_{ij}}\right)\cosh\left(\frac{\beta}{R_{ij}}\right)-2\sinh\left(\frac{\beta}{R_{ij}}\right)\right]. \tag{2.55}$$

We introduce the following designation into (2.55):

$$\kappa=\frac{n_i|z_i|e}{\varepsilon_S\varepsilon_0}. \tag{2.56}$$

With regard to (2.56), Equation 2.55 acquires the following form:

$$\operatorname{div} E^0=2\kappa\left[\operatorname{csch}\left(\frac{\beta}{R_{ij}}\right)\cosh\left(\frac{\beta}{R_{ij}}\right)-2\sinh\left(\frac{\beta}{R_{ij}}\right)\right]. \tag{2.57}$$

We take into account that electric field intensity for electrostatic fields has the following dependence through potential gradient:

$$E^0=-\operatorname{grad}\Psi^0(R), \tag{2.58}$$

where $\Psi^0(R)$ is the potential. In this case, with regard to (2.58), Equation 2.57 is as follows:

$$\Delta\Psi^0(R)=-2\kappa\left[\operatorname{csch}\left(\frac{\beta}{R_{ij}}\right)\cosh\left(\frac{\beta}{R_{ij}}\right)-2\sinh\left(\frac{\beta}{R_{ij}}\right)\right], \tag{2.59}$$

with boundary conditions:

$$\frac{d\Psi^0(R)}{dR} \to 0 \quad \text{with } R \to \infty,$$
$$\Psi^0(R) \to U_{sp} + U_{vv} + U_x \quad \text{with } R \to a, \tag{2.60}$$

where in (2.59)

$\Delta$ is the Laplacian operator

$\Psi^0(R)$ is the supramolecular interactions potential

In (2.60)

$U_{sp}$ is the supramolecular interactions (various interactions of ion–dipole and dipole–dipole)

$U_{vv}$ is the van der Waals forces (orientation interaction [Keesom effect], induction interaction [Debye effect], and dispersion interaction [London effect] in an electrolyte solution)

$U_x$ is the short-range forces (supramolecular weak hydrogen bonds associated with manifestation of Pauli principle, and short-range repulsion forces)

$a$ is the minimal distance of interaction of electrostatic forces between ions, nm

We will dwell on the structure of supramolecular interactions potential and the introduced boundary conditions in more detail. Proceeding from Equation 2.59 and boundary conditions (2.60), it is possible to distinguish three characteristic zones of host–guest aggregates with increase in their concentrations, which sequentially overlap:

1. The first zone of concentrations is dominated by long-range forces of electrostatic interaction, $\psi^0_{el}$. Host–guest aggregates are formed in initial water network, which depend on coordination numbers of ion hydration There are dipoles in an electrolyte solution, which are not bonded into ion hydration shells. The structure of water H-bonds is slightly disrupted.
2. In the second zone of increase in ion concentrations, steric effects are significantly manifested, eigenstructure of water H-bonds is almost destroyed, and structural transition of the electrolyte solution starts to be largely defined by supermolecules formation. With $r = a$, the influence of electrostatic forces between ions reduces to zero, because ionic or ion-aqueous clusters and fragments of ion hydration spheres are intensively formed, and the solution is quantitatively described by interactions $U_{sp}$, $U_{vv}$, and $U_x$.
3. In the third area of increase in ion concentrations, water molecules are already not enough for their entire intraspheric hydration. Formation of supramolecular assemblies in a host solution occurs to an appreciable extent with the formation of aqueous polymer melts of crystalline hydrates or salts, hydroxides, and acids. In case of contact of fragments of

ion hydration spheres with each other, formation of weak chemical bond occurs, which appears with the participation of interparticle hydrogen bond and overlap of electron clouds, which significantly affects physical and chemical properties of such solvent as water. Water structure is completely destroyed. From our point of view, supramolecular host–guest structure interchanges its places, and zone of coordination chemistry regularities appears, where water-host acts as a ligand-guest. Interactions $U_x$ dominate.

To date, the intermolecular interactions theory has achieved significant success [54–61]. Quantitative description of $\Psi^0$, $U_{sp}$, $U_{vv}$, and $U_x$ in concentrated electrolyte solutions remains unsatisfactory.

## 2.7 SOLVING THE EQUATION FOR POTENTIAL IN THE ABSENCE OF EXTERNAL FIELDS

In Equation 2.59, the Laplacian operator for scalar fields with spherical (central) symmetry is represented by the following expression:

$$\Delta\Psi^0(R)=\frac{1}{R}\frac{\partial^2}{\partial R^2}\left[R\Psi^0(R)\right]. \tag{2.61}$$

Taking into account (2.61) and (2.59), we obtain the following equation in spherical coordinates:

$$\frac{1}{R}\frac{\partial^2}{\partial R^2}\left[R\Psi^0(R)\right]=-2\kappa\left[\operatorname{csch}\left(\frac{\beta}{R}\right)\cosh\left(\frac{\beta}{R}\right)-2\sinh\left(\frac{\beta}{R}\right)\right], \tag{2.62}$$

or

$$\frac{\partial}{\partial R}\left[R\Psi^0(R)\right]=-2\kappa\left[\operatorname{csch}\left(\frac{\beta}{R}\right)\cosh\left(\frac{\beta}{R}\right)-2\sinh\left(\frac{\beta}{R}\right)\right]R\mathrm{d}R. \tag{2.63}$$

We introduce substitutions:

$$R=\frac{\beta}{x};\quad \mathrm{d}R=-\frac{\beta \mathrm{d}x}{x^2},\quad x=\frac{\beta}{R}. \tag{2.64}$$

We substitute (2.64) into (2.63), and, after some transformations, we obtain

$$\frac{\partial}{\partial R}\left[R\Psi^0(R)\right]=2\kappa\beta^2\int\left[\operatorname{csch}(x)\cosh(x)-2\sinh(x)\right]\frac{\mathrm{d}x}{x^3}. \tag{2.65}$$

We consider the first expression in square brackets under the integral sign in (2.65). We will perform integration in the following manner. For hyperbolic functions csch($x$) and ch($x$), we use expansions into series:

$$\operatorname{csch}(x) \approx \frac{1}{x} - \frac{x}{6} + \frac{7x^3}{360} - \cdots, \quad x^2 < \frac{\pi^2}{4}, \tag{2.66}$$

$$\cosh(x) \approx 1 + \frac{x^2}{2!} + \frac{x^4}{4!} + \cdots, \quad x^2 < \infty. \tag{2.67}$$

We multiply (2.66) and (2.67), and after some transformations, we have

$$\operatorname{csch}(x)\operatorname{ch}(x) = \frac{1}{x} + \frac{x}{3} - \frac{x^3}{45} - \frac{x^5}{6\cdot 4!} + \frac{7x^7}{360\cdot 4!} + \cdots. \tag{2.68}$$

We substitute (2.68) into the first expression in square brackets under integral sign (2.65) and obtain

$$\int \operatorname{csch}(x)\cosh(x)\frac{dx}{x^3} = \int \left(\frac{1}{x} + \frac{x}{3} - \frac{x^3}{45} - \frac{x^5}{6\cdot 4!} + \frac{7x^7}{360\cdot 4!}\right)\frac{dx}{x^3}$$

$$= \int \left(\frac{1}{x^4} + \frac{1}{3x^2} - \frac{1}{45} - \frac{x^2}{6\cdot 4!} + \frac{7x^4}{360\cdot 4!}\right)dx. \tag{2.69}$$

We integrate (2.69), and the first summand under integral (2.65) has the following form:

$$\int \operatorname{csch}(x)\cosh(x)\frac{dx}{x^3} = -\frac{1}{3x^3} - \frac{1}{3x} - \frac{x}{45} - \frac{x^3}{432} + \frac{7x^5}{43{,}200} + C_1, \tag{2.70}$$

where $C_1$ is the integration constant.

We use the first boundary condition of (2.60), in which case $d\Psi^0(R)/dR \to 0$ with $R \to \infty$. We find constant $C_1$ from it and obtain

$$C_1 = \frac{1}{3x^3} + \frac{1}{3x}. \tag{2.71}$$

Such condition is necessary to ensure that the expression in (2.70) with $x \to 0$ does not turn into $\infty$. We also take into account that, according to the presented estimates in [62], the last two summands in (2.70) are of low significance during subsequent integration. We substitute (2.71) into (2.70), and (2.70) is as follows:

$$\int \operatorname{csch}(x)\cosh(x)\frac{dx}{x^3} = -\frac{x}{45}. \tag{2.72}$$

Now we proceed to the integral of the right side of (2.65) $\int(\sinh x/x^3)\mathrm{d}x$ and integrate it in parts. We assume

$$U = \sinh x, \quad \mathrm{d}U = \cosh x\mathrm{d}x, \quad V = \int\frac{\mathrm{d}x}{x^3} = -\frac{1}{2x^2}.$$

For the integral under consideration, we have

$$\int\frac{\sinh x}{x^3}\mathrm{d}x = -\frac{\sinh x}{2x^2} + \frac{1}{2}\int\frac{\cosh x}{x^2}\mathrm{d}x$$

$$= -\frac{\sinh x}{2x^2} + \frac{1}{2}\left(-\frac{\cosh x}{x} + x + \frac{x^3}{3\cdot 3!} + \frac{x^5}{5\cdot 5!} + \cdots\right) + C_2. \quad (2.73)$$

To obtain integration results, we use tables of integrals [63].

We use the first boundary condition of (2.60). We find constant $C_2$ from it and obtain the following:

$$C_2 = \frac{\sinh x}{2x^2} + \frac{\cosh x}{2x}. \quad (2.74)$$

We also take into account that, according to the presented estimates in [62], the last two summands in (2.73) are of low significance during subsequent integration. We substitute (2.74) into (2.73), and (2.73) is as follows:

$$\int\frac{\sinh x}{x^3}\mathrm{d}x = \frac{x}{2}. \quad (2.75)$$

We substitute (2.75) and (2.72) into (2.65), and after some transformations, (2.65) is as follows:

$$\frac{\partial}{\partial R}\left[R\Psi^0(R)\right] = -\frac{92}{45}\kappa\beta^2 x. \quad (2.76)$$

We develop Laplacian operator (2.76) further, take into account the substitutions in (2.64), and proceed to the second stage of integration:

$$R\Psi^0(R) = \frac{92}{45}\kappa\beta^3\int\frac{\mathrm{d}x}{x} = \frac{92}{45}\kappa\beta^3\ln\beta_R + C, \quad (2.77)$$

where $\beta_R = \beta/R$.

We use the second boundary condition of (2.60) and find integration constant from Equation 2.77:

$$C = -\frac{92}{45}\kappa\beta^3 \ln\beta_a + a\left(U_{sp} + U_{vv} + U_x\right), \tag{2.78}$$

where $\beta_a = \beta/a$.

We substitute (2.78) into (2.77) and obtain the final form of supramolecular interactions potential, which takes into account all kinds of interactions in an aqueous host solution:

$$\Psi^0 = \kappa\beta^2\Psi_{el}^0 + \frac{a}{R}\left(U_{sp} + U_{vv} + U_x\right), \tag{2.79}$$

where $\Psi_{el}^0$ is the electrostatic part of supramolecular interactions potential in an electrolyte solution and is equal to

$$\Psi_{el}^0 = -\frac{92}{45}\beta_R \ln\left(\frac{R}{a}\right) \cong -2\beta_R \ln\left(\frac{R}{a}\right). \tag{2.80}$$

Figures 2.4 and 2.5 show the run of the curve of the electrostatic part of supramolecular interactions potential in 1–2 electrolyte solution according to formula (2.80).

It is clear from the figures that there are three concentration areas, which correspond to the experimentally verified structural transition from dilute solutions to concentrated ones, shown in Figure 1.1. In Figure 2.4, the curve is displayed in a

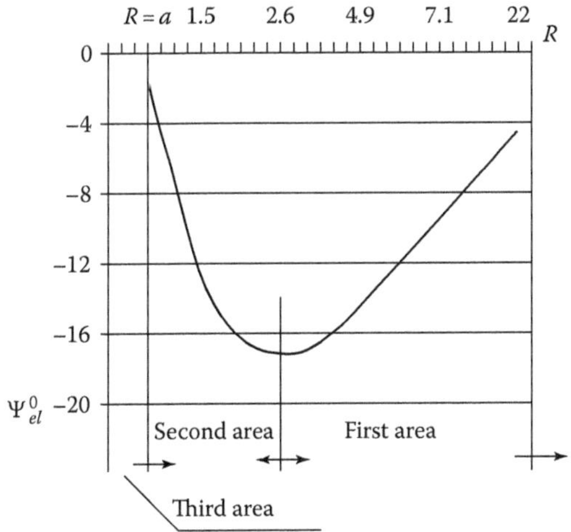


**FIGURE 2.4** Electrostatic interactions in various concentration areas of supramolecular host–guest structure.

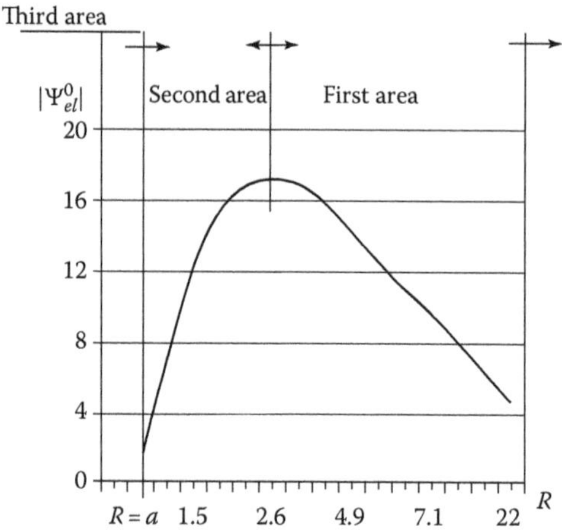


**FIGURE 2.5** The same as in Figure 2.4, but only $\Psi^0_{el}$ is taken in terms of absolute value $\left|\Psi^0_{el}\right|$.

conventional manner, $Y$ axis is going down by the increasing negative values, and for better perception Figure 2.5 presents the dependence of module $\left|\Psi(R)\right|$. The experimental data to calculate the points of the curves were taken from [21].

As it is seen from the run of the curves, in case of increase of guest concentration in a host solution (or reduction of radius between ions), the value of electrostatic forces of supermolecular interactions in terms of absolute value dramatically increases to a conventional boarder with the second area of ion concentrations in the solution. Intensive complex formation takes place; interactions appear not only between ions but also between complexes. Water dipoles are enough for all of these processes.

When approaching the second area, electrostatic interactions slow down their course and then dramatically decrease in terms of absolute value. This is the area of various supermolecular interactions, and electrostatic fields decrease because of their strong shielding up to the border of the first area.

The third area begins, in accordance with the second boundary condition of (2.60), in case of $R = a$. With this value of $R$, complete shielding of electrostatic ion fields has occurred, and various supramolecular assemblies with their interactions appear. In the beginning of the area, with $R \to \infty$ tendency, the curve asymptotically approaches axis $R$, as it is shown in enlarged scale in Figure 2.6.

Figures 2.4 and 2.5 demonstrate the distribution only of electrostatic interactions, when guest concentration increases in a host solution, so the diagrams show only the border of the third area. Distribution of all supermolecular and supramolecular interactions according to formula (2.79) will be demonstrated in the Chapter 3.

Dependence of $\psi$ on $R$, as it is required by formula (2.80), is not really demonstrative for analysis, structural transition boundaries have to be enumerated from right to left, so Figure 2.7 presents $\psi$–$m$ curve (mol/kg $H_2O$).

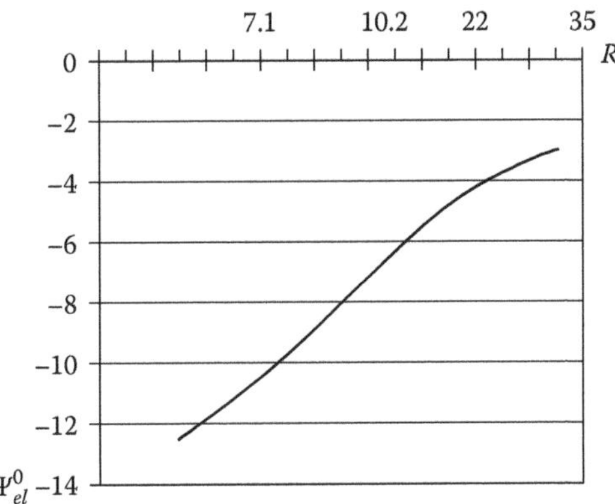


**FIGURE 2.6** Behavior of electrostatic curve in case of large distances between ions.

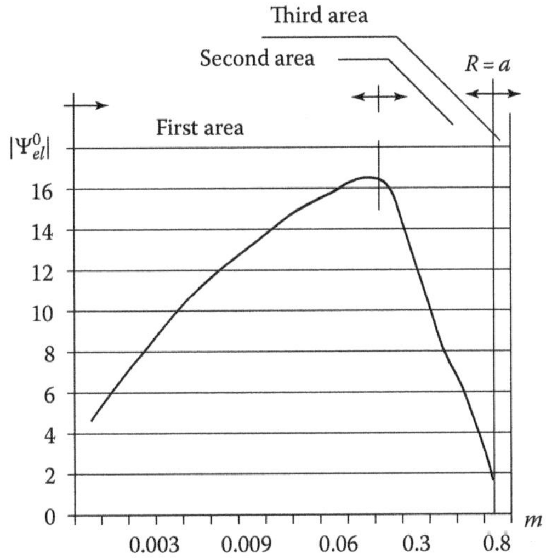


**FIGURE 2.7** Electrostatic part of supramolecular interactions potential.

## REFERENCES

1. Harned G., Owen B. 1952. *Physical Chemistry of Solutions of Electrolytes*. Publishing house of the of foreign lit., Moscow, Russia, p. 628.
2. Robinson R., Stokes R. 1963. *Solutions of Electrolytes*. 1, 2. Publishing house of the of foreign lit., Moscow, Russia, p. 1.
3. Semenchenko V.K. 1941. *Physical Theory of Solutions*. Gostekhteorizdat, Moscow, Russia, p. 368.
4. Frank H.S., Thompson W. 1959. In: Hamer W.J. (ed.). *Structure of the Electrolyte Solutions*, Vol. 31. Wiley, New York, pp. 113–120.
5. Huckel E. 1925. Zur Theorie konzentrierterer wasseriger Losungen starker Elektrolyte. *Phys. Z.* 26: 93–147.

6. Gronwall T.H., La Mer V.K. 1926. Über den einfluß der sogenannten höheren glieder in der debye-hückelschen theorie der lösungen starker elektrolyte. *Science* 64: 122–130.
7. Nakajima M.J. 1950. *Electrochem. Soc. Jpn.* 18: 106–130.
8. Ed. by G.I. Mikulin. 1968. *Questions of Physical Chemistry of Solutions of Electrolytes.* Chemistry, Leningrad, Russia, p. 420.
9. Mikulin G.I. 1955. To the electrostatic theory of strong solutions of electrolytes. *Ukr. Chem. J.* 21 (4): 435–448.
10. Guggenheim E.A. 1959. The accurate numerical solution of the Poisson–Boltzmann equation. *Trans. Faraday Soc.* 55: 1714–1720.
11. Falkenhagen G., Kelbg G. 1962. *New Problems of the Modern Electrochemistry.* Inostr. litas., Moscow, Russia, p. 130.
12. Aseyev G.G. 1988. *Electrostatic, Solvation and Chemical Interactions in Strong Solutions of Electrolytes.* KHNPO "Carbonate". Dep. in ONIITEKHIM, Cherkassy, November 29, 1988, 1153-hp88, Kharkov, Ukraine, p. 181.
13. Aseyev G.G. 1989. *Theory Development Strong Solutions of Electrolytes.* Ser. Industry-Wide Issues, Vol. 8 (286). NIITEKHIM, Moscow, Russia, p. 57.
14. Aseyev G.G. 1988. *Theoretical Dependence of the Activity Coefficients of Concentrated Solutions of Electrolytes, Based on the Potential of Interionic Interaction.* KHNPO "Carbonate". Dep. in ONIITEKHIM, Cherkassy, June 5, 1989, 528-hp89, Kharkov, Ukraine, p. 48.
15. Aseyev G.G. 1990. *Research of the Equation of Calculation of Activity Coefficients of Strong Solutions of Electrolytes.* KHNPO "Carbonate". Dep. in ONIITEKHIM, Cherkassy, January 23, 1990, 98-hp90, Kharkov, Ukraine, p. 158.
16. Aseyev G.G., Yudina L.V. 1990. *Average Ionic Activity Coefficients of the Strong Electrolytes in the Aqueous Solutions, Depending on Electrostatic, Salvation and Chemical Interactions.* KHNPO "Carbonate". Dep. in ONIITEKHIM, Cherkassy, November 1, 1990, 688-hp90, Kharkov, Ukraine, p. 207.
17. Aseyev G.G. 1990. *Dependence of Activity of Water in Strong Solutions of Electrolytes from Electrostatic, Salvation and Chemical Interactions.* KHNPO "Carbonate". Dep. in ONIITEKHIM, Cherkassy, March 15, 1990, 202-hp90, Kharkov, Ukraine, p. 31.
18. Aseyev G.G., Yudina L.V. 1991. *Aktivnost of Water in Strong Solutions of Electrolytes and Interionic Interactions.* KHNPO "Carbonate". Dep. in ONIITEKHIM, Cherkassy, November 18, 1991, 133-hp91, Kharkov, Ukraine, p. 54.
19. Aseyev G.G. 1991. *The Theory of the Irreversible Processes Which Are Flowing Past in Solutions of Electrolytes.* Ser. Industry-Wide Issues, Vol. 2 (304). NIITEKHIM, Moscow, Russia, p. 72.
20. Aseyev G.G. 1991. *Non-Equilibrium Phenomena in Strong Solutions of Electrolytes (Electrical Conductance, Viscosity, Diffusion).* (*Reviews on Thermal Properties of Substances*), Vol. 5 (91). IVT of Academy of Sciences of the USSR, Moscow, Russia, p. 82.
21. Aseyev G.G. 1998. *Elektrolites: Interparticle Interactions. Theory, Calculation Methods, and Experimental Data.* Begell House, Inc., New York, p. 754.
22. Aseyev G.G., Ryshchenko I.M., Savenkov A.S. 2004. Stable processes of thermodynamics of multicomponent strong solutions of electrolytes. *Quest. Chem. Chem. Technol.* 4: 61–66.
23. Barthel J., Buchner R., Munsterer M. 1996. *Electrolyte Data Collection. Part 2: Dielectric Properties of Nonaqueous Electrolyte Solutions.* Chemistry Data Series, Vol. XII. Dechema, Frankfurt, Germany, p. 387.
24. Barthel J., Buchner R., Munsterer M. 1995. *Electrolyte Data Collection. Part 2: Dielectric Properties of Water and Aqueous Electrolyte Solutions.* Chemistry Data Series, Vol. XII(2). Dechema, Frankfurt, Germany, p. 365.
25. Akhadov Ya.Yu. 1977. *Dielectric Properties of Binary Solutions.* Science, Moscow, Russia, p. 399.

26. Hasted J.B. 1973. *Aqueous Dielectrics*. Chapman and Hall, London, U.K., p. 302.
27. Barthel J., Krienke H., Kunz W.H. 1988. *Physical Chemistry of Electrolyte Solutions: Modern Aspects*. Springer, New York, p. 401.
28. Lyashchenko A.K., Karataeva I.M. 2010. The activity of the water and the dielectric constant of water solutions of electrolytes. *J. Phys. Chem.* 84 (2): 376–384.
29. Lyashchenko A.K., Zasetsky A.Yu. 1998. Complex dielectric permittivity and relaxation parameters of concentrated aqueous electrolyte solutions in millimeter and centimeter wavelength ranges. *J. Mol. Liq.* 77: 61–65.
30. Lileev A.S., Balakaeva I.V., Lyashchenko A.K. 2001. Dielectric properties, hydration and ionic association in binary and multicomponent formate water-salt systems. *J. Mol. Liq.* 87: 11–20.
31. Lileev AS, Lyaschenko A.K., Harkin V.S. 1992. Dielectric properties of aqueous solutions of yttrium and copper". *J. Inorg. Chem.* 37 (10): 2287–2291.
32. Lyashchenko A.K. Ed. by W. Coffey. 1994. Structure and structure-sensitive properties of aqueous solutions of electrolytes and nonelectrolytes in relaxation phenomena in condensed matter. *Advances in Chem. Phys.* LXXXVII: 379–426.
33. Lyashchenko A.K. 2001. Concentration transition from water-electrolyte to electrolyte water solvents and ionic clusters in solutions. *J. Mol. Liq.* 91: 21–31.
34. Lyashchenko A.K., Karatayeva I.M. 2007. Communication of activity of water with a statistical dielectric constant of strong solutions of electrolytes. *Dokl. Akad. Sciences.* 414 (3): 357–359.
35. Loginova D.V., Lileev A.S., Lyashchenko A.K. 2002. Temperature dependence of dielectric properties of aqueous solutions of chloride of a potassium. *Zhurn. Inorg. Chem.* 47 (9): 1558–1565.
36. Filimonova Z.A., Lileev A.S., Lyashchenko A.K. 2002. Complex inductivity and relaxation of aqueous solutions of nitrates of alkali metals. *Zhurn. Inorg. Chem.* 47 (12): 2055–2061.
37. Lileev A.S., Filimonova Z.A., Lyashchenko A.K. 2003. Dielectric permittivity and relaxation in aqueous solutions of sulfates and nitrates of alkali metals in temperature range 288–313 K. *J. Mol. Liquids* 103–104: 299–308.
38. Loginova D.V., Lileev A.S., Lyashchenko A.K. 2006. Microwave dielectric properties of aqueous solutions of fluorides of a potassium and cesium. *Zhurn. Phys. Chem.* 80 (10): 1830–1838.
39. Zasetsky A.Yu., Lyashchenko A.K. 1999. *Quasioptical Technique of Measurement of a Complex Inductivity of Aqueous Solutions of Electrolytes in the Millimetric Range of Lengths of Waves and Relaxational Characteristics of Solutions*. Dep. in VINITI of 06.07.99, 2181-B 29, Moscow, Russia, p. 62.
40. Lehn J.-M. 1995. *Supramolecular Chemistry: Concepts and Perspectives*. VCH, Moscow, Russia, p. 334.
41. Stid Dzh.B., Atwood J.L. 2007. *Supramolecular Chemistry*. In 2 t. Akademkniga, Moscow, Russia.
42. Landau L.D., Lifshits E.M. 1976. *Theoretical Physics. Part I. V: Statistical Physics*. Nauka, Moscow, Russia, p. 584.
43. Landau L.D., Lifshits E.M. 1988. *Theoretical Physics. II: Field Theory*. Nauka, Moscow, Russia, p. 542.
44. Sivukhin D.V. 2005. *General Course of Physics. II: Thermodynamics and Molecular Physics*. FIZMATLIT, Moscow, Russia, p. 544.
45. Anselm A.I. 1973. *Bases of a Statistical Physics and Thermodynamics*. Nauka, Moscow, Russia, p. 424.
46. Savukov V.V. 2006. *Specification of the Axiomatic Principles of a Statistical Physics*. Balt. State Technical University "Voenmekh", St. Petersburg, Russia, p. 322.
47. Akhiyezer A.I., Peletminsky S.V. 1977. *Methods of a Statistical Physics*. Nauka, Moscow, Russia, p. 367.

48. Kompaneets A.S. 1975. *Course of Theoretical Physics. 2: Statistical Laws*. Education, Moscow, Russia, p. 480.
49. Semenchenko V.K. 1966. *Selected Heads of Theoretical Physics*. Education, Moscow, Russia, p. 396.
50. Vonsovsky S.V. 2005. *The Modern Naturally-Scientific Picture of the World*. Publishing House of Gumanit University, Yekaterinburg, Russia, p. 680.
51. Kitaigorodsky A.I. 1979. Not valence interactions of atoms in organic crystals and molecules. *Usp. Fiz. Nauk* 127 (3): 391–419.
52. Yukhnovsky I.R. 1988. *Electrolytes*. Naukova Dumka, Kiev, Ukraine, p. 168.
53. Matveev A.N. 1983. *Electricity and Magnetism*. Higher School, Moscow, Russia, p. 463.
54. Prezhdo V.V., Kraynov N. 1994. *Molecular Interactions and Electric Properties of Molecules*. Publishing house "Osnova", Kharkov, Ukraine, p. 240.
55. Simkin B. Yu., Sheykhet I. I. 1989. *Quantum Chemical and Statistical Theory of Solutions. Computing Methods and Their Application*. Chemistry, Moscow, Russia, p. 256.
56. Shakhparonov M. I. 1976. *Introduction in the Modern Theory of Solutions (Intermolecular Interactions. Structure. Prime liquids)*. The highest school, Moscow, Russia, p. 296.
57. Eremin V.V., Kargov S.I., Uspenskaya I.A. et al. 2005. *Fundamentals of Physical chemistry. Theory and tasks*. Examination, Moscow, Russia, p. 480.
58. Girshfelder Dzh. Curtice Ch., Byrd R. 1961. *Molecular Theory of Gases and Liquids*. Publishing house of the of foreign lit. Moscow, Russia, p. 930.
59. Kaplan I.G. 1982. *Introduction to the Theory of Intermolecular Interactions*. Nauka, Moscow, Russia, p. 312.
60. Barash Yu. S. 1988. *Van-der-Waals Forces*. Nauka, Moscow, Russia, p. 344.
61. Smirnova N.A. 1987. *Molecular Theories of Solutions*. Chemistry, Leningrad, Russia, p. 336.
62. Aseyev G.G. 2010. Thermodynamics of interactions of electrostatic forces in strong solutions of electrolytes. *Zhurn. Obshch. Chem.* 80 (11): 1767–1773.
63. Dvait G.B. 1977. *Tables of Integrals and Other Mathematical Equations*. Nauka, Moscow, Russia, p. 228.

# 3 Thermodynamics of Supermolecular and Supramolecular Interactions

## 3.1 SUPERMOLECULAR–THERMODYNAMIC INTERACTIONS OF ACTIVITY COEFFICIENTS

Activity coefficients are a real physical quantity, and being defined by various methods, they give results coinciding well with each other, which indicates the self-consistency of these ideas. Salem demonstrated a comparison of activity coefficients obtained by various methods in different examples in aqueous solutions calculated from solution vapor pressure data, from cryoscopy, and from measuring electromotive forces of galvanic element [1]. At the same time, numerical values of the dissociation degree obtained, for example, from osmotic pressure, saturated vapor pressure, or electrical conductivity measuring tests do not coincide with one another. Therefore, the thermodynamic method of describing various interactions in electrolyte solutions is more reliable. When describing electrolyte solution properties from the thermodynamic point of view, it is absolutely not important in what degree the electrolyte is dissociated because this description does not reveal the mechanism of interparticle (super- and supramolecular) interactions.

The activity represents the effective concentration that a real system should have had in order to perform the same activities as the ideal system does. The activity theory tries to keep usual formulations of the thermodynamic correlations by means of common thermodynamic potentials, introducing the notion about the effective activity depending on surrounding conditions and concentrations. However, in this case, pure thermodynamics is unable to determine the type of functional dependence $\gamma = \gamma(c)$. This target can be reached by several ways: first of all, one can study the laws, followed by the changes in activity coefficients at change of concentration, temperature, etc., experimentally. Substitution of these data into empirical dependencies in thermodynamic formulas allows deducing a number of correlations and, thus, building more or less empirical real system thermodynamics of the type being studied. This was the way taken by Lewis, Bronsted, their learners, and their successors. Their works demonstrated that the activity theory provides an extremely convenient tool for processing experimental data and finding new empirical regularities. Results obtained by them demonstrate that the activity coefficient is a real physical quantity whose value does not depend on the determination method and represents the function of temperature and concentration.

The second way is an attempt of disclosing a physical meaning of activity coefficients by means of quantum chemistry and statistical mechanics methods, by means of finding regularities followed by that quantity. However, the Schrödinger equation or statistical sum for a complete ensemble of particles in a solution in volume unit cannot be solved yet. Reviews under these methods are not discussed in this chapter; some information about these methods is given in Section 1.2.

That is why we will take an opportunity to consider interparticle interactions in electrolyte solutions based on activity coefficients using the supramolecular–thermodynamic method.

During interaction of a cation with surrounding anions and vice versa, the energy of electrostatic interactions $F_{el}$ decreases for a value equal to the product of its charge $|z|e$ by potential $\Psi^0$. If we used this argument for 1–1 electrolytes to every ion in the solution, then every ion would be accounted twice: first as a neutral and the second time as the one interacting with the central one.

Then for 1–1 and 2–2 electrolytes, one can put down change of energy in electrostatic interactions of ion $i$ with the rest of the ions:

$$F_{el} = \frac{\kappa\beta^2\Psi^0_{el}|z_i|e}{2}. \tag{3.1}$$

If we used this argument for 1–2 and 2–1 electrolytes, then it would be necessary to divide by 3, while for 1–3 and 1–4, 2–3 and 2–4, and 3–3 electrolytes by 6, 10, and 16, respectively. These arguments are correct for a formed electrolyte structure in the solution, that is, starting from the average values of solution concentrations. For the dilute solutions area, when the ordered ion structure with its complexes is absent, these values of coefficients are lower than the recommended ones, and then with increase of electrolyte concentration, they acquire the stated values.

Instead of κ, we will substitute expression (2.56) into (3.1). Then

$$F_{el} = \frac{1}{\varepsilon_S\varepsilon_0}\cdot\frac{\beta^2 n_i(z_i e)^2}{\chi}\Psi^0_{el}, \tag{3.2}$$

where $\chi$ is a constant depending on the electrolyte type:

| Type of Electrolyte | 1–1<br>2–2 | 1–2<br>2–1 | 1–3 | 1–4<br>2–3 | 2–4<br>3–3 |
|---|---|---|---|---|---|
| $\chi$ | 2 | 3 | 6 | 10 | 16 |

Then, the contribution of electrostatic interactions with other ions for 1 mol of ions is expressed as

$$\bar{F}_{el} = \frac{1}{\varepsilon_S\varepsilon_0}\cdot\frac{N_A\beta^2 n_i(z_i e)^2}{\chi}\Psi^0_{el}, \tag{3.3}$$

where $N_A$ is Avogadro's number equal to $6.0220943\cdot 10^{23}$ mol$^{-1}$.

Let us assume that in the absence of interparticle interactions, the solution behaves as ideal. Then, the partial energy of 1 mol of ions $\overline{F}_i$ will be expressed as follows:

$$\overline{F}_i^0 = \overline{F}_{id,i} + \overline{F}_{el} + \overline{F}_{sp} + \overline{F}_{vv} + \overline{F}_x, \tag{3.4}$$

where

$\overline{F}_{id,i}$ is the partial energy of the ideal solution

$\overline{F}_{sp}, \overline{F}_{vv}$, and $\overline{F}_x$ are the partial energy of supermolecular forces, van der Waals forces, and chemical interactions, respectively, for 1 mol of ions from (2.79):

$$\overline{F}_{sp} + \overline{F}_{vv} + \overline{F}_x = N_A n_i n_{hi} \frac{a}{R}\left(U_{sp} + U_{vv} + U_x\right), \tag{3.5}$$

where $n_{hi}$ are the coordination numbers (CN) of ion hydration, whose systematized values are given in Table 3.2 and [2–8].

Taking into account (3.3) and (3.5), expression (3.4) for the electrolyte solution will be presented as follows:

$$\overline{F}_i^0 + R_0 T \ln f_i + R_0 T \ln N_i = \overline{F}_i^0 + R_0 T \ln N_i$$
$$- \frac{1}{\varepsilon_S \varepsilon_0} \cdot \frac{N_A \beta^2 \Psi_{el}^0 c_i (z_i e_i)^2}{\chi} + N_A n_i n_{hi} \frac{a}{R}\left(U_{sp} + U_{vv} + U_x\right), \tag{3.6}$$

where

$\overline{F}_i^0$ is the standard state energy

$R_0$ is the gas constant equal to $8.31441 \cdot 10^3$, J/(mol K)

$T$ is the temperature, K

$f_i$ is the ion activity coefficient being one of the most important energy and thermodynamic characteristics of the electrolyte solution

$N_i$ is the mole fraction

From (3.6), after several transformations, we obtain the following (the fundamental relationship used is $R_0 = kN_A$):

$$\ln f_i = -\frac{1}{\varepsilon_S \varepsilon_0} \cdot \frac{\beta^2 \Psi_{el}^0 n_i (z_i e_i)^2}{\chi k T} + \frac{a n_i n_{hi}}{RkT}\left(U_{sp} + U_{vv} + U_x\right). \tag{3.7}$$

Deviations of real solutes from the ideal state are convenient to be expressed using a coefficient called the molar activity coefficient $f_\pm$:

$$f_\pm = \left(f_i^{\nu_i} f_j^{\nu_j}\right)^{1/\nu_k}, \tag{3.8}$$

where $v_k$ is the number of ions into which an electrolyte molecule is split in the solvent and index $k$ is the component number.

Then, from (3.8), one can go to the average electrolyte activity coefficient dissociated into the $k$ types of ions:

$$\ln f_{\pm} = \frac{1}{\nu_k}\sum_{i=1}^{k} \nu_i \ln f_i. \tag{3.9}$$

If during dissociation of one electrolyte molecule $\nu_k$ of ions is formed, from which $\nu_i$ are the ions of type $i$.

By combining Equations 3.7 and 3.9, we obtain

$$\ln f_{\pm} = \frac{1}{\nu_k}\sum_{i} \nu_i \left[\frac{an_i n_{hi}}{RkT}\left(U_{sp} + U_{vv} + U_x\right) - \frac{1}{\varepsilon_S \varepsilon_0}\cdot\frac{\beta^2 \Psi^0_{el} n_i (z_i e_i)^2}{\chi kT}\right]. \tag{3.10}$$

In (3.10), we will take into account that concentration $c_i$ of ions $i$-type in mol/dm³ of the solution is connected with a number of these ions in m³ $n_i$ by the following ratio:

$$c_i = \frac{n_i}{N_A}, \quad n_i = N_A c_i, \tag{3.11}$$

and apparent equations that

$$\sum_{i=1}^{2} \nu_i = \nu_i + \nu_j = \nu_+ + \nu_- = \nu_k, \quad \sum_{i=1}^{2} \nu_i z_i^2 = \nu_i z_i^2 + \nu_j z_j^2 = \nu_k \left|z_i z_j\right|,$$
$$\sum_{i=1}^{2} c_i = c_i + c_j = \nu_+ c_k + \nu_- c_k = \nu_k c_k \equiv \left(\left|z_i\right| + \left|z_j\right|\right) c_k, \tag{3.12}$$

where

$\nu_+$ and $\nu_-$ are the number of cations and anions to which an electrolyte molecule is split in the solution

$c_k$ is the electrolyte weight contents, mol/dm³

Then, taking into account (3.11) and (3.12), expression (3.10) will be presented as follows:

$$\ln f_{\pm} = \sum_{i}\left[\frac{N_A a c_k \nu_i n_{hi}}{RkT}\left(U_{sp} + U_{vv} + U_x\right)\right] + \frac{1}{\varepsilon_S \varepsilon_0}\cdot\frac{\beta^2 N_A c_k \Psi^0_{el}\left|z_i z_j\right| e^2}{\chi kT}. \tag{3.13}$$

Let us depict (3.13) as follows:

$$\ln f_{\pm} = \sum_i \left[ \frac{N_A a c_k \nu_i n_{hi}}{RkT} \left( U_{sp} + U_{vv} + U_x \right) \right] + \frac{\Omega_g c_k}{\chi \varepsilon_S T} \Psi_{el}^0, \quad (3.14)$$

where

$$\Omega_g = \frac{1}{\varepsilon_0} \cdot \frac{N_A \beta^2 |z_i z_j| e^2}{k}. \quad (3.15)$$

When substituting fundamental physical constants and all constants into Equation 3.15, it will be as follows:

$$\Omega_g = 1.26525 \cdot 10^{23} \beta^2 \nu_k |z_i z_j|. \quad (3.16)$$

When substituting all fundamental physical constants into the equation for β (2.48), we will obtain

$$\beta = \frac{1.67337 \cdot 10^{-5} |z_i z_j|}{\varepsilon_S T}. \quad (3.17)$$

Further transformations of (3.14) to the form convenient for practical calculations will be made in the next sections, and now we will determine a few ratios required for the narration to follow.

## 3.2 GUEST ACTIVITY FUNCTION

The equations discussed in the previous sections contain all variables and are general enough to serve as a thermodynamic basis for consideration of all electrolyte solutions. Using chemical potentials and introducing corresponding limitations, it is possible to apply thermodynamics to solutions quite strictly. In particular, the activity is widely used in solution thermodynamics.

The activity $a_i$ of a clear chemical substance or one of the solution components in general form is defined by the following equation:

$$\mu_i = \mu_i^0 + RT \ln a_i, \quad (3.18)$$

where $\mu_i^0$ is the chemical potential of the substance in some conventionally chosen standard state. We have to demonstrate that the value $\mu_i^0$ depends on the choice of concentration units (molarity, molality, etc.), in which $a_i$ is expressed.

The connection among activities of separate ions and salt is set purely formally from the electrical neutrality condition in a solution:

$$\nu^{+}z^{+} = \nu^{-}\left|z^{-}\right|, \tag{3.19}$$

and then

$$\mu_{\pm} = \nu^{+}\mu_{+} + \nu^{-}\mu_{-}. \tag{3.20}$$

Since the electrical neutrality condition is kept in a standard state, then

$$\mu_{\pm}^{0} = \nu^{+}\mu_{\pm}^{0} + \nu^{-}\mu_{-}^{0}. \tag{3.21}$$

Then, from (3.18) through (3.21), it follows that

$$RT\ln a_{\pm} = \nu^{+}RT\ln a_{+} + \nu^{-}RT\ln a_{-}, \tag{3.22}$$

or

$$a_{\pm} = a_{+}^{\nu^{+}} a_{-}^{\nu^{-}}, \tag{3.23}$$

that is, electrolyte activity is expressed through the product of activities of ions into which such electrolyte can split. However, using formula (3.23), it is impossible to distinguish the activity of every of ions. Since chemical potential served as a primary message for describing the equilibrium state in a solution, then these formulas are true for any concentration scale (in molality, molarity concentration, and molar fraction scales):

$$a_{\pm}(m) = m_{\pm}\gamma_{\pm}, \quad a_{\pm}(c) = c_{\pm}f_{\pm}, \quad a_{\pm}(N) = N_{\pm}y_{\pm}, \tag{3.24}$$

where $\gamma$, $f$, and $y$ are called molality, molarity, and rational activity coefficients, respectively.

Ion concentrations are connected with the electrolyte concentration by the following ratios:

$$\begin{aligned} m_{+} &= \nu^{+}m_{\pm}, \quad m_{-} = \nu^{-}m_{\pm}, \\ c_{+} &= \nu^{+}c_{\pm}, \quad c_{-} = \nu^{-}c_{\pm}, \\ N_{+} &= \nu^{+}N_{\pm}, \quad N_{-} = \nu^{-}N_{\pm}. \end{aligned} \tag{3.25}$$

Combining expressions (3.24) and (3.25) with (3.23), we obtain

$$a_{\pm} = a_{+}^{\nu^{+}} a_{-}^{\nu^{-}} = (\gamma_{+} m_{+})^{\nu^{+}} (\gamma_{-} m_{+})^{\nu^{-}} = (\gamma_{+} \nu_{+} m_{\pm})^{\nu^{+}} (\gamma_{-} \nu_{-} m_{\pm})^{\nu^{-}} = (\nu_{+})^{\nu^{+}} (\nu_{-})^{\nu^{-}} (m_{\pm} \gamma_{\pm})^{\nu}. \tag{3.26}$$

In expression (3.26) $\nu = \nu^{+} + \nu^{-}$, and $\gamma_{\pm} = \left(\gamma_{+}^{\nu^{+}} \gamma_{-}^{\nu^{-}}\right)^{1/\nu}$ is the ion-average ion activity coefficient. It is most frequently used in actual practice and in publications, and Harned and Owen called it the practical activity coefficient. We can obtain similar expressions for $N_{\pm}$ and $y_{\pm}$ as well. The rational ion activity coefficient $y_{\pm}$ has not become that widely used.

Expressing the equation for chemical potential in (3.18) form, we did not set a goal to determine $\mu_{\pm}^{0}$, that is, we did not set properties for the standard state. We will now determine the standard state for the electrolyte solution according to Salem [1].

In every concentration scale, the standard state is chosen so that the average ion activity coefficient in such scale tends to one (or $f_{\pm} = y_{\pm} = 1$) at concentration tending toward zero. This requirement stays in force at any temperature and pressure values. Such requirement, at first glance, may appear strange because $\ln a_{\pm} = \ln \gamma_{\pm} m_{\pm}$, and if $\gamma_{\pm} \to 1$ and $m_{\pm} \to 0$, then in the end it will be within limits of the infinite dilute solution $\lim \ln(\gamma_{\pm} m_{\pm}) = -\infty$. However, it is clear that in the standard state,

$$\mu_{\pm} = \mu_{\pm}^{0}, \tag{3.27}$$

which is possible only if $a_{\pm} = 1$. Therefore, such requirement means only that it goes not about a real solution but about some assumed state of the solution in which the average ion activity coefficient is equal to 1 and, accordingly, average molality, molarity, and normality are also equal to one, but these are already not dilute solutions. One should avoid a mistaken notion about the standard state as a state of the infinite dilute solution, for at infinite dilution, as shown earlier, the chemical potential turns to $-\infty$.

Then, the connection among activities and activity coefficients in different concentration scales can be obtained, if it is accounted that chemical potential $m_{\pm}$ does not depend on the choice of the concentration scale, and using (3.18), (3.24), and (3.26), we have

$$\mu_{\pm} = \mu_{\pm}^{0}(m) + RT \ln a_{\pm}(m) = \mu_{\pm}^{0}(c) + RT \ln a_{\pm}(c) = \mu_{\pm}^{0}(N) + RT \ln a_{\pm}(N). \tag{3.28}$$

Combining Equations 3.18, 3.24, 3.26, and 3.29, we obtain

$$\mu_{\pm} = \mu_{\pm}^{0}(m) + \nu RT \ln(\gamma_{\pm} m_{\pm}) = \mu_{\pm}^{0}(c) + \nu RT \ln(f_{\pm} c_{\pm}) = \mu_{\pm}^{0}(N) + RT \ln(y_{\pm} N_{\pm}). \tag{3.29}$$

If following condition (3.27), it is possible to introduce ultimate values $N_{\pm}/m_{\pm}$ and $c_{\pm}/m_{\pm}$ at infinite dilution and to obtain from Equation 3.29 the ratio

$$\mu_{\pm}^{0}(N) = \mu_{\pm}^{0}(m) + \nu RT \ln \frac{1000}{M_{H_2O}} = \mu_{\pm}^{0}(c) + \nu RT \ln \frac{1000\rho_0}{M_{H_2O}}, \quad (3.30)$$

where

$M_{H_2O}$ is the molecular weight of solvent (water)
$\rho_0$ is its density

The appearance of $\rho_0$ in the last term of this equation points to a substantial difference between weight and volume concentration units. The phase concentration (composition) in the sense this expression is being used when introducing main thermodynamic correlations is an *independent* variable. When the process takes place at the constant composition and, accordingly, at the same condition, partial derivation of equations is made, then concentrations $m$, $N$, etc., expressed in weight units, stay constant, whereas concentrations $c$, expressed in volume units, can change and in practice they usually do. For constancy of $c$, two conditions are usually required: constant composition and constant volume. Consequently, if $\mu_{\pm}^{0}$ (and other thermodynamic functions expressed through concentrations $c$) must be differentiated under $T$ or $P$ at a constant composition, it is necessary to remember that $c$ is a variable value. Those mistakes that might happen if this circumstance is disregarded make the usage of volume concentration units undesirable except in cases when temperature and pressure values are constant. But we are forced to use dependence $f_{\pm}(c)$ (volume concentration unit) because the contemporary thermodynamic theory offers dependence of a number of ions in the solution through $n_i(c)$, and if dependence $n_i(m)$ (weight concentration unit) is used, then, in any case, solution density appears and the task becomes sharply more complicated.

By combining Equations 3.29 and 3.30, we obtain

$$\ln f_{\pm} = \ln \gamma_{\pm} + \ln \frac{m_{\pm}}{N_{\pm}} \cdot \frac{M_{H_2O}}{1000} = \ln y_{\pm} + \ln \frac{c_{\pm}}{N_{\pm}} \cdot \frac{M_{H_2O}}{1000\rho_0}. \quad (3.31)$$

The ratio between concentrations at any dilutions is presented by common expression

$$N_{\pm} = \frac{m_{\pm}}{\nu m + 1000/M_{H_2O}} = \frac{c_{\pm}}{\nu c + \left(1000\rho - cM_{H_2O}\right)/M_E}, \quad (3.32)$$

where

$m$ and $c$ are the stoichiometric molality and molarity of the electrolyte
$M_E$ is its molecular weight
$\rho$ is the solution density

Substituting (3.32) into (3.31), we obtain a very important ratio:

$$\ln f_{\pm} = \ln \gamma_{\pm} + \ln\left(1 + \frac{\nu m M_{H_2O}}{1000}\right), \tag{3.33}$$

connecting dependence of activity coefficients $f_{\pm}(c)$ (volume concentration unit) with dependence of activity coefficients $\gamma_{\pm}(m)$ (weight concentration unit). Let us take into account that $M_{H_2O} = 18.02$. After several transformations, we will obtain

$$\ln \gamma_{\pm} = \ln f_{\pm} - \ln\left(1 + \frac{\nu m}{55.51}\right). \tag{3.34}$$

It is necessary to notice that in some physical chemistry textbooks and monographs, $\gamma_{\pm} \equiv f_{\pm}$ is declared. This equation is true only for dilute solutions.

## 3.3 MINIMUM INTERACTION DISTANCE OF ELECTROSTATIC FORCES AMONG IONS IN THE HOST SOLUTION

According to Bockris in (3.14), value $a$ can be calculated by the ratio

$$a = \xi(a_i + a_j), \tag{3.35}$$

where

- $\xi$ assumes values from 0.5 to 1.5 depending on electrolyte thermodynamic properties
- $a_i$ and $a_j$ are the distance from the ion center to the oxygen atom in the water molecule (nm) of simple ions, which is defined by the x-ray crystal structure analysis or, in case of absence of such data, is defined as effective radius:

$$a_{i,j} = r_{i,j} + r_w, \tag{3.36}$$

where

- $r_{i,j}$ is the ionic radius of a cation or anion, nm
- $r_w$ is the water molecule radius assumed as 0.138 nm

Ionic radius, conditional characteristics of ions used for approximate evaluation of internuclear distances in ionic crystals. As a result, it is considered that the distance between the nearest cation and anion is equal to the sum of their ionic radius. Values of ionic radius are consistently connected with the position of elements in the Mendeleev periodic system. Several systems of values of ionic radius are offered. These systems are usually based on the following observation: the difference of internuclear distances A–X and B–X in ionic crystals of AX and BX composition, where A and B are metal and X is nonmetal, practically does not change in case of substitution of X for another nonmetal analogous to it (e.g., substitution of chlorine for bromine) and if CN of analogous ions in the salts being compared are the same. Therefore, it implies that ionic radius possesses the additivity property, that is, experimentally defined internuclear distances can be considered as the sum of the corresponding *radius* of ions. The division of this sum into summands is always based on more or less optional assumptions. The systems of ionic radius, offered by different authors, vary mainly by usage of various basic assumptions.

There are several systems of atomic ionic radius, differed by values of atomic radius of individual ions but leading to almost the same internuclear distances. The work on determining ion atomic radius was done for the first time in the 1920s by Goldschmidt, who relied, on one hand, upon internuclear distances in crystals measured by x-ray structural analysis methods and, on the other hand, upon values of atomic radius $F^-$ and $O^{2-}$ defined by the refractometry method (0.133 and 0.132 nm, respectively). The majority of other systems also relies upon internuclear distances in crystals defined by diffraction methods and upon some *benchmark* value of atomic radius of a definite ion. In the most widely known Pauling system, such benchmark value is an atomic radius $O^{2-}$ (0.140 nm). In the Belov and Bokiy system, considered as one of the most reliable, the atomic radius $O^{2-}$ is assumed to be equal to 0.136 nm. The system of atomic and ionic radiuses according to Belov and Bokiy is presented in Table 3.1.

Calculated values $a_{i,j}$ under (3.36) for different ions are presented in Table 3.2. This table also contains CN of ion hydration grouped from [2–8].

For polyatomic and complex ions, the effective ion radius is calculated in the same way as it was done in (3.36):

$$a_{i,j} = r_{ki,j} + r_w, \tag{3.37}$$

where $r_{ki,j}$ are thermodynamic radius of polyatomic and complex ions, nm, whose grouped data from [9–13] are given in Table 3.3. Calculated values under (3.36) of effective radius, nm, are also presented there. Table 3.4 presents CN and effective radius, nm, of some polyatomic and complex ions.*

## 3.4 DISTANCE BETWEEN IONS IN THE HOST SOLUTION

Value $R$ in (3.14) is calculated as follows. If addressing [14], then Shakhparonov provides formula (2.31) for calculation of $R$, having a misprint. If one assumes ions as point and considers that they are located on a sphere with diameter $R$ and calculates it through the geometric volume, then we will have $R = ((6/\pi)V)^{1/3}$, where $V$ is the volume. On the other hand, the molar volume of particles in a

* If one tries to use a theoretical formula to calculate hydrate numbers and the hydrated ion radius of Tanganov [14], advertised by him in other publications, we will obtain a negative value of the hydrate number, and then this negative number for radius calculation goes under the square root.

**TABLE 3.1**
**Atomic and Ionic Radii of Elements (nm) According to Belov and Bokiy for Calculation According to Formulae (3.35) and (3.36)**

| | Ion Charge | | | | | | | | | | |
|---|---|---|---|---|---|---|---|---|---|---|---|
| **Element** | **−3** | **−2** | **−1** | **0** | **+1** | **+2** | **+3** | **+4** | **+5** | **+6** | **+7** |
| H | | | 0.136 | 0.028 | 0.000 | | | | | | |
| He | | | | 0.122 | | | | | | | |
| Li | | | | 0.155 | 0.068 | | | | | | |
| Be | | | | 0.113 | | 0.034 | | | | | |
| B | | | | 0.091 | | | 0.020 | | | | |
| C | | | | 0.077 | | | | 0.020 | | | |
| N | 0.148 | | | 0.055 | | | | | 0.015 | | |
| O | | 0.136 | | 0.059 | | | | | | | |
| F | | | 0.133 | 0.064 | | | | | | | |
| Ne | | | | 0.160 | | | | | | | |
| Na | | | | 0.189 | 0.098 | | | | | | |
| Mg | | | | 0.160 | | 0.074 | | | | | |
| Al | | | | 0.143 | | | 0.057 | | | | |
| Si | | | | 0.134 | | | | 0.039 | | | |
| P | 0.186 | | | 0.110 | | | | | 0.035 | | |
| S | | 0.182 | | 0.104 | | | | | | 0.029 | |
| Cl | | | 0.181 | 0.099 | | | | | | | 0.026 |
| Ar | | | | 0.192 | | | | | | | |
| K | | | | 0.236 | 0.133 | | | | | | |
| Ca | | | | 0.197 | | 0.104 | | | | | |
| Se | | | | 0.164 | | | 0.083 | | | | |
| Ti | | | | 0.146 | | 0.078 | 0.069 | 0.064 | | | |
| V | | | | 0.134 | | 0.072 | 0.067 | 0.061 | 0.040 | | |
| Cr | | | | 0.127 | | 0.083 | 0.064 | | | 0.035 | |
| Mn | | | | 0.130 | | 0.091 | 0.070 | 0.052 | | | 0.046 |
| Fe | | | | 0.126 | | 0.080 | 0.067 | | | | |
| Co | | | | 0.125 | | 0.078 | 0.064 | | | | |
| Ni | | | | 0.124 | | 0.074 | | | | | |
| Cu | | | | 0.128 | 0.098 | 0.080 | | | | | |
| Zn | | | | 0.139 | | 0.083 | | | | | |
| Ga | | | | 0.139 | | | 0.062 | | | | |
| Ge | | | | 0.139 | | 0.065 | | 0.044 | | | |
| As | 0.191 | | | 0.148 | | | 0.069 | | 0.047 | | |
| Se | | 0.193 | | 0.160 | | | | 0.069 | | 0.035 | |
| Br | | | 0.196 | 0.114 | | | | | | | 0.039 |
| Ru | | | | 0.198 | | | | | | | |
| Rb | | | | 0.248 | 0.149 | | | | | | |
| Sr | | | | 0.215 | | 0.120 | | | | | |
| Y | | | | 0.181 | | | 0.097 | | | | |
| Zr | | | | 0.160 | | | | 0.082 | | | |

*(Continued)*

**TABLE 3.1** (*Continued*)
**Atomic and Ionic Radii of Elements (nm) According to Belov and Bokiy for Calculation According to Formulae (3.35) and (3.36)**

| | Ion Charge | | | | | | | | | | |
|---|---|---|---|---|---|---|---|---|---|---|---|
| **Element** | **−3** | **−2** | **−1** | **0** | **+1** | **+2** | **+3** | **+4** | **+5** | **+6** | **+7** |
| Nb | | | | 0.145 | | | | 0.067 | 0.066 | | |
| Mo | | | | 0.139 | | | | 0.068 | | 0.065 | |
| Tc | | | | 0.136 | | | | | | | |
| Ru | | | | 0.133 | | | | 0.062 | | | |
| Rh | | | | 0.134 | | | 0.075 | 0.065 | | | |
| Pd | | | | 0.137 | | | | 0.064 | | | |
| Ag | | | | 0.144 | 0.113 | | | | | | |
| Cd | | | | 0.156 | | 0.099 | | | | | |
| In | | | | 0.166 | 0.136 | | 0.092 | | | | |
| Sn | | | | 0.158 | | 0.102 | | 0.067 | | | |
| Sb | 0.208 | | | 0.161 | | | 0.090 | | 0.062 | | |
| Te | | 0.211 | | 0.170 | | | | 0.089 | | 0.056 | |
| I | | | 0.220 | 0.133 | | | | | | | 0.050 |
| Xe | | | | 0.218 | | | | | | | |
| Cs | | | | 0.267 | 0.165 | | | | | | |
| Ba | | | | 0.221 | | 0.138 | | | | | |
| La | | | | 0.187 | | | 0.104 | 0.090 | | | |
| Ce | | | | 0.183 | | | 0.102 | 0.088 | | | |
| Pr | | | | 0.182 | | | 0.100 | | | | |
| Nd | | | | 0.182 | | | 0.099 | | | | |
| Pm | | | | | | | 0.098 | | | | |
| Sm | | | | 0.181 | | | 0.097 | | | | |
| Eu | | | | 0.202 | | | 0.097 | | | | |
| Gd | | | | 0.179 | | | 0.094 | | | | |
| Tb | | | | 0.177 | | | 0.089 | | | | |
| Dy | | | | 0.177 | | | 0.088 | | | | |
| Ho | | | | 0.176 | | | 0.086 | | | | |
| Er | | | | 0.175 | | | 0.085 | | | | |
| Tm | | | | 0.174 | | | 0.085 | | | | |
| Yb | | | | 0.193 | | | 0.081 | | | | |
| Lu | | | | 0.174 | | | 0.080 | | | | |
| Hf | | | | 0.159 | | | | 0.082 | | | |
| Ta | | | | 0.146 | | | | | 0.066 | | |
| W | | | | 0.140 | | | | 0.068 | | 0.065 | |
| Re | | | | 0.137 | | | | | | | 0.052 |
| Os | | | | 0.135 | | | | 0.065 | | | |
| Ir | | | | 0.136 | | | | 0.065 | | | |
| Pt | | | | 0.138 | | | | 0.064 | | | |
| Au | | | | 0.144 | 0.137 | | | | | | |
| Hg | | | | 0.160 | | 0.112 | | | | | |
| Tl | | | | 0.171 | 0.136 | | 0.105 | | | | |

(*Continued*)

**TABLE 3.1 (*Continued*)**
**Atomic and Ionic Radii of Elements (nm) According to Belov and Bokiy for Calculation According to Formulae (3.35) and (3.36)**

| | Ion Charge | | | | | | | | | | |
|---|---|---|---|---|---|---|---|---|---|---|---|
| **Element** | **−3** | **−2** | **−1** | **0** | **+1** | **+2** | **+3** | **+4** | **+5** | **+6** | **+7** |
| Pb | | | | 0.175 | | 0.126 | | 0.076 | | | |
| Bi | | | | 0.182 | | | 0.120 | | 0.074 | | |
| Fr | | | | 0.280 | | | | | | | |
| Ra | | | | 0.235 | | 0.144 | | | | | |
| Ac | | | | 0.203 | | | 0.111 | | | | |
| Th | | | | 0.180 | | | 0.108 | 0.095 | | | |
| Pa | | | | 0.162 | | | 0.106 | 0.091 | | | |
| U | | | | 0.153 | | | 0.102 | 0.088 | | | |
| Np | | | | 0.150 | | | 0.102 | 0.086 | | | |
| Pu | | | | 0.162 | | | 0.101 | 0.085 | | | |
| Am | | | | | | | 0.100 | 0.085 | | | |

**TABLE 3.2**
**CN of Ion Hydration, $n_{hi}$, and Distances from Ion Center to Oxygen Atom of Water Molecule $a_i$ (nm), for Ions in a Solution for Calculation According to Formula (3.36)**

| Ion | $n_{hi}$ | $a_i$ | Ion | $n_{hi}$ | $a_i$ | Ion | $n_{hi}$ | $a_i$ |
|---|---|---|---|---|---|---|---|---|
| $Al^{3+}$ | 6 | 0.195 | $H^-$ | 6 | 0.274 | $Ti^{3+}$ | 6 | 0.207 |
| $Ac^{3+}$ | 8 | 0.249 | $Hg^{2+}$ | 8 | 0.250 | $Ti^{4+}$ | 6 | 0.202 |
| $Am^{3+}$ | 8 | 0.238 | $K^+$ | 4 | 0.271 | $Tm^{3+}$ | 6 | 0.223 |
| $Ba^{2+}$ | 8 | 0.276 | $La^{3+}$ | 8 | 0.242 | $U^{3+}$ | 6 | 0.240 |
| $Be^{2+}$ | 4 | 0.172 | $Li^+$ | 4 | 0.206 | $U^{4+}$ | 6 | 0.226 |
| $Ca^{2+}$ | 6 | 0.242 | $Lu^{3+}$ | 6 | 0.218 | $Y^{3+}$ | 8 | 0.235 |
| $Cd^{2+}$ | 6 | 0.237 | $Mg^{2+}$ | 6 | 0.212 | $Yb^{2+}$ | 6 | 0.234 |
| $Ce^{3+}$ | 9 | 0.240 | $Mo^{6+}$ | 6 | 0.203 | $Yb^{3+}$ | 6 | 0.219 |
| $Cr^{3+}$ | 6 | 0.202 | $Na^+$ | 4 | 0.236 | $Zn^{2+}$ | 6 | 0.221 |
| $Co^{2+}$ | 6 | 0.216 | $Nd^{3+}$ | 6 | 0.237 | $Zr^{4+}$ | 7 | 0.220 |
| $Cs^+$ | 8 | 0.303 | $Np^{3+}$ | 6 | 0.240 | $As^{3-}$ | 6 | 0.329 |
| $Cu^{2+}$ | 6 | 0.218 | $Ni^{2+}$ | 6 | 0.212 | $Br^-$ | 6 | 0.334 |
| $Dy^{3+}$ | 8 | 0.226 | $Pb^{2+}$ | 8 | 0.264 | $Cl^-$ | 6 | 0.319 |
| $Er^{3+}$ | 8 | 0.223 | $Pr^{3+}$ | 6 | 0.238 | $F^-$ | 4 | 0.271 |
| $Eu^{3+}$ | 6 | 0.235 | $Rb^+$ | 6 | 0.287 | $I^-$ | 6 | 0.358 |
| $Fe^{2+}$ | 6 | 0.218 | $Sm^{3+}$ | 6 | 0.235 | $O^{2-}$ | 3 | 0.274 |
| $Fe^{3+}$ | 6 | 0.205 | $Mo^{6+}$ | 6 | 0.203 | $S^{2-}$ | 6 | 0.320 |
| $Ga^{3+}$ | 6 | 0.200 | $Sn^{2+}$ | 6 | 0.240 | $Sb^{3-}$ | 6 | 0.346 |
| $Gd^{3+}$ | 8 | 0.232 | $Sc^{2+}$ | 6 | 0.258 | $Se^{2-}$ | 6 | 0.331 |
| $Ge^{4+}$ | 4 | 0.182 | $Tb^{3+}$ | 6 | 0.227 | $Te^{2-}$ | 6 | 0.349 |

**TABLE 3.3**
**Thermochemical and Effective Radii of Polyatomic Ions, $r_i$ and $a_i$ (nm), for Calculation According to Formula (3.37)**

| Ion | $r_i$ | $a_i$ | Ion | $r_i$ | $a_i$ |
|---|---|---|---|---|---|
| $NH^+_-$ | 0.143 | 0.281 | $IO^{3-}$ | 0.182 | 0.320 |
| $PH^+_4$ | 0.180 | 0.318 | $IO^{4-}$ | 0.249 | 0.387 |
| $AsO_4^{2-}$ | 0.248 | 0.386 | $MnO^{4-}$ | 0.240 | 0.378 |
| $BrO^-$ | 0.191 | 0.329 | $MoO_4^{2-}$ | 0.254 | 0.392 |
| $BO^-$ | 0.191 | 0.329 | $NH_4^-$ | 0.143 | 0.281 |
| $CH_3COO^-$ | 0.159 | 0.297 | $NO_2^-$ | 0.155 | 0.293 |
| $ClO_3^-$ | 0.200 | 0.338 | $NO_3^-$ | 0.189 | 0.327 |
| $ClO_4^-$ | 0.236 | 0.374 | $OH^-$ | 0.140 | 0.278 |
| $CN^-$ | 0.192 | 0.330 | $PO_4^{3-}$ | 0.238 | 0.376 |
| $CNO^-$ | 0.159 | 0.297 | $ReO_4^-$ | 0.280 | 0.418 |
| $CNS^-$ | 0.195 | 0.333 | $SCN^-$ | 0.195 | 0.333 |
| $CO_3^{2-}$ | 0.185 | 0.323 | $SbO_4^{3-}$ | 0.260 | 0.398 |
| $CrO_4^{2-}$ | 0.240 | 0.378 | $SeO_4^{2-}$ | 0.243 | 0.381 |
| $HF_2^-$ | 0.155 | 0.293 | $SiO_4^{4-}$ | 0.240 | 0.378 |
| $HCO_2^-$ | 0.158 | 0.296 | $SO_3^{2-}$ | 0.200 | 0.338 |
| $HCO_3^-$ | 0.163 | 0.301 | $SO_4^{2-}$ | 0.230 | 0.368 |
| $HS^-$ | 0.195 | 0.333 | $TeO_4^{2-}$ | 0.254 | 0.392 |
| $HSO_4^-$ | 0.206 | 0.344 | $WO_4^{2-}$ | 0.257 | 0.395 |

**TABLE 3.4**
**CN of Ion Hydration, $n_{hi}$, and Distances from Ion Center to Oxygen Atom of Water Molecule, $a_i$ (nm), for Some Polyatomic Ions in a Solution for Calculation According to Formula (3.37)**

| Ion | $CH_3COO^-$ | $ClO_4^-$ | $Cr_2O_4^{2-}$ | $CO_3^{2-}$ | $HS^-$ | $NO_3^-$ | $SO_4^{2-}$ | $OH^-$ |
|---|---|---|---|---|---|---|---|---|
| $n_{hi}$ | 8 | 12 | 4 | 9 | 6 | 5 | 8 | 6 |
| $a_i$ | 0.297 | 0.374 | 0.378 | 0.323 | 0.333 | 0.327 | 0.368 | 0.278 |

solution is determined as $V \cong M_k/\rho_k N_A$, where $M_k$ is the molecular weight of the dilute electrolyte and $\rho_k$ is the solution density, kg/m³. Let us substitute the last expression to the previous one and we will obtain an expression for calculation of $R$ in m:

$$R = \left( \frac{6}{\pi} \frac{M_k}{\rho_k N_A} \right)^{1/3}. \tag{3.38}$$

Let us substitute all constants to (3.38) and we will obtain an expression for calculation of $R$ in m:

$$R = 1.47 \cdot 10^{-9} \left( \frac{M_k}{\rho_k} \right)^{1/3}. \tag{3.39}$$

The solution density $\rho_k$, kg/m$^3$, required for calculation of $R$ in (3.39) for a big quantity of binary electrolytes in a wide range of temperatures and concentrations can be found in [15], where the data from the world scientific literature at that time, considerably added with proper experimental researches, are presented. Experimental density data in [15] are processed by regression analysis methods, and different density calculation coefficients as well as the equation are obtained to allow calculation of the density not only of the binary electrolyte solution but also of the multicomponent solution with high accuracy [2,3,15–17].

The density of the multicomponent solution is calculated by the following formula:

$$\rho_k = \rho_0 + \sum_{i=1}^{k} B_i (A_{1i} + A_{2i}t + A_{3i}B_i), \tag{3.40}$$

where

$\rho_0$ is the water density, kg/m$^3$

$k$ is the number of solution components

$B_i$ is the electrolyte weight content in the solution, %

$A_{ij}$ are the coefficients obtained based on processing proper experimental data and data from the world scientific literature [2,3,15–17], using regression analysis methods, which are given in Table 3.5

$t$ is the temperature, °C

Coefficients provided here and in [15,16] are much more accurate than those in [17].

Water density on the saturation line with relative error 0.0048% is calculated by the following formulas:

- At temperature from 0°C to 150°C:

$$\rho_0 = 999.810745 + R_1 t^* + R_2 t^{*1.5} + R_3 t^{*2} + R_4 t^{*2.5} + R_5 t^{*3} + R_6 t^{*3.5} + R_7 t^{*4} + R_8 t^{*4.5}, \tag{3.41}$$

where

$t^* = 0.01t$

$R_1 = 15.910174$

$R_2 = -73.840671$

$R_3 = 141.693443$

$R_4 = -297.162815$

$R_5 = 160.758796$

$R_6 = 146.727674$

$R_7 = -194.580793$

$R_8 = 58.847919$

**TABLE 3.5**
**Coefficients of Equation 3.40 for Density Calculation**

| Electrolyte | $A_{1i}$ | $A_{2i}$ | $A_{3i}$ | Range $B$ (%) from $B_{min}$ to $B_{max}$ | Range $t$ (°C) from $t_{min}$ to $t_{max}$ | $S_\rho$ (kg/m³) | Δ (%) |
|---|---|---|---|---|---|---|---|
| $AgClO_4$ | 7.9760 | −0.0067 | 0.0835 | 035 | 0–100 | 2.20 | 0.13 |
| | 5.3002 | −0.0077 | 0.1597 | 35–65 | 0–100 | 5.27 | 0.25 |
| | −3.1825 | −0.0076 | 0.2901 | 65–80 | 0–90 | 7.14 | 0.25 |
| $AgNO_3$ | 8.0942 | −0.0069 | 0.0962 | 0–45 | 0–100 | 4.91 | 0.25 |
| | 4.1303 | −0.0066 | 0.1876 | 45–65 | 0–100 | 5.35 | 0.25 |
| | −2.3869 | −0.0084 | 0.2916 | 65–75 | 40–100 | 4.66 | 0.16 |
| $Al(NO_3)_3$ | 6.9571 | −0.0005 | 0.0621 | 0–50 | 0–100 | 4.88 | 0.27 |
| $Al_2(SO_4)_3$ | 10.1132 | 0.0021 | 0.0816 | 0–30 | 0–100 | 2.47 | 0.16 |
| $BaBr_2$ | 8.7019 | −0.0162 | 0.1111 | 0–50 | 0–100 | 5.54 | 0.32 |
| $BaCl_2$ | 8.6677 | −0.0066 | 0.1059 | 0–30 | 0–100 | 1.89 | 0.14 |
| $Ba(ClO_3)_2$ | 7.3631 | 0.0075 | 0.0518 | 0–40 | 0–60 | 4.21 | 0.26 |
| $Ba(ClO_4)_2$ | 7.7966 | −0.0201 | 0.0889 | 0–45 | 0–100 | 4.13 | 0.25 |
| | 6.7478 | −0.0133 | 0.1091 | 45–65 | 0–100 | 5.70 | 0.27 |
| $BaI_2$ | 9.0029 | −0.0114 | 0.0965 | 0–35 | 0–100 | 3.57 | 0.24 |
| | 7.4748 | −0.0099 | 0.1438 | 35–60 | 0–100 | 5.06 | 0.25 |
| $Ba(NO_2)_2$ | 9.0089 | −0.0226 | 0.0843 | 0–45 | 0–100 | 5.08 | 0.28 |
| $Ba(NO_3)_2$ | 7.0275 | 0.0051 | 0.0922 | 0–20 | 0–100 | 2.89 | 0.21 |
| $Ba(OH)_2$ | 12.2031 | −0.0244 | 0.1019 | 0–40 | 30–100 | 4.93 | 0.29 |
| $BaS$ | 11.4614 | −0.0125 | 0.0848 | 0–30 | 20–100 | 2.99 | 0.18 |
| $CaBr_2$ | 8.2102 | 0.0052 | 0.0848 | 0–50 | 0–100 | 5.16 | 0.25 |
| | 4.9164 | 0.0073 | 0.1577 | 50–70 | 0–100 | 4.73 | 0.22 |
| $Ca(CH_3COO)_2$ | 5.1867 | 0.0314 | 0.0945 | 2–13 | 10–100 | 3.25 | 0.25 |
| | 5.2558 | 0.0147 | 0.0134 | 13–24 | 10–100 | 1.99 | 0.15 |
| | 5.1247 | 0.0251 | 0.0511 | 2–24 | 10–100 | 5.29 | 0.37 |
| $CaCl_2$ | 8.5647 | 0.0008 | 0.0260 | 0–60 | 0–100 | 4.34 | 0.22 |
| | 6.3626 | 0.0172 | 0.0119 | 4–34 | 125–280 | 4.04 | 0.27 |
| | −3.1723 | 0.0555 | 0.0474 | 4–34 | 280–340 | 5.50 | 0.40 |
| $Ca(ClO_4)_2$ | 7.0068 | −0.0049 | 0.0597 | 0–60 | 0–100 | 3.99 | 0.24 |
| $CaCrO_4$ | 7.6755 | −0.0046 | 0.0786 | 0–20 | 25–80 | 0.87 | 0.07 |
| $CaCr_2O_7$ | 7.6845 | −0.0182 | 0.0852 | 0–50 | 0–100 | 5.28 | 0.29 |
| $CaI_2$ | 9.5813 | −0.0310 | 0.0833 | 0–30 | 0–100 | 4.12 | 0.27 |
| | 8.0856 | −0.0213 | 0.1148 | 30–45 | 0–100 | 3.60 | 0.19 |
| $Ca(NO_2)_2$ | 7.8732 | −0.0174 | 0.0456 | 0–55 | 0–100 | 3.95 | 0.24 |
| $Ca(NO_3)_2$ | 8.0889 | −0.0078 | 0.0388 | 0–70 | 0–100 | 4.67 | 0.28 |
| $CdBr_2$ | 10.2046 | −0.0102 | 0.0543 | 0–45 | 0–100 | 2.25 | 0.14 |
| $CdCl_2$ | 9.0569 | −0.0132 | 0.0916 | 0–50 | 0–100 | 3.43 | 0.19 |
| $CdI_2$ | 8.7932 | −0.0132 | 0.0962 | 0–45 | 0–100 | 4.11 | 0.27 |
| $Cd(NO_3)_2$ | 8.3096 | −0.0235 | 0.1038 | 0–60 | 0–100 | 4.72 | 0.27 |

*(Continued)*

**TABLE 3.5 (*Continued*)**
**Coefficients of Equation 3.40 for Density Calculation**

| Electrolyte | $A_{1i}$ | $A_{2i}$ | $A_{3i}$ | Range $B$ (%) from $B_{min}$ to $B_{max}$ | Range $t$ (°C) from $t_{min}$ to $t_{max}$ | $S_\rho$ (kg/m³) | Δ (%) |
|---|---|---|---|---|---|---|---|
| $CdSO_4$ | 9.8600 | −0.0150 | 0.1182 | 0–40 | 0–100 | 3.55 | 0.22 |
| $CoBr_2$ | 8.6719 | −0.0218 | 0.1228 | 0–35 | 0–100 | 4.94 | 0.30 |
| | 7.1494 | −0.0122 | 0.1565 | 35–60 | 0–100 | 4.85 | 0.25 |
| $Co(CH_3COO)_2$ | 6.6332 | 0.0361 | 0.1233 | 2–13 | 10–100 | 3.08 | 0.23 |
| | 8.0664 | 0.0210 | 0.0426 | 13–24 | 10–100 | 1.97 | 0.15 |
| | 6.8485 | 0.0270 | 0.0373 | 4–24 | 10–100 | 5.63 | 0.41 |
| $CoCl_2$ | 9.2511 | −0.0122 | 0.0851 | 0–40 | 0–100 | 5.39 | 0.31 |
| $Co(ClO_4)_2$ | 7.7366 | −0.0127 | 0.0635 | 0–33 | 15–55 | 0.71 | 0.04 |
| $CoI_2$ | 9.2710 | −0.0236 | 0.0875 | 0–30 | 0–100 | 3.53 | 0.21 |
| | 5.4124 | −0.0192 | 0.2050 | 30–55 | 0–100 | 3.81 | 0.20 |
| | 6.6584 | −0.0160 | 0.1773 | 55–70 | 30–100 | 3.08 | 0.13 |
| $Co(NO_3)_2$ | 10.1195 | −0.0371 | 0.1299 | 0–20 | 0–100 | 5.40 | 0.40 |
| | 11.8491 | −0.0225 | −0.0012 | 20–40 | 0–100 | 4.87 | 0.29 |
| | 7.0816 | −0.0160 | 0.1134 | 40–60 | 0–100 | 1.20 | 0.10 |
| $CoSO_4$ | 10.7788 | −0.0175 | 0.0870 | 0–35 | 0–100 | 4.45 | 0.27 |
| $CrCl_3$ | 8.8493 | −0.0226 | 0.0670 | 0–30 | 0–100 | 4.54 | 0.29 |
| $Cr(NO_3)_3$ | 8.2549 | −0.0157 | 0.0678 | 0–45 | 0–100 | 4.88 | 0.30 |
| $Cr_2(SO_4)_3$ | 9.3510 | −0.0160 | 0.1057 | 0–50 | 0–100 | 5.35 | 0.28 |
| CsBr | 9.6569 | −0.0059 | 0.0308 | 0–30 | 0–100 | 2.77 | 0.19 |
| | 8.0404 | −0.0058 | 0.0905 | 30–50 | 0–100 | 2.42 | 0.14 |
| $Cs_2CO_3$ | 10.6810 | −0.0099 | 0.0428 | 0–30 | 0–100 | 4.67 | 0.28 |
| | 7.7480 | −0.0063 | 0.1371 | 30–60 | 0–100 | 2.68 | 0.14 |
| CsCl | 7.3120 | 0.0026 | 0.0680 | 0–25 | 0–100 | 3.26 | 0.23 |
| | 6.5303 | −0.0036 | 0.1056 | 25–55 | 0–100 | 3.81 | 0.21 |
| | 4.0636 | −0.0044 | 0.1531 | 55–65 | 0–100 | 3.62 | 0.16 |
| CsF | 9.4797 | −0.0072 | 0.0718 | 0–35 | 0–50 | 3.91 | 0.22 |
| | 6.2247 | −0.0070 | 0.1682 | 35–55 | 0–50 | 3.27 | 0.17 |
| CsI | 8.0480 | −0.0082 | 0.0926 | 0–50 | 0–100 | 3.43 | 0.20 |
| $CsIO_3$ | 7.9941 | −0.0045 | 0.2469 | 0–2 | 48–100 | 0.17 | 0.01 |
| $CsNO_3$ | 7.2336 | −0.0020 | 0.0596 | 0–30 | 0–100 | 1.53 | 0.11 |
| | 5.9275 | 0.0039 | 0.0890 | 30–50 | 0–100 | 4.44 | 0.29 |
| $Cs_2SO_4$ | 9.8397 | −0.0062 | 0.0487 | 0–30 | 0–100 | 2.21 | 0.15 |
| | 8.1125 | −0.0056 | 0.1078 | 30–55 | 0–100 | 2.69 | 0.14 |
| | 2.7442 | −0.0053 | 0.2033 | 55–65 | 0–100 | 4.15 | 0.18 |
| $CuCl_2$ | 8.9282 | −0.0194 | 0.0858 | 0–45 | 0–100 | 4.52 | 0.25 |
| $Cu(ClO_4)_2$ | 7.6216 | −0.0125 | 0.0758 | 0–50 | 15–55 | 3.52 | 0.18 |
| $Cu(NO_3)_2$ | 8.8108 | −0.0192 | 0.0795 | 0–55 | 0–100 | 5.18 | 0.29 |
| $CuSO_4$ | 10.7672 | −0.0150 | 0.0635 | 0–25 | 0–100 | 2.90 | 0.21 |
| $FeCl_2$ | 10.0388 | −0.0103 | 0.0456 | 0–30 | 0–100 | 3.32 | 0.22 |
| $FeCl_3$ | 9.6221 | −0.0158 | 0.0380 | 0–45 | 0–100 | 4.13 | 0.26 |

(*Continued*)

**TABLE 3.5 (*Continued*)**
**Coefficients of Equation 3.40 for Density Calculation**

| Electrolyte | $A_{1i}$ | $A_{2i}$ | $A_{3i}$ | Range $B$ (%) from $B_{min}$ to $B_{max}$ | Range $t$ (°C) from $t_{min}$ to $t_{max}$ | $S_\rho$ (kg/m³) | Δ (%) |
|---|---|---|---|---|---|---|---|
| $Fe(ClO_4)_2$ | 7.5652 | −0.0131 | 0.0631 | 0–40 | 15–45 | 1.35 | 0.09 |
| $Fe(NO_3)_3$ | 7.7019 | −0.0129 | 0.0760 | 0–50 | 0–100 | 5.38 | 0.27 |
| $FeSO_4$ | 9.7417 | −0.0097 | 0.0751 | 0–30 | 0–100 | 3.23 | 0.21 |
| $H_3BO_3$ | 5.0490 | −0.0255 | 0.0143 | 4–12 | 20–80 | 2.02 | 0.15 |
| HCl | 5.5136 | −0.0091 | −0.0020 | 0–35 | 0–100 | 3.53 | 0.28 |
| $HClO_4$ | 7.7179 | −0.0031 | −0.0914 | 0–20 | 0–100 | 2.90 | 0.21 |
| | 2.3660 | −0.0081 | 0.1758 | 20–30 | 0–100 | 3.19 | 0.22 |
| | 6.5439 | −0.0139 | 0.0431 | 30–60 | 0–100 | 4.66 | 0.26 |
| HCOOH | 2.2853 | −0.00004 | 0.0027 | 0–70 | 20–260 | 3.39 | 0.27 |
| $H_2C_2O_4$ | 7.6482 | −0.0072 | −0.5101 | 0–9 | 20–160 | 2.97 | 0.24 |
| | 3.1880 | 0.2990 | 0.0842 | 5–15 | 15–85 | 2.78 | 0.21 |
| | 3.3695 | 0.0128 | 0.0103 | 15–35 | 15–85 | 2.10 | 0.14 |
| HF | 4.1635 | −0.0177 | −0.0141 | 7–40 | 23–27 | 0.75 | 0.05 |
| $HNO_3$ | 5.7225 | −0.0127 | 0.0165 | 0–50 | 0–100 | 4.01 | 0.24 |
| | 7.6699 | −0.0157 | −0.0207 | 50–100 | 0–50 | 4.88 | 0.27 |
| | 6.5474 | −0.0089 | −0.0097 | 50–100 | 50–100 | 4.78 | 0.27 |
| $H_3PO_4$ | 5.0915 | −0.0032 | 0.0369 | 0–90 | 25–100 | 3.86 | 0.19 |
| | 4.4435 | 0.0002 | 0.0425 | 0–90 | 100–260 | 4.16 | 0.25 |
| $H_4P_2O_7$ | 11.3082 | 0.0114 | 0.1134 | 13–36 | 20–90 | 5.97 | 0.35 |
| $H_2SO_4$ | 6.3787 | −0.0076 | 0.0363 | 10–80 | 0–90 | 3.48 | 0.21 |
| $H_2SiF_6$ | 9.9657 | −0.0409 | −0.0173 | 0–30 | 0–100 | 3.64 | 0.23 |
| | 8.8068 | −0.0482 | 0.0278 | 30–40 | 0–100 | 5.13 | 0.34 |
| $HgCl_2$ | 6.0738 | 0.0014 | 0.2074 | 0–11 | 0–50 | 2.26 | 0.16 |
| $KAl(SO_4)_2$ | 8.0925 | 0.0021 | 0.2219 | 0–20 | 0–80 | 4.02 | 0.29 |
| | 8.9816 | 0.0302 | 0.0595 | 4–25 | 25–90 | 4.36 | 0.31 |
| | 11.8002 | 0.0496 | 0.0293 | 25–40 | 25–90 | 5.40 | 0.34 |
| KBr | 7.0554 | −0.0049 | 0.0602 | 0–50 | 0–100 | 3.23 | 0.18 |
| $KBrO_3$ | 6.1591 | 0.0141 | 0.0647 | 0–35 | 0–100 | 3.65 | 0.26 |
| $K_2CO_3$ | 8.8376 | −0.0054 | 0.0444 | 0–50 | 0–100 | 2.89 | 0.18 |
| | 7.7303 | 0.0081 | 0.0392 | 10–50 | 100–250 | 2.57 | 0.14 |
| | 2.2514 | 0.0307 | 0.0285 | 10–50 | 250–330 | 5.92 | 0.40 |
| $K_2C_2O_4$ | 7.1533 | 0.0253 | 0.0707 | 2–22 | 10–100 | 4.61 | 0.31 |
| KCl | 6.2665 | −0.0041 | 0.0265 | 0–30 | 0–100 | 2.76 | 0.19 |
| | 5.0868 | 0.0130 | −0.0170 | 0–7 | 25–300 | 1.96 | 0.16 |
| $KClO_3$ | 5.8859 | 0.0056 | 0.0246 | 0–35 | 0–100 | 1.55 | 0.11 |
| $KClO_4$ | 4.1721 | 0.0422 | −0.0450 | 0–14 | 40–100 | 3.65 | 0.23 |
| $K_2CrO_4$ | 8.3737 | −0.0134 | 0.0461 | 0–40 | 0–100 | 3.05 | 0.18 |
| $K_2Cr_2O_7$ | 6.8627 | −0.0026 | 0.0495 | 0–35 | 0–100 | 2.65 | 0.20 |

(*Continued*)

**TABLE 3.5 (*Continued*)**
**Coefficients of Equation 3.40 for Density Calculation**

| Electrolyte | $A_{1i}$ | $A_{2i}$ | $A_{3i}$ | Range $B$ (%) from $B_{min}$ to $B_{max}$ | Range $t$ (°C) from $t_{min}$ to $t_{max}$ | $S_\rho$ (kg/m³) | Δ (%) |
|---|---|---|---|---|---|---|---|
| KF | 8.5289 | 0.0036 | 0.0344 | 0–45 | 0–100 | 3.33 | 0.19 |
| $K_3Fe(CN)_6$ | 5.1129 | 0.0018 | 0.0233 | 0–50 | 0–100 | 3.55 | 0.22 |
| $K_4Fe(CN)_6$ | 6.3414 | 0.0038 | 0.0290 | 0–45 | 0–100 | 2.71 | 0.19 |
| $KHCO_3$ | 8.2975 | −0.0147 | −0.0047 | 0–30 | 0–100 | 3.27 | 0.23 |
| $KH_2PO_4$ | 6.8038 | −0.0154 | 0.0742 | 0–30 | 0–100 | 4.06 | 0.29 |
| $K_2HPO_4$ | 8.4533 | −0.0040 | 0.0485 | 0–55 | 0–100 | 2.93 | 0.19 |
| $KHSO_4$ | 7.4113 | −0.0232 | 0.0438 | 0–35 | 0–100 | 4.35 | 0.26 |
| KI | 6.8940 | −0.0070 | 0.0841 | 0–55 | 0–100 | 4.63 | 0.27 |
| $KIO_3$ | 8.0363 | −0.0044 | 0.1040 | 1–8 | 24–100 | 0.44 | 0.03 |
| | 7.7267 | 0.0338 | 0.2102 | 2–15 | 10–100 | 4.86 | 0.31 |
| $KMnO_4$ | 6.4094 | −0.0072 | 0.1470 | 0–7 | 0–25 | 0.41 | 0.03 |
| $KNO_2$ | 6.0226 | 0.0127 | 0.0368 | 0–70 | 0–60 | 4.13 | 0.20 |
| | 5.0375 | 0.0015 | 0.0413 | 0–70 | 60–100 | 5.02 | 0.26 |
| $KNO_3$ | 6.4235 | −0.0066 | 0.0205 | 0–30 | 0–100 | 1.77 | 0.13 |
| | 7.8763 | 0.0091 | −0.2318 | 1–13 | 100–280 | 3.84 | 0.33 |
| KOH | 9.9519 | −0.0067 | 0.0088 | 0–50 | 0–100 | 2.78 | 0.17 |
| | 7.1344 | 0.0145 | 0.0072 | 1–13 | 55–250 | 2.80 | 0.17 |
| *$K_2O$–$Al_2O_3$* | | | | | | | |
| $K_2O/Al_2O_3$ = 14.97 | 10.7816 | −0.0139 | 0.0713 | 0–35 | 25–100 | 3.54 | 0.19 |
| | 6.9395 | 0.0255 | 0.0672 | 0–35 | 100–150 | 4.60 | 0.26 |
| $K_2O/Al_2O_3$ = 6.97 | 10.7861 | −0.0146 | 0.0993 | 0–35 | 25–100 | 3.89 | 0.24 |
| | 6.9552 | 0.0229 | 0.1012 | 0–35 | 100–150 | 4.30 | 0.27 |
| $K_3PO_4$ | 10.1582 | −0.0241 | 0.0362 | 0–30 | 0–100 | 4.76 | 0.30 |
| | 8.7045 | −0.0130 | 0.0821 | 30–60 | 0–100 | 3.76 | 0.19 |
| $K_4P_2O_7$ | 6.7944 | 0.0009 | 0.0799 | 5–38 | 20–90 | 4.71 | 0.33 |
| KSCN | 5.2451 | −0.0068 | 0.0163 | 0–65 | 0–100 | 3.13 | 0.19 |
| $K_2SO_4$ | 7.7902 | −0.0070 | 0.0581 | 0–18 | 0–100 | 1.78 | 0.13 |
| | 7.2513 | 0.0021 | 0.0762 | 0–10 | 25–300 | 1.37 | 0.10 |
| $LaCl_3$ | 8.8201 | −0.0005 | 0.1031 | 0–50 | 0–90 | 5.82 | 0.28 |
| LiBr | 6.8940 | −0.0022 | 0.0711 | 0–45 | 0–100 | 4.21 | 0.27 |
| | 4.4025 | −0.0031 | 0.1293 | 45–65 | 0–100 | 3.74 | 0.19 |
| LiCNS | 3.3051 | −0.0089 | 0.0167 | 0–36 | 0–100 | 4.25 | 0.28 |
| LiCl | 5.3062 | 0.0042 | 0.0225 | 0–50 | 0–100 | 2.67 | 0.17 |
| $LiClO_4$ | 6.0424 | −0.0040 | 0.0274 | 0–50 | 0–100 | 3.79 | 0.24 |
| LiI | 6.8186 | −0.0071 | 0.1026 | 0–60 | 0–100 | 4.45 | 0.25 |
| $LiNO_3$ | 6.1832 | −0.0071 | 0.0302 | 0–55 | 0–100 | 3.33 | 0.20 |

(*Continued*)

**TABLE 3.5 (*Continued*)**
**Coefficients of Equation 3.40 for Density Calculation**

| Electrolyte | $A_{1i}$ | $A_{2i}$ | $A_{3i}$ | Range $B$ (%) from $B_{min}$ to $B_{max}$ | Range $t$ (°C) from $t_{min}$ to $t_{max}$ | $S_\rho$ (kg/m³) | Δ (%) |
|---|---|---|---|---|---|---|---|
| LiOH | 12.0687 | −0.0006 | −0.0883 | 0–10 | 0–100 | 2.91 | 0.24 |
| | 10.1476 | 0.0231 | −0.2947 | 0–7 | 55–250 | 1.95 | 0.15 |
| $Li_2SO_4$ | 7.6439 | 0.0095 | 0.0201 | 0–21 | 100–300 | 2.59 | 0.15 |
| | 8.6040 | −0.0018 | 0.0277 | 0–25 | 0–100 | 2.25 | 0.16 |
| $MgBr_2$ | 8.3963 | 0.0003 | 0.0756 | 0–30 | 0–100 | 4.85 | 0.32 |
| | 7.3258 | 0.0002 | 0.0948 | 30–50 | 0–100 | 5.48 | 0.32 |
| $Mg(BrO_3)_2$ | 9.6252 | −0.1040 | 0.0964 | 0–35 | 0–100 | 4.75 | 0.23 |
| $MgCl_2$ | 7.9592 | 0.0059 | 0.0357 | 0–40 | 0–140 | 2.57 | 0.17 |
| $Mg(ClO_4)_2$ | 7.0378 | −0.0077 | 0.0602 | 0–50 | 0–100 | 2.86 | 0.16 |
| $MgCrO_4$ | 10.2181 | −0.0143 | 0.0588 | 0–35 | 0–100 | 2.08 | 0.15 |
| $MgI_2$ | 8.0019 | −0.0178 | 0.1070 | 0–40 | 0–100 | 3.54 | 0.22 |
| | 5.7374 | −0.0129 | 0.1617 | 40–60 | 0–100 | 4.80 | 0.23 |
| $Mg(NO_3)_2$ | 7.0424 | −0.0009 | 0.0510 | 0–40 | 0–100 | 2.84 | 0.20 |
| | 7.8231 | 0.0023 | 0.0336 | 55–72 | 90–150 | 2.06 | 0.12 |
| $MgSO_4$ | 9.6408 | −0.0085 | 0.0603 | 0–27 | 0–100 | 1.90 | 0.13 |
| | 9.8994 | 0.0041 | 0.0082 | 0–10 | 25–150 | 0.75 | 0.05 |
| $MnCl_2$ | 8.1013 | −0.0089 | 0.0708 | 0–40 | 0–100 | 3.10 | 0.20 |
| $Mn(ClO_4)_2$ | 7.3094 | −0.0121 | 0.0602 | 0–42 | 15–55 | 1.30 | 0.07 |
| $Mn(NO_3)_2$ | 9.2369 | −0.0479 | 0.0864 | 0–20 | 0–100 | 5.02 | 0.31 |
| | 8.1501 | −0.0225 | 0.0688 | 20–40 | 0–100 | 4.88 | 0.28 |
| | 7.7549 | −0.0090 | 0.0631 | 40–60 | 0–100 | 4.13 | 0.22 |
| $MnSO_4$ | 9.4271 | −0.0162 | 0.1010 | 0–40 | 0–100 | 3.80 | 0.23 |
| $NH_3$ | -3.9701 | −0.0111 | 0.0184 | 0–50 | 0–100 | 2.92 | 0.25 |
| $NH_4Al(SO_4)_2$ | 12.9953 | 0.0607 | 0.2308 | 1–14 | 20–100 | 5.30 | 0.36 |
| $NH_4Br$ | 6.2905 | −0.0263 | 0.0593 | 0–20 | 0–100 | 4.27 | 0.32 |
| | 5.7279 | −0.0068 | 0.0305 | 20–55 | 0–100 | 5.49 | 0.38 |
| $NH_4CH_3COO$ | 1.9363 | −0.0012 | −0.0063 | 5–25 | 25–80 | 0.44 | 0.03 |
| | 2.1247 | 0.0079 | 0.0020 | 10–55 | 10–100 | 3.64 | 0.27 |
| | 2.0589 | 0.0116 | 0.0113 | 5–30 | 10–100 | 2.94 | 0.22 |
| | 2.2095 | 0.0055 | 0.0070 | 30–55 | 10–100 | 2.12 | 0.17 |
| $(NH_4)_2C_2O_4$ | 4.8765 | −0.0025 | −0.0116 | 2–12 | 20–80 | 0.35 | 0.02 |
| | 3.4919 | 0.0117 | 0.0700 | 2–20 | 25–100 | 3.24 | 0.24 |
| $NH_4Cl$ | 2.9307 | 0.00003 | −0.0020 | 0–40 | 0–100 | 1.57 | 0.13 |
| $NH_4ClO_4$ | 4.8630 | −0.0194 | 0.0347 | 0–35 | 0–100 | 2.72 | 0.17 |
| $(NH_4)_2CrO_4$ | 5.8466 | −0.0107 | 0.0218 | 0–40 | 0–100 | 3.2 | 0.19 |
| $(NH_4)_2Cr_2O_7$ | 4.2916 | −0.0056 | 0.0652 | 0–45 | 0–100 | 4.66 | 0.33 |
| $NH_4F$ | 3.5953 | −0.0034 | −0.0142 | 0–50 | 0–100 | 3.54 | 0.23 |
| $NH_4HCO_3$ | 5.4689 | 0.0120 | −0.0482 | 0–25 | 0–100 | 4.22 | 0.32 |
| $NH_4H_2PO_4$ | 5.0458 | −0.0062 | 0.0229 | 0–50 | 0–100 | 4.65 | 0.33 |

(*Continued*)

**TABLE 3.5 (*Continued*)**
**Coefficients of Equation 3.40 for Density Calculation**

| Electrolyte | $A_{1i}$ | $A_{2i}$ | $A_{3i}$ | Range $B$ (%) from $B_{min}$ to $B_{max}$ | Range $t$ (°C) from $t_{min}$ to $t_{max}$ | $S_\rho$ (kg/m³) | Δ (%) |
|---|---|---|---|---|---|---|---|
| $NH_4HSO_4$ | 5.7645 | −0.0132 | 0.0211 | 0–65 | 0–100 | 3.85 | 0.23 |
| $NH_4I$ | 6.2062 | −0.0143 | 0.0635 | 0–45 | 0–100 | 4.47 | 0.29 |
| | 5.5231 | −0.0069 | 0.0699 | 45–60 | 0–100 | 3.84 | 0.21 |
| $NH_4NO_3$ | 4.1148 | −0.0101 | 0.0165 | 0–70 | 0–100 | 4.15 | 0.26 |
| $(NH_4)_2SO_4$ | 5.8719 | −0.0012 | −0.0020 | 0–50 | 0–100 | 1.89 | 0.13 |
| $NaBO_2$ | 14.1061 | −0.0124 | −0.1558 | 0–20 | 25–100 | 4.05 | 0.28 |
| | 12.6538 | 0.0012 | −0.1303 | 0–20 | 100–250 | 4.15 | 0.33 |
| | 1.7858 | 0.0535 | −0.2343 | 0–20 | 250–305 | 5.82 | 0.53 |
| $Na_2B_4O_7$ | 8.5344 | 0.0038 | 0.0435 | 0–35 | 30–80 | 1.37 | 0.09 |
| | 14.4953 | −0.0665 | 0.0066 | 0–35 | 80–100 | 5.22 | 0.28 |
| | 8.1130 | 0.0227 | 0.1161 | 2–20 | 25–100 | 4.33 | 0.29 |
| NaBr | 7.3632 | −0.0030 | 0.0776 | 0–45 | 0–110 | 3.94 | 0.23 |
| $NaBrO_3$ | 7.2067 | 0.0030 | 0.0670 | 0–50 | 0–100 | 5.27 | 0.35 |
| $Na_2CO_3$ | 12.1498 | −0.0056 | −0.1033 | 0–15 | 90–270 | 2.38 | 0.19 |
| | 10.2647 | −0.0090 | 0.0391 | 0–30 | 0–100 | 2.54 | 0.18 |
| NaCl | 7.2181 | −0.0057 | 0.0188 | 0–22 | 0–100 | 1.25 | 0.09 |
| $NaClO_3$ | 6.5170 | −0.0002 | 0.0422 | 0–55 | 0–90 | 4.75 | 0.29 |
| | 5.5352 | 0.0048 | 0.0554 | 55–65 | 0–90 | 4.02 | 0.21 |
| $NaClO_4$ | 6.5184 | −0.0216 | 0.0616 | 0–55 | 0–100 | 4.74 | 0.27 |
| | 3.2579 | −0.0154 | 0.1146 | 55–70 | 0–100 | 3.20 | 0.16 |
| $Na_2CrO_4$ | 8.9951 | −0.0057 | 0.0382 | 0–45 | 0–150 | 3.57 | 0.22 |
| | 7.7831 | 0.0049 | 0.0301 | 0–45 | 150–280 | 3.34 | 0.22 |
| $Na_2Cr_2O_7$ | 6.6089 | −0.0142 | 0.0826 | 0–60 | 0–100 | 2.99 | 0.17 |
| NaF | 10.6223 | −0.0069 | 0.0181 | 0–4 | 0–80 | 0.58 | 0.05 |
| | 7.4249 | 0.0322 | −0.2292 | 0–1.5 | 25–325 | 1.34 | 0.11 |
| $NaHCO_3$ | 7.6141 | 0.0426 | −0.1793 | 0–10 | 0–100 | 3.19 | 0.22 |
| $NaH_2PO_4$ | 6.9265 | −0.0016 | 0.0462 | 0–40 | 0–100 | 3.50 | 0.23 |
| | 5.8335 | 0.0008 | 0.0658 | 40–60 | 10–100 | 2.88 | 0.16 |
| $Na_2HPO_4$ | 8.5871 | 0.0142 | 0.0438 | 0–50 | 10–100 | 5.43 | 0.36 |
| NaHS | 5.9615 | 0.0021 | 0.0095 | 0–30 | 0–100 | 2.81 | 0.20 |
| $NaHSO_4$ | 7.4884 | −0.0140 | 0.0590 | 0–40 | 0–100 | 4.39 | 0.28 |
| NaI | 7.3218 | −0.0075 | 0.0915 | 0–45 | 0–100 | 5.05 | 0.35 |
| | 4.6918 | −0.0093 | 0.1472 | 45–60 | 0–100 | 5.11 | 0.26 |
| $NaIO_3$ | 8.6039 | −0.0076 | 0.0969 | 1–8 | 24–100 | 0.40 | 0.03 |
| $Na_2MoO_4$ | 8.6123 | −0.0122 | 0.0636 | 0–40 | 15–115 | 5.28 | 0.37 |
| | 7.2930 | 0.0065 | 0.0406 | 0–40 | 115–280 | 5.52 | 0.41 |
| $NaNO_2$ | 6.3692 | −0.0242 | 0.0543 | 0–30 | 0–100 | 3.40 | 0.26 |
| | 5.5488 | −0.0138 | 0.0587 | 30–50 | 0–100 | 2.98 | 0.17 |
| $NaNO_3$ | 6.6910 | −0.0064 | 0.0329 | 0–50 | 0–100 | 3.75 | 0.24 |

(*Continued*)

**TABLE 3.5 (*Continued*)**
**Coefficients of Equation 3.40 for Density Calculation**

| Electrolyte | $A_{1i}$ | $A_{2i}$ | $A_{3i}$ | Range $B$ (%) from $B_{min}$ to $B_{max}$ | Range $t$ (°C) from $t_{min}$ to $t_{max}$ | $S_\rho$ (kg/m³) | Δ (%) |
|---|---|---|---|---|---|---|---|
| NaOH | 11.3404 | −0.0079 | −0.0095 | 0–50 | 0–100 | 4.15 | 0.24 |
| | 9.9294 | 0.0075 | −0.0162 | 0–28 | 100–200 | 3.03 | 0.18 |
| | 7.8024 | 0.0216 | −0.0493 | 0–28 | 200–300 | 4.70 | 0.29 |
| *$Na_2O$–$Al_2O_3$* | | | | | | | |
| $Na_2O/Al_2O_3$ = 14.95 | 14.3247 | −0.0062 | 0.0366 | 0–25 | 25–150 | 5.48 | 0.38 |
| | 15.2391 | −0.0220 | 0.0376 | 0–25 | 25–105 | 3.93 | 0.24 |
| | 10.0201 | 0.0281 | 0.0391 | 0–25 | 105–150 | 3.82 | 0.25 |
| $Na_2O/Al_2O_3$ = 6.95 | 15.4737 | −0.0021 | 0.0197 | 0–25 | 25–150 | 5.00 | 0.34 |
| | 16.2173 | −0.0166 | 0.0271 | 0–25 | 25–105 | 1.98 | 0.12 |
| | 11.6179 | 0.0299 | 0.0107 | 0–25 | 105–150 | 3.31 | 0.20 |
| $Na_2O/Al_2O_3$ = 3.48 | 16.5367 | −0.0040 | 0.0974 | 0–25 | 25–150 | 5.54 | 0.38 |
| | 17.4161 | −0.0213 | 0.1063 | 0–25 | 25–105 | 2.71 | 0.16 |
| | 12.3081 | 0.0314 | 0.0864 | 0–25 | 105–150 | 3.82 | 0.22 |
| $Na_2O/Al_2O_3$ = 2.49 | 18.0024 | −0.0042 | 0.1213 | 0–25 | 25–150 | 5.72 | 0.38 |
| | 18.8640 | −0.0217 | 0.1322 | 0–25 | 25–105 | 2.45 | 0.14 |
| | 13.8106 | 0.0314 | 0.1068 | 0–25 | 105–150 | 3.94 | 0.22 |
| *$SiO_2/Na_2O$* | | | | | | | |
| $SiO_2/Na_2O$ = 2.06 | 33.3772 | −0.0112 | 0.3814 | 0–20 | 15–100 | 5.73 | 0.33 |
| $SiO_2/Na_2O$ = 2.44 | 33.6988 | −0.0069 | 0.0308 | 0–16 | 15–100 | 2.44 | 0.12 |
| $SiO_2/Na_2O$ = 3.34 | 33.6251 | −0.0293 | 0.3558 | 0–12 | 15–100 | 5.66 | 0.42 |
| $Na_3PO_4$ | 10.0248 | 0.0070 | 0.0637 | 0–32 | 10–100 | 3.97 | 0.27 |
| $Na_4P_2O_7$ | 9.0116 | 0.0053 | 0.0502 | 0–30 | 0–100 | 2.08 | 0.15 |
| $Na_2S$ | 10.4034 | −0.0040 | −0.0154 | 0–27 | 0–100 | 2.93 | 0.19 |
| NaSCN | 6.2979 | −0.0174 | 0.0019 | 0–40 | 0–100 | 4.05 | 0.25 |
| | 5.2242 | −0.0055 | 0.0145 | 40–60 | 0–100 | 4.83 | 0.30 |
| $Na_2SO_3$ | 11.1333 | −0.0258 | −0.0078 | 0–20 | 0–100 | 2.70 | 0.19 |
| $Na_2SO_4$ | 8.8058 | −0.0066 | 0.0568 | 0–30 | 0–100 | 2.15 | 0.13 |
| | 8.8486 | 0.0061 | −0.2988 | 0–4 | 80–300 | 1.74 | 0.12 |
| $Na_2S_2O_3$ | 7.4342 | −0.0163 | 0.0662 | 0–60 | 0–100 | 4.51 | 0.24 |
| $Na_2S_2O_8$ | 7.9457 | 0.0187 | 0.1090 | 2–34 | 10–100 | 2.67 | 0.18 |
| $Na_2SiO_3$ | 8.4869 | −0.0037 | 0.1289 | 0–35 | 0–100 | 4.98 | 0.33 |
| $Na_2WO_4$ | 12.3856 | −0.0218 | −0.0203 | 0 ≤ 25 | 20–100 | 4.33 | 0.29 |
| | 7.8388 | −0.0117 | 0.1336 | >25–40 | 20–100 | 4.64 | 0.28 |
| | 9.8602 | 0.0067 | −0.0334 | 0–25 | 100–280 | 3.98 | 0.30 |
| | 7.1409 | 0.0017 | 0.1149 | 25–40 | 100–280 | 3.85 | 0.24 |

(*Continued*)

**TABLE 3.5 (*Continued*)**
**Coefficients of Equation 3.40 for Density Calculation**

| Electrolyte | $A_{1i}$ | $A_{2i}$ | $A_{3i}$ | Range $B$ (%) from $B_{min}$ to $B_{max}$ | Range $t$ (°C) from $t_{min}$ to $t_{max}$ | $S_\rho$ (kg/m³) | Δ (%) |
|---|---|---|---|---|---|---|---|
| $NdCl_3$ | 8.7320 | −0.0028 | 0.1214 | 0–50 | 0–100 | 5.48 | 0.29 |
| $NiCl_2$ | 9.4441 | −0.0170 | 0.0979 | 0–30 | 0–100 | 4.74 | 0.32 |
| $Ni(ClO_4)_2$ | 7.8071 | −0.0121 | 0.0697 | 0–42 | 15–55 | 1.87 | 0.10 |
| $Ni(NO_3)_2$ | 8.6918 | −0.0594 | 0.2036 | 0–15 | 0–100 | 4.63 | 0.31 |
| | 8.7727 | −0.0331 | 0.1114 | 15–3 | 0–100 | 5.04 | 0.32 |
| | 9.3756 | −0.0229 | 0.0740 | 30–50 | 0–100 | 4.25 | 0.25 |
| $NiSO_4$ | 10.6549 | 0.0006 | 0.0654 | 0–30 | 0–180 | 3.31 | 0.23 |
| $Pb(CH_3COO)_2$ | 6.2086 | 0.0261 | 0.0917 | 2–24 | 10–100 | 5.18 | 0.36 |
| $Pb(NO_3)_2$ | 9.4070 | −0.0153 | 0.0624 | 0–30 | 0–100 | 1.86 | 0.12 |
| RbBr | 7.2846 | −0.0033 | 0.0853 | 0–50 | 0–100 | 4.80 | 0.25 |
| $Rb_2CO_3$ | 10.1837 | −0.0065 | 0.0384 | 0–30 | 0–100 | 4.96 | 0.33 |
| | 7.2776 | −0.0055 | 0.1373 | 30–60 | 0–100 | 3.20 | 0.16 |
| RbCl | 11.4984 | −0.0283 | −0.1194 | 0–15 | 0–100 | 4.59 | 0.34 |
| | 8.0192 | −0.0090 | 0.0532 | 15–40 | 0–100 | 4.56 | 0.29 |
| RbF | 9.8085 | −0.0099 | 0.0653 | 0–30 | 0–50 | 2.95 | 0.21 |
| | 7.7124 | 0.0078 | 0.1329 | 30–55 | 0–55 | 3.94 | 0.21 |
| RbI | 6.3856 | −0.0044 | 0.1593 | 0–45 | 0–60 | 4.46 | 0.28 |
| | -3.8653 | −0.0066 | 0.3885 | 45–55 | 0–60 | 4.50 | 0.21 |
| | 7.4991 | 0.0206 | 0.1111 | 5–30 | 10–90 | 3.74 | 0.24 |
| | 7.7137 | 0.0138 | 0.0886 | 30–50 | 10–90 | 4.06 | 0.20 |
| $RbNO_2$ | 8.3990 | −0.0063 | 0.0292 | 0–50 | 0–80 | 4.26 | 0.30 |
| | 2.0419 | −0.0071 | 0.1570 | 50–65 | 0–80 | 5.29 | 0.28 |
| | 3.3986 | −0.0082 | 0.1334 | 65–86 | 0–80 | 2.99 | 0.12 |
| $RbNO_3$ | 6.4819 | 0.0041 | 0.0584 | 0–45 | 0–100 | 4.17 | 0.28 |
| | 5.0127 | 0.0059 | 0.0875 | 45–60 | 0–100 | 5.28 | 0.30 |
| RbOH | 9.6277 | −0.0030 | 0.0977 | 0–30 | 0–100 | 3.42 | 0.24 |
| | 10.8714 | −0.0015 | 0.0514 | 30–50 | 0–100 | 2.13 | 0.12 |
| $Rb_2SO_4$ | 10.2610 | −0.0064 | 0.0028 | 0–30 | 0–100 | 2.43 | 0.17 |
| $SrCl_2$ | 8.8279 | −0.0053 | 0.0782 | 0–35 | 0–100 | 3.52 | 0.22 |
| $Sr(ClO_4)_2$ | 7.5533 | 0.0019 | 0.0466 | 0–30 | 0–100 | 4.45 | 0.30 |
| | 6.9054 | −0.0074 | 0.0835 | 30–55 | 0–100 | 4.30 | 0.24 |
| | 4.2329 | −0.0088 | 0.1319 | 55–70 | 0–100 | 4.93 | 0.23 |
| $SrI_2$ | 8.2021 | −0.0092 | 0.1030 | 0–40 | 0–100 | 2.51 | 0.16 |
| | 4.9963 | −0.0088 | 0.1833 | 40–60 | 0–100 | 4.68 | 0.23 |
| $Sr(NO_3)_2$ | 7.6378 | −0.0124 | 0.0851 | 0–45 | 0–100 | 2.96 | 0.20 |
| $Tl_2SO_4$ | 9.9517 | −0.0044 | −0.0807 | 0–16 | 0–95 | 1.64 | 0.12 |
| $ZnBr_2$ | 8.3300 | −0.0115 | 0.0915 | 0–50 | 0–100 | 4.52 | 0.23 |
| | 4.7266 | −0.0131 | 0.1687 | 50–70 | 0–100 | 4.33 | 0.19 |
| | 0.1474 | −0.0113 | 0.2339 | 70–80 | 0–100 | 4.56 | 0.16 |

(*Continued*)

**TABLE 3.5 (*Continued*)**
**Coefficients of Equation 3.40 for Density Calculation**

| Electrolyte | $A_{1i}$ | $A_{2i}$ | $A_{3i}$ | Range $B$ (%) from $B_{min}$ to $B_{max}$ | Range $t$ (°C) from $t_{min}$ to $t_{max}$ | $S_\rho$ (kg/m³) | Δ (%) |
|---|---|---|---|---|---|---|---|
| $Zn(CH_3COO)_2$ | 5.5210 | 0.0316 | 0.0493 | 2–24 | 10–100 | 5.13 | 0.35 |
| | 5.6845 | 0.0384 | 0.0847 | 2–13 | 10–100 | 3.25 | 0.23 |
| | 5.3244 | 0.0209 | 0.0277 | 13–24 | 10–100 | 1.97 | 0.15 |
| $ZnCl_2$ | 9.0154 | −0.0127 | 0.0499 | 0–55 | 0–100 | 4.77 | 0.28 |
| | 4.8146 | −0.0140 | 0.1297 | 55–70 | 0–100 | 3.67 | 0.16 |
| $Zn(ClO_4)_2$ | 7.6286 | −0.0125 | 0.0739 | 0–48 | 15–55 | 3.05 | 0.16 |
| $ZnI_2$ | 8.5777 | −0.0263 | 0.1078 | 0–40 | 0–100 | 4.10 | 0.25 |
| | 6.2867 | −0.0218 | 0.1604 | 40–60 | 0–100 | 2.30 | 0.11 |
| | 2.9302 | −0.0211 | 0.2157 | 60–75 | 0–100 | 2.80 | 0.11 |
| $Zn(NO_3)_2$ | 8.4240 | −0.0196 | 0.1086 | 0–30 | 0–100 | 4.10 | 0.29 |
| $ZnSO_4$ | 9.2356 | −0.0065 | 0.1184 | 0–30 | 0–100 | 4.20 | 0.27 |

*Notes:* $S_\rho$ (kg/m³), mean squared error of calculation; Δ (%), mean relative error of calculation.

- At temperature from 150°C to 350°C:

$$\rho_0 = 1014.554664 - R_1 t^{*1.5} + R_2 t^{*2} + R_3 t^{*2.5} + R_4 t^{*4} + R_5 t^{*5} + R_6 t^{*6}, \quad (3.42)$$

where

$R_1 = -85.134917$
$R_2 = -8.171228$
$R_3 = 51.282976$
$R_4 = -22.685588$
$R_5 = 8.11453$
$R_6 = 0.987312$

The water-specific volume on the saturation line with relative error 0.0026% in the temperature interval 0°C–180°C can be calculated by dependence

$$V_0 = 1.000199 + R_1 t + R_2 t^{1.5} + R_3 t^2 + R_4 t^{2.5} + R_5 t^3 + R_6 t^{3.5} + R_7 t^4, \quad (3.43)$$

where

$V_0$ is the water-specific volume, cm³
$R_1 = -1.5616 \cdot 10^{-4}$
$R_2 = 6.479 \cdot 10^{-5}$
$R_3 = -1.16457 \cdot 10^{-5}$
$R_4 = 3.38531 \cdot 10^{-6}$
$R_5 = -4.12981 \cdot 10^{-7}$
$R_6 = 2.36493 \cdot 10^{-8}$
$R_7 = -5.1287 \cdot 10^{-10}$

The water-specific volume ($V \cdot 10^3$ m$^3$/kg) on the saturation line in the temperature interval 0°C ≤ $t$ ≤ 500°C and pressure values 1 ≤ $P$ ≤ 1000 MPa can be calculated by the following equations:

- In the temperature interval 0°C ≤ $t$ ≤ 175°C and pressure values 1 ≤ $P$ ≤ 10 MPa, or in the temperature interval 175°C < $t$ ≤ 200°C and pressure values 2.5 ≤ $P$ ≤ 10 MPa, or in the temperature interval 200°C < $t$ ≤ 250°C and pressure values 5 ≤ $P$ ≤ 10 MPa, or in the temperature interval 250°C < $t$ ≤ 300°C and pressure values $P$ = 10 MPa with mean square error 0.009 m$^3$/kg and relative error 0.6%:

$$V_0 \cdot 10^3 = 1.610954 + R_1\left(\frac{T^*}{P^*}\right) - R_2(T^{*0.7}) + R_3(T^{*2}), \tag{3.44}$$

where
$T^* = T/647.27$
$P^* = P/22.15$
$T$ is the temperature, K
$R_1 = 0.000561$
$R_2 = 1.650281$
$R_3 = 1.627441$

- In the temperature interval 200°C ≤ $t$ < 300°C and pressure values $P$ = 1 MPa, or in the temperature interval 250°C ≤ $t$ ≤ 300°C and pressure values 1 < $P$ ≤ 2.5 MPa, or in the temperature interval 300°C < $t$ ≤ 500°C and pressure values 1 ≤ $P$ ≤ 10 MPa with mean square error 0.147 m$^3$/kg and relative error 0.09%:

$$V_0 \cdot 10^3 = -133.68217 + \frac{R_1}{P^*} + \frac{R_2(T^5)}{P^*} + \frac{R_3}{P^*} + R_4 T^{*0.7} + \frac{R_5 T^*}{P^{*0.5}}$$
$$+ R_6\left(\frac{1}{P^{*T^*}}\right) + R_7 T^{*(1/P^*)} + R_8 T^*, \tag{3.45}$$

where
$R_1 = -9.588874$
$R_2 = 23.34165$
$R_3 = -1.178347$
$R_4 = 340.241186$
$R_5 = 1.626891$
$R_6 = 0.696686$
$R_7 = -0.014187$
$R_8 = -213.999632$

- In the temperature interval $0°C \leq t \leq 300°C$ and pressure values $10 \leq P \leq 100$ MPa with mean square error 0.010 $m^3/kg$ and relative error 0.09%:

$$V_0 \cdot 10^3 = 2.346719 + \frac{R_1}{P^*} + R_2 T^* + R_3 T^{*0.7} + R_4 T^{*4} + \frac{R_5 T^*}{P^{*0.5}} + R_6 T^{*5}, \quad (3.46)$$

where
$R_1 = -0.052553$
$R_2 = 8.547468$
$R_3 = -9.138168$
$R_4 = -2.617144$
$R_5 = 0.390176$
$R_6 = 2.184545$

- In the temperature interval $300°C < t \leq 350°C$ and pressure values $17.5 \leq P \leq 100$ MPa with mean square error 0.007 $m^3/kg$ and relative error 0.41%:

$$V_0 \cdot 10^3 = -2.26703 + R_1 T^{*2} + R_2 \left( \frac{1}{P^{*T^*}} \right) + R_3 T^{*(1/P^*)} + R_4 \exp\left( \frac{T^*}{P^*} \right), \quad (3.47)$$

where
$R_1 = 0.495533$
$R_2 = 0.822158$
$R_3 = 3.157102$
$R_4 = -0.155522$

- In the temperature interval $400°C \leq t \leq 500°C$ and pressure values $17.5 \leq P \leq 50$ MPa with mean square error 0.137 $m^3/kg$ and relative error 0.85%:

$$V_0 = -595.562918 + \frac{R_1}{P^*} + R_2 T^* + R_3 T^{*3} + R_4 T^{*(1/P^*)} + R_5 \left( \frac{1}{P^*} \right)^{T^*}$$

$$+ \frac{R_6 T^*}{P^{*0.5}} + \frac{R_7 T^*}{P^*} + \frac{R_8 T^{*2}}{P^*}, \quad (3.48)$$

where
$R_1 = 599.003573$
$R_2 = -340.295244$
$R_3 = 71.604246$
$R_4 = 894.882219$
$R_5 = -248.375903$
$R_6 = -84.650427$
$R_7 = 43.705264$
$R_8 = -334.788341$

## 3.5 DIELECTRIC PERMITTIVITY OF HOST–GUEST SOLUTION

The study of interparticle interactions of supermolecules and supramolecular assemblies in the host–guest structure, where electrolyte-dissolved ions interact with one another and with water molecules, is of special interest. These interactions and their energy vary depending on the type of particles and the nature of forces acting among them. The nature of interparticle, interionic, ion-molecular, and other forces is very complicated.

All the main types of interactions in the electrolyte solution also happen among dissolved ions and solvent molecules [18–50]. However, since the solvent is usually always in excess, there is a big number of its molecules per one or more electrolyte-dissolved ions that is taking place. This influence has many peculiarities appearing in certain cases and conditions. The influence of the medium on the energy of pair-wise interactions among dissolved ions and molecules is usually studied for estimation of solvent influence on behavior of particles in the solution. The estimation of influence of the solvent dielectric permittivity value $\varepsilon_S$ on electrostatic, solvation, and other interactions of ions (or other particles) serves as a specific example. The value $\varepsilon_S$ in dilute solutions is just approximately equal to water dielectric permittivity, but it cannot be accepted for average and high concentration solutions when the value $\varepsilon_S$ sharply decreases with increase of electrolyte concentration in the solution. This decrease of static dielectric constant values integrates changes of water during hydration and complex formation processes. It includes changes of water composition both during ion hydration and the formation of ion and hydrate complexes in the solution [51].

Real values $\varepsilon_S$ of concentrated electrolyte solutions were unavailable for quite a long time. It is explained by high electrical conductivity of strong electrolytes, which made the direct measurements of $\varepsilon_S$ at low frequencies impossible. Achievements of strong high frequency (SHF) radio electronics have changed such situation. Measurements of dielectric permittivity ($\varepsilon'$) and losses ($\varepsilon''$) of electrolyte solutions into SHF can be made using contemporary radio-technical means $\varepsilon_S$ that can be found from these data. At present, only some (two or three) laboratories of the Russian Federation (including the Institute of General and Inorganic Chemistry of the Russian Academy of Sciences, and the School of Liashchenko) and abroad are conducting systematic SHF researches of aqueous and nonaqueous electrolyte solutions in a wide range of concentrations [51–57].

Thermodynamic water activity ($a_w$) also reflects total change of water composition in electrolyte solutions and is connected with solution dielectric permittivity $\varepsilon_S$. The authors of works [51,54] provide the formula of this connection:

$$a_w = 1 + b\left(\frac{1}{\varepsilon_{H_2O}} - \frac{1}{\varepsilon_S}\right), \tag{3.49}$$

where

$b$ is the constant > 0 and defines peculiarities of solutions of different composition

$\varepsilon_{H_2O}$ is the water dielectric permittivity

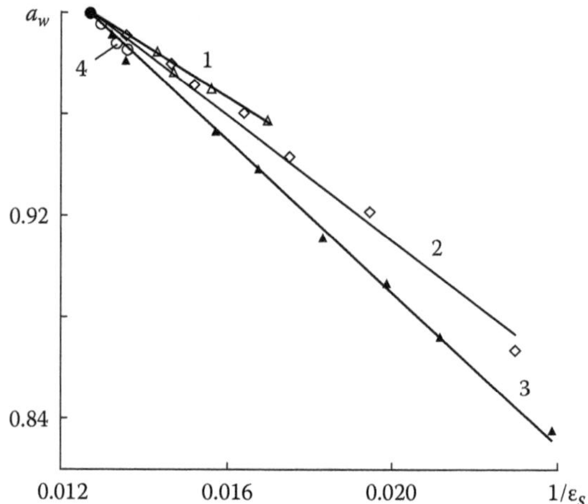


**FIGURE 3.1** Dependence of water activity ($a_w$) from $1/\varepsilon_S$ alkali-metal sulfate solutions at 298 K: 1—$Na_2SO_4$, 2—$Li_2SO_4$, 3—$Cs_2SO_4$, and 4—$K_2SO_4$.

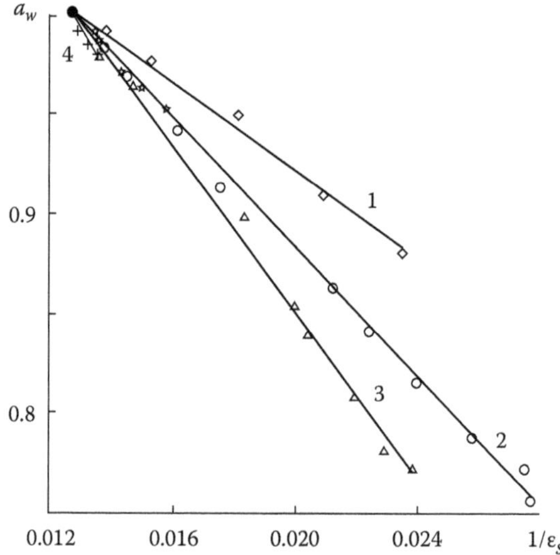


**FIGURE 3.2** Dependence of water activity ($a_w$) from $1/\varepsilon_S$ alkali-metal sulfate solutions at 298 K: 1—$Na_2SO_4$, 2—$Li_2SO_4$, 3—$Cs_2SO_4$, and 4—$K_2SO_4$.

To this formula, the authors [51,54] provide dependence diagrams $a_w$ from $1/\varepsilon_S$, some of which are shown in Figures 3.1 and 3.2.

The dependencies under consideration are straight line. As a result, deviations from linearity do not exceed errors of definitions $\varepsilon_S$ from experimental data. The stated dependencies $a_w$ from $1/\varepsilon_S$ at 298 K are available for a big number of electrolytes. This regularity can be observed in a wide area of solution concentrations. In some cases, these concentrations reach high values, for example, for solutions

$NaNO_3$ up to 10*m* (mol/kg $H_2O$) and for solutions $LiNO_3$ up to 7.38*m*. In some solutions, they comprise the whole area with measurements $\varepsilon_S$, for example, KCl (up to 4.6*m*), NaCl (up to 5.6*m*), KF (up to 5*m*), and CsF (up to 6*m*). In other cases, the area under consideration is smaller than the studied concentration measurement range $\varepsilon_S$. It can be seen from solution LiCl [51]. Linear dependence can be observed up to 5.7 mol/kg $H_2O$. Deviation from linear dependence $a_w$ from $1/\varepsilon_S$ is available at rather large concentrations. Ion-aqueous and ionic clusters of complex structure, where several ions [52] are available, are found in the same range.

For electrolytes with different types of valency, the linear dependencies in various concentration ranges (at maximum deviation $a_w$ 1.0%–1.5%) for sulfates K, Na, Li, and Cs are 0.46*m*, 1.29*m*, 2.77*m*, and 4.27*m*, respectively. They can be found within the whole concentration range for which measurements $\varepsilon_S$ are made. The border corresponds to ~2*m* for solutions $Cu(NO_3)_2$, and 1.6*m* or a bit more for solutions $Y(NO_3)_3$. Minimum concentration limit can be seen for solutions $MgCl_2$ (~1.4–1.5*m*). Molecular research methods demonstrate that in the same concentration areas where deviation from linear dependence $a_w$ from $1/\varepsilon_S$ can be seen, complex ionic and (or) ion-aqueous groupings in solutions $Cu(NO_3)_2$ (1.5–3.1*m*), $MgCl_2$ (1.5–2.5*m*), and $Y(NO_3)_3$ (1.3–2.48*m*) begin to appear. Dependencies obtained in whole establish a tie between a dielectric constant, structural and molecular-kinetic changes of water in ion hydration shells, and thermodynamic characteristics of solutions [51,54]

Hence, linear dependence $a_w$ from $1/\varepsilon_S$ comprises a large concentration area of electrolyte solutions in water. Taking into account dependence (3.29), given the description and structure of Figures 3.1 and 3.2, the task can be resolved vice versa, determining $1/\varepsilon_S$ based on data $a_w$ and coefficient *b*. After several transformations, dependence (3.49) takes the following form:

$$\frac{1}{\varepsilon_S} = \frac{1}{b}(1 - a_w) + \frac{1}{\varepsilon_{H_2O}}. \tag{3.50}$$

It can be seen from Figures 3.1 and 3.2 that constant *b* is nothing but a slope angle tangent of the presented straight lines and it is expressed as

$$b \equiv \text{tg}\,\alpha = \frac{(1 - a_{w1})}{\left((1/\varepsilon_{S1}) - (1/\varepsilon_{H_2O})\right)}, \tag{3.51}$$

where $\varepsilon_{S1}$ is one experimental value $\varepsilon_S$, whose concentration corresponds to the water activity value $a_{w1}$.

Having found constant *b* from (3.51), proceed to (3.50), and in the concentration range depending on $a_w$, values $1/\varepsilon_S$ are calculated. The error of calculation lies in the range of measurement errors $\varepsilon_S$ and $a_w$. For a series of electrolytes, the author of [51] provides tables of constants *b* and $a_w$. Taking into account water activity, shown in [58], tables of constants *b* and $a_w$ were revised and new electrolytes from [51] were added, which are not available in [59]. In Table 3.6, constant *b* is shown for calculation at 25°C, and in Table 3.7 for the temperature range.

**TABLE 3.6**
**Values of Thermodynamic Data to Calculate Dielectric Permittivity of an Electrolyte Solution at 25°C According to Formula (3.50)**

| Ion | $Ba^{2+}$ | $Ca^{2+}$ | $Cs^{+}$ | $Cu^{2+}$ | $H^{+}$ | $K^{+}$ | $Li^{+}$ | $Mg^{2+}$ | $Na^{+}$ | $Ni^{2+}$ | $Rb^{+}$ |
|---|---|---|---|---|---|---|---|---|---|---|---|
| $Br^{-}$ | | | | | | | | | | | |
| $b$ | | | | | | 13.540 | 18.460 | | | | |
| $m_{max}$ | | | | | | 3.390 | 5.710 | | | | |
| $a_w$ | | | | | | 0.889 | 0.628 | | | | |
| $\varepsilon_S$ | | | | | | 47.640 | 30.410 | | | | |
| $Cl^{-}$ | | | | | | | | | | | |
| $b$ | | 12.460 | 20.000 | | | 12.640 | 26.180 | 21.740 | 13.740 | | 16.630 |
| $m_{max}$ | | 1.200 | 5.660 | | | 4.610 | 14.500 | 4.700 | 5.620 | | 6.580 |
| $a_w$ | | 0.926 | 0.826 | | | 0.851 | 0.215 | 0.481 | 0.769 | | 0.820 |
| $\varepsilon_S$ | | 53.500 | 46.600 | | | 40.800 | 23.400 | 27.300 | 33.800 | | 42.400 |
| $F^{-}$ | | | | | | | | | | | |
| $b$ | | | 28.080 | | | 24.260 | | | | | |
| $m_{max}$ | | | 6.000 | | | 9.190 | | | | | |
| $a_w$ | | | 0.741 | | | 0.518 | | | | | |
| $\varepsilon_S$ | | | 45.500 | | | 30.640 | | | | | |
| $I^{-}$ | | | | | | | | | | | |
| $b$ | | | | | | 13.360 | | | | | |
| $m_{max}$ | | | | | | 5.040 | | | | | |
| $a_w$ | | | | | | 0.849 | | | | | |
| $\varepsilon_S$ | | | | | | 41.590 | | | | | |
| $NO_3^{-}$ | | | | | | | | | | | |
| $b$ | 6.680 | | 21.610 | 15.890 | | 30.710 | 25.80 | | 15.280 | | |
| $m_{max}$ | 0.340 | | 0.750 | 4.870 | | 2.030 | 8.30 | | 9.990 | | |
| $a_w$ | 0.988 | | 0.980 | 0.604 | | 0.953 | 0.630 | | 0.772 | | |
| $\varepsilon_S$ | 68.500 | | 73.000 | 26.500 | | 70.000 | 36.9 | | 36.100 | | |
| $OH^{-}$ | | | | | | | | | | | |
| $b$ | | | | | | 15.490 | | | | | |
| $m_{max}$ | | | | | | 3.090 | | | | | |
| $a_w$ | | | | | | 0.866 | | | | | |
| $\varepsilon_S$ | | | | | | 46.700 | | | | | |
| $SO_4^{2-}$ | | | | | | | | | | | |
| $b$ | | | 14.850 | | 12.470 | 18.820 | 12.80 | | 12.760 | 10.84 | |
| $m_{max}$ | | | 4.270 | | 0.510 | 0.460 | 2.77 | | 1.290 | 2.04 | |
| $a_w$ | | | 0.835 | | 0.984 | 0.983 | 0.867 | | 0.945 | 0.921 | |
| $\varepsilon_S$ | | | 41.900 | | 52.000 | 73.300 | 43.2 | | 58.700 | 50.0 | |

Despite a seemed simplicity of formula (3.50), calculations $1/\varepsilon_S$ under it involve a certain difficulty. If the calculation is done in the temperature range, it is necessary to have information about water dielectric permittivity $\varepsilon_{H_2O}$. In Table 3.8, water dielectric permittivity [60] is given. The data in Table 3.8 are given at a pitch of 10°C, and in case their interpolation drops to a pitch of 5°C, errors will occur. One can

**TABLE 3.7**
**Values of Thermodynamic Data to Calculate Dielectric Permittivity of an Electrolyte Solution at Various Temperatures According to Formula (3.50)**

Temperature (K)

| Ion | 283 | 288 | 308 | 313 | 283 | 293 | 303 | 313 | 323 |
|---|---|---|---|---|---|---|---|---|---|
| | $K^+$ | | | | $Na^+$ | | | | |
| $Cl^-$ | | | | | | | | | |
| $b$ | 12.80 | 14.390 | 16.760 | 18.760 | 13.230 | 15.05 | 15.200 | 13.520 | 13.020 |
| $m_{max}$ | 3.950 | 4.610 | 4.610 | 4.610 | 5.580 | 5.59 | 5.600 | 3.210 | 3.220 |
| $a_w$ | 0.871 | 0.848 | 0.851 | 0.851 | 0.775 | 0.770 | 0.770 | 0.875 | 0.874 |
| $\varepsilon_S$ | 45.500 | 43.400 | 45.000 | 46.300 | 34.600 | 36.0 | 36.200 | 43.600 | 41.700 |

| Ion | 288 | 308 | 288 | 308 |
|---|---|---|---|---|
| | $Cs^+$ | | $K^+$ | |
| $F^-$ | | | | |
| $b$ | | | 29.100 | 23.630 |
| $m_{max}$ | 0.500 | 6.0 | 0.500 | 5.000 |
| $a_w$ | 0.741 | | 0.773 | 0.800 |
| $\varepsilon_S$ | 45.500 | 42.9 | 50.700 | 45.200 |

| Ion | 283 | 313 | 283 | 313 |
|---|---|---|---|---|
| | $Cs^+$ | | $Ba^{2+}$ | |
| $NO_3^-$ | | | | |
| $b$ | 10.200 | 5.140 | 11.590 | 9.340 |
| $m_{max}$ | 0.750 | 0.760 | 0.340 | 0.340 |
| $a_w$ | 0.972 | 0.987 | 0.987 | 0.987 |
| $\varepsilon_S$ | 68.300 | 79.150 | 76.600 | 66.300 |

| Ion | 283 | 313 | 283 | 313 | 283 | 313 |
|---|---|---|---|---|---|---|
| | $K^+$ | | $Li^+$ | | $Na^+$ | |
| $NO_3^-$ | | | | | | |
| $b$ | 23.090 | 63.260 | 20.470 | 24.810 | 16.060 | 16.910 |
| $m_{max}$ | 2.000 | 2.000 | 8.340 | 8.340 | 10.000 | 10.000 |
| $a_w$ | 0.901 | 0.933 | 0.621 | 0.622 | 0.777 | 0.754 |
| $\varepsilon_S$ | 61.670 | 67.920 | 32.900 | 34.580 | 38.750 | 35.420 |

| Ion | 293 | 303 | 293 | 303 |
|---|---|---|---|---|
| | $H^+$ | | $Ni^{2+}$ | |
| $SO_4^{2-}$ | | | | |
| $b$ | 1.380 | 1.730 | 8.580 | 6.180 |
| $m_{max}$ | 0.510 | 0.510 | 2.040 | 2.040 |
| $a_w$ | 0.991 | 0.988 | 0.941 | 0.955 |
| $\varepsilon_S$ | 52.700 | 50.900 | 51.000 | 49.800 |

offer a water dielectric permittivity calculation formula in the temperature range 0°C–100°C at atmospheric pressure with error of ±0.05ε [61]:

$$\varepsilon_{H_2O} = 87.740 - 0.40008t + 9.398 \cdot 10^{-4} t^2 - 1.410 \cdot 10^{-6} t^3. \tag{3.52}$$

The calculation of water activity in the concentration and temperature range also involves a certain difficulty in formula (3.50) (Table 3.9). Some publications contain data about vapor pressure over electrolyte solutions and some about water activity systematized in [58]. These works frequently provide only experimental

**TABLE 3.8**
**Static Dielectric Permittivity of Water and Aqueous Vapor in Saturation State**

| $T$ | $\varepsilon'$ | $\varepsilon''$ | $T$ | $\varepsilon'$ | $\varepsilon''$ | $T$ | $\varepsilon'$ | $\varepsilon''$ |
|---|---|---|---|---|---|---|---|---|
| 273.16 | 87.90 | 1.000 | 413.15 | 46.13 | 1.018 | 553.15 | 22.82 | 1.250 |
| 283.15 | 83.97 | 1.000 | 423.15 | 44.03 | 1.022 | 563.15 | 21.48 | 1.296 |
| 293.15 | 80.22 | 1.000 | 433.15 | 42.02 | 1.028 | 573.15 | 20.14 | 1.350 |
| 303.15 | 76.63 | 1.000 | 443.15 | 40.09 | 1.035 | 583.15 | 18.79 | 1.417 |
| 313.15 | 73.20 | 1.001 | 453.15 | 38.24 | 1.043 | 593.15 | 17.44 | 1.501 |
| 323.15 | 69.91 | 1.001 | 463.15 | 36.46 | 1.053 | 603.15 | 16.05 | 1.609 |
| 333.15 | 66.77 | 1.001 | 473.15 | 34.74 | 1.064 | 613.15 | 14.61 | 1.755 |
| 343.15 | 63.77 | 1.002 | 483.15 | 33.09 | 1.076 | 623.15 | 13.04 | 1.966 |
| 353.15 | 60.90 | 1.003 | 493.15 | 31.50 | 1.091 | 633.15 | 11.23 | 2.311 |
| 363.15 | 58.15 | 1.004 | 503.15 | 29.95 | 1.109 | 643.15 | 8.67 | 3.108 |
| 373.15 | 55.53 | 1.006 | 513.15 | 28.46 | 1.129 | 644.15 | 8.28 | 3.279 |
| 383.15 | 53.02 | 1.008 | 523.15 | 27.00 | 1.153 | 645.15 | 7.80 | 3.512 |
| 393.15 | 50.62 | 1.011 | 533.15 | 25.58 | 1.180 | 646.15 | 7.15 | 3.883 |
| 403.15 | 48.33 | 1.014 | 543.15 | 24.19 | 1.212 | | | |

data of vapor pressure over electrolyte solutions, when water activity can be found from the thermodynamic ratio

$$a_w = \frac{P}{P_0}, \tag{3.53}$$

where

$P$ is the water vapor pressure over solution, Pa
$P_0$ is the water vapor pressure over clear water, Pa

Water vapor pressure over clear water is calculated with mean square error 39.54 Pa and mean relative error 0.006% using the following formula [2,58]:

$$P_0 = P_k \exp B, \tag{3.54}$$

where $P_k$ is the critical pressure equal to $2.2064 \cdot 10^7$ Pa, and value $B$ makes

$$B = \frac{T_k A}{T}, \tag{3.55}$$

where

$T_k$ is the critical temperature equal to 647.14 K
$T$ is the temperature, K

**TABLE 3.9**
**Calculated Values of Thermodynamic Quantities Included in Formulae (4.84) and (4.85)**

| $m$ | $c_k$ | $R$ (nm) | $R_{H_2O}$ (nm) | $-\Psi_\gamma^0$ | $\chi$ | $-E_{sp}$ | $E_H$ | $U_{ot}$ | $B$ | $\varepsilon_S$ | $\gamma_\pm$ |
|---|---|---|---|---|---|---|---|---|---|---|---|
| *$CaCl_2$; $\xi$ = 1.18; $a$ = 0.66198* | | | | | | | | | | | |
| 0.001 | 0.001 | 0.7064 | 0.3854 | 0.1169 | 0.1518 | 0.00293 | 0.0000 | 0.0000 | 0.0000 | 78.2 | 0.887 |
| 0.002 | 0.002 | 0.7064 | 0.3854 | 0.1519 | 0.2337 | 0.00586 | 0.0000 | 0.0000 | 0.0000 | 78.2 | 0.854 |
| 0.003 | 0.003 | 0.7064 | 0.3853 | 0.1810 | 0.2939 | 0.00878 | 0.0000 | 0.0000 | 0.0000 | 78.2 | 0.827 |
| 0.004 | 0.004 | 0.7063 | 0.3853 | 0.2037 | 0.3480 | 0.01171 | 0.0000 | 0.0000 | 0.0000 | 78.2 | 0.806 |
| 0.005 | 0.005 | 0.7063 | 0.3853 | 0.2233 | 0.3966 | 0.01465 | 0.0000 | 0.0000 | 0.0000 | 78.2 | 0.788 |
| 0.006 | 0.006 | 0.7063 | 0.3853 | 0.2396 | 0.4445 | 0.01759 | 0.0000 | 0.0000 | 0.0000 | 78.1 | 0.773 |
| 0.007 | 0.007 | 0.7063 | 0.3853 | 0.2509 | 0.4949 | 0.02053 | 0.0000 | 0.0000 | 0.0000 | 78.1 | 0.762 |
| 0.008 | 0.008 | 0.7062 | 0.3853 | 0.2611 | 0.5432 | 0.02346 | 0.0000 | 0.0000 | 0.0000 | 78.1 | 0.752 |
| 0.009 | 0.009 | 0.7062 | 0.3853 | 0.2824 | 0.5648 | 0.02640 | 0.0000 | 0.0000 | 0.0000 | 78.1 | 0.734 |
| 0.010 | 0.010 | 0.7062 | 0.3852 | 0.3197 | 0.5540 | 0.02934 | 0.0000 | 0.0000 | 0.0000 | 78.1 | 0.705 |
| 0.020 | 0.020 | 0.7059 | 0.3851 | 0.3436 | 1.0281 | 0.05882 | 0.0000 | 0.0000 | 0.0000 | 78.1 | 0.668 |
| 0.030 | 0.030 | 0.7057 | 0.3850 | 0.3672 | 1.4387 | 0.08846 | 0.0000 | 0.0000 | 0.0000 | 78.0 | 0.633 |
| 0.040 | 0.040 | 0.7054 | 0.3848 | 0.3838 | 1.8300 | 0.11824 | 0.0000 | 0.0000 | 0.0000 | 78.0 | 0.604 |
| 0.050 | 0.050 | 0.7051 | 0.3847 | 0.3921 | 2.2322 | 0.14816 | 0.0000 | 0.0000 | 0.0000 | 77.9 | 0.581 |
| 0.060 | 0.060 | 0.7049 | 0.3845 | 0.3930 | 2.6649 | 0.17824 | 0.0000 | 0.0000 | 0.0000 | 77.9 | 0.563 |
| 0.070 | 0.070 | 0.7046 | 0.3844 | 0.4071 | 3.0000 | 0.20859 | 0.0056 | 0.0141 | 0.5334 | 77.8 | 0.549 |
| 0.080 | 0.080 | 0.7044 | 0.3843 | 0.4638 | 3.0000 | 0.23898 | 0.0064 | 0.0807 | 2.6576 | 77.7 | 0.538 |
| 0.090 | 0.090 | 0.7041 | 0.3841 | 0.5202 | 3.0000 | 0.26952 | 0.0072 | 0.1372 | 4.0076 | 77.7 | 0.522 |
| 0.100 | 0.099 | 0.7062 | 0.3853 | 0.5996 | 3.0000 | 0.29221 | 0.0079 | 0.1961 | 5.2346 | 77.6 | 0.500 |
| 0.200 | 0.198 | 0.7036 | 0.3839 | 1.1965 | 3.0000 | 0.60307 | 0.0161 | 1.0538 | 13.7806 | 76.6 | 0.477 |
| 0.300 | 0.297 | 0.7011 | 0.3825 | 2.0204 | 3.0000 | 0.96258 | 0.0255 | 2.1948 | 18.1796 | 73.3 | 0.459 |
| 0.400 | 0.395 | 0.6992 | 0.3814 | 3.0303 | 3.0000 | 1.35292 | 0.0355 | 3.5683 | 21.2042 | 70.3 | 0.449 |

*(Continued)*

**TABLE 3.9 (*Continued*)**
**Calculated Values of Thermodynamic Quantities Included in Formulae (4.84) and (4.85)**

| $m$ | $c_k$ | $R$ (nm) | $R_{H_2O}$ (nm) | $-\Psi_\gamma^0$ | $\chi$ | $-E_{sp}$ | $E_H$ | $U_{ot}$ | $B$ | $\varepsilon_S$ | $\gamma_\pm$ |
|---|---|---|---|---|---|---|---|---|---|---|---|
| 0.500 | 0.493 | 0.6971 | 0.3803 | 4.1946 | 3.0000 | 1.78327 | 0.0464 | 5.1528 | 23.4460 | 67.6 | 0.447 |
| 0.600 | 0.590 | 0.6953 | 0.3793 | 5.5388 | 3.0000 | 2.24320 | 0.0579 | 6.9574 | 25.3661 | 65.2 | 0.450 |
| 0.700 | 0.687 | 0.6934 | 0.3783 | 6.9994 | 3.0000 | 2.74219 | 0.0702 | 8.9253 | 26.8489 | 63.0 | 0.457 |
| 0.800 | 0.783 | 0.6916 | 0.3773 | 8.6431 | 3.0000 | 3.27243 | 0.0832 | 11.1110 | 28.2234 | 60.9 | 0.466 |
| 0.900 | 0.879 | 0.6898 | 0.3763 | 10.3338 | 3.0000 | 3.84015 | 0.0968 | 13.3927 | 29.2278 | 59.1 | 0.481 |
| 1.000 | 0.975 | 0.6879 | 0.3753 | 12.0795 | 3.0000 | 4.45091 | 0.1113 | 15.7963 | 29.9968 | 57.3 | 0.509 |
| 1.250 | 1.208 | 0.6842 | 0.3733 | 18.2705 | 3.0000 | 6.16916 | 0.1518 | 23.7608 | 33.0831 | 52.6 | 0.553 |
| 1.500 | 1.437 | 0.6808 | 0.3714 | 25.1505 | 3.0000 | 8.13555 | 0.1971 | 32.6807 | 35.0489 | 48.7 | 0.615 |
| 1.750 | 1.663 | 0.6773 | 0.3695 | 32.7898 | 3.0000 | 10.46058 | 0.2494 | 42.7260 | 36.2029 | 45.0 | 0.694 |
| 2.000 | 1.886 | 0.6738 | 0.3676 | 39.7525 | 3.0000 | 13.16679 | 0.3091 | 52.4821 | 35.8846 | 41.6 | 0.794 |
| 2.250 | 2.106 | 0.6705 | 0.3658 | 42.3303 | 3.0000 | 16.16749 | 0.3738 | 58.1520 | 32.8815 | 38.8 | 0.917 |
| 2.500 | 2.323 | 0.6672 | 0.3640 | 40.8624 | 3.0000 | 19.93176 | 0.4540 | 60.5317 | 28.1820 | 35.6 | 1.067 |
| 2.750 | 2.520 | 0.6655 | 0.3631 | 40.9200 | 3.0000 | 23.63976 | 0.5343 | 64.3887 | 25.4720 | 33.0 | 1.252 |
| 3.000 | 2.731 | 0.6623 | 0.3613 | 5.7517 | 3.0000 | 28.23811 | 0.6289 | 33.9004 | 11.3938 | 30.7 | 1.476 |
| 3.250 | 2.940 | 0.6592 | 0.3596 | 0.0000 | 0.0000 | 33.41017 | 0.7333 | 33.3982 | 9.62700 | 28.6 | 1.750 |
| 3.500 | 3.125 | 0.6576 | 0.3588 | 0.0000 | 0.0000 | 38.45505 | 0.8379 | 38.5241 | 9.71802 | 26.7 | 2.083 |
| 3.750 | 3.305 | 0.6561 | 0.3579 | 0.0000 | 0.0000 | 43.69677 | 0.9455 | 43.8460 | 9.80155 | 25.1 | 2.485 |
| 4.000 | 3.506 | 0.6531 | 0.3563 | 0.0000 | 0.0000 | 50.40524 | 1.0753 | 50.6138 | 9.94901 | 23.7 | 2.969 |
| 4.250 | 3.681 | 0.6515 | 0.3554 | 0.0000 | 0.0000 | 56.84815 | 1.2040 | 57.1172 | 10.0276 | 22.3 | 3.548 |
| 4.500 | 3.851 | 0.6501 | 0.3547 | 0.0000 | 0.0000 | 63.43114 | 1.3343 | 63.7576 | 10.1003 | 21.1 | 4.234 |
| 4.750 | 4.046 | 0.6471 | 0.353 | 0.0000 | 0.0000 | 71.70678 | 1.4874 | 72.0652 | 10.2409 | 20.1 | 5.040 |
| 5.000 | 4.211 | 0.6457 | 0.3523 | 0.0000 | 0.0000 | 79.14622 | 1.6306 | 79.5427 | 10.3113 | 19.2 | 5.977 |
| 5.250 | 4.373 | 0.6443 | 0.3515 | 0.0000 | 0.0000 | 86.91974 | 1.7786 | 87.3444 | 10.3803 | 18.3 | 7.054 |

| | | | | | | | | | | | |
|---|---|---|---|---|---|---|---|---|---|---|---|
| 5.500 | 4.532 | 0.6429 | 0.3507 | 0.0000 | 0.0000 | 95.05552 | 1.9320 | 95.4969 | 10.4478 | 17.6 | 8.275 |
| 5.750 | 4.689 | 0.6414 | 0.3499 | 0.0000 | 0.0000 | 103.3810 | 2.0869 | 103.8305 | 10.5165 | 16.9 | 9.640 |
| 6.000 | 4.843 | 0.6400 | 0.3492 | 0.0000 | 0.0000 | 112.0975 | 2.2478 | 112.5413 | 10.5831 | 16.3 | 11.14 |
| 6.250 | 4.995 | 0.6386 | 0.3484 | 0.0000 | 0.0000 | 119.9690 | 2.3894 | 120.4179 | 10.6525 | 15.9 | 12.77 |
| 6.500 | 5.144 | 0.6373 | 0.3477 | 0.0000 | 0.0000 | 128.0784 | 2.5342 | 128.5209 | 10.7196 | 15.5 | 14.52 |
| 6.750 | 5.291 | 0.6359 | 0.3469 | 0.0000 | 0.0000 | 136.3205 | 2.6796 | 136.7496 | 10.7871 | 15.1 | 16.40 |
| 7.000 | 5.436 | 0.6345 | 0.3462 | 0.0000 | 0.0000 | 144.8462 | 2.8286 | 145.2437 | 10.8537 | 14.8 | 18.27 |
| 7.250 | 5.579 | 0.6332 | 0.3454 | 0.0000 | 0.0000 | 153.2961 | 2.9742 | 153.6427 | 10.9194 | 14.5 | 19.89 |
| 7.460 | 5.668 | 0.6332 | 0.3454 | 0.0000 | 0.0000 | 158.0849 | 3.0672 | 158.4136 | 10.9170 | 14.3 | 21.26 |
| 7.500 | 5.684 | 0.6332 | 0.3454 | 0.0000 | 0.0000 | 159.0876 | 3.0866 | 159.4631 | 10.9205 | 14.2 | 22.68 |
| 7.750 | 5.823 | 0.6318 | 0.3447 | 0.0000 | 0.0000 | 167.7599 | 3.2336 | 168.0681 | 10.9865 | 14.0 | 24.33 |
| 8.000 | 5.960 | 0.6305 | 0.3440 | 0.0000 | 0.0000 | 176.6622 | 3.3832 | 176.9024 | 11.0525 | 13.7 | 26.15 |
| 8.250 | 6.095 | 0.6292 | 0.3433 | 0.0000 | 0.0000 | 185.7899 | 3.5354 | 185.9583 | 11.1180 | 13.5 | 28.08 |
| 8.500 | 6.228 | 0.6279 | 0.3426 | 0.0000 | 0.0000 | 195.1383 | 3.6902 | 195.2294 | 11.1829 | 13.3 | 30.06 |
| 8.750 | 6.321 | 0.6279 | 0.3426 | 0.0000 | 0.0000 | 201.4274 | 3.8091 | 201.4736 | 11.1804 | 13.0 | 32.07 |
| 9.000 | 6.452 | 0.6266 | 0.3418 | 0.0000 | 0.0000 | 211.2888 | 3.9698 | 211.2459 | 11.2481 | 12.8 | 34.14 |
| 9.250 | 6.580 | 0.6253 | 0.3412 | 0.0000 | 0.0000 | 221.4968 | 4.1359 | 221.3574 | 11.3130 | 12.6 | 36.27 |
| 9.500 | 6.667 | 0.6253 | 0.3411 | 0.0000 | 0.0000 | 228.0816 | 4.2586 | 227.8876 | 11.3113 | 12.4 | 38.48 |
| 9.750 | 6.793 | 0.6241 | 0.3404 | 0.0000 | 0.0000 | 238.5653 | 4.4263 | 238.2701 | 11.3785 | 12.2 | 40.76 |
| 10.000 | 6.917 | 0.6228 | 0.3398 | 0.0000 | 0.0000 | 249.2473 | 4.5961 | 248.8470 | 11.4445 | 12.0 | 43.11 |
| *$Cs_2SO_4$; ξ = 1; a = 0.671* | | | | | | | | | | | |
| 0.100 | 0.099 | 0.8497 | 0.3842 | 1.9106 | 3.000 | 0.2150 | 0.0090 | 1.3541 | 31.6934 | 76.7 | 0.464 |
| 0.200 | 0.199 | 0.8429 | 0.3811 | 3.8530 | 3.000 | 0.4535 | 0.0186 | 3.3571 | 38.1849 | 76.1 | 0.390 |
| 0.400 | 0.395 | 0.8347 | 0.3774 | 8.2390 | 3.000 | 0.9721 | 0.0386 | 8.0450 | 44.0012 | 74.1 | 0.317 |
| 0.600 | 0.585 | 0.8284 | 0.3746 | 13.1707 | 3.000 | 1.5360 | 0.0596 | 13.4025 | 47.4958 | 72.2 | 0.279 |
| 0.800 | 0.765 | 0.8243 | 0.3727 | 18.6837 | 3.000 | 2.1123 | 0.0808 | 19.3950 | 50.7528 | 70.4 | 0.256 |
| 1.000 | 0.950 | 0.8170 | 0.3694 | 24.5446 | 3.000 | 2.8079 | 0.1045 | 25.8736 | 52.3518 | 68.8 | 0.240 |

*(Continued)*

**TABLE 3.9 (*Continued*)**
**Calculated Values of Thermodynamic Quantities Included in Formulae (4.84) and (4.85)**

| $m$ | $c_k$ | $R$ (nm) | $R_{H_2O}$ (nm) | $-\Psi_\gamma^0$ | $\chi$ | $-E_{sp}$ | $E_H$ | $U_{ot}$ | $B$ | $\varepsilon_S$ | $\gamma_\pm$ |
|---|---|---|---|---|---|---|---|---|---|---|---|
| 2.000 | 1.766 | 0.7961 | 0.3600 | 63.3111 | 3.000 | 6.6577 | 0.2285 | 68.2335 | 63.1134 | 61.6 | 0.200 |
| 3.000 | 2.450 | 0.7821 | 0.3536 | 121.3994 | 3.000 | 11.2166 | 0.3644 | 130.6871 | 75.8049 | 55.5 | 0.180 |
| 4.000 | 3.038 | 0.7709 | 0.3486 | 202.8001 | 3.000 | 16.4798 | 0.5119 | 217.2490 | 89.7152 | 50.4 | 0.180 |
| *CsCl; ξ = 1.075; a = 0.66865* | | | | | | | | | | | |
| 0.002 | 0.002 | 0.8116 | 0.3853 | 0.0505 | 0.0758 | 0.0007 | 0.00000 | 0.00000 | 0.00000 | 78.2 | 0.950 |
| 0.003 | 0.003 | 0.8116 | 0.3853 | 0.0617 | 0.0931 | 0.0011 | 0.00000 | 0.00000 | 0.00000 | 78.2 | 0.939 |
| 0.004 | 0.004 | 0.8115 | 0.3853 | 0.0698 | 0.1096 | 0.0015 | 0.00000 | 0.00000 | 0.00000 | 78.2 | 0.931 |
| 0.005 | 0.005 | 0.8115 | 0.3853 | 0.0769 | 0.1245 | 0.0019 | 0.00000 | 0.00000 | 0.00000 | 78.2 | 0.924 |
| 0.006 | 0.006 | 0.8114 | 0.3853 | 0.0830 | 0.1384 | 0.0023 | 0.00000 | 0.00000 | 0.00000 | 78.2 | 0.918 |
| 0.007 | 0.007 | 0.8114 | 0.3852 | 0.0892 | 0.1505 | 0.0026 | 0.00000 | 0.00000 | 0.00000 | 78.2 | 0.912 |
| 0.008 | 0.008 | 0.8113 | 0.3852 | 0.0910 | 0.1686 | 0.0030 | 0.00000 | 0.00000 | 0.00000 | 78.2 | 0.910 |
| 0.009 | 0.009 | 0.8113 | 0.3852 | 0.1005 | 0.1720 | 0.0034 | 0.00000 | 0.00000 | 0.00000 | 78.1 | 0.901 |
| 0.010 | 0.010 | 0.8112 | 0.3852 | 0.1180 | 0.1627 | 0.0038 | 0.00000 | 0.00000 | 0.00000 | 78.1 | 0.885 |
| 0.020 | 0.020 | 0.8108 | 0.3850 | 0.1424 | 0.2711 | 0.0077 | 0.00000 | 0.00000 | 0.00000 | 78.0 | 0.860 |
| 0.030 | 0.030 | 0.8103 | 0.3847 | 0.1628 | 0.3570 | 0.0116 | 0.00000 | 0.00000 | 0.00000 | 77.9 | 0.839 |
| 0.040 | 0.040 | 0.8099 | 0.3845 | 0.1814 | 0.4288 | 0.0155 | 0.00000 | 0.00000 | 0.00000 | 77.7 | 0.820 |
| 0.050 | 0.050 | 0.8094 | 0.3843 | 0.1956 | 0.5000 | 0.0195 | 0.00000 | 0.00000 | 0.00000 | 77.6 | 0.805 |
| 0.060 | 0.060 | 0.8090 | 0.3841 | 0.2075 | 0.5678 | 0.0235 | 0.00000 | 0.00000 | 0.00000 | 77.5 | 0.792 |
| 0.070 | 0.070 | 0.8085 | 0.3839 | 0.2170 | 0.6366 | 0.0276 | 0.00000 | 0.00000 | 0.00000 | 77.3 | 0.781 |
| 0.080 | 0.079 | 0.8115 | 0.3853 | 0.2227 | 0.7155 | 0.0306 | 0.00000 | 0.00000 | 0.00000 | 77.2 | 0.774 |
| 0.090 | 0.089 | 0.8107 | 0.3849 | 0.2351 | 0.7660 | 0.0347 | 0.00000 | 0.00000 | 0.00000 | 77.0 | 0.761 |
| 0.100 | 0.100 | 0.8072 | 0.3833 | 0.2615 | 0.7645 | 0.0400 | 0.00000 | 0.00000 | 0.00000 | 76.9 | 0.737 |
| 0.200 | 0.197 | 0.8068 | 0.3831 | 0.2705 | 1.5567 | 0.0804 | 0.00000 | 0.00000 | 0.00000 | 75.6 | 0.699 |

| | | | | | | | | | | | |
|---|---|---|---|---|---|---|---|---|---|---|---|
| 0.300 | 0.296 | 0.8021 | 0.3808 | 0.3297 | 2.0000 | 0.1268 | 0.01730 | 0.03897 | 0.47614 | 74.3 | 0.663 |
| 0.400 | 0.391 | 0.8003 | 0.3800 | 0.4608 | 2.0000 | 0.1723 | 0.02334 | 0.16678 | 1.51056 | 73.1 | 0.633 |
| 0.500 | 0.489 | 0.7960 | 0.3779 | 0.5999 | 2.0000 | 0.2254 | 0.02999 | 0.31890 | 2.24774 | 71.9 | 0.610 |
| 0.600 | 0.583 | 0.7936 | 0.3768 | 0.7508 | 2.0000 | 0.2773 | 0.03654 | 0.48880 | 2.82787 | 70.8 | 0.592 |
| 0.700 | 0.680 | 0.7897 | 0.3749 | 0.9103 | 2.0000 | 0.3374 | 0.04373 | 0.68082 | 3.29105 | 69.7 | 0.578 |
| 0.800 | 0.772 | 0.7875 | 0.3739 | 1.0846 | 2.0000 | 0.3949 | 0.05070 | 0.88806 | 3.70227 | 68.6 | 0.566 |
| 0.900 | 0.863 | 0.7853 | 0.3729 | 1.2706 | 2.0000 | 0.4549 | 0.05788 | 1.11086 | 4.05708 | 67.6 | 0.555 |
| 1.000 | 0.953 | 0.7831 | 0.3718 | 1.4687 | 2.0000 | 0.5176 | 0.06525 | 1.34593 | 4.36033 | 66.6 | 0.543 |
| 1.250 | 1.180 | 0.7764 | 0.3686 | 2.0076 | 2.0000 | 0.6961 | 0.08526 | 2.02577 | 5.02229 | 64.2 | 0.529 |
| 1.500 | 1.403 | 0.7700 | 0.3656 | 2.6145 | 2.0000 | 0.8960 | 0.10675 | 2.79478 | 5.53421 | 62.0 | 0.516 |
| 1.750 | 1.626 | 0.7632 | 0.3624 | 3.2305 | 2.0000 | 1.1214 | 0.12974 | 3.60009 | 5.86569 | 60.2 | 0.505 |
| 2.000 | 1.832 | 0.7587 | 0.3602 | 3.8965 | 2.0000 | 1.3396 | 0.15196 | 4.45263 | 6.19374 | 58.5 | 0.496 |
| 2.250 | 2.037 | 0.7539 | 0.3579 | 4.5714 | 2.0000 | 1.5803 | 0.17546 | 5.34095 | 6.43437 | 57.1 | 0.490 |
| 2.500 | 2.262 | 0.7465 | 0.3544 | 5.1171 | 2.0000 | 1.8886 | 0.20291 | 6.16752 | 6.42511 | 55.9 | 0.486 |
| 2.750 | 2.465 | 0.7416 | 0.3521 | 5.6025 | 2.0000 | 2.1643 | 0.22744 | 6.90403 | 6.41652 | 55.1 | 0.482 |
| 3.000 | 2.667 | 0.7366 | 0.3497 | 6.0084 | 2.0000 | 2.4586 | 0.25263 | 7.58106 | 6.34322 | 54.4 | 0.479 |
| 3.250 | 2.840 | 0.7341 | 0.3485 | 6.4924 | 2.0000 | 2.6990 | 0.27413 | 8.28783 | 6.39075 | 53.7 | 0.477 |
| 3.500 | 3.039 | 0.7291 | 0.3462 | 6.8250 | 2.0000 | 3.0342 | 0.30120 | 8.93447 | 6.27015 | 53.1 | 0.476 |
| 3.750 | 3.239 | 0.7240 | 0.3438 | 7.1301 | 2.0000 | 3.4050 | 0.33019 | 9.58718 | 6.13755 | 52.3 | 0.475 |
| 4.000 | 3.403 | 0.7215 | 0.3426 | 7.6539 | 2.0000 | 3.7012 | 0.35478 | 10.38848 | 6.18945 | 51.5 | 0.474 |
| 4.250 | 3.602 | 0.7165 | 0.3402 | 7.8671 | 2.0000 | 4.1298 | 0.38662 | 11.00627 | 6.01750 | 50.7 | 0.474 |
| 4.500 | 3.762 | 0.7140 | 0.3390 | 8.3553 | 2.0000 | 4.4679 | 0.41335 | 11.81364 | 6.04129 | 49.9 | 0.474 |
| 4.750 | 3.919 | 0.7115 | 0.3378 | 8.8449 | 2.0000 | 4.8231 | 0.44098 | 12.63848 | 6.05816 | 49.1 | 0.474 |
| 5.000 | 4.118 | 0.7064 | 0.3354 | 8.8546 | 2.0000 | 5.3535 | 0.47786 | 13.15154 | 5.81756 | 48.3 | 0.475 |
| 5.250 | 4.272 | 0.7040 | 0.3342 | 9.1846 | 2.0000 | 5.7483 | 0.50705 | 13.85676 | 5.77663 | 47.5 | 0.476 |
| 5.500 | 4.425 | 0.7014 | 0.3330 | 9.4438 | 2.0000 | 6.1640 | 0.53716 | 14.51336 | 5.71117 | 46.8 | 0.478 |
| 5.750 | 4.576 | 0.6989 | 0.3319 | 9.7054 | 2.0000 | 6.6092 | 0.56904 | 15.19990 | 5.64630 | 46.0 | 0.480 |
| 6.000 | 4.725 | 0.6965 | 0.3307 | 9.8893 | 2.0000 | 7.0729 | 0.60169 | 15.82640 | 5.55999 | 45.3 | 0.482 |

*(Continued)*

**TABLE 3.9 (*Continued*)**
**Calculated Values of Thermodynamic Quantities Included in Formulae (4.84) and (4.85)**

| $m$ | $c_k$ | $R$ (nm) | $R_{H_2O}$ (nm) | $-\Psi_\gamma^0$ | $\chi$ | $-E_{sp}$ | $E_H$ | $U_{ot}$ | $B$ | $\varepsilon_S$ | $\gamma_\pm$ |
|---|---|---|---|---|---|---|---|---|---|---|---|
| 6.250 | 4.874 | 0.6939 | 0.3295 | 9.9267 | 2.0000 | 7.5586 | 0.63511 | 16.32563 | 5.43354 | 44.6 | 0.483 |
| 6.500 | 5.020 | 0.6915 | 0.3283 | 9.9810 | 2.0000 | 8.0796 | 0.67081 | 16.87668 | 5.31805 | 43.8 | 0.485 |
| 6.750 | 5.166 | 0.6890 | 0.3271 | 10.1535 | 2.0000 | 8.6894 | 0.71266 | 17.62848 | 5.22874 | 42.7 | 0.487 |
| 7.000 | 5.311 | 0.6865 | 0.3260 | 10.1624 | 2.0000 | 9.3344 | 0.75624 | 18.24808 | 5.10060 | 41.7 | 0.488 |
| 7.250 | 5.478 | 0.6831 | 0.3243 | 9.4758 | 2.0000 | 10.1531 | 0.80870 | 18.33891 | 4.79346 | 40.6 | 0.490 |
| 7.500 | 5.630 | 0.6803 | 0.3230 | 8.7580 | 2.0000 | 10.9422 | 0.85955 | 18.37061 | 4.51769 | 39.6 | 0.492 |
| 7.750 | 5.783 | 0.6775 | 0.3217 | 7.8586 | 2.0000 | 11.8902 | 0.92091 | 18.36894 | 4.21628 | 38.2 | 0.494 |
| 8.000 | 5.863 | 0.6775 | 0.3217 | 9.1698 | 2.0000 | 12.4946 | 0.96763 | 20.24697 | 4.42296 | 36.9 | 0.495 |
| 8.250 | 6.014 | 0.6747 | 0.3203 | 7.3911 | 2.0000 | 13.5567 | 1.03517 | 19.47371 | 3.97651 | 35.7 | 0.497 |
| 8.500 | 6.165 | 0.6719 | 0.3190 | 4.6180 | 2.0000 | 14.6450 | 1.10262 | 17.73250 | 3.39945 | 34.6 | 0.499 |
| 8.750 | 6.317 | 0.6691 | 0.3177 | 0.7131 | 2.0000 | 15.8036 | 1.17301 | 14.92464 | 2.68946 | 33.6 | 0.500 |
| 9.000 | 6.389 | 0.6691 | 0.3177 | 0.7925 | 2.0000 | 16.4471 | 1.22072 | 15.61066 | 2.70313 | 32.7 | 0.502 |
| 9.250 | 6.540 | 0.6663 | 0.3163 | 0.0000 | 0.0000 | 17.7030 | 1.29529 | 16.00817 | 2.61239 | 31.8 | 0.503 |
| 9.500 | 6.691 | 0.6635 | 0.3150 | 0.0000 | 0.0000 | 19.0281 | 1.37260 | 17.26672 | 2.65907 | 31.0 | 0.505 |
| 9.750 | 6.758 | 0.6635 | 0.3150 | 0.0000 | 0.0000 | 19.7291 | 1.42309 | 17.92588 | 2.66265 | 30.2 | 0.506 |
| 10.000 | 6.908 | 0.6608 | 0.3137 | 0.0000 | 0.0000 | 21.1603 | 1.50483 | 19.28599 | 2.70906 | 29.4 | 0.508 |
| 10.500 | 7.122 | 0.6580 | 0.3124 | 0.0000 | 0.0000 | 23.4404 | 1.64318 | 21.44475 | 2.75867 | 28.0 | 0.510 |
| 11.000 | 7.332 | 0.6552 | 0.3111 | 0.0000 | 0.0000 | 25.8680 | 1.78766 | 23.74477 | 2.80767 | 26.7 | 0.512 |
| *CsF; ξ = 1; a = 0.574* | | | | | | | | | | | |
| 0.100 | 0.099 | 0.7830 | 0.3847 | 0.2122 | 1.5265 | 0.02887 | 0.00000 | 0.00000 | 0.00000 | 77.7 | 0.783 |
| 0.200 | 0.199 | 0.7779 | 0.3822 | 0.3325 | 2.0000 | 0.06065 | 0.00849 | 0.09884 | 2.45955 | 77.0 | 0.746 |
| 0.400 | 0.398 | 0.7704 | 0.3785 | 0.7000 | 2.0000 | 0.13008 | 0.01764 | 0.49961 | 5.98534 | 75.6 | 0.721 |
| 0.600 | 0.592 | 0.7653 | 0.3760 | 1.1028 | 2.0000 | 0.20412 | 0.02709 | 0.96712 | 7.54689 | 74.2 | 0.716 |

| | | | | | | | | | | | |
|---|---|---|---|---|---|---|---|---|---|---|---|
| 0.800 | 0.790 | 0.7581 | 0.3725 | 1.5477 | 2.0000 | 0.29179 | 0.03753 | 1.49768 | 8.43599 | 72.8 | 0.717 |
| 1.000 | 0.981 | 0.7530 | 0.3700 | 2.0352 | 2.0000 | 0.38262 | 0.04812 | 2.08214 | 9.14567 | 71.5 | 0.724 |
| 2.000 | 1.916 | 0.7283 | 0.3579 | 5.2874 | 2.0000 | 0.98274 | 0.11058 | 5.98829 | 11.44694 | 64.9 | 0.786 |
| 3.000 | 2.791 | 0.7089 | 0.3483 | 10.2772 | 2.0000 | 1.81939 | 0.18703 | 11.88211 | 13.42924 | 59.0 | 0.878 |
| 4.000 | 3.676 | 0.6887 | 0.3384 | 17.1096 | 2.0000 | 3.05837 | 0.28513 | 20.01747 | 14.83990 | 54.0 | 1.000 |
| 5.000 | 4.481 | 0.6739 | 0.3311 | 25.9840 | 2.0000 | 4.54550 | 0.39374 | 30.43245 | 16.33782 | 49.8 | 1.140 |
| 6.000 | 5.288 | 0.6592 | 0.3239 | 36.5508 | 2.0000 | 6.51404 | 0.52356 | 43.02968 | 17.37249 | 46.2 | 1.340 |
| *$CsNO_3$; ξ = 1; a = 0.63* | | | | | | | | | | | |
| 0.100 | 0.099 | 0.8497 | 0.3842 | 0.2811 | 1.0312 | 0.02586 | 0.00000 | 0.00000 | 0.00000 | 77.5 | 0.733 |
| 0.200 | 0.198 | 0.8443 | 0.3818 | 0.3619 | 1.6476 | 0.05409 | 0.00000 | 0.00000 | 0.00000 | 76.7 | 0.655 |
| 0.400 | 0.390 | 0.8383 | 0.3791 | 0.6241 | 2.0000 | 0.11294 | 0.01896 | 0.15436 | 1.72109 | 75.2 | 0.561 |
| 0.600 | 0.577 | 0.8322 | 0.3763 | 0.9712 | 2.0000 | 0.17678 | 0.02895 | 0.44926 | 3.27989 | 73.9 | 0.501 |
| 0.800 | 0.762 | 0.8254 | 0.3732 | 1.3394 | 2.0000 | 0.24808 | 0.03951 | 0.79545 | 4.25564 | 72.7 | 0.458 |
| 1.000 | 0.943 | 0.8191 | 0.3704 | 1.7231 | 2.0000 | 0.32490 | 0.05041 | 1.17029 | 4.90751 | 71.6 | 0.422 |
| *$Cu(NO_3)_2$; ξ = 1.35; a = 0.73575* | | | | | | | | | | | |
| 0.003 | 0.003 | 0.8307 | 0.3853 | 0.1855 | 0.4513 | 0.00364 | 0.00000 | 0.00000 | 0.00000 | 78.4 | 0.827 |
| 0.004 | 0.004 | 0.8307 | 0.3853 | 0.2106 | 0.5309 | 0.00485 | 0.00000 | 0.00000 | 0.00000 | 78.3 | 0.806 |
| 0.005 | 0.005 | 0.8306 | 0.3853 | 0.2315 | 0.6046 | 0.00607 | 0.00000 | 0.00000 | 0.00000 | 78.3 | 0.788 |
| 0.006 | 0.006 | 0.8306 | 0.3853 | 0.2495 | 0.6730 | 0.00729 | 0.00000 | 0.00000 | 0.00000 | 78.3 | 0.773 |
| 0.007 | 0.007 | 0.8305 | 0.3852 | 0.2628 | 0.7465 | 0.00851 | 0.00000 | 0.00000 | 0.00000 | 78.2 | 0.762 |
| 0.008 | 0.008 | 0.8305 | 0.3852 | 0.2751 | 0.8146 | 0.00973 | 0.00000 | 0.00000 | 0.00000 | 78.2 | 0.751 |
| 0.009 | 0.009 | 0.8304 | 0.3852 | 0.2971 | 0.8498 | 0.01095 | 0.00000 | 0.00000 | 0.00000 | 78.2 | 0.734 |
| 0.010 | 0.010 | 0.8304 | 0.3852 | 0.3373 | 0.8315 | 0.01217 | 0.00000 | 0.00000 | 0.00000 | 78.2 | 0.704 |
| 0.020 | 0.020 | 0.8299 | 0.3849 | 0.3784 | 1.4906 | 0.02448 | 0.00000 | 0.00000 | 0.00000 | 78.0 | 0.667 |
| 0.030 | 0.030 | 0.8294 | 0.3847 | 0.4194 | 2.0241 | 0.03691 | 0.00000 | 0.00000 | 0.00000 | 77.9 | 0.632 |
| 0.040 | 0.040 | 0.8289 | 0.3845 | 0.4539 | 2.5072 | 0.04948 | 0.00000 | 0.00000 | 0.00000 | 77.7 | 0.603 |
| 0.050 | 0.050 | 0.8284 | 0.3842 | 0.4799 | 2.9743 | 0.06216 | 0.00000 | 0.00000 | 0.00000 | 77.5 | 0.580 |

*(Continued)*

**TABLE 3.9 (*Continued*)**
**Calculated Values of Thermodynamic Quantities Included in Formulae (4.84) and (4.85)**

| $m$ | $c_k$ | $R$ (nm) | $R_{H_2O}$ (nm) | $-\Psi_\gamma^0$ | $\chi$ | $-E_{sp}$ | $E_H$ | $U_{ot}$ | $B$ | $\varepsilon_S$ | $\gamma_\pm$ |
|---|---|---|---|---|---|---|---|---|---|---|---|
| 0.060 | 0.060 | 0.8279 | 0.3840 | 0.5739 | 3.0000 | 0.07501 | 0.00407 | 0.07149 | 3.70959 | 77.3 | 0.561 |
| 0.070 | 0.069 | 0.8314 | 0.3856 | 0.6860 | 3.0000 | 0.08459 | 0.00465 | 0.16930 | 7.68832 | 77.2 | 0.548 |
| 0.080 | 0.079 | 0.8304 | 0.3852 | 0.7862 | 3.0000 | 0.09768 | 0.00536 | 0.26147 | 10.32122 | 77.0 | 0.537 |
| 0.090 | 0.090 | 0.8264 | 0.3833 | 0.8711 | 3.0000 | 0.11426 | 0.00617 | 0.33243 | 11.38643 | 76.8 | 0.521 |
| 0.100 | 0.100 | 0.8259 | 0.3831 | 0.9729 | 3.0000 | 0.12765 | 0.00688 | 0.40149 | 12.33236 | 76.7 | 0.497 |
| 0.200 | 0.199 | 0.8225 | 0.3815 | 2.0495 | 3.0000 | 0.26541 | 0.01412 | 1.56165 | 23.37349 | 75.0 | 0.472 |
| 0.300 | 0.296 | 0.8200 | 0.3804 | 3.2554 | 3.0000 | 0.40987 | 0.02161 | 2.86653 | 28.03922 | 73.3 | 0.452 |
| 0.400 | 0.394 | 0.8159 | 0.3784 | 4.5388 | 3.0000 | 0.57266 | 0.02972 | 4.28168 | 30.45685 | 71.7 | 0.439 |
| 0.500 | 0.493 | 0.8111 | 0.3762 | 5.8996 | 3.0000 | 0.75554 | 0.03849 | 5.80833 | 31.89806 | 70.1 | 0.433 |
| 0.600 | 0.587 | 0.8087 | 0.3751 | 7.4859 | 3.0000 | 0.93387 | 0.04715 | 7.56730 | 33.92513 | 68.5 | 0.432 |
| 0.700 | 0.686 | 0.8039 | 0.3729 | 9.0723 | 3.0000 | 1.15233 | 0.05711 | 9.37199 | 34.68928 | 66.9 | 0.434 |
| 0.800 | 0.778 | 0.8017 | 0.3719 | 10.9928 | 3.0000 | 1.35714 | 0.06669 | 11.50278 | 36.46148 | 65.3 | 0.439 |
| 0.900 | 0.869 | 0.7995 | 0.3708 | 13.0846 | 3.0000 | 1.57469 | 0.07669 | 13.82622 | 38.10699 | 63.8 | 0.447 |
| 1.000 | 0.968 | 0.7947 | 0.3686 | 14.9427 | 3.0000 | 1.85152 | 0.08851 | 15.98980 | 38.18800 | 62.3 | 0.463 |
| 1.250 | 1.198 | 0.7874 | 0.3653 | 21.0948 | 3.0000 | 2.55645 | 0.11874 | 22.88288 | 40.73455 | 58.6 | 0.489 |
| 1.500 | 1.425 | 0.7803 | 0.3619 | 28.0279 | 3.0000 | 3.38846 | 0.15296 | 30.69511 | 42.41822 | 55.1 | 0.524 |
| 1.750 | 1.651 | 0.7730 | 0.3586 | 35.9628 | 3.0000 | 4.40076 | 0.19294 | 39.69427 | 43.48735 | 51.5 | 0.567 |
| 2.000 | 1.858 | 0.7683 | 0.3564 | 45.5020 | 3.0000 | 5.43001 | 0.23356 | 50.32191 | 45.54358 | 48.5 | 0.619 |
| 2.250 | 2.081 | 0.7612 | 0.3531 | 51.4190 | 3.0000 | 6.78366 | 0.28331 | 57.64812 | 43.01185 | 45.6 | 0.679 |
| 2.500 | 2.283 | 0.7564 | 0.3508 | 58.9558 | 3.0000 | 8.16778 | 0.33445 | 66.62723 | 42.11031 | 42.9 | 0.749 |
| 2.750 | 2.482 | 0.7516 | 0.3486 | 63.8512 | 3.0000 | 9.75916 | 0.39179 | 73.16872 | 39.47649 | 40.4 | 0.828 |
| 3.000 | 2.679 | 0.7469 | 0.3464 | 60.8202 | 3.0000 | 11.50584 | 0.45277 | 71.93767 | 33.58467 | 38.2 | 0.917 |
| 3.250 | 2.874 | 0.7421 | 0.3442 | 47.7092 | 3.0000 | 13.51654 | 0.52137 | 60.78103 | 24.64250 | 36.0 | 0.918 |

| | | | | | | | | | | | |
|---|---|---|---|---|---|---|---|---|---|---|---|
| 3.500 | 3.067 | 0.7375 | 0.3421 | 17.0540 | 3.0000 | 15.71424 | 0.59423 | 32.46970 | 11.55012 | 34.2 | 1.130 |
| 3.750 | 3.228 | 0.7351 | 0.3410 | 0.0000 | 0.0000 | 17.70188 | 0.66279 | 17.45077 | 5.56547 | 32.4 | 1.255 |
| 4.000 | 3.418 | 0.7304 | 0.3388 | 0.0000 | 0.0000 | 20.35643 | 0.74697 | 20.13663 | 5.69829 | 30.9 | 1.393 |
| 4.250 | 3.572 | 0.7282 | 0.3378 | 0.0000 | 0.0000 | 22.64929 | 0.82302 | 22.46805 | 5.77054 | 29.5 | 1.545 |
| 4.500 | 3.760 | 0.7235 | 0.3356 | 0.0000 | 0.0000 | 25.76147 | 0.91728 | 25.59930 | 5.89913 | 28.2 | 1.711 |
| 4.750 | 3.909 | 0.7212 | 0.3345 | 0.0000 | 0.0000 | 28.38595 | 1.00081 | 28.25195 | 5.96706 | 27.0 | 1.893 |
| 5.000 | 4.055 | 0.7190 | 0.3335 | 0.0000 | 0.0000 | 31.14578 | 1.08734 | 31.03513 | 6.03325 | 26.0 | 2.090 |
| 5.250 | 4.241 | 0.7143 | 0.3313 | 0.0000 | 0.0000 | 34.97195 | 1.19615 | 34.86036 | 6.16044 | 25.0 | 2.304 |
| 5.500 | 4.383 | 0.7121 | 0.3303 | 0.0000 | 0.0000 | 38.04418 | 1.28853 | 37.94589 | 6.22491 | 24.1 | 2.534 |
| 5.750 | 4.524 | 0.7098 | 0.3293 | 0.0000 | 0.0000 | 41.29232 | 1.38448 | 41.50864 | 6.33744 | 23.3 | 3.781 |
| 6.000 | 4.663 | 0.7076 | 0.3282 | 0.0000 | 0.0000 | 44.62963 | 1.48133 | 44.54319 | 6.35613 | 22.6 | 3.046 |
| 6.250 | 4.799 | 0.7054 | 0.3272 | 0.0000 | 0.0000 | 48.08551 | 1.58039 | 47.99895 | 6.41994 | 22.0 | 3.329 |
| 6.500 | 4.934 | 0.7031 | 0.3261 | 0.0000 | 0.0000 | 51.64219 | 1.68038 | 51.55237 | 6.48491 | 21.4 | 3.631 |
| 6.750 | 5.068 | 0.7009 | 0.3251 | 0.0000 | 0.0000 | 55.37205 | 1.78361 | 55.27370 | 6.55062 | 20.8 | 3.952 |
| 7.000 | 5.200 | 0.6987 | 0.3241 | 0.0000 | 0.0000 | 59.26170 | 1.88993 | 59.15197 | 6.61589 | 20.3 | 4.303 |
| 7.250 | 5.331 | 0.6964 | 0.3230 | 0.0000 | 0.0000 | 63.32788 | 1.99942 | 63.20122 | 6.68167 | 19.8 | 4.674 |
| 7.500 | 5.408 | 0.6965 | 0.3231 | 0.0000 | 0.0000 | 65.81138 | 2.07827 | 65.69136 | 6.68144 | 19.3 | 5.042 |
| 7.750 | 5.536 | 0.6943 | 0.3220 | 0.0000 | 0.0000 | 70.14629 | 2.19287 | 69.98296 | 6.74594 | 18.8 | 5.363 |
| 7.840 | 5.617 | 0.6920 | 0.3210 | 0.0000 | 0.0000 | 72.99242 | 2.25856 | 72.81321 | 6.81463 | 18.7 | 5.618 |
| *$H_2SO_4$; ξ = 0.995; a = 0.63879* | | | | | | | | | | | |
| 0.001 | 0.001 | 0.6779 | 0.3854 | 0.3854 | 0.0983 | 0.0047 | 0.0000 | 0.0000 | 0.00000 | 73.8 | 0.801 |
| 0.002 | 0.002 | 0.6779 | 0.3854 | 0.3854 | 0.1503 | 0.0094 | 0.0000 | 0.0000 | 0.00000 | 73.8 | 0.745 |
| 0.003 | 0.003 | 0.6778 | 0.3854 | 0.3854 | 0.1883 | 0.0142 | 0.0000 | 0.0000 | 0.00000 | 73.7 | 0.701 |
| 0.004 | 0.004 | 0.6778 | 0.3853 | 0.3853 | 0.2204 | 0.0189 | 0.0000 | 0.0000 | 0.00000 | 73.7 | 0.666 |
| 0.005 | 0.005 | 0.6778 | 0.3853 | 0.3853 | 0.2496 | 0.0236 | 0.0000 | 0.0000 | 0.00000 | 73.7 | 0.636 |
| 0.006 | 0.006 | 0.6778 | 0.3853 | 0.3853 | 0.2779 | 0.0284 | 0.0000 | 0.0000 | 0.00000 | 73.7 | 0.612 |
| 0.007 | 0.007 | 0.6778 | 0.3853 | 0.3853 | 0.3073 | 0.0331 | 0.0000 | 0.0000 | 0.00000 | 73.7 | 0.594 |
| 0.008 | 0.008 | 0.6777 | 0.3853 | 0.3853 | 0.3348 | 0.0379 | 0.0000 | 0.0000 | 0.00000 | 73.7 | 0.577 |

(*Continued*)

**TABLE 3.9 (*Continued*)**
**Calculated Values of Thermodynamic Quantities Included in Formulae (4.84) and (4.85)**

| $m$ | $c_k$ | $R$ (nm) | $R_{H_2O}$ (nm) | $-\Psi_\gamma^0$ | $\chi$ | $-E_{sp}$ | $E_H$ | $U_{ot}$ | $B$ | $\varepsilon_S$ | $\gamma_\pm$ |
|---|---|---|---|---|---|---|---|---|---|---|---|
| 0.009 | 0.009 | 0.6777 | 0.3853 | 0.3853 | 0.3482 | 0.0427 | 0.0000 | 0.0000 | 0.00000 | 73.6 | 0.551 |
| 0.010 | 0.010 | 0.6777 | 0.3853 | 0.3853 | 0.3378 | 0.0474 | 0.0000 | 0.0000 | 0.00000 | 73.6 | 0.506 |
| 0.020 | 0.020 | 0.6775 | 0.3851 | 0.3851 | 0.6135 | 0.0952 | 0.0000 | 0.0000 | 0.00000 | 73.5 | 0.451 |
| 0.030 | 0.030 | 0.6772 | 0.3850 | 0.3850 | 0.8412 | 0.1434 | 0.0000 | 0.0000 | 0.00000 | 73.3 | 0.401 |
| 0.040 | 0.040 | 0.6770 | 0.3849 | 0.3849 | 1.0449 | 0.1918 | 0.0000 | 0.0000 | 0.00000 | 73.2 | 0.361 |
| 0.050 | 0.050 | 0.6768 | 0.3848 | 0.3848 | 1.2477 | 0.2406 | 0.0000 | 0.0000 | 0.00000 | 73.1 | 0.330 |
| 0.060 | 0.060 | 0.6766 | 0.3846 | 0.3846 | 1.4627 | 0.2899 | 0.0000 | 0.0000 | 0.00000 | 72.9 | 0.306 |
| 0.070 | 0.070 | 0.6764 | 0.3845 | 0.3845 | 1.6979 | 0.3394 | 0.0000 | 0.0000 | 0.00000 | 72.8 | 0.290 |
| 0.080 | 0.080 | 0.6761 | 0.3844 | 0.3844 | 1.9462 | 0.3892 | 0.0000 | 0.0000 | 0.00000 | 72.6 | 0.276 |
| 0.090 | 0.090 | 0.6759 | 0.3843 | 0.3843 | 2.1546 | 0.4396 | 0.0000 | 0.0000 | 0.00000 | 72.5 | 0.257 |
| 0.100 | 0.099 | 0.6780 | 0.3854 | 0.3854 | 2.3258 | 0.4770 | 0.0000 | 0.0000 | 0.00000 | 72.4 | 0.231 |
| 0.200 | 0.198 | 0.6758 | 0.3842 | 0.3842 | 3.0000 | 0.9883 | 0.0194 | 0.9391 | 10.20101 | 71.0 | 0.201 |
| 0.300 | 0.296 | 0.6744 | 0.3834 | 0.3834 | 3.0000 | 1.5211 | 0.0297 | 2.2046 | 15.65847 | 69.7 | 0.177 |
| 0.400 | 0.394 | 0.6726 | 0.3824 | 0.3824 | 3.0000 | 2.0903 | 0.0405 | 3.5617 | 18.55428 | 68.4 | 0.160 |
| 0.500 | 0.490 | 0.6717 | 0.3819 | 0.3819 | 3.0000 | 2.6659 | 0.0515 | 5.0539 | 20.73351 | 67.2 | 0.148 |
| 0.600 | 0.587 | 0.6700 | 0.3809 | 0.3809 | 3.0000 | 3.2923 | 0.0631 | 6.5848 | 22.04395 | 66.0 | 0.140 |
| 0.700 | 0.678 | 0.6702 | 0.3810 | 0.3810 | 3.0000 | 3.8638 | 0.0741 | 8.4033 | 23.95086 | 64.9 | 0.134 |
| 0.800 | 0.773 | 0.6687 | 0.3801 | 0.3801 | 3.0000 | 4.5319 | 0.0863 | 10.1243 | 24.77090 | 63.8 | 0.130 |
| 0.900 | 0.868 | 0.6671 | 0.3792 | 0.3792 | 3.0000 | 5.2402 | 0.0991 | 11.8899 | 25.34249 | 62.7 | 0.126 |
| 1.000 | 0.961 | 0.6659 | 0.3785 | 0.3785 | 3.0000 | 5.9516 | 0.1120 | 13.7501 | 25.94583 | 61.7 | 0.123 |
| 1.500 | 1.417 | 0.6600 | 0.3752 | 0.3752 | 3.0000 | 10.3967 | 0.1904 | 27.3704 | 30.37253 | 54.4 | 0.121 |
| 2.000 | 1.848 | 0.6557 | 0.3727 | 0.3727 | 3.0000 | 15.7661 | 0.2831 | 45.6975 | 34.11667 | 48.4 | 0.122 |
| 2.500 | 2.280 | 0.6498 | 0.3694 | 0.3694 | 3.0000 | 22.4441 | 0.3921 | 58.8323 | 31.71470 | 43.9 | 0.127 |

| | | | | | | | | | | | |
|---|---|---|---|---|---|---|---|---|---|---|---|
| 3.000 | 2.686 | 0.6454 | 0.3669 | 0.3669 | 3.0000 | 31.0599 | 0.5316 | 74.8891 | 29.77598 | 38.6 | 0.135 |
| 3.500 | 3.081 | 0.6411 | 0.3645 | 0.3645 | 3.0000 | 41.1341 | 0.6897 | 67.3495 | 20.63984 | 34.6 | 0.147 |
| 4.000 | 3.444 | 0.6382 | 0.3628 | 0.3628 | 0.0000 | 51.5876 | 0.8530 | 49.1076 | 12.16810 | 31.6 | 0.161 |
| 4.500 | 3.821 | 0.6338 | 0.3603 | 0.3603 | 0.0000 | 64.6238 | 1.0464 | 62.0736 | 12.53922 | 29.0 | 0.178 |
| 5.000 | 4.163 | 0.6309 | 0.3586 | 0.3586 | 0.0000 | 79.3303 | 1.2666 | 76.6859 | 12.79767 | 26.3 | 0.198 |
| 5.500 | 4.495 | 0.6280 | 0.3570 | 0.3570 | 0.0000 | 95.1983 | 1.4988 | 92.4487 | 13.03808 | 24.2 | 0.220 |
| 6.000 | 4.819 | 0.6250 | 0.3553 | 0.3553 | 0.0000 | 112.2613 | 1.7425 | 109.3943 | 13.26973 | 22.6 | 0.245 |
| 6.500 | 5.135 | 0.6222 | 0.3537 | 0.3537 | 0.0000 | 130.4228 | 1.9961 | 127.4268 | 13.49368 | 21.2 | 0.272 |
| 7.000 | 5.445 | 0.6193 | 0.3520 | 0.3520 | 0.0000 | 150.1107 | 2.2649 | 146.9680 | 13.71585 | 20.0 | 0.301 |
| 7.500 | 5.709 | 0.6178 | 0.3512 | 0.3512 | 0.0000 | 168.4171 | 2.5231 | 165.1355 | 13.83451 | 18.9 | 0.333 |
| 8.000 | 6.006 | 0.6149 | 0.3496 | 0.3496 | 0.0000 | 190.7873 | 2.8177 | 187.3276 | 14.05276 | 18.0 | 0.367 |
| 8.500 | 6.254 | 0.6135 | 0.3488 | 0.3488 | 0.0000 | 210.1509 | 3.0819 | 206.5407 | 14.16614 | 17.2 | 0.404 |
| 9.000 | 6.541 | 0.6106 | 0.3471 | 0.3471 | 0.0000 | 234.4719 | 3.3896 | 230.6644 | 14.38455 | 16.5 | 0.443 |
| 9.500 | 6.776 | 0.6092 | 0.3463 | 0.3463 | 0.0000 | 255.2099 | 3.6634 | 251.2360 | 14.49633 | 15.9 | 0.484 |
| 10.000 | 7.054 | 0.6064 | 0.3447 | 0.3447 | 0.0000 | 281.4675 | 3.9832 | 277.2780 | 14.71428 | 15.3 | 0.528 |
| 10.500 | 7.279 | 0.6050 | 0.3439 | 0.3439 | 0.0000 | 302.5667 | 4.2512 | 298.2105 | 14.82774 | 14.9 | 0.574 |
| 11.000 | 7.499 | 0.6035 | 0.3431 | 0.3431 | 0.0000 | 323.6228 | 4.5141 | 319.1017 | 14.94216 | 14.5 | 0.622 |
| 11.500 | 7.713 | 0.6021 | 0.3423 | 0.3423 | 0.0000 | 344.7350 | 4.7745 | 340.0488 | 15.05473 | 14.2 | 0.673 |
| 12.000 | 7.922 | 0.6007 | 0.3415 | 0.3415 | 0.0000 | 366.3290 | 5.0378 | 361.4717 | 15.16690 | 13.9 | 0.726 |
| 12.500 | 8.128 | 0.5993 | 0.3407 | 0.3407 | 0.0000 | 388.1447 | 5.2992 | 383.1150 | 15.28195 | 13.6 | 0.781 |
| 13.000 | 8.329 | 0.5979 | 0.3399 | 0.3399 | 0.0000 | 410.4975 | 5.5646 | 405.2883 | 15.39539 | 13.3 | 0.838 |
| 13.500 | 8.526 | 0.5965 | 0.3391 | 0.3391 | 0.0000 | 432.7611 | 5.8250 | 427.3744 | 15.50870 | 13.1 | 0.896 |
| 14.000 | 8.721 | 0.5951 | 0.3383 | 0.3383 | 0.0000 | 455.8700 | 6.0917 | 450.2963 | 15.62513 | 12.9 | 0.955 |
| 14.500 | 8.912 | 0.5937 | 0.3375 | 0.3375 | 0.0000 | 478.9502 | 6.3544 | 473.1904 | 15.74052 | 12.7 | 1.016 |
| 15.000 | 9.100 | 0.5923 | 0.3367 | 0.3367 | 0.0000 | 502.5185 | 6.6197 | 496.5672 | 15.85633 | 12.5 | 1.077 |
| 15.500 | 9.220 | 0.5923 | 0.3367 | 0.3367 | 0.0000 | 515.3813 | 6.7894 | 509.3309 | 15.85735 | 12.3 | 1.139 |
| 16.000 | 9.402 | 0.5909 | 0.3359 | 0.3359 | 0.0000 | 536.2190 | 7.0132 | 530.0127 | 15.97460 | 12.2 | 1.201 |
| 16.500 | 9.581 | 0.5896 | 0.3352 | 0.3352 | 0.0000 | 557.4083 | 7.2386 | 551.0417 | 16.09127 | 12.1 | 1.264 |

*(Continued)*

TABLE 3.9 (*Continued*)
Calculated Values of Thermodynamic Quantities Included in Formulae (4.84) and (4.85)

| $m$ | $c_k$ | $R$ (nm) | $R_{H_2O}$ (nm) | $-\Psi_\gamma^0$ | $\chi$ | $-E_{sp}$ | $E_H$ | $U_{ot}$ | $B$ | $\varepsilon_S$ | $\gamma_\pm$ |
|---|---|---|---|---|---|---|---|---|---|---|---|
| 17.000 | 9.758 | 0.5882 | 0.3344 | 0.3344 | 0.0000 | 578.7543 | 7.4624 | 572.2265 | 16.20887 | 12.0 | 1.327 |
| 17.500 | 9.864 | 0.5882 | 0.3344 | 0.3344 | 0.0000 | 589.3140 | 7.5982 | 582.7103 | 16.21066 | 12.0 | 1.389 |
| 18.000 | 10.036 | 0.5868 | 0.3336 | 0.3336 | 0.0000 | 611.0761 | 7.8228 | 604.3056 | 16.32885 | 11.9 | 1.451 |
| 18.500 | 10.135 | 0.5868 | 0.3336 | 0.3336 | 0.0000 | 621.1525 | 7.9518 | 614.3086 | 16.32979 | 11.8 | 1.514 |
| 19.000 | 10.303 | 0.5854 | 0.3328 | 0.3328 | 0.0000 | 643.0409 | 8.1732 | 636.0294 | 16.44931 | 11.7 | 1.576 |
| 19.500 | 10.470 | 0.5840 | 0.3320 | 0.3320 | 0.0000 | 665.4768 | 8.3975 | 658.2930 | 16.57026 | 11.6 | 1.639 |
| 20.000 | 10.560 | 0.5841 | 0.3320 | 0.3320 | 0.0000 | 675.1393 | 8.5200 | 667.8836 | 16.57003 | 11.6 | 1.701 |
| 20.500 | 10.723 | 0.5827 | 0.3313 | 0.3313 | 0.0000 | 697.5528 | 8.7399 | 690.1265 | 16.69113 | 11.5 | 1.764 |
| 21.000 | 10.808 | 0.5827 | 0.3313 | 0.3313 | 0.0000 | 706.9246 | 8.8577 | 699.4287 | 16.69109 | 11.5 | 1.828 |
| 21.500 | 10.968 | 0.5813 | 0.3305 | 0.3305 | 0.0000 | 729.9177 | 9.0802 | 722.2466 | 16.81328 | 11.4 | 1.892 |
| 22.000 | 11.049 | 0.5813 | 0.3305 | 0.3305 | 0.0000 | 739.0269 | 9.1933 | 731.2890 | 16.81428 | 11.3 | 1.958 |
| 22.500 | 11.205 | 0.5800 | 0.3297 | 0.3297 | 0.0000 | 761.8865 | 9.4107 | 753.9766 | 16.93543 | 11.3 | 2.024 |
| 23.000 | 11.282 | 0.5800 | 0.3297 | 0.3297 | 0.0000 | 770.8593 | 9.5211 | 762.8836 | 16.93683 | 11.2 | 2.091 |
| 23.500 | 11.436 | 0.5786 | 0.3289 | 0.3289 | 0.0000 | 794.3591 | 9.7416 | 786.2064 | 17.05953 | 11.2 | 2.157 |
| 24.000 | 11.508 | 0.5786 | 0.3289 | 0.3289 | 0.0000 | 801.9461 | 9.8351 | 793.7421 | 17.05927 | 11.1 | 2.224 |
| 24.500 | 11.660 | 0.5773 | 0.3282 | 0.3282 | 0.0000 | 823.3099 | 10.0252 | 814.9563 | 17.18318 | 11.1 | 2.289 |
| 25.000 | 11.729 | 0.5773 | 0.3282 | 0.3282 | 0.0000 | 829.2238 | 10.0971 | 820.8364 | 17.18382 | 11.1 | 2.351 |
| 25.500 | 11.796 | 0.5773 | 0.3282 | 0.3282 | 0.0000 | 835.0739 | 10.1684 | 826.6506 | 17.18429 | 11.1 | 2.408 |
| 26.500 | 12.009 | 0.5759 | 0.3274 | 0.3274 | 0.0000 | 862.4730 | 10.4267 | 853.8335 | 17.30959 | 11.1 | 2.455 |
| 26.000 | 11.945 | 0.5759 | 0.3274 | 0.3274 | 0.0000 | 856.8659 | 10.3591 | 848.2979 | 17.30964 | 11.1 | 2.493 |
| 27.000 | 12.155 | 0.5746 | 0.3267 | 0.3267 | 0.0000 | 884.2297 | 10.6146 | 875.4386 | 17.43353 | 11.1 | 2.518 |
| 27.500 | 12.216 | 0.5746 | 0.3267 | 0.3267 | 0.0000 | 889.6268 | 10.6793 | 880.7875 | 17.43381 | 11.0 | 2.532 |

| | | | | | | | | | | | |
|---|---|---|---|---|---|---|---|---|---|---|---|
| *$K_2SO_4$; ξ = 1; a = 0.639* | | | | | | | | | | | |
| 0.001 | 0.001 | 0.8210 | 0.3854 | 0.1236 | 0.4731 | 0.00121 | 0.00000 | 0.00000 | 0.00000 | 78.3 | 0.882 |
| 0.002 | 0.002 | 0.8210 | 0.3853 | 0.1635 | 0.7152 | 0.00242 | 0.00000 | 0.00000 | 0.00000 | 78.3 | 0.847 |
| 0.003 | 0.003 | 0.8209 | 0.3853 | 0.1975 | 0.8897 | 0.00363 | 0.00000 | 0.00000 | 0.00000 | 78.3 | 0.817 |
| 0.004 | 0.004 | 0.8209 | 0.3853 | 0.2261 | 1.0358 | 0.00485 | 0.00000 | 0.00000 | 0.00000 | 78.3 | 0.793 |
| 0.005 | 0.005 | 0.8208 | 0.3853 | 0.2505 | 1.1685 | 0.00606 | 0.00000 | 0.00000 | 0.00000 | 78.3 | 0.773 |
| 0.006 | 0.006 | 0.8208 | 0.3853 | 0.2717 | 1.2925 | 0.00727 | 0.00000 | 0.00000 | 0.00000 | 78.3 | 0.756 |
| 0.007 | 0.007 | 0.8207 | 0.3852 | 0.2877 | 1.4265 | 0.00849 | 0.00000 | 0.00000 | 0.00000 | 78.2 | 0.743 |
| 0.008 | 0.008 | 0.8207 | 0.3852 | 0.3028 | 1.5485 | 0.00971 | 0.00000 | 0.00000 | 0.00000 | 78.2 | 0.731 |
| 0.009 | 0.009 | 0.8206 | 0.3852 | 0.3305 | 1.5957 | 0.01092 | 0.00000 | 0.00000 | 0.00000 | 78.2 | 0.710 |
| 0.010 | 0.010 | 0.8206 | 0.3852 | 0.3818 | 1.5370 | 0.01215 | 0.00000 | 0.00000 | 0.00000 | 78.2 | 0.674 |
| 0.020 | 0.020 | 0.8201 | 0.3849 | 0.4396 | 2.6788 | 0.02440 | 0.00000 | 0.00000 | 0.00000 | 78.1 | 0.628 |
| 0.030 | 0.030 | 0.8196 | 0.3847 | 0.5907 | 3.0000 | 0.03675 | 0.00177 | 0.08892 | 10.63152 | 78.0 | 0.583 |
| 0.040 | 0.040 | 0.8192 | 0.3845 | 0.7888 | 3.0000 | 0.04918 | 0.00236 | 0.23194 | 20.75642 | 77.9 | 0.545 |
| 0.050 | 0.050 | 0.8187 | 0.3843 | 0.9892 | 3.0000 | 0.06174 | 0.00296 | 0.38671 | 27.61976 | 77.8 | 0.514 |
| 0.060 | 0.060 | 0.8182 | 0.3841 | 1.1909 | 3.0000 | 0.07440 | 0.00356 | 0.55159 | 32.75172 | 77.8 | 0.490 |
| 0.070 | 0.069 | 0.8217 | 0.3857 | 1.3939 | 3.0000 | 0.08384 | 0.00406 | 0.72647 | 37.78005 | 77.7 | 0.471 |
| 0.080 | 0.079 | 0.8207 | 0.3852 | 1.5982 | 3.0000 | 0.09668 | 0.00467 | 0.90928 | 41.15349 | 77.6 | 0.456 |
| 0.090 | 0.090 | 0.8168 | 0.3834 | 1.8035 | 3.0000 | 0.11300 | 0.00538 | 1.07711 | 42.33297 | 77.5 | 0.432 |
| 0.100 | 0.100 | 0.8164 | 0.3832 | 2.0103 | 3.0000 | 0.12607 | 0.00599 | 1.20561 | 42.54377 | 77.4 | 0.394 |
| 0.200 | 0.197 | 0.8158 | 0.3829 | 4.2123 | 3.0000 | 0.25315 | 0.01200 | 3.41435 | 60.12046 | 76.1 | 0.350 |
| 0.300 | 0.296 | 0.8109 | 0.3806 | 6.6200 | 3.0000 | 0.39881 | 0.01856 | 5.84582 | 66.59441 | 74.9 | 0.310 |
| 0.400 | 0.394 | 0.8069 | 0.3787 | 9.1951 | 3.0000 | 0.55235 | 0.02531 | 8.46587 | 70.70364 | 73.8 | 0.278 |
| 0.500 | 0.488 | 0.8050 | 0.3779 | 11.8615 | 3.0000 | 0.70072 | 0.03188 | 11.19049 | 74.20346 | 73.0 | 0.255 |
| 0.600 | 0.581 | 0.8029 | 0.3768 | 14.7711 | 3.0000 | 0.85720 | 0.03867 | 14.18104 | 77.50915 | 72.0 | 0.236 |
| 0.692 | 0.672 | 0.7983 | 0.3747 | 17.5854 | 3.0000 | 1.03352 | 0.04580 | 17.11197 | 78.97157 | 71.1 | 0.223 |

*(Continued)*

**TABLE 3.9 (*Continued*)**
**Calculated Values of Thermodynamic Quantities Included in Formulae (4.84) and (4.85)**

| $m$ | $c_k$ | $R$ (nm) | $R_{H_2O}$ (nm) | $-\Psi_\gamma^0$ | $\chi$ | $-E_{sp}$ | $E_H$ | $U_{ot}$ | $B$ | $\varepsilon_S$ | $\gamma_\pm$ |
|---|---|---|---|---|---|---|---|---|---|---|---|
| *KBr; $\xi = 1.065$; $a = 0.644325$* | | | | | | | | | | | |
| 0.003 | 0.003 | 0.7230 | 0.3853 | 0.0593 | 0.0641 | 0.00139 | 0.00000 | 0.00000 | 0.00000 | 78.4 | 0.941 |
| 0.004 | 0.004 | 0.7229 | 0.3853 | 0.0674 | 0.0753 | 0.00185 | 0.00000 | 0.00000 | 0.00000 | 78.4 | 0.933 |
| 0.005 | 0.005 | 0.7229 | 0.3853 | 0.0744 | 0.0854 | 0.00232 | 0.00000 | 0.00000 | 0.00000 | 78.3 | 0.926 |
| 0.006 | 0.006 | 0.7229 | 0.3853 | 0.0804 | 0.0948 | 0.00278 | 0.00000 | 0.00000 | 0.00000 | 78.3 | 0.920 |
| 0.007 | 0.007 | 0.7229 | 0.3853 | 0.0853 | 0.1041 | 0.00324 | 0.00000 | 0.00000 | 0.00000 | 78.3 | 0.915 |
| 0.008 | 0.008 | 0.7228 | 0.3853 | 0.0870 | 0.1169 | 0.00371 | 0.00000 | 0.00000 | 0.00000 | 78.3 | 0.913 |
| 0.009 | 0.009 | 0.7228 | 0.3853 | 0.0953 | 0.1201 | 0.00418 | 0.00000 | 0.00000 | 0.00000 | 78.3 | 0.905 |
| 0.010 | 0.010 | 0.7228 | 0.3852 | 0.1115 | 0.1142 | 0.00464 | 0.00000 | 0.00000 | 0.00000 | 78.2 | 0.890 |
| 0.020 | 0.020 | 0.7225 | 0.3851 | 0.1327 | 0.1928 | 0.00932 | 0.00000 | 0.00000 | 0.00000 | 78.1 | 0.867 |
| 0.030 | 0.030 | 0.7222 | 0.3849 | 0.1498 | 0.2572 | 0.01404 | 0.00000 | 0.00000 | 0.00000 | 77.9 | 0.848 |
| 0.040 | 0.040 | 0.7219 | 0.3848 | 0.1649 | 0.3127 | 0.01879 | 0.00000 | 0.00000 | 0.00000 | 77.8 | 0.831 |
| 0.050 | 0.050 | 0.7216 | 0.3846 | 0.1755 | 0.3695 | 0.02358 | 0.00000 | 0.00000 | 0.00000 | 77.6 | 0.818 |
| 0.060 | 0.060 | 0.7213 | 0.3845 | 0.1839 | 0.4248 | 0.02841 | 0.00000 | 0.00000 | 0.00000 | 77.5 | 0.807 |
| 0.070 | 0.070 | 0.7211 | 0.3843 | 0.1899 | 0.4818 | 0.03327 | 0.00000 | 0.00000 | 0.00000 | 77.4 | 0.798 |
| 0.080 | 0.080 | 0.7208 | 0.3842 | 0.1934 | 0.5425 | 0.03816 | 0.00000 | 0.00000 | 0.00000 | 77.2 | 0.791 |
| 0.090 | 0.089 | 0.7232 | 0.3855 | 0.2034 | 0.5944 | 0.04182 | 0.00000 | 0.00000 | 0.00000 | 77.1 | 0.780 |
| 0.100 | 0.099 | 0.7226 | 0.3852 | 0.2241 | 0.6008 | 0.04679 | 0.00000 | 0.00000 | 0.00000 | 76.9 | 0.760 |
| 0.200 | 0.198 | 0.7198 | 0.3837 | 0.2116 | 1.3289 | 0.09729 | 0.00000 | 0.00000 | 0.00000 | 75.5 | 0.729 |
| 0.300 | 0.295 | 0.7187 | 0.3831 | 0.2220 | 2.0000 | 0.14878 | 0.01584 | 0.01041 | 0.13894 | 74.2 | 0.701 |
| 0.400 | 0.395 | 0.7149 | 0.3811 | 0.3050 | 2.0000 | 0.20831 | 0.02180 | 0.11718 | 1.13611 | 72.9 | 0.678 |
| 0.500 | 0.491 | 0.7136 | 0.3803 | 0.3992 | 2.0000 | 0.26604 | 0.02767 | 0.24145 | 1.84431 | 71.7 | 0.661 |
| 0.600 | 0.587 | 0.7118 | 0.3794 | 0.4993 | 2.0000 | 0.32766 | 0.03381 | 0.38068 | 2.37997 | 70.5 | 0.648 |
| 0.700 | 0.683 | 0.7098 | 0.3783 | 0.6050 | 2.0000 | 0.39330 | 0.04022 | 0.53353 | 2.80408 | 69.3 | 0.638 |

| | | | | | | | | | | | |
|---|---|---|---|---|---|---|---|---|---|---|---|
| 0.800 | 0.778 | 0.7080 | 0.3774 | 0.7179 | 2.0000 | 0.46145 | 0.04680 | 0.69893 | 3.15674 | 68.2 | 0.630 |
| 0.900 | 0.872 | 0.7063 | 0.3765 | 0.8366 | 2.0000 | 0.53182 | 0.05353 | 0.87361 | 3.44988 | 67.2 | 0.623 |
| 1.000 | 0.966 | 0.7045 | 0.3755 | 0.9619 | 2.0000 | 0.60661 | 0.06055 | 1.05726 | 3.69094 | 66.1 | 0.615 |
| 1.250 | 1.194 | 0.7010 | 0.3736 | 1.3174 | 2.0000 | 0.80013 | 0.07858 | 1.58377 | 4.26023 | 63.6 | 0.607 |
| 1.500 | 1.419 | 0.6973 | 0.3717 | 1.7096 | 2.0000 | 1.01443 | 0.09793 | 2.16960 | 4.68296 | 61.3 | 0.601 |
| 1.750 | 1.640 | 0.6937 | 0.3698 | 2.1474 | 2.0000 | 1.24940 | 0.11861 | 2.82352 | 5.03207 | 59.1 | 0.597 |
| 2.000 | 1.859 | 0.6900 | 0.3678 | 2.6162 | 2.0000 | 1.50956 | 0.14082 | 3.53366 | 5.30417 | 57.0 | 0.594 |
| 2.250 | 2.075 | 0.6863 | 0.3658 | 3.1078 | 2.0000 | 1.79303 | 0.16441 | 4.29182 | 5.51788 | 55.1 | 0.593 |
| 2.500 | 2.290 | 0.6826 | 0.3638 | 3.5975 | 2.0000 | 2.10534 | 0.18964 | 5.07686 | 5.65880 | 53.3 | 0.593 |
| 2.750 | 2.503 | 0.6789 | 0.3619 | 4.0791 | 2.0000 | 2.44611 | 0.21648 | 5.88400 | 5.74550 | 51.6 | 0.595 |
| 3.000 | 2.693 | 0.6770 | 0.3609 | 4.7272 | 2.0000 | 2.75594 | 0.24170 | 6.82654 | 5.97016 | 50.0 | 0.596 |
| 3.250 | 2.902 | 0.6733 | 0.3589 | 5.1637 | 2.0000 | 3.15344 | 0.27168 | 7.64369 | 5.94710 | 48.5 | 0.599 |
| 3.500 | 3.086 | 0.6714 | 0.3579 | 5.8043 | 2.0000 | 3.50787 | 0.29941 | 8.62405 | 6.08852 | 47.0 | 0.602 |
| 3.750 | 3.293 | 0.6677 | 0.3559 | 6.0583 | 2.0000 | 3.96969 | 0.33272 | 9.31948 | 5.92071 | 45.7 | 0.605 |
| 4.000 | 3.471 | 0.6658 | 0.3549 | 6.6106 | 2.0000 | 4.37008 | 0.36292 | 10.25643 | 5.97374 | 44.4 | 0.609 |
| 4.250 | 3.676 | 0.6621 | 0.3529 | 6.5392 | 2.0000 | 4.90188 | 0.39975 | 10.69438 | 5.65503 | 43.1 | 0.613 |
| 4.500 | 3.849 | 0.6602 | 0.3519 | 6.8805 | 2.0000 | 5.35487 | 0.43271 | 11.47007 | 5.60314 | 42.0 | 0.617 |
| 4.750 | 4.020 | 0.6584 | 0.3509 | 7.0758 | 2.0000 | 5.83093 | 0.46679 | 12.12313 | 5.48981 | 40.9 | 0.622 |
| 5.000 | 4.188 | 0.6565 | 0.3499 | 7.1247 | 2.0000 | 6.32767 | 0.50190 | 12.64929 | 5.32742 | 39.8 | 0.627 |
| 5.250 | 4.354 | 0.6546 | 0.3489 | 6.9734 | 2.0000 | 6.85106 | 0.53836 | 13.00048 | 5.10449 | 38.8 | 0.632 |
| 5.500 | 4.557 | 0.6509 | 0.3469 | 5.1745 | 2.0000 | 7.57490 | 0.58417 | 11.89505 | 4.30418 | 37.9 | 0.637 |
| *KCl; $\xi = 0.98$; $a = 0.5782$* | | | | | | | | | | | |
| 0.001 | 0.001 | 0.6187 | 0.3854 | 0.0337 | 0.0260 | 0.00086 | 0.00000 | 0.00000 | 0.00000 | 78.2 | 0.966 |
| 0.002 | 0.002 | 0.6187 | 0.3854 | 0.0495 | 0.0354 | 0.00172 | 0.00000 | 0.00000 | 0.00000 | 78.2 | 0.950 |
| 0.003 | 0.003 | 0.6186 | 0.3854 | 0.0592 | 0.0445 | 0.00258 | 0.00000 | 0.00000 | 0.00000 | 78.2 | 0.940 |
| 0.004 | 0.004 | 0.6186 | 0.3854 | 0.0668 | 0.0526 | 0.00344 | 0.00000 | 0.00000 | 0.00000 | 78.2 | 0.932 |
| 0.005 | 0.005 | 0.6186 | 0.3853 | 0.0724 | 0.0606 | 0.00430 | 0.00000 | 0.00000 | 0.00000 | 78.2 | 0.926 |

(*Continued*)

**TABLE 3.9 (*Continued*)**
**Calculated Values of Thermodynamic Quantities Included in Formulae (4.84) and (4.85)**

| $m$ | $c_k$ | $R$ (nm) | $R_{H_2O}$ (nm) | $-\Psi_\gamma^0$ | $\chi$ | $-E_{sp}$ | $E_H$ | $U_{ot}$ | $B$ | $\varepsilon_S$ | $\gamma_\pm$ |
|---|---|---|---|---|---|---|---|---|---|---|---|
| 0.006 | 0.006 | 0.6186 | 0.3853 | 0.0780 | 0.0677 | 0.00516 | 0.00000 | 0.00000 | 0.00000 | 78.1 | 0.920 |
| 0.007 | 0.007 | 0.6186 | 0.3853 | 0.0837 | 0.0736 | 0.00602 | 0.00000 | 0.00000 | 0.00000 | 78.1 | 0.914 |
| 0.008 | 0.008 | 0.6186 | 0.3853 | 0.0849 | 0.0828 | 0.00688 | 0.00000 | 0.00000 | 0.00000 | 78.1 | 0.912 |
| 0.009 | 0.009 | 0.6185 | 0.3853 | 0.0929 | 0.0852 | 0.00774 | 0.00000 | 0.00000 | 0.00000 | 78.1 | 0.904 |
| 0.010 | 0.010 | 0.6185 | 0.3853 | 0.1087 | 0.0811 | 0.00861 | 0.00000 | 0.00000 | 0.00000 | 78.1 | 0.889 |
| 0.020 | 0.020 | 0.6184 | 0.3852 | 0.1259 | 0.1405 | 0.01728 | 0.00000 | 0.00000 | 0.00000 | 77.9 | 0.866 |
| 0.030 | 0.030 | 0.6182 | 0.3851 | 0.1390 | 0.1917 | 0.02600 | 0.00000 | 0.00000 | 0.00000 | 77.8 | 0.847 |
| 0.040 | 0.040 | 0.6181 | 0.3850 | 0.1501 | 0.2375 | 0.03477 | 0.00000 | 0.00000 | 0.00000 | 77.6 | 0.830 |
| 0.050 | 0.050 | 0.6179 | 0.3849 | 0.1579 | 0.2833 | 0.04360 | 0.00000 | 0.00000 | 0.00000 | 77.5 | 0.816 |
| 0.060 | 0.060 | 0.6178 | 0.3848 | 0.1623 | 0.3322 | 0.05248 | 0.00000 | 0.00000 | 0.00000 | 77.3 | 0.805 |
| 0.070 | 0.070 | 0.6176 | 0.3847 | 0.1655 | 0.3815 | 0.06142 | 0.00000 | 0.00000 | 0.00000 | 77.2 | 0.795 |
| 0.080 | 0.080 | 0.6175 | 0.3846 | 0.1637 | 0.4424 | 0.07041 | 0.00000 | 0.00000 | 0.00000 | 77.1 | 0.789 |
| 0.090 | 0.090 | 0.6173 | 0.3845 | 0.1683 | 0.4858 | 0.07945 | 0.00000 | 0.00000 | 0.00000 | 76.9 | 0.778 |
| 0.100 | 0.100 | 0.6172 | 0.3844 | 0.1862 | 0.4898 | 0.08855 | 0.00000 | 0.00000 | 0.00000 | 76.8 | 0.757 |
| 0.200 | 0.198 | 0.6177 | 0.3848 | 0.1380 | 1.4286 | 0.17782 | 0.00000 | 0.00000 | 0.00000 | 75.4 | 0.724 |
| 0.300 | 0.296 | 0.6169 | 0.3843 | 0.1548 | 2.0000 | 0.27226 | 0.01658 | 0.05742 | 0.73200 | 74.1 | 0.695 |
| 0.400 | 0.395 | 0.6152 | 0.3832 | 0.2119 | 2.0000 | 0.37444 | 0.02261 | 0.17755 | 1.65967 | 72.9 | 0.670 |
| 0.500 | 0.493 | 0.6141 | 0.3825 | 0.2739 | 2.0000 | 0.47954 | 0.02879 | 0.31475 | 2.31102 | 71.7 | 0.652 |
| 0.600 | 0.587 | 0.6142 | 0.3826 | 0.3485 | 2.0000 | 0.57940 | 0.03481 | 0.46503 | 2.82401 | 70.6 | 0.638 |
| 0.700 | 0.684 | 0.6130 | 0.3818 | 0.4179 | 2.0000 | 0.69227 | 0.04133 | 0.62694 | 3.20616 | 69.6 | 0.627 |
| 0.800 | 0.781 | 0.6117 | 0.3811 | 0.4894 | 2.0000 | 0.81067 | 0.04809 | 0.79914 | 3.51239 | 68.5 | 0.618 |
| 0.900 | 0.873 | 0.6116 | 0.3810 | 0.5804 | 2.0000 | 0.92134 | 0.05463 | 0.98471 | 3.81049 | 67.5 | 0.610 |
| 1.000 | 0.969 | 0.6104 | 0.3802 | 0.6664 | 2.0000 | 1.05073 | 0.06191 | 1.18314 | 4.03948 | 66.3 | 0.602 |
| 1.250 | 1.206 | 0.6078 | 0.3786 | 0.9043 | 2.0000 | 1.39294 | 0.08098 | 1.73609 | 4.53162 | 63.7 | 0.592 |

| | | | | | | | | | | | |
|---|---|---|---|---|---|---|---|---|---|---|---|
| 1.500 | 1.432 | 0.6065 | 0.3778 | 1.2046 | 2.0000 | 1.73857 | 0.10041 | 2.35749 | 4.96307 | 61.2 | 0.584 |
| 1.750 | 1.664 | 0.6040 | 0.3762 | 1.5148 | 2.0000 | 2.15238 | 0.12268 | 3.05749 | 5.26827 | 58.7 | 0.578 |
| 2.000 | 1.883 | 0.6027 | 0.3754 | 1.9101 | 2.0000 | 2.56116 | 0.14499 | 3.83895 | 5.59669 | 56.5 | 0.573 |
| 2.250 | 2.098 | 0.6014 | 0.3746 | 2.3436 | 2.0000 | 2.99344 | 0.16833 | 4.68633 | 5.88487 | 54.4 | 0.571 |
| 2.500 | 2.323 | 0.5989 | 0.3731 | 2.6890 | 2.0000 | 3.50776 | 0.19472 | 5.52620 | 5.99901 | 52.5 | 0.570 |
| 2.750 | 2.532 | 0.5976 | 0.3723 | 3.1490 | 2.0000 | 3.99580 | 0.22033 | 6.45506 | 6.19297 | 50.8 | 0.569 |
| 3.000 | 2.738 | 0.5963 | 0.3715 | 3.6228 | 2.0000 | 4.51199 | 0.24709 | 7.42824 | 6.35475 | 49.2 | 0.570 |
| 3.250 | 2.959 | 0.5938 | 0.3699 | 3.8702 | 2.0000 | 5.15029 | 0.27832 | 8.29257 | 6.29811 | 47.6 | 0.571 |
| 3.500 | 3.160 | 0.5925 | 0.3691 | 4.3035 | 2.0000 | 5.73570 | 0.30785 | 9.29151 | 6.37983 | 46.2 | 0.572 |
| 3.750 | 3.357 | 0.5913 | 0.3683 | 4.7160 | 2.0000 | 6.34012 | 0.33811 | 10.29140 | 6.43407 | 44.9 | 0.575 |
| 4.000 | 3.552 | 0.5900 | 0.3676 | 5.0827 | 2.0000 | 6.98196 | 0.36985 | 11.27950 | 6.44652 | 43.6 | 0.577 |
| 4.250 | 3.745 | 0.5888 | 0.3668 | 5.3583 | 2.0000 | 7.65472 | 0.40272 | 12.20805 | 6.40774 | 42.4 | 0.580 |
| 4.500 | 3.935 | 0.5875 | 0.3660 | 5.5344 | 2.0000 | 8.34965 | 0.43636 | 13.05839 | 6.32577 | 41.3 | 0.583 |
| 4.750 | 4.123 | 0.5862 | 0.3652 | 5.5663 | 2.0000 | 9.07417 | 0.47102 | 13.79637 | 6.19143 | 40.2 | 0.588 |
| 5.000 | 4.308 | 0.5850 | 0.3644 | 5.4629 | 2.0000 | 9.82475 | 0.50664 | 14.42408 | 6.01801 | 39.2 | 0.593 |
| *KF; ξ = 0.96; a = 0.52032* | | | | | | | | | | | |
| 0.001 | 0.001 | 0.5693 | 0.3854 | 0.0335 | 0.0376 | 0.00080 | 0.00000 | 0.00000 | 0.00000 | 78.3 | 0.966 |
| 0.002 | 0.002 | 0.5693 | 0.3854 | 0.0490 | 0.0515 | 0.00161 | 0.00000 | 0.00000 | 0.00000 | 78.3 | 0.950 |
| 0.003 | 0.003 | 0.5693 | 0.3854 | 0.0585 | 0.0647 | 0.00241 | 0.00000 | 0.00000 | 0.00000 | 78.3 | 0.940 |
| 0.004 | 0.004 | 0.5693 | 0.3854 | 0.0664 | 0.0760 | 0.00322 | 0.00000 | 0.00000 | 0.00000 | 78.3 | 0.932 |
| 0.005 | 0.005 | 0.5693 | 0.3854 | 0.0726 | 0.0870 | 0.00403 | 0.00000 | 0.00000 | 0.00000 | 78.3 | 0.926 |
| 0.006 | 0.006 | 0.5693 | 0.3853 | 0.0779 | 0.0972 | 0.00483 | 0.00000 | 0.00000 | 0.00000 | 78.3 | 0.920 |
| 0.007 | 0.007 | 0.5693 | 0.3853 | 0.0828 | 0.1067 | 0.00564 | 0.00000 | 0.00000 | 0.00000 | 78.3 | 0.915 |
| 0.008 | 0.008 | 0.5692 | 0.3853 | 0.0844 | 0.1196 | 0.00644 | 0.00000 | 0.00000 | 0.00000 | 78.3 | 0.912 |
| 0.009 | 0.009 | 0.5692 | 0.3853 | 0.0924 | 0.1230 | 0.00725 | 0.00000 | 0.00000 | 0.00000 | 78.3 | 0.904 |
| 0.010 | 0.010 | 0.5692 | 0.3853 | 0.1085 | 0.1163 | 0.00806 | 0.00000 | 0.00000 | 0.00000 | 78.3 | 0.889 |

(Continued)

**TABLE 3.9 (*Continued*)**
**Calculated Values of Thermodynamic Quantities Included in Formulae (4.84) and (4.85)**

| $m$ | $c_k$ | $R$ (nm) | $R_{H_2O}$ (nm) | $-\Psi_\gamma^0$ | $\chi$ | $-E_{sp}$ | $E_H$ | $U_{ot}$ | $B$ | $\varepsilon_S$ | $\gamma_\pm$ |
|---|---|---|---|---|---|---|---|---|---|---|---|
| 0.020 | 0.020 | 0.5691 | 0.3852 | 0.1256 | 0.2010 | 0.01614 | 0.00000 | 0.00000 | 0.00000 | 78.2 | 0.867 |
| 0.030 | 0.030 | 0.5690 | 0.3852 | 0.1384 | 0.2735 | 0.02424 | 0.00000 | 0.00000 | 0.00000 | 78.2 | 0.849 |
| 0.040 | 0.040 | 0.5689 | 0.3851 | 0.1498 | 0.3367 | 0.03236 | 0.00000 | 0.00000 | 0.00000 | 78.2 | 0.832 |
| 0.050 | 0.050 | 0.5688 | 0.3850 | 0.1572 | 0.4011 | 0.04052 | 0.00000 | 0.00000 | 0.00000 | 78.1 | 0.819 |
| 0.060 | 0.060 | 0.5687 | 0.3849 | 0.1616 | 0.4680 | 0.04869 | 0.00000 | 0.00000 | 0.00000 | 78.1 | 0.808 |
| 0.070 | 0.070 | 0.5686 | 0.3849 | 0.1647 | 0.5352 | 0.05688 | 0.00000 | 0.00000 | 0.00000 | 78.1 | 0.799 |
| 0.080 | 0.079 | 0.5708 | 0.3864 | 0.1660 | 0.6256 | 0.06293 | 0.00000 | 0.00000 | 0.00000 | 78.0 | 0.793 |
| 0.090 | 0.090 | 0.5683 | 0.3847 | 0.1688 | 0.6715 | 0.07334 | 0.00000 | 0.00000 | 0.00000 | 78.0 | 0.782 |
| 0.100 | 0.100 | 0.5682 | 0.3846 | 0.1854 | 0.6787 | 0.08160 | 0.00000 | 0.00000 | 0.00000 | 78.0 | 0.762 |
| 0.200 | 0.199 | 0.5681 | 0.3846 | 0.1390 | 1.8296 | 0.16332 | 0.00000 | 0.00000 | 0.00000 | 77.6 | 0.733 |
| 0.300 | 0.299 | 0.5667 | 0.3836 | 0.1893 | 2.0000 | 0.24954 | 0.01313 | 0.09100 | 1.46524 | 77.3 | 0.707 |
| 0.400 | 0.397 | 0.5664 | 0.3834 | 0.2541 | 2.0000 | 0.33356 | 0.01752 | 0.20965 | 2.52943 | 77.0 | 0.687 |
| 0.500 | 0.497 | 0.5651 | 0.3825 | 0.3155 | 2.0000 | 0.42459 | 0.02214 | 0.33978 | 3.24459 | 76.6 | 0.673 |
| 0.600 | 0.593 | 0.5651 | 0.3825 | 0.3828 | 2.0000 | 0.50861 | 0.02652 | 0.47511 | 3.78708 | 76.3 | 0.662 |
| 0.700 | 0.694 | 0.5634 | 0.3814 | 0.4411 | 2.0000 | 0.60682 | 0.03135 | 0.61959 | 4.17706 | 76.0 | 0.655 |
| 0.800 | 0.789 | 0.5634 | 0.3814 | 0.5097 | 2.0000 | 0.69330 | 0.03581 | 0.76621 | 4.52275 | 75.7 | 0.650 |
| 0.900 | 0.890 | 0.5618 | 0.3803 | 0.5665 | 2.0000 | 0.79653 | 0.04079 | 0.91902 | 4.76211 | 75.3 | 0.647 |
| 1.000 | 0.984 | 0.5617 | 0.3802 | 0.6366 | 2.0000 | 0.88559 | 0.04533 | 1.07364 | 5.00669 | 75.0 | 0.644 |
| 1.250 | 1.234 | 0.5586 | 0.3781 | 0.8114 | 2.0000 | 1.16811 | 0.05875 | 1.52489 | 5.48650 | 73.4 | 0.644 |
| 1.500 | 1.474 | 0.5569 | 0.3770 | 1.0434 | 2.0000 | 1.45721 | 0.07262 | 2.04461 | 5.95145 | 71.3 | 0.646 |
| 1.750 | 1.712 | 0.5553 | 0.3759 | 1.3543 | 2.0000 | 1.78478 | 0.08813 | 2.68563 | 6.44120 | 68.7 | 0.652 |
| 2.000 | 1.947 | 0.5538 | 0.3749 | 1.7115 | 2.0000 | 2.13466 | 0.10452 | 3.39847 | 6.87321 | 66.2 | 0.661 |
| 2.250 | 2.200 | 0.5506 | 0.3727 | 2.0249 | 2.0000 | 2.57115 | 0.12364 | 4.15211 | 7.09846 | 64.0 | 0.671 |

| | | | | | | | | | | | |
|---|---|---|---|---|---|---|---|---|---|---|---|
| 2.500 | 2.434 | 0.5490 | 0.3717 | 2.4126 | 2.0000 | 2.97591 | 0.14185 | 4.95063 | 7.37727 | 62.0 | 0.682 |
| 2.750 | 2.666 | 0.5475 | 0.3706 | 2.8162 | 2.0000 | 3.40182 | 0.16075 | 5.78776 | 7.61062 | 60.3 | 0.694 |
| 3.000 | 2.897 | 0.5460 | 0.3696 | 3.2247 | 2.0000 | 3.85242 | 0.18044 | 6.65497 | 7.79631 | 58.7 | 0.708 |
| 3.250 | 3.127 | 0.5444 | 0.3685 | 3.6383 | 2.0000 | 4.33042 | 0.20101 | 7.55607 | 7.94604 | 57.2 | 0.724 |
| 3.500 | 3.355 | 0.5429 | 0.3675 | 4.1206 | 2.0000 | 4.85123 | 0.22322 | 8.56814 | 8.11378 | 55.6 | 0.741 |
| 3.750 | 3.582 | 0.5413 | 0.3664 | 4.6187 | 2.0000 | 5.40789 | 0.24664 | 9.63241 | 8.25520 | 54.0 | 0.760 |
| 4.000 | 3.808 | 0.5398 | 0.3654 | 5.0847 | 2.0000 | 5.99031 | 0.27080 | 10.69028 | 8.34458 | 52.6 | 0.779 |
| 4.250 | 4.033 | 0.5383 | 0.3644 | 5.5237 | 2.0000 | 6.60450 | 0.29593 | 11.75345 | 8.39527 | 51.3 | 0.801 |
| 4.500 | 4.257 | 0.5368 | 0.3633 | 6.1226 | 2.0000 | 7.31216 | 0.32476 | 13.06694 | 8.50497 | 49.6 | 0.824 |
| 4.750 | 4.480 | 0.5353 | 0.3623 | 6.7483 | 2.0000 | 8.08145 | 0.35579 | 14.46793 | 8.59570 | 47.9 | 0.848 |
| 5.000 | 4.702 | 0.5338 | 0.3613 | 7.2875 | 2.0000 | 8.88815 | 0.38789 | 15.81767 | 8.61974 | 46.4 | 0.873 |
| 5.250 | 4.882 | 0.5338 | 0.3613 | 8.5579 | 2.0000 | 9.51543 | 0.41528 | 17.72684 | 9.02300 | 45.0 | 0.900 |
| 5.500 | 5.101 | 0.5323 | 0.3603 | 9.0077 | 2.0000 | 10.39696 | 0.44979 | 19.06802 | 8.96117 | 43.6 | 0.934 |
| 5.750 | 5.319 | 0.5308 | 0.3593 | 9.3489 | 2.0000 | 11.34115 | 0.48638 | 20.36595 | 8.85098 | 42.3 | 0.974 |
| 6.000 | 5.537 | 0.5293 | 0.3583 | 9.4462 | 2.0000 | 12.34092 | 0.52462 | 21.47147 | 8.65126 | 41.1 | 1.013 |
| 6.250 | 5.754 | 0.5278 | 0.3573 | 9.2180 | 2.0000 | 13.37629 | 0.56371 | 22.28136 | 8.35500 | 39.9 | 1.048 |
| 6.500 | 5.921 | 0.5278 | 0.3573 | 10.5616 | 2.0000 | 14.13806 | 0.59582 | 24.39302 | 8.65388 | 38.9 | 1.082 |
| 6.750 | 6.136 | 0.5264 | 0.3563 | 9.8913 | 2.0000 | 15.29254 | 0.63884 | 24.87170 | 8.22952 | 37.8 | 1.115 |
| 7.000 | 6.350 | 0.5249 | 0.3553 | 8.9884 | 2.0000 | 16.62948 | 0.68871 | 25.29451 | 7.76345 | 36.5 | 1.150 |
| 7.250 | 6.564 | 0.5234 | 0.3543 | 7.2981 | 2.0000 | 18.04205 | 0.74074 | 25.00542 | 7.13560 | 35.2 | 1.190 |
| 7.500 | 6.721 | 0.5235 | 0.3543 | 8.5444 | 2.0000 | 19.07436 | 0.78321 | 27.28163 | 7.36306 | 34.1 | 1.229 |
| 7.750 | 6.933 | 0.5220 | 0.3534 | 5.3734 | 2.0000 | 20.57065 | 0.83733 | 25.59293 | 6.46082 | 33.1 | 1.271 |
| 8.000 | 7.145 | 0.5206 | 0.3524 | 0.8954 | 2.0000 | 22.14712 | 0.89368 | 22.67583 | 5.36343 | 32.2 | 1.314 |
| 8.250 | 7.296 | 0.5206 | 0.3524 | 0.9714 | 2.0000 | 22.99620 | 0.92794 | 23.60871 | 5.37795 | 31.6 | 1.361 |
| 8.500 | 7.505 | 0.5191 | 0.3514 | 0.0000 | 0.0000 | 24.19711 | 0.96811 | 23.83911 | 5.20511 | 31.4 | 1.409 |
| 8.750 | 7.715 | 0.5177 | 0.3505 | 0.0000 | 0.0000 | 25.44784 | 1.00937 | 25.08943 | 5.25415 | 31.1 | 1.457 |
| 9.000 | 7.860 | 0.5177 | 0.3504 | 0.0000 | 0.0000 | 26.15012 | 1.03719 | 25.80497 | 5.25905 | 30.8 | 1.508 |
| 9.250 | 8.068 | 0.5163 | 0.3495 | 0.0000 | 0.0000 | 27.45033 | 1.07946 | 27.08534 | 5.30385 | 30.6 | 1.532 |

*(Continued)*

**TABLE 3.9 (*Continued*)**
**Calculated Values of Thermodynamic Quantities Included in Formulae (4.84) and (4.85)**

| $m$ | $c_k$ | $R$ (nm) | $R_{H_2O}$ (nm) | $-\Psi_\gamma^0$ | $\chi$ | $-E_{sp}$ | $E_H$ | $U_{ot}$ | $B$ | $\varepsilon_S$ | $\gamma_\pm$ |
|---|---|---|---|---|---|---|---|---|---|---|---|
| 9.500 | 8.276 | 0.5149 | 0.3485 | 0.0000 | 0.0000 | 28.79657 | 1.12275 | 28.44492 | 5.35533 | 30.3 | 1.610 |
| 9.750 | 8.415 | 0.5149 | 0.3485 | 0.0000 | 0.0000 | 29.52505 | 1.15116 | 29.18628 | 5.35930 | 30.1 | 1.667 |
| 10.000 | 8.622 | 0.5135 | 0.3476 | 0.0000 | 0.0000 | 30.93436 | 1.19579 | 30.60228 | 5.40956 | 29.8 | 1.743 |
| 10.500 | 8.963 | 0.5121 | 0.3466 | 0.0000 | 0.0000 | 33.14572 | 1.27033 | 32.81213 | 5.45987 | 29.3 | 1.851 |
| 11.000 | 9.299 | 0.5107 | 0.3457 | 0.0000 | 0.0000 | 35.43730 | 1.34661 | 35.10369 | 5.51030 | 28.9 | 1.972 |
| 11.500 | 9.631 | 0.5093 | 0.3447 | 0.0000 | 0.0000 | 37.81469 | 1.42468 | 37.48075 | 5.56103 | 28.4 | 2.104 |
| 12.000 | 9.958 | 0.5079 | 0.3438 | 0.0000 | 0.0000 | 40.26189 | 1.50409 | 39.92413 | 5.61083 | 28.0 | 2.241 |
| 12.500 | 10.281 | 0.5066 | 0.3429 | 0.0000 | 0.0000 | 42.79771 | 1.58537 | 42.45261 | 5.66028 | 27.5 | 2.383 |
| 13.000 | 10.601 | 0.5052 | 0.3420 | 0.0000 | 0.0000 | 45.42879 | 1.66859 | 45.07159 | 5.70974 | 27.1 | 2.527 |
| 13.500 | 10.918 | 0.5038 | 0.3410 | 0.0000 | 0.0000 | 48.15007 | 1.75354 | 47.77665 | 5.75923 | 26.7 | 2.674 |
| 14.000 | 11.231 | 0.5025 | 0.3401 | 0.0000 | 0.0000 | 50.95985 | 1.84029 | 50.56555 | 5.80808 | 26.3 | 2.822 |
| 14.500 | 11.541 | 0.5011 | 0.3392 | 0.0000 | 0.0000 | 53.86479 | 1.92890 | 53.44479 | 5.85678 | 26.0 | 2.970 |
| 15.000 | 11.848 | 0.4998 | 0.3383 | 0.0000 | 0.0000 | 56.85894 | 2.01914 | 56.40939 | 5.90541 | 25.6 | 3.119 |
| 15.500 | 12.056 | 0.4998 | 0.3383 | 0.0000 | 0.0000 | 58.67698 | 2.08366 | 58.21938 | 5.90615 | 25.2 | 3.262 |
| 16.000 | 12.356 | 0.4985 | 0.3374 | 0.0000 | 0.0000 | 61.81154 | 2.17665 | 61.31050 | 5.95403 | 24.9 | 3.388 |
| 16.500 | 12.654 | 0.4971 | 0.3365 | 0.0000 | 0.0000 | 65.05973 | 2.27189 | 64.50901 | 6.00202 | 24.6 | 3.506 |
| 17.000 | 12.950 | 0.4958 | 0.3356 | 0.0000 | 0.0000 | 68.41638 | 2.36917 | 67.81663 | 6.05068 | 24.2 | 3.638 |
| 17.500 | 13.139 | 0.4958 | 0.3356 | 0.0000 | 0.0000 | 70.33839 | 2.43570 | 69.72827 | 6.05131 | 23.9 | 3.806 |
| *KI; ξ = 1.1; a = 0.6919* | | | | | | | | | | | |
| 0.001 | 0.001 | 0.8079 | 0.3854 | 0.0343 | 0.0447 | 0.00022 | 0.00000 | 0.00000 | 0.00000 | 78.3 | 0.966 |
| 0.002 | 0.002 | 0.8078 | 0.3853 | 0.0497 | 0.0617 | 0.00044 | 0.00000 | 0.00000 | 0.00000 | 78.3 | 0.951 |
| 0.003 | 0.003 | 0.8078 | 0.3853 | 0.0600 | 0.0766 | 0.00065 | 0.00000 | 0.00000 | 0.00000 | 78.3 | 0.941 |
| 0.004 | 0.004 | 0.8077 | 0.3853 | 0.0683 | 0.0897 | 0.00087 | 0.00000 | 0.00000 | 0.00000 | 78.3 | 0.933 |

| | | | | | | | | | | | |
|---|---|---|---|---|---|---|---|---|---|---|---|
| 0.005 | 0.005 | 0.8077 | 0.3853 | 0.0756 | 0.1013 | 0.00109 | 0.00000 | 0.00000 | 0.00000 | 78.3 | 0.926 |
| 0.006 | 0.006 | 0.8076 | 0.3853 | 0.0819 | 0.1123 | 0.00131 | 0.00000 | 0.00000 | 0.00000 | 78.3 | 0.920 |
| 0.007 | 0.007 | 0.8076 | 0.3852 | 0.0870 | 0.1231 | 0.00153 | 0.00000 | 0.00000 | 0.00000 | 78.3 | 0.915 |
| 0.008 | 0.008 | 0.8075 | 0.3852 | 0.0890 | 0.1380 | 0.00175 | 0.00000 | 0.00000 | 0.00000 | 78.2 | 0.913 |
| 0.009 | 0.009 | 0.8075 | 0.3852 | 0.0975 | 0.1416 | 0.00197 | 0.00000 | 0.00000 | 0.00000 | 78.2 | 0.905 |
| 0.010 | 0.010 | 0.8074 | 0.3852 | 0.1140 | 0.1345 | 0.00219 | 0.00000 | 0.00000 | 0.00000 | 78.2 | 0.890 |
| 0.020 | 0.020 | 0.8070 | 0.3850 | 0.1365 | 0.2241 | 0.00439 | 0.00000 | 0.00000 | 0.00000 | 78.2 | 0.868 |
| 0.030 | 0.030 | 0.8066 | 0.3848 | 0.1548 | 0.2960 | 0.00661 | 0.00000 | 0.00000 | 0.00000 | 78.2 | 0.850 |
| 0.040 | 0.040 | 0.8061 | 0.3845 | 0.1712 | 0.3566 | 0.00884 | 0.00000 | 0.00000 | 0.00000 | 78.1 | 0.834 |
| 0.050 | 0.050 | 0.8057 | 0.3843 | 0.1843 | 0.4138 | 0.01109 | 0.00000 | 0.00000 | 0.00000 | 78.1 | 0.821 |
| 0.060 | 0.060 | 0.8052 | 0.3841 | 0.1940 | 0.4705 | 0.01335 | 0.00000 | 0.00000 | 0.00000 | 78.1 | 0.811 |
| 0.070 | 0.070 | 0.8048 | 0.3839 | 0.2025 | 0.5254 | 0.01563 | 0.00000 | 0.00000 | 0.00000 | 78.0 | 0.802 |
| 0.080 | 0.080 | 0.8044 | 0.3837 | 0.2073 | 0.5860 | 0.01793 | 0.00000 | 0.00000 | 0.00000 | 78.0 | 0.796 |
| 0.090 | 0.089 | 0.8069 | 0.3849 | 0.2192 | 0.6291 | 0.01962 | 0.00000 | 0.00000 | 0.00000 | 77.9 | 0.785 |
| 0.100 | 0.099 | 0.8062 | 0.3846 | 0.2397 | 0.6366 | 0.02193 | 0.00000 | 0.00000 | 0.00000 | 77.9 | 0.767 |
| 0.200 | 0.198 | 0.8018 | 0.3825 | 0.2513 | 1.2004 | 0.04537 | 0.00000 | 0.00000 | 0.00000 | 77.6 | 0.738 |
| 0.300 | 0.295 | 0.7994 | 0.3813 | 0.2585 | 1.7354 | 0.06900 | 0.00000 | 0.00000 | 0.00000 | 77.3 | 0.713 |
| 0.400 | 0.390 | 0.7975 | 0.3804 | 0.2978 | 2.0000 | 0.09283 | 0.01640 | 0.02180 | 0.28104 | 76.9 | 0.693 |
| 0.500 | 0.488 | 0.7931 | 0.3783 | 0.3660 | 2.0000 | 0.12014 | 0.02083 | 0.09457 | 0.95975 | 76.6 | 0.678 |
| 0.600 | 0.581 | 0.7912 | 0.3774 | 0.4369 | 2.0000 | 0.14560 | 0.02503 | 0.17543 | 1.48126 | 76.3 | 0.668 |
| 0.700 | 0.673 | 0.7891 | 0.3764 | 0.5056 | 2.0000 | 0.17173 | 0.02927 | 0.25744 | 1.85936 | 75.9 | 0.660 |
| 0.800 | 0.770 | 0.7849 | 0.3744 | 0.5683 | 2.0000 | 0.20311 | 0.03400 | 0.34274 | 2.13113 | 75.6 | 0.655 |
| 0.900 | 0.860 | 0.7830 | 0.3735 | 0.6354 | 2.0000 | 0.23090 | 0.03833 | 0.42910 | 2.36657 | 75.3 | 0.650 |
| 1.000 | 0.950 | 0.7808 | 0.3725 | 0.7002 | 2.0000 | 0.26017 | 0.04277 | 0.51453 | 2.54290 | 74.9 | 0.645 |
| 1.250 | 1.176 | 0.7743 | 0.3694 | 0.8485 | 2.0000 | 0.34067 | 0.05442 | 0.73411 | 2.85157 | 74.1 | 0.641 |
| 1.500 | 1.399 | 0.7678 | 0.3663 | 0.9804 | 2.0000 | 0.42831 | 0.06649 | 0.94700 | 3.01075 | 73.4 | 0.639 |
| 1.750 | 1.622 | 0.7611 | 0.3631 | 1.3082 | 2.0000 | 0.55039 | 0.08290 | 1.38743 | 3.53785 | 69.5 | 0.638 |
| 2.000 | 1.828 | 0.7566 | 0.3609 | 1.6974 | 2.0000 | 0.67310 | 0.09935 | 1.89443 | 4.03065 | 66.1 | 0.640 |

*(Continued)*

**TABLE 3.9 (*Continued*)**
**Calculated Values of Thermodynamic Quantities Included in Formulae (4.84) and (4.85)**

| $m$ | $c_k$ | $R$ (nm) | $R_{H_2O}$ (nm) | $-\Psi_\gamma^0$ | $\chi$ | $-E_{sp}$ | $E_H$ | $U_{ot}$ | $B$ | $\varepsilon_S$ | $\gamma_\pm$ |
|---|---|---|---|---|---|---|---|---|---|---|---|
| 2.250 | 2.031 | 0.7520 | 0.3587 | 2.1173 | 2.0000 | 0.80849 | 0.11687 | 2.44367 | 4.41976 | 63.2 | 0.642 |
| 2.500 | 2.231 | 0.7475 | 0.3566 | 2.5527 | 2.0000 | 0.95633 | 0.13537 | 3.01985 | 4.71555 | 60.7 | 0.644 |
| 2.750 | 2.429 | 0.7428 | 0.3544 | 3.1362 | 2.0000 | 1.13178 | 0.15681 | 3.77178 | 5.08426 | 57.7 | 0.648 |
| 3.000 | 2.624 | 0.7383 | 0.3522 | 3.8205 | 2.0000 | 1.33005 | 0.18045 | 4.64501 | 5.44123 | 54.9 | 0.652 |
| 3.250 | 2.819 | 0.7337 | 0.3500 | 4.4699 | 2.0000 | 1.54763 | 0.20544 | 5.50121 | 5.66039 | 52.4 | 0.656 |
| 3.500 | 3.013 | 0.7290 | 0.3478 | 5.0613 | 2.0000 | 1.78502 | 0.23180 | 6.31930 | 5.76249 | 50.3 | 0.661 |
| 3.750 | 3.175 | 0.7268 | 0.3467 | 5.8602 | 2.0000 | 1.98726 | 0.25532 | 7.31240 | 6.05402 | 48.4 | 0.666 |
| 4.000 | 3.366 | 0.7221 | 0.3445 | 6.2475 | 2.0000 | 2.25821 | 0.28379 | 7.95908 | 5.92831 | 46.8 | 0.672 |
| 4.250 | 3.524 | 0.7198 | 0.3433 | 6.8667 | 2.0000 | 2.48386 | 0.30860 | 8.79430 | 6.02372 | 45.3 | 0.677 |
| 4.500 | 3.713 | 0.7152 | 0.3412 | 6.8678 | 2.0000 | 2.79088 | 0.33916 | 9.08852 | 5.66437 | 44.0 | 0.683 |
| 5.000 | 4.016 | 0.7106 | 0.3390 | 7.4162 | 2.0000 | 3.29701 | 0.39172 | 10.12898 | 5.46580 | 41.8 | 0.699 |
| 6.000 | 4.639 | 0.6991 | 0.3335 | 4.7820 | 2.0000 | 4.53665 | 0.50936 | 8.69301 | 3.60753 | 38.3 | 0.732 |
| *$KNO_3$; $\xi = 1$; $a = 0.598$* | | | | | | | | | | | |
| 0.001 | 0.001 | 0.6848 | 0.3854 | 0.0353 | 0.0446 | 0.00032 | 0.00000 | 0.00000 | 0.00000 | 78.4 | 0.965 |
| 0.002 | 0.002 | 0.6848 | 0.3854 | 0.0516 | 0.0610 | 0.00064 | 0.00000 | 0.00000 | 0.00000 | 78.4 | 0.949 |
| 0.003 | 0.003 | 0.6847 | 0.3854 | 0.0629 | 0.0752 | 0.00096 | 0.00000 | 0.00000 | 0.00000 | 78.3 | 0.938 |
| 0.004 | 0.004 | 0.6847 | 0.3853 | 0.0722 | 0.0875 | 0.00128 | 0.00000 | 0.00000 | 0.00000 | 78.3 | 0.929 |
| 0.005 | 0.005 | 0.6847 | 0.3853 | 0.0794 | 0.0995 | 0.00160 | 0.00000 | 0.00000 | 0.00000 | 78.3 | 0.922 |
| 0.006 | 0.006 | 0.6847 | 0.3853 | 0.0856 | 0.1109 | 0.00192 | 0.00000 | 0.00000 | 0.00000 | 78.3 | 0.916 |
| 0.007 | 0.007 | 0.6847 | 0.3853 | 0.0918 | 0.1208 | 0.00224 | 0.00000 | 0.00000 | 0.00000 | 78.2 | 0.910 |
| 0.008 | 0.008 | 0.6846 | 0.3853 | 0.0937 | 0.1354 | 0.00256 | 0.00000 | 0.00000 | 0.00000 | 78.2 | 0.908 |
| 0.009 | 0.009 | 0.6846 | 0.3853 | 0.1033 | 0.1383 | 0.00289 | 0.00000 | 0.00000 | 0.00000 | 78.2 | 0.899 |
| 0.010 | 0.010 | 0.6846 | 0.3853 | 0.1220 | 0.1303 | 0.00321 | 0.00000 | 0.00000 | 0.00000 | 78.1 | 0.882 |

| | | | | | | | | | | | |
|---|---|---|---|---|---|---|---|---|---|---|---|
| 0.020 | 0.020 | 0.6844 | 0.3851 | 0.1495 | 0.2149 | 0.00645 | 0.00000 | 0.00000 | 0.00000 | 77.9 | 0.855 |
| 0.030 | 0.030 | 0.6841 | 0.3850 | 0.1719 | 0.2833 | 0.00972 | 0.00000 | 0.00000 | 0.00000 | 77.6 | 0.833 |
| 0.040 | 0.040 | 0.6839 | 0.3849 | 0.1938 | 0.3388 | 0.01303 | 0.00000 | 0.00000 | 0.00000 | 77.4 | 0.812 |
| 0.050 | 0.050 | 0.6837 | 0.3847 | 0.2112 | 0.3928 | 0.01637 | 0.00000 | 0.00000 | 0.00000 | 77.1 | 0.795 |
| 0.060 | 0.060 | 0.6834 | 0.3846 | 0.2253 | 0.4464 | 0.01974 | 0.00000 | 0.00000 | 0.00000 | 76.9 | 0.781 |
| 0.070 | 0.070 | 0.6832 | 0.3845 | 0.2383 | 0.4977 | 0.02314 | 0.00000 | 0.00000 | 0.00000 | 76.7 | 0.768 |
| 0.080 | 0.080 | 0.6830 | 0.3844 | 0.2450 | 0.5593 | 0.02658 | 0.00000 | 0.00000 | 0.00000 | 76.4 | 0.760 |
| 0.090 | 0.089 | 0.6853 | 0.3857 | 0.2620 | 0.6024 | 0.02914 | 0.00000 | 0.00000 | 0.00000 | 76.2 | 0.745 |
| 0.100 | 0.099 | 0.6848 | 0.3854 | 0.2978 | 0.5944 | 0.03265 | 0.00000 | 0.00000 | 0.00000 | 75.9 | 0.716 |
| 0.200 | 0.198 | 0.6825 | 0.3841 | 0.3263 | 1.2018 | 0.06852 | 0.00000 | 0.00000 | 0.00000 | 73.6 | 0.669 |
| 0.300 | 0.296 | 0.6811 | 0.3833 | 0.3562 | 1.6921 | 0.10470 | 0.00000 | 0.00000 | 0.00000 | 72.8 | 0.624 |
| 0.400 | 0.393 | 0.6798 | 0.3826 | 0.4111 | 2.0000 | 0.14176 | 0.01599 | 0.01161 | 0.15344 | 72.1 | 0.583 |
| 0.500 | 0.490 | 0.6782 | 0.3817 | 0.5240 | 2.0000 | 0.18069 | 0.02022 | 0.10445 | 1.09178 | 71.4 | 0.550 |
| 0.600 | 0.586 | 0.6768 | 0.3809 | 0.6410 | 2.0000 | 0.22047 | 0.02451 | 0.21013 | 1.81218 | 70.8 | 0.523 |
| 0.700 | 0.681 | 0.6756 | 0.3802 | 0.7625 | 2.0000 | 0.26111 | 0.02885 | 0.32655 | 2.39221 | 70.1 | 0.500 |
| 0.800 | 0.771 | 0.6756 | 0.3802 | 0.8937 | 2.0000 | 0.29817 | 0.03295 | 0.45335 | 2.90827 | 69.5 | 0.480 |
| 0.900 | 0.865 | 0.6741 | 0.3793 | 1.0212 | 2.0000 | 0.34131 | 0.03745 | 0.58043 | 3.27617 | 68.9 | 0.460 |
| 1.000 | 0.958 | 0.6727 | 0.3786 | 1.1519 | 2.0000 | 0.38515 | 0.04199 | 0.70260 | 3.53710 | 68.4 | 0.437 |
| 1.250 | 1.186 | 0.6698 | 0.3769 | 1.4751 | 2.0000 | 0.49609 | 0.05332 | 1.07028 | 4.24284 | 67.2 | 0.410 |
| 1.500 | 1.409 | 0.6671 | 0.3754 | 1.7937 | 2.0000 | 0.61050 | 0.06476 | 1.42713 | 4.65835 | 66.3 | 0.381 |
| 1.750 | 1.629 | 0.6643 | 0.3738 | 2.1080 | 2.0000 | 0.73098 | 0.07648 | 1.78523 | 4.93403 | 65.5 | 0.354 |
| 2.000 | 1.846 | 0.6615 | 0.3723 | 2.4179 | 2.0000 | 0.85758 | 0.08849 | 2.14786 | 5.13061 | 64.7 | 0.330 |
| 2.250 | 2.059 | 0.6588 | 0.3707 | 2.7209 | 2.0000 | 0.98857 | 0.10066 | 2.52204 | 5.29605 | 63.9 | 0.312 |
| 2.500 | 2.256 | 0.6574 | 0.3700 | 3.0454 | 2.0000 | 1.10676 | 0.11192 | 2.91245 | 5.50083 | 63.3 | 0.297 |
| 2.750 | 2.463 | 0.6547 | 0.3684 | 3.2532 | 2.0000 | 1.24085 | 0.12379 | 3.20595 | 5.47449 | 63.0 | 0.284 |
| 3.000 | 2.652 | 0.6532 | 0.3676 | 3.4829 | 2.0000 | 1.35696 | 0.13441 | 3.50240 | 5.50786 | 62.7 | 0.271 |
| 3.250 | 2.853 | 0.6506 | 0.3661 | 3.6476 | 2.0000 | 1.49755 | 0.14638 | 3.75474 | 5.42200 | 62.4 | 0.258 |
| 3.500 | 3.034 | 0.6492 | 0.3654 | 3.8562 | 2.0000 | 1.61700 | 0.15698 | 4.03259 | 5.42988 | 62.2 | 0.246 |

*(Continued)*

**TABLE 3.9 (*Continued*)**
**Calculated Values of Thermodynamic Quantities Included in Formulae (4.84) and (4.85)**

| $m$ | $c_k$ | $R$ (nm) | $R_{H_2O}$ (nm) | $-\Psi_\gamma^0$ | $\chi$ | $-E_{sp}$ | $E_H$ | $U_{ot}$ | $B$ | $\varepsilon_S$ | $\gamma_\pm$ |
|---|---|---|---|---|---|---|---|---|---|---|---|
| *KOH; $\xi = 0.96$; $a = 0.52704$* | | | | | | | | | | | |
| 0.001 | 0.001 | 0.5627 | 0.3854 | 0.0337 | 0.0318 | 0.0008 | 0.00000 | 0.0000 | 0.0000 | 75.6 | 0.966 |
| 0.002 | 0.002 | 0.5627 | 0.3854 | 0.0484 | 0.0442 | 0.0017 | 0.00000 | 0.0000 | 0.0000 | 75.6 | 0.951 |
| 0.003 | 0.003 | 0.5627 | 0.3854 | 0.0580 | 0.0553 | 0.0026 | 0.00000 | 0.0000 | 0.0000 | 75.6 | 0.941 |
| 0.004 | 0.004 | 0.5627 | 0.3854 | 0.0656 | 0.0651 | 0.0035 | 0.00000 | 0.0000 | 0.0000 | 75.6 | 0.933 |
| 0.005 | 0.005 | 0.5627 | 0.3854 | 0.0722 | 0.0740 | 0.0044 | 0.00000 | 0.0000 | 0.0000 | 75.6 | 0.926 |
| 0.006 | 0.006 | 0.5627 | 0.3853 | 0.0778 | 0.0825 | 0.0053 | 0.00000 | 0.0000 | 0.0000 | 75.6 | 0.920 |
| 0.007 | 0.007 | 0.5627 | 0.3853 | 0.0823 | 0.0910 | 0.0062 | 0.00000 | 0.0000 | 0.0000 | 75.6 | 0.915 |
| 0.008 | 0.008 | 0.5627 | 0.3853 | 0.0836 | 0.1024 | 0.0071 | 0.00000 | 0.0000 | 0.0000 | 75.6 | 0.913 |
| 0.009 | 0.009 | 0.5627 | 0.3853 | 0.0915 | 0.1053 | 0.0080 | 0.00000 | 0.0000 | 0.0000 | 75.6 | 0.905 |
| 0.010 | 0.010 | 0.5627 | 0.3853 | 0.1072 | 0.0997 | 0.0089 | 0.00000 | 0.0000 | 0.0000 | 75.6 | 0.890 |
| 0.020 | 0.020 | 0.5625 | 0.3852 | 0.1218 | 0.1755 | 0.0179 | 0.00000 | 0.0000 | 0.0000 | 75.5 | 0.869 |
| 0.030 | 0.030 | 0.5624 | 0.3852 | 0.1334 | 0.2401 | 0.0268 | 0.00000 | 0.0000 | 0.0000 | 75.5 | 0.851 |
| 0.040 | 0.040 | 0.5623 | 0.3851 | 0.1430 | 0.2985 | 0.0358 | 0.00000 | 0.0000 | 0.0000 | 75.5 | 0.835 |
| 0.050 | 0.050 | 0.5622 | 0.3850 | 0.1481 | 0.3593 | 0.0449 | 0.00000 | 0.0000 | 0.0000 | 75.5 | 0.823 |
| 0.060 | 0.060 | 0.5621 | 0.3850 | 0.1509 | 0.4228 | 0.0539 | 0.00000 | 0.0000 | 0.0000 | 75.4 | 0.813 |
| 0.070 | 0.070 | 0.5620 | 0.3849 | 0.1526 | 0.4874 | 0.0630 | 0.00000 | 0.0000 | 0.0000 | 75.4 | 0.804 |
| 0.080 | 0.079 | 0.5643 | 0.3864 | 0.1518 | 0.5863 | 0.0697 | 0.00000 | 0.0000 | 0.0000 | 75.3 | 0.799 |
| 0.090 | 0.090 | 0.5618 | 0.3847 | 0.1524 | 0.6262 | 0.0813 | 0.00000 | 0.0000 | 0.0000 | 75.3 | 0.789 |
| 0.100 | 0.100 | 0.5617 | 0.3847 | 0.1647 | 0.6435 | 0.0904 | 0.00000 | 0.0000 | 0.0000 | 75.3 | 0.772 |
| 0.200 | 0.199 | 0.5616 | 0.3846 | 0.1070 | 2.0000 | 0.1809 | 0.00919 | 0.0030 | 0.0690 | 74.9 | 0.749 |
| 0.300 | 0.300 | 0.5596 | 0.3833 | 0.1556 | 2.0000 | 0.2790 | 0.01401 | 0.1166 | 1.7607 | 74.6 | 0.730 |
| 0.400 | 0.398 | 0.5595 | 0.3832 | 0.2231 | 2.0000 | 0.3779 | 0.01897 | 0.2651 | 2.9552 | 73.1 | 0.718 |

| | | | | | | | | | | | |
|---|---|---|---|---|---|---|---|---|---|---|---|
| 0.500 | 0.499 | 0.5580 | 0.3821 | 0.2881 | 2.0000 | 0.4898 | 0.02436 | 0.4331 | 3.7577 | 71.8 | 0.713 |
| 0.600 | 0.596 | 0.5578 | 0.3820 | 0.3694 | 2.0000 | 0.5969 | 0.02967 | 0.6184 | 4.4060 | 70.4 | 0.712 |
| 0.700 | 0.698 | 0.5561 | 0.3808 | 0.4419 | 2.0000 | 0.7234 | 0.03561 | 0.8177 | 4.8539 | 69.2 | 0.714 |
| 0.800 | 0.794 | 0.5560 | 0.3807 | 0.5364 | 2.0000 | 0.8383 | 0.04124 | 1.0305 | 5.2827 | 68.0 | 0.718 |
| 0.900 | 0.896 | 0.5544 | 0.3797 | 0.6165 | 2.0000 | 0.9765 | 0.04762 | 1.2557 | 5.5744 | 66.8 | 0.725 |
| 1.000 | 0.991 | 0.5543 | 0.3796 | 0.7256 | 2.0000 | 1.0996 | 0.05358 | 1.5045 | 5.9357 | 65.7 | 0.739 |
| 1.250 | 1.245 | 0.5509 | 0.3773 | 0.9523 | 2.0000 | 1.4854 | 0.07102 | 2.1350 | 6.3542 | 63.0 | 0.759 |
| 1.500 | 1.489 | 0.5491 | 0.3761 | 1.2466 | 2.0000 | 1.8810 | 0.08904 | 2.8542 | 6.7759 | 60.5 | 0.789 |
| 1.750 | 1.730 | 0.5476 | 0.3750 | 1.5687 | 2.0000 | 2.3031 | 0.10804 | 3.6374 | 7.1167 | 58.3 | 0.829 |
| 2.000 | 1.971 | 0.5458 | 0.3738 | 1.9074 | 2.0000 | 2.7685 | 0.12859 | 4.4834 | 7.3702 | 56.1 | 0.875 |
| 2.250 | 2.209 | 0.5442 | 0.3727 | 2.3692 | 2.0000 | 3.3002 | 0.15189 | 5.5110 | 7.6694 | 53.6 | 0.919 |
| 2.500 | 2.446 | 0.5426 | 0.3716 | 2.8538 | 2.0000 | 3.8797 | 0.17692 | 6.6062 | 7.8930 | 51.2 | 0.964 |
| 2.750 | 2.682 | 0.5410 | 0.3705 | 3.3134 | 2.0000 | 4.5001 | 0.20329 | 7.7196 | 8.0267 | 49.2 | 1.015 |
| 3.000 | 2.917 | 0.5393 | 0.3694 | 3.7348 | 2.0000 | 5.1680 | 0.23128 | 8.8408 | 8.0803 | 47.3 | 1.069 |
| 3.250 | 3.150 | 0.5378 | 0.3683 | 4.0863 | 2.0000 | 5.8740 | 0.26049 | 9.9327 | 8.0600 | 45.6 | 1.130 |
| 3.500 | 3.382 | 0.5362 | 0.3672 | 4.3936 | 2.0000 | 6.6541 | 0.29242 | 11.0522 | 7.9893 | 43.9 | 1.195 |
| 3.750 | 3.613 | 0.5346 | 0.3661 | 4.6413 | 2.0000 | 7.5302 | 0.32793 | 12.2070 | 7.8685 | 42.1 | 1.267 |
| 4.000 | 3.842 | 0.5331 | 0.3651 | 4.6665 | 2.0000 | 8.4545 | 0.36496 | 13.1856 | 7.6368 | 40.4 | 1.343 |
| 4.250 | 4.071 | 0.5316 | 0.3641 | 4.3148 | 2.0000 | 9.4382 | 0.40377 | 13.8472 | 7.2491 | 38.9 | 1.427 |
| 4.500 | 4.298 | 0.5301 | 0.3630 | 3.5745 | 2.0000 | 10.4812 | 0.44451 | 14.1768 | 6.7415 | 37.5 | 1.515 |
| 4.750 | 4.525 | 0.5286 | 0.3620 | 2.2440 | 2.0000 | 11.6201 | 0.48845 | 14.0106 | 6.0631 | 36.2 | 1.611 |
| 5.000 | 4.751 | 0.5271 | 0.3610 | 0.1577 | 2.0000 | 12.8260 | 0.53442 | 13.1497 | 5.2011 | 34.9 | 1.707 |
| 5.250 | 4.975 | 0.5257 | 0.3600 | 0.0000 | 0.0000 | 14.0738 | 0.58144 | 14.2639 | 5.1855 | 33.8 | 1.819 |
| 5.500 | 5.199 | 0.5242 | 0.3590 | 0.0000 | 0.0000 | 15.3990 | 0.63072 | 15.6215 | 5.2354 | 32.7 | 1.959 |
| 5.750 | 5.378 | 0.5242 | 0.3590 | 0.0000 | 0.0000 | 16.4121 | 0.67217 | 16.6839 | 5.2466 | 31.8 | 2.129 |
| 6.000 | 5.599 | 0.5228 | 0.3580 | 0.0000 | 0.0000 | 17.8269 | 0.72387 | 18.1321 | 5.2948 | 30.9 | 2.301 |
| 6.250 | 5.819 | 0.5214 | 0.3571 | 0.0000 | 0.0000 | 19.3118 | 0.77752 | 19.6387 | 5.3390 | 30.0 | 2.463 |
| 6.500 | 6.038 | 0.5200 | 0.3561 | 0.0000 | 0.0000 | 20.9428 | 0.83614 | 21.2783 | 5.3792 | 29.1 | 2.615 |

*(Continued)*

**TABLE 3.9 (*Continued*)**
**Calculated Values of Thermodynamic Quantities Included in Formulae (4.84) and (4.85)**

| $m$ | $c_k$ | $R$ (nm) | $R_{H_2O}$ (nm) | $-\Psi_\gamma^0$ | $\chi$ | $-E_{sp}$ | $E_H$ | $U_{ot}$ | $B$ | $\varepsilon_S$ | $\gamma_\pm$ |
|---|---|---|---|---|---|---|---|---|---|---|---|
| 6.750 | 6.207 | 0.5200 | 0.3561 | 0.0000 | 0.0000 | 22.1764 | 0.88532 | 22.5277 | 5.3787 | 28.3 | 2.770 |
| 7.000 | 6.423 | 0.5186 | 0.3551 | 0.0000 | 0.0000 | 23.9340 | 0.94758 | 24.2904 | 5.4185 | 27.5 | 2.942 |
| 7.250 | 6.639 | 0.5172 | 0.3542 | 0.0000 | 0.0000 | 25.7476 | 1.01092 | 26.1126 | 5.4600 | 26.8 | 3.139 |
| 7.500 | 6.855 | 0.5158 | 0.3532 | 0.0000 | 0.0000 | 27.6568 | 1.07682 | 28.0252 | 5.5013 | 26.1 | 3.340 |
| 7.750 | 7.014 | 0.5158 | 0.3532 | 0.0000 | 0.0000 | 29.0597 | 1.13149 | 29.4420 | 5.5002 | 25.4 | 3.552 |
| 8.000 | 7.227 | 0.5145 | 0.3523 | 0.0000 | 0.0000 | 31.1958 | 1.20470 | 31.5751 | 5.5402 | 24.7 | 3.784 |
| 8.250 | 7.440 | 0.5131 | 0.3514 | 0.0000 | 0.0000 | 33.4412 | 1.28081 | 33.8181 | 5.5812 | 24.1 | 4.045 |
| 8.500 | 7.593 | 0.5131 | 0.3514 | 0.0000 | 0.0000 | 35.0069 | 1.34071 | 35.3941 | 5.5803 | 23.5 | 4.309 |
| 8.750 | 7.803 | 0.5118 | 0.3505 | 0.0000 | 0.0000 | 37.3386 | 1.41851 | 37.7174 | 5.6205 | 22.9 | 4.587 |
| 9.000 | 8.013 | 0.5105 | 0.3496 | 0.0000 | 0.0000 | 39.7748 | 1.49889 | 40.1435 | 5.6612 | 22.4 | 4.888 |
| 9.250 | 8.160 | 0.5105 | 0.3496 | 0.0000 | 0.0000 | 41.4335 | 1.56130 | 41.8131 | 5.6609 | 21.9 | 5.224 |
| 9.500 | 8.368 | 0.5091 | 0.3487 | 0.0000 | 0.0000 | 43.9710 | 1.64367 | 44.3302 | 5.7009 | 21.4 | 5.521 |
| 9.750 | 8.511 | 0.5091 | 0.3487 | 0.0000 | 0.0000 | 45.6628 | 1.70678 | 46.0294 | 5.7006 | 21.0 | 5.885 |
| 10.000 | 8.717 | 0.5078 | 0.3478 | 0.0000 | 0.0000 | 48.3514 | 1.79295 | 48.7210 | 5.7439 | 20.6 | 6.391 |
| 10.500 | 9.060 | 0.5066 | 0.3469 | 0.0000 | 0.0000 | 52.5896 | 1.93467 | 52.9440 | 5.7846 | 19.9 | 7.158 |
| 11.000 | 9.398 | 0.5053 | 0.3460 | 0.0000 | 0.0000 | 56.9856 | 2.07981 | 57.3254 | 5.8262 | 19.3 | 8.051 |
| 11.500 | 9.731 | 0.5040 | 0.3451 | 0.0000 | 0.0000 | 61.4878 | 2.22647 | 61.8179 | 5.8689 | 18.8 | 9.115 |
| 12.000 | 10.059 | 0.5027 | 0.3443 | 0.0000 | 0.0000 | 66.2229 | 2.37922 | 66.5320 | 5.9109 | 18.3 | 10.268 |
| 12.500 | 10.382 | 0.5015 | 0.3434 | 0.0000 | 0.0000 | 71.0978 | 2.53467 | 71.3851 | 5.9532 | 17.8 | 11.590 |
| 13.000 | 10.701 | 0.5003 | 0.3426 | 0.0000 | 0.0000 | 76.0894 | 2.69173 | 76.3477 | 5.9955 | 17.3 | 13.012 |
| 13.500 | 11.017 | 0.4990 | 0.3417 | 0.0000 | 0.0000 | 81.2067 | 2.85047 | 81.4345 | 6.0388 | 16.9 | 14.614 |
| 14.000 | 11.328 | 0.4978 | 0.3409 | 0.0000 | 0.0000 | 86.5839 | 3.01610 | 86.7685 | 6.0810 | 16.5 | 16.318 |
| 14.500 | 11.635 | 0.4966 | 0.3401 | 0.0000 | 0.0000 | 91.4815 | 3.16268 | 91.6407 | 6.1248 | 16.3 | 18.203 |

| | | | | | | | | | | | |
|---|---|---|---|---|---|---|---|---|---|---|---|
| 15.000 | 11.939 | 0.4954 | 0.3393 | 0.0000 | 0.0000 | 96.4411 | 3.30900 | 96.5696 | 6.1689 | 16.0 | 20.193 |
| 15.500 | 12.240 | 0.4942 | 0.3385 | 0.0000 | 0.0000 | 101.6590 | 3.46176 | 101.7486 | 6.2129 | 15.8 | 22.368 |
| 16.000 | 12.448 | 0.4942 | 0.3385 | 0.0000 | 0.0000 | 104.7563 | 3.56725 | 104.8492 | 6.2129 | 15.6 | 24.655 |
| 16.500 | 12.741 | 0.4930 | 0.3376 | 0.0000 | 0.0000 | 110.0759 | 3.72040 | 110.1220 | 6.2567 | 15.4 | 27.112 |
| 17.000 | 13.032 | 0.4919 | 0.3368 | 0.0000 | 0.0000 | 115.4563 | 3.87300 | 115.4509 | 6.3010 | 15.2 | 29.660 |
| 17.500 | 13.319 | 0.4907 | 0.3360 | 0.0000 | 0.0000 | 120.8375 | 4.02370 | 120.7786 | 6.3449 | 15.0 | 32.329 |
| 18.000 | 13.508 | 0.4907 | 0.3361 | 0.0000 | 0.0000 | 124.0898 | 4.13211 | 124.0146 | 6.3440 | 14.8 | 35.058 |
| 18.500 | 13.789 | 0.4896 | 0.3353 | 0.0000 | 0.0000 | 129.8048 | 4.29063 | 129.6588 | 6.3877 | 14.6 | 37.857 |
| 19.000 | 14.068 | 0.4884 | 0.3345 | 0.0000 | 0.0000 | 135.6833 | 4.45204 | 135.4586 | 6.4314 | 14.5 | 40.684 |
| 19.500 | 14.245 | 0.4884 | 0.3345 | 0.0000 | 0.0000 | 138.9325 | 4.55858 | 138.6796 | 6.4305 | 14.3 | 43.534 |
| 20.000 | 14.518 | 0.4873 | 0.3337 | 0.0000 | 0.0000 | 145.0164 | 4.72359 | 144.6725 | 6.4740 | 14.1 | 46.386 |
| *$Li_2SO_4$; ξ = 1.135; a = 0.65149* | | | | | | | | | | | |
| 0.005 | 0.005 | 0.7041 | 0.3853 | 0.2339 | 0.4503 | 0.01298 | 0.00000 | 0.00000 | 0.0000 | 78.4 | 0.7810 |
| 0.006 | 0.006 | 0.7041 | 0.3853 | 0.2521 | 0.5025 | 0.01559 | 0.00000 | 0.00000 | 0.0000 | 78.3 | 0.7649 |
| 0.007 | 0.007 | 0.7040 | 0.3853 | 0.2653 | 0.5569 | 0.01819 | 0.00000 | 0.00000 | 0.0000 | 78.3 | 0.7529 |
| 0.008 | 0.008 | 0.7040 | 0.3853 | 0.2775 | 0.6082 | 0.02079 | 0.00000 | 0.00000 | 0.0000 | 78.3 | 0.7418 |
| 0.009 | 0.009 | 0.7040 | 0.3853 | 0.3007 | 0.6326 | 0.02341 | 0.00000 | 0.00000 | 0.0000 | 78.3 | 0.7228 |
| 0.010 | 0.010 | 0.7040 | 0.3853 | 0.3448 | 0.6127 | 0.02601 | 0.00000 | 0.00000 | 0.0000 | 78.3 | 0.6898 |
| 0.020 | 0.020 | 0.7037 | 0.3851 | 0.3802 | 1.1202 | 0.05228 | 0.00000 | 0.00000 | 0.0000 | 78.0 | 0.6482 |
| 0.030 | 0.030 | 0.7034 | 0.3850 | 0.4165 | 1.5418 | 0.07876 | 0.00000 | 0.00000 | 0.0000 | 77.8 | 0.6084 |
| 0.040 | 0.040 | 0.7032 | 0.3848 | 0.4466 | 1.9275 | 0.10547 | 0.00000 | 0.00000 | 0.0000 | 77.6 | 0.5745 |
| 0.050 | 0.050 | 0.7029 | 0.3847 | 0.4676 | 2.3192 | 0.13248 | 0.00000 | 0.00000 | 0.0000 | 77.4 | 0.5473 |
| 0.060 | 0.060 | 0.7027 | 0.3845 | 0.4801 | 2.7244 | 0.15965 | 0.00000 | 0.00000 | 0.0000 | 77.2 | 0.5257 |
| 0.070 | 0.070 | 0.7024 | 0.3844 | 0.5113 | 3.0000 | 0.18705 | 0.00497 | 0.02366 | 1.0053 | 77.0 | 0.5099 |
| 0.080 | 0.080 | 0.7022 | 0.3843 | 0.5874 | 3.0000 | 0.21468 | 0.00570 | 0.09972 | 3.6963 | 76.8 | 0.4961 |
| 0.090 | 0.090 | 0.7019 | 0.3841 | 0.6658 | 3.0000 | 0.24269 | 0.00644 | 0.16415 | 5.3881 | 76.6 | 0.4758 |

*(Continued)*

**TABLE 3.9 (*Continued*)**
**Calculated Values of Thermodynamic Quantities Included in Formulae (4.84) and (4.85)**

| $m$ | $c_k$ | $R$ (nm) | $R_{H_2O}$ (nm) | $-\Psi_\gamma^0$ | $\chi$ | $-E_{sp}$ | $E_H$ | $U_{ot}$ | $B$ | $\varepsilon_S$ | $\gamma_\pm$ |
|---|---|---|---|---|---|---|---|---|---|---|---|
| 0.100 | 0.099 | 0.7040 | 0.3853 | 0.7668 | 3.0000 | 0.26358 | 0.00706 | 0.21607 | 6.4711 | 76.4 | 0.4437 |
| 0.200 | 0.199 | 0.7003 | 0.3832 | 1.5979 | 3.0000 | 0.55815 | 0.01471 | 1.25096 | 17.9805 | 74.5 | 0.4061 |
| 0.300 | 0.297 | 0.6990 | 0.3825 | 2.5688 | 3.0000 | 0.86206 | 0.02258 | 2.43794 | 22.8191 | 72.7 | 0.3729 |
| 0.400 | 0.395 | 0.6971 | 0.3815 | 3.6046 | 3.0000 | 1.18859 | 0.03088 | 3.72500 | 25.4960 | 71.1 | 0.3469 |
| 0.500 | 0.493 | 0.6950 | 0.3804 | 4.7506 | 3.0000 | 1.54285 | 0.03972 | 5.16566 | 27.4880 | 69.4 | 0.3280 |
| 0.600 | 0.591 | 0.6929 | 0.3792 | 5.9120 | 3.0000 | 1.91867 | 0.04893 | 6.65591 | 28.7535 | 68.0 | 0.3142 |
| 0.700 | 0.688 | 0.6910 | 0.3782 | 7.2374 | 3.0000 | 2.31712 | 0.05861 | 8.34293 | 30.0908 | 66.4 | 0.3042 |
| 0.800 | 0.785 | 0.6890 | 0.3771 | 8.5739 | 3.0000 | 2.73911 | 0.06868 | 10.07065 | 30.9944 | 65.0 | 0.2964 |
| 0.900 | 0.882 | 0.6870 | 0.3760 | 9.9836 | 3.0000 | 3.19170 | 0.07932 | 11.90385 | 31.7241 | 63.6 | 0.2895 |
| 1.000 | 0.978 | 0.6852 | 0.3750 | 11.5018 | 3.0000 | 3.66509 | 0.09035 | 13.86649 | 32.4422 | 62.3 | 0.2829 |
| 1.250 | 1.213 | 0.6814 | 0.3729 | 15.7552 | 3.0000 | 4.92775 | 0.11943 | 19.34516 | 34.2398 | 59.1 | 0.2770 |
| 1.500 | 1.445 | 0.6777 | 0.3709 | 20.2487 | 3.0000 | 6.34867 | 0.15129 | 25.22539 | 35.2455 | 56.2 | 0.2729 |
| 1.750 | 1.675 | 0.6739 | 0.3688 | 25.0520 | 3.0000 | 7.98705 | 0.18708 | 31.63671 | 35.7463 | 53.3 | 0.2710 |
| 2.000 | 1.902 | 0.6702 | 0.3668 | 29.2384 | 3.0000 | 9.79638 | 0.22566 | 37.60799 | 35.2281 | 50.7 | 0.2715 |
| 2.250 | 2.127 | 0.6666 | 0.3648 | 33.3602 | 3.0000 | 11.92074 | 0.27006 | 43.83249 | 34.3080 | 47.9 | 0.2744 |
| 2.500 | 2.351 | 0.6629 | 0.3628 | 34.5004 | 3.0000 | 14.25918 | 0.31762 | 47.29394 | 31.4745 | 45.5 | 0.2795 |
| 2.750 | 2.552 | 0.6611 | 0.3618 | 38.7543 | 3.0000 | 16.49321 | 0.36437 | 53.76996 | 31.1929 | 43.3 | 0.2860 |
| 3.000 | 2.771 | 0.6576 | 0.3599 | 32.7481 | 3.0000 | 19.36356 | 0.42076 | 50.61190 | 25.4260 | 41.1 | 0.2925 |
| 3.140 | 2.891 | 0.6557 | 0.3589 | 26.8492 | 3.0000 | 21.05969 | 0.45376 | 46.39983 | 21.6149 | 40.0 | 0.2976 |
| 3.165 | 2.908 | 0.6557 | 0.3589 | 27.5535 | 3.0000 | 21.28194 | 0.45856 | 47.33281 | 21.8187 | 39.8 | 0.3006 |
| *LiBr; $\xi = 1.111$; $a = 0.59994$* | | | | | | | | | | | |
| 0.001 | 0.001 | 0.6510 | 0.3854 | 0.0341 | 0.0293 | 0.00049 | 0.00000 | 0.00000 | 0.0000 | 78.4 | 0.966 |
| 0.002 | 0.002 | 0.6509 | 0.3854 | 0.0492 | 0.0405 | 0.00099 | 0.00000 | 0.00000 | 0.0000 | 78.4 | 0.951 |

| 0.003 | 0.003 | 0.6509 | 0.3854 | 0.0582 | 0.0514 | 0.00148 | 0.00000 | 0.00000 | 0.0000 | 78.4 | 0.942 |
|---|---|---|---|---|---|---|---|---|---|---|---|
| 0.004 | 0.004 | 0.6509 | 0.3853 | 0.0662 | 0.0602 | 0.00197 | 0.00000 | 0.00000 | 0.0000 | 78.4 | 0.934 |
| 0.005 | 0.005 | 0.6509 | 0.3853 | 0.0721 | 0.0692 | 0.00247 | 0.00000 | 0.00000 | 0.0000 | 78.3 | 0.928 |
| 0.006 | 0.006 | 0.6509 | 0.3853 | 0.0769 | 0.0778 | 0.00296 | 0.00000 | 0.00000 | 0.0000 | 78.3 | 0.923 |
| 0.007 | 0.007 | 0.6508 | 0.3853 | 0.0818 | 0.0853 | 0.00346 | 0.00000 | 0.00000 | 0.0000 | 78.3 | 0.918 |
| 0.008 | 0.008 | 0.6508 | 0.3853 | 0.0835 | 0.0957 | 0.00395 | 0.00000 | 0.00000 | 0.0000 | 78.3 | 0.916 |
| 0.009 | 0.009 | 0.6508 | 0.3853 | 0.0917 | 0.0979 | 0.00445 | 0.00000 | 0.00000 | 0.0000 | 78.3 | 0.908 |
| 0.010 | 0.010 | 0.6508 | 0.3853 | 0.1056 | 0.0945 | 0.00494 | 0.00000 | 0.00000 | 0.0000 | 78.3 | 0.895 |
| 0.020 | 0.020 | 0.6506 | 0.3852 | 0.1229 | 0.1630 | 0.00992 | 0.00000 | 0.00000 | 0.0000 | 78.2 | 0.875 |
| 0.030 | 0.030 | 0.6504 | 0.3851 | 0.1360 | 0.2213 | 0.01492 | 0.00000 | 0.00000 | 0.0000 | 78.1 | 0.859 |
| 0.040 | 0.040 | 0.6502 | 0.3849 | 0.1470 | 0.2739 | 0.01996 | 0.00000 | 0.00000 | 0.0000 | 77.9 | 0.845 |
| 0.050 | 0.050 | 0.6500 | 0.3848 | 0.1535 | 0.3285 | 0.02502 | 0.00000 | 0.00000 | 0.0000 | 77.8 | 0.835 |
| 0.060 | 0.060 | 0.6498 | 0.3847 | 0.1589 | 0.3815 | 0.03010 | 0.00000 | 0.00000 | 0.0000 | 77.7 | 0.826 |
| 0.070 | 0.070 | 0.6497 | 0.3846 | 0.1619 | 0.4383 | 0.03523 | 0.00000 | 0.00000 | 0.0000 | 77.6 | 0.819 |
| 0.080 | 0.080 | 0.6495 | 0.3845 | 0.1625 | 0.4998 | 0.04038 | 0.00000 | 0.00000 | 0.0000 | 77.5 | 0.814 |
| 0.090 | 0.090 | 0.6493 | 0.3844 | 0.1669 | 0.5496 | 0.04557 | 0.00000 | 0.00000 | 0.0000 | 77.4 | 0.806 |
| 0.100 | 0.100 | 0.6491 | 0.3843 | 0.1775 | 0.5789 | 0.05086 | 0.00000 | 0.00000 | 0.0000 | 77.2 | 0.793 |
| 0.200 | 0.199 | 0.6483 | 0.3838 | 0.1446 | 1.4684 | 0.10315 | 0.00000 | 0.00000 | 0.0000 | 76.2 | 0.775 |
| 0.300 | 0.298 | 0.6469 | 0.3830 | 0.1635 | 2.0000 | 0.15849 | 0.01362 | 0.04861 | 0.7542 | 75.1 | 0.763 |
| 0.400 | 0.394 | 0.6469 | 0.3830 | 0.2291 | 2.0000 | 0.21262 | 0.01828 | 0.15804 | 1.8278 | 74.0 | 0.756 |
| 0.500 | 0.491 | 0.6457 | 0.3823 | 0.2962 | 2.0000 | 0.27137 | 0.02320 | 0.28521 | 2.5989 | 72.9 | 0.758 |
| 0.600 | 0.589 | 0.6440 | 0.3813 | 0.3646 | 2.0000 | 0.33500 | 0.02839 | 0.41945 | 3.1225 | 71.9 | 0.761 |
| 0.700 | 0.687 | 0.6423 | 0.3803 | 0.4365 | 2.0000 | 0.40233 | 0.03381 | 0.56600 | 3.5382 | 70.8 | 0.768 |
| 0.800 | 0.779 | 0.6422 | 0.3802 | 0.5254 | 2.0000 | 0.46345 | 0.03894 | 0.72468 | 3.9340 | 69.7 | 0.776 |
| 0.900 | 0.875 | 0.6408 | 0.3794 | 0.6097 | 2.0000 | 0.53481 | 0.04462 | 0.89482 | 4.2387 | 68.6 | 0.789 |
| 1.000 | 0.972 | 0.6392 | 0.3784 | 0.7135 | 2.0000 | 0.61582 | 0.05096 | 1.10550 | 4.5858 | 67.1 | 0.812 |
| 1.250 | 1.210 | 0.6358 | 0.3764 | 0.9253 | 2.0000 | 0.81200 | 0.06608 | 1.54688 | 4.9481 | 65.1 | 0.845 |
| 1.500 | 1.446 | 0.6326 | 0.3745 | 2.8375 | 2.0000 | 1.28858 | 0.10318 | 3.96463 | 8.1225 | 50.3 | 0.895 |

(Continued)

**TABLE 3.9 (*Continued*)**
**Calculated Values of Thermodynamic Quantities Included in Formulae (4.84) and (4.85)**

| $m$ | $c_k$ | $R$ (nm) | $R_{H_2O}$ (nm) | $-\Psi_\gamma^0$ | $\chi$ | $-E_{sp}$ | $E_H$ | $U_{ot}$ | $B$ | $\varepsilon_S$ | $\gamma_\pm$ |
|---|---|---|---|---|---|---|---|---|---|---|---|
| 1.750 | 1.668 | 0.6310 | 0.3736 | 2.0189 | 2.0000 | 1.35074 | 0.10727 | 3.28477 | 6.4728 | 56.1 | 0.962 |
| 2.000 | 1.901 | 0.6277 | 0.3716 | 2.6216 | 2.0000 | 1.67861 | 0.13107 | 4.27605 | 6.8962 | 52.9 | 1.038 |
| 2.250 | 2.117 | 0.6260 | 0.3706 | 3.4158 | 2.0000 | 2.00138 | 0.15492 | 5.44818 | 7.4339 | 50.1 | 1.114 |
| 2.500 | 2.348 | 0.6226 | 0.3686 | 4.1033 | 2.0000 | 2.40633 | 0.18307 | 6.59008 | 7.6092 | 47.5 | 1.194 |
| 2.750 | 2.559 | 0.6208 | 0.3676 | 4.9490 | 2.0000 | 2.78178 | 0.20974 | 7.86626 | 7.9278 | 45.5 | 1.285 |
| 3.000 | 2.790 | 0.6174 | 0.3655 | 5.3593 | 2.0000 | 3.25482 | 0.24105 | 8.80142 | 7.7182 | 43.6 | 1.385 |
| 3.250 | 2.996 | 0.6157 | 0.3645 | 6.0892 | 2.0000 | 3.68562 | 0.27055 | 10.02116 | 7.8295 | 42.0 | 1.501 |
| 3.500 | 3.200 | 0.6139 | 0.3635 | 6.7669 | 2.0000 | 4.15172 | 0.30199 | 11.22152 | 7.8545 | 40.4 | 1.626 |
| 3.750 | 3.430 | 0.6105 | 0.3614 | 6.2890 | 2.0000 | 4.73727 | 0.33840 | 11.38556 | 7.1118 | 39.1 | 1.770 |
| 4.000 | 3.631 | 0.6087 | 0.3604 | 6.3742 | 2.0000 | 5.26539 | 0.37266 | 12.05650 | 6.8386 | 37.8 | 1.925 |
| 4.250 | 3.829 | 0.6070 | 0.3594 | 6.0915 | 2.0000 | 5.79843 | 0.40669 | 12.36909 | 6.4288 | 36.7 | 2.103 |
| 4.500 | 4.026 | 0.6052 | 0.3583 | 5.3830 | 2.0000 | 6.36244 | 0.44209 | 12.28391 | 5.8733 | 35.7 | 2.294 |
| 4.750 | 4.221 | 0.6035 | 0.3573 | 4.2331 | 2.0000 | 6.96379 | 0.47938 | 11.79653 | 5.2016 | 34.7 | 2.512 |
| 5.000 | 4.415 | 0.6017 | 0.3562 | 2.4660 | 2.0000 | 7.61171 | 0.51902 | 10.72900 | 4.3696 | 33.8 | 2.731 |
| 5.250 | 4.607 | 0.5999 | 0.3552 | 0.0000 | 0.0000 | 8.40741 | 0.56789 | 9.11238 | 3.3918 | 32.4 | 3.003 |
| 5.500 | 4.797 | 0.5982 | 0.3541 | 0.0000 | 0.0000 | 9.20247 | 0.61583 | 9.97908 | 3.4252 | 31.3 | 3.359 |
| 5.750 | 4.986 | 0.5964 | 0.3531 | 0.0000 | 0.0000 | 10.07091 | 0.66763 | 10.93207 | 3.4612 | 30.2 | 3.821 |
| 6.000 | 5.175 | 0.5947 | 0.3520 | 0.0000 | 0.0000 | 10.94763 | 0.71876 | 11.88552 | 3.4954 | 29.3 | 4.310 |
| 6.250 | 5.362 | 0.5929 | 0.3510 | 0.0000 | 0.0000 | 11.89527 | 0.77360 | 12.89400 | 3.5231 | 28.3 | 4.803 |
| 6.500 | 5.549 | 0.5911 | 0.3500 | 0.0000 | 0.0000 | 12.84870 | 0.82754 | 13.89209 | 3.5484 | 27.6 | 5.262 |
| 6.750 | 5.734 | 0.5894 | 0.3489 | 0.0000 | 0.0000 | 13.87357 | 0.88511 | 14.95552 | 3.5716 | 26.8 | 5.751 |
| 7.000 | 5.919 | 0.5876 | 0.3479 | 0.0000 | 0.0000 | 14.95651 | 0.94504 | 16.08104 | 3.5968 | 26.1 | 6.326 |
| 7.250 | 6.104 | 0.5858 | 0.3468 | 0.0000 | 0.0000 | 16.01761 | 1.00224 | 17.19777 | 3.6271 | 25.5 | 7.031 |

| | | | | | | | | | | | |
|---|---|---|---|---|---|---|---|---|---|---|---|
| 7.500 | 6.231 | 0.5859 | 0.3468 | 0.0000 | 0.0000 | 16.75975 | 1.04876 | 17.99646 | 3.6272 | 24.9 | 7.739 |
| 7.750 | 6.413 | 0.5841 | 0.3458 | 0.0000 | 0.0000 | 17.88967 | 1.10862 | 19.16996 | 3.6551 | 24.4 | 8.522 |
| 8.000 | 6.595 | 0.5823 | 0.3447 | 0.0000 | 0.0000 | 19.15363 | 1.17535 | 20.47651 | 3.6825 | 23.8 | 9.440 |
| 8.250 | 6.776 | 0.5806 | 0.3437 | 0.0000 | 0.0000 | 20.47044 | 1.24401 | 21.84318 | 3.7115 | 23.2 | 10.554 |
| 8.500 | 6.894 | 0.5806 | 0.3437 | 0.0000 | 0.0000 | 21.21133 | 1.28906 | 22.64680 | 3.7136 | 22.8 | 11.674 |
| 8.750 | 7.073 | 0.5788 | 0.3427 | 0.0000 | 0.0000 | 22.58622 | 1.35930 | 24.05849 | 3.7412 | 22.3 | 12.904 |
| 9.000 | 7.253 | 0.5770 | 0.3416 | 0.0000 | 0.0000 | 24.03425 | 1.43216 | 25.54558 | 3.7704 | 21.9 | 14.334 |
| 9.250 | 7.432 | 0.5753 | 0.3406 | 0.0000 | 0.0000 | 25.40904 | 1.49933 | 26.97337 | 3.8028 | 21.5 | 16.055 |
| 9.500 | 7.542 | 0.5753 | 0.3406 | 0.0000 | 0.0000 | 26.27473 | 1.55042 | 27.88168 | 3.8013 | 21.1 | 17.514 |
| 9.750 | 7.759 | 0.5726 | 0.3390 | 0.0000 | 0.0000 | 28.22289 | 1.64006 | 29.85251 | 3.8475 | 20.7 | 19.465 |
| 10.000 | 7.866 | 0.5726 | 0.3390 | 0.0000 | 0.0000 | 29.00113 | 1.68517 | 30.72875 | 3.8544 | 20.5 | 22.311 |
| 10.500 | 8.253 | 0.5683 | 0.3365 | 0.0000 | 0.0000 | 32.68910 | 1.85451 | 34.45782 | 3.9275 | 19.8 | 27.177 |
| 11.000 | 8.548 | 0.5662 | 0.3352 | 0.0000 | 0.0000 | 35.45037 | 1.98731 | 37.29786 | 3.9671 | 19.3 | 33.147 |
| 11.500 | 8.841 | 0.5641 | 0.3340 | 0.0000 | 0.0000 | 38.48779 | 2.13162 | 40.41902 | 4.0081 | 18.7 | 41.107 |
| 12.000 | 9.132 | 0.5620 | 0.3327 | 0.0000 | 0.0000 | 41.09132 | 2.24816 | 43.11913 | 4.0542 | 18.5 | 50.232 |
| 12.500 | 9.420 | 0.5599 | 0.3315 | 0.0000 | 0.0000 | 43.84459 | 2.36976 | 45.97029 | 4.1004 | 18.2 | 61.784 |
| 13.000 | 9.706 | 0.5578 | 0.3302 | 0.0000 | 0.0000 | 46.77684 | 2.49756 | 48.97898 | 4.1453 | 17.9 | 74.854 |
| 13.500 | 9.991 | 0.5557 | 0.3290 | 0.0000 | 0.0000 | 49.85432 | 2.62928 | 52.13079 | 4.1910 | 17.7 | 90.867 |
| *LiCl; ξ = 0.934; a = 0.49035* | | | | | | | | | | | |
| 0.001 | 0.001 | 0.5126 | 0.3854 | 0.0333 | 0.0215 | 0.00127 | 0.00000 | 0.00000 | 0.00000 | 77.6 | 0.966 |
| 0.002 | 0.002 | 0.5126 | 0.3854 | 0.0476 | 0.0301 | 0.00254 | 0.00000 | 0.00000 | 0.00000 | 77.5 | 0.951 |
| 0.003 | 0.003 | 0.5125 | 0.3854 | 0.0569 | 0.0379 | 0.00381 | 0.00000 | 0.00000 | 0.00000 | 77.5 | 0.941 |
| 0.004 | 0.004 | 0.5125 | 0.3854 | 0.0641 | 0.0450 | 0.00509 | 0.00000 | 0.00000 | 0.00000 | 77.4 | 0.933 |
| 0.005 | 0.005 | 0.5125 | 0.3854 | 0.0693 | 0.0522 | 0.00636 | 0.00000 | 0.00000 | 0.00000 | 77.4 | 0.927 |
| 0.006 | 0.006 | 0.5125 | 0.3854 | 0.0734 | 0.0592 | 0.00764 | 0.00000 | 0.00000 | 0.00000 | 77.3 | 0.922 |
| 0.007 | 0.007 | 0.5125 | 0.3854 | 0.0775 | 0.0655 | 0.00892 | 0.00000 | 0.00000 | 0.00000 | 77.3 | 0.917 |
| 0.008 | 0.008 | 0.5125 | 0.3853 | 0.0794 | 0.0732 | 0.01020 | 0.00000 | 0.00000 | 0.00000 | 77.2 | 0.914 |
| 0.009 | 0.009 | 0.5125 | 0.3853 | 0.0858 | 0.0764 | 0.01148 | 0.00000 | 0.00000 | 0.00000 | 77.2 | 0.907 |

(*Continued*)

**TABLE 3.9 (*Continued*)**
**Calculated Values of Thermodynamic Quantities Included in Formulae (4.84) and (4.85)**

| $m$ | $c_k$ | $R$ (nm) | $R_{H_2O}$ (nm) | $-\Psi_\gamma^0$ | $\chi$ | $-E_{sp}$ | $E_H$ | $U_{ot}$ | $B$ | $\varepsilon_S$ | $\gamma_\pm$ |
|---|---|---|---|---|---|---|---|---|---|---|---|
| 0.010 | 0.010 | 0.5125 | 0.3853 | 0.1000 | 0.0730 | 0.01277 | 0.00000 | 0.00000 | 0.00000 | 77.2 | 0.893 |
| 0.020 | 0.020 | 0.5124 | 0.3853 | 0.1107 | 0.1318 | 0.02557 | 0.00000 | 0.00000 | 0.00000 | 77.1 | 0.872 |
| 0.030 | 0.030 | 0.5123 | 0.3852 | 0.1160 | 0.1885 | 0.03841 | 0.00000 | 0.00000 | 0.00000 | 77.1 | 0.856 |
| 0.040 | 0.040 | 0.5123 | 0.3852 | 0.1204 | 0.2422 | 0.05129 | 0.00000 | 0.00000 | 0.00000 | 77.0 | 0.841 |
| 0.050 | 0.050 | 0.5122 | 0.3851 | 0.5708 | 0.0638 | 0.06419 | 0.00000 | 0.00000 | 0.00000 | 76.9 | 0.529 |
| 0.060 | 0.060 | 0.5121 | 0.3851 | 0.1192 | 0.3666 | 0.07713 | 0.00000 | 0.00000 | 0.00000 | 76.9 | 0.820 |
| 0.070 | 0.070 | 0.5121 | 0.3850 | 0.1144 | 0.4452 | 0.09011 | 0.00000 | 0.00000 | 0.00000 | 76.9 | 0.813 |
| 0.080 | 0.080 | 0.5120 | 0.3850 | 0.1084 | 0.5363 | 0.10311 | 0.00000 | 0.00000 | 0.00000 | 76.8 | 0.807 |
| 0.090 | 0.089 | 0.5138 | 0.3863 | 0.1084 | 0.6453 | 0.11271 | 0.00000 | 0.00000 | 0.00000 | 76.8 | 0.799 |
| 0.100 | 0.099 | 0.5136 | 0.3861 | 0.1140 | 0.6775 | 0.12578 | 0.00000 | 0.00000 | 0.00000 | 76.7 | 0.784 |
| 0.200 | 0.199 | 0.5120 | 0.3850 | 0.0745 | 2.0000 | 0.25847 | 0.00921 | 0.06310 | 1.44843 | 76.2 | 0.765 |
| 0.300 | 0.297 | 0.5121 | 0.3851 | 0.1147 | 2.0000 | 0.38757 | 0.01382 | 0.21285 | 3.25542 | 75.8 | 0.751 |
| 0.400 | 0.397 | 0.5110 | 0.3842 | 0.1493 | 2.0000 | 0.52715 | 0.01867 | 0.37508 | 4.24663 | 75.3 | 0.743 |
| 0.500 | 0.494 | 0.5110 | 0.3842 | 0.1909 | 2.0000 | 0.65948 | 0.02337 | 0.54515 | 4.93176 | 74.9 | 0.741 |
| 0.600 | 0.590 | 0.5111 | 0.3843 | 0.2346 | 2.0000 | 0.79158 | 0.02806 | 0.72382 | 5.45209 | 74.4 | 0.744 |
| 0.700 | 0.689 | 0.5103 | 0.3837 | 0.2699 | 2.0000 | 0.93823 | 0.03309 | 0.91094 | 5.81927 | 73.9 | 0.749 |
| 0.800 | 0.784 | 0.5103 | 0.3837 | 0.3151 | 2.0000 | 1.07347 | 0.03787 | 1.09809 | 6.12952 | 73.5 | 0.755 |
| 0.900 | 0.884 | 0.5092 | 0.3829 | 0.3452 | 2.0000 | 1.23069 | 0.04313 | 1.29678 | 6.35513 | 73.1 | 0.765 |
| 1.000 | 0.978 | 0.5093 | 0.3829 | 0.3775 | 2.0000 | 1.35646 | 0.04755 | 1.47595 | 6.56097 | 73.3 | 0.782 |
| 1.250 | 1.217 | 0.5083 | 0.3822 | 0.5213 | 2.0000 | 1.77038 | 0.06171 | 2.05838 | 7.05115 | 70.6 | 0.806 |
| 1.500 | 1.454 | 0.5074 | 0.3815 | 0.6853 | 2.0000 | 2.21539 | 0.07677 | 2.70459 | 7.44660 | 68.0 | 0.842 |
| 1.750 | 1.689 | 0.5064 | 0.3808 | 0.9126 | 2.0000 | 2.72479 | 0.09389 | 3.48699 | 7.85075 | 64.9 | 0.889 |
| 2.000 | 1.922 | 0.5055 | 0.3801 | 1.1397 | 2.0000 | 3.24918 | 0.11132 | 4.28735 | 8.14079 | 62.5 | 0.942 |

| | | | | | | | | | | | |
|---|---|---|---|---|---|---|---|---|---|---|---|
| 2.250 | 2.153 | 0.5046 | 0.3794 | 1.3668 | 2.0000 | 3.79255 | 0.12922 | 5.10101 | 8.34445 | 60.5 | 0.993 |
| 2.500 | 2.383 | 0.5036 | 0.3787 | 1.6190 | 2.0000 | 4.38486 | 0.14852 | 5.98751 | 8.52183 | 58.5 | 1.047 |
| 2.750 | 2.596 | 0.5036 | 0.3787 | 1.9816 | 2.0000 | 4.91421 | 0.16647 | 6.92548 | 8.79360 | 56.8 | 1.107 |
| 3.000 | 2.821 | 0.5027 | 0.3780 | 2.2674 | 2.0000 | 5.55680 | 0.18717 | 7.89926 | 8.92088 | 55.1 | 1.173 |
| 3.250 | 3.045 | 0.5018 | 0.3773 | 2.5144 | 2.0000 | 6.21720 | 0.20819 | 8.85570 | 8.99136 | 53.7 | 1.248 |
| 3.500 | 3.267 | 0.5008 | 0.3766 | 2.7752 | 2.0000 | 6.92118 | 0.23045 | 9.86916 | 9.05242 | 52.3 | 1.329 |
| 3.750 | 3.488 | 0.4999 | 0.3759 | 3.0101 | 2.0000 | 7.66294 | 0.25367 | 10.89750 | 9.08059 | 50.9 | 1.421 |
| 4.000 | 3.687 | 0.4999 | 0.3759 | 3.5477 | 2.0000 | 8.32595 | 0.27561 | 12.15070 | 9.31906 | 49.5 | 1.519 |
| 4.250 | 3.904 | 0.4990 | 0.3752 | 3.8555 | 2.0000 | 9.18414 | 0.30228 | 13.36901 | 9.34878 | 48.0 | 1.631 |
| 4.500 | 4.121 | 0.4980 | 0.3744 | 4.0331 | 2.0000 | 10.06071 | 0.32914 | 14.47457 | 9.29590 | 46.7 | 1.750 |
| 4.750 | 4.336 | 0.4970 | 0.3737 | 4.2181 | 2.0000 | 11.01900 | 0.35840 | 15.67007 | 9.24192 | 45.3 | 1.884 |
| 5.000 | 4.524 | 0.4971 | 0.3737 | 4.9632 | 2.0000 | 11.84146 | 0.38518 | 17.28817 | 9.48742 | 44.0 | 2.020 |
| 5.250 | 4.736 | 0.4961 | 0.3730 | 4.9435 | 2.0000 | 12.83215 | 0.41497 | 18.31511 | 9.32957 | 42.9 | 2.184 |
| 5.500 | 4.947 | 0.4952 | 0.3723 | 4.8219 | 2.0000 | 13.90184 | 0.44693 | 19.33138 | 9.14291 | 41.7 | 2.396 |
| 5.750 | 5.157 | 0.4942 | 0.3716 | 4.5102 | 2.0000 | 15.02169 | 0.48013 | 20.21876 | 8.90149 | 40.7 | 2.661 |
| 6.000 | 5.336 | 0.4942 | 0.3716 | 5.1133 | 2.0000 | 15.91130 | 0.50854 | 21.78814 | 9.05653 | 39.7 | 2.934 |
| 6.250 | 5.543 | 0.4933 | 0.3709 | 4.5255 | 2.0000 | 17.16103 | 0.54530 | 22.50718 | 8.72461 | 38.6 | 3.199 |
| 6.500 | 5.750 | 0.4923 | 0.3702 | 3.5500 | 2.0000 | 18.47062 | 0.58344 | 22.88542 | 8.29138 | 37.6 | 3.448 |
| 6.750 | 5.922 | 0.4923 | 0.3702 | 3.9968 | 2.0000 | 19.45505 | 0.61453 | 24.36468 | 8.38073 | 36.8 | 3.705 |
| 7.000 | 6.126 | 0.4914 | 0.3695 | 2.4262 | 2.0000 | 20.83722 | 0.65434 | 24.21898 | 7.82376 | 35.9 | 3.995 |
| 7.250 | 6.329 | 0.4905 | 0.3688 | 0.3363 | 2.0000 | 22.27640 | 0.69549 | 23.61582 | 7.17757 | 35.0 | 4.334 |
| 7.500 | 6.495 | 0.4905 | 0.3688 | 0.3440 | 2.0000 | 23.32744 | 0.72825 | 24.72526 | 7.17664 | 34.3 | 4.678 |
| 7.750 | 6.696 | 0.4895 | 0.3681 | 0.0000 | 0.0000 | 24.83908 | 0.77092 | 25.93281 | 7.11056 | 33.5 | 5.045 |
| 8.000 | 6.857 | 0.4895 | 0.3681 | 0.0000 | 0.0000 | 26.00427 | 0.80712 | 27.14640 | 7.10949 | 32.8 | 5.452 |
| 8.250 | 7.057 | 0.4886 | 0.3674 | 0.0000 | 0.0000 | 27.61874 | 0.85208 | 28.80405 | 7.14556 | 32.1 | 5.913 |
| 8.500 | 7.214 | 0.4886 | 0.3674 | 0.0000 | 0.0000 | 28.71966 | 0.88607 | 29.95407 | 7.14579 | 31.5 | 6.381 |
| 8.750 | 7.411 | 0.4877 | 0.3667 | 0.0000 | 0.0000 | 30.40109 | 0.93246 | 31.67041 | 7.17941 | 30.9 | 6.874 |
| 9.000 | 7.608 | 0.4867 | 0.3660 | 0.0000 | 0.0000 | 32.15174 | 0.98033 | 33.45510 | 7.21366 | 30.3 | 7.410 |

*(Continued)*

TABLE 3.9 (*Continued*)
**Calculated Values of Thermodynamic Quantities Included in Formulae (4.84) and (4.85)**

| $m$ | $c_k$ | $R$ (nm) | $R_{H_2O}$ (nm) | $-\Psi_\gamma^0$ | $\chi$ | $-E_{sp}$ | $E_H$ | $U_{ot}$ | $B$ | $\varepsilon_S$ | $\gamma_\pm$ |
|---|---|---|---|---|---|---|---|---|---|---|---|
| 9.250 | 7.760 | 0.4867 | 0.3660 | 0.0000 | 0.0000 | 33.30780 | 1.01555 | 34.66045 | 7.21435 | 29.8 | 8.009 |
| 9.500 | 7.955 | 0.4858 | 0.3653 | 0.0000 | 0.0000 | 35.10667 | 1.06411 | 36.48169 | 7.24693 | 29.3 | 8.540 |
| 9.750 | 8.103 | 0.4858 | 0.3653 | 0.0000 | 0.0000 | 36.39910 | 1.10330 | 37.81498 | 7.24495 | 28.8 | 9.190 |
| 10.000 | 8.297 | 0.4849 | 0.3646 | 0.0000 | 0.0000 | 38.30148 | 1.15404 | 39.76698 | 7.28390 | 28.3 | 10.093 |
| 10.500 | 8.634 | 0.4839 | 0.3639 | 0.0000 | 0.0000 | 41.36832 | 1.23904 | 42.88752 | 7.31661 | 27.5 | 11.443 |
| 11.000 | 8.966 | 0.4830 | 0.3632 | 0.0000 | 0.0000 | 44.66722 | 1.32993 | 46.23632 | 7.34884 | 26.7 | 13.003 |
| 11.500 | 9.293 | 0.4821 | 0.3625 | 0.0000 | 0.0000 | 47.85460 | 1.41647 | 49.47961 | 7.38384 | 26.1 | 14.803 |
| 12.000 | 9.616 | 0.4811 | 0.3618 | 0.0000 | 0.0000 | 51.29485 | 1.50934 | 52.96101 | 7.41708 | 25.4 | 16.713 |
| 12.500 | 9.936 | 0.4802 | 0.3611 | 0.0000 | 0.0000 | 54.54099 | 1.59520 | 56.25162 | 7.45389 | 25.0 | 18.803 |
| 13.000 | 10.251 | 0.4793 | 0.3604 | 0.0000 | 0.0000 | 57.96513 | 1.68536 | 59.70794 | 7.48867 | 24.5 | 20.989 |
| 13.500 | 10.502 | 0.4793 | 0.3604 | 0.0000 | 0.0000 | 60.32540 | 1.75394 | 62.11845 | 7.48635 | 24.1 | 23.351 |
| 14.000 | 10.808 | 0.4784 | 0.3597 | 0.0000 | 0.0000 | 63.74115 | 1.84233 | 65.55830 | 7.52183 | 23.7 | 25.818 |
| 14.500 | 11.112 | 0.4774 | 0.3590 | 0.0000 | 0.0000 | 67.09962 | 1.92774 | 68.94096 | 7.55948 | 23.4 | 28.468 |
| 15.000 | 11.412 | 0.4765 | 0.3583 | 0.0000 | 0.0000 | 70.50361 | 2.01353 | 72.36337 | 7.59670 | 23.1 | 31.225 |
| 15.500 | 11.709 | 0.4756 | 0.3576 | 0.0000 | 0.0000 | 73.99084 | 2.10060 | 75.86494 | 7.63418 | 22.8 | 34.158 |
| 16.000 | 11.934 | 0.4756 | 0.3576 | 0.0000 | 0.0000 | 76.08817 | 2.16015 | 77.99944 | 7.63258 | 22.6 | 37.197 |
| 16.500 | 12.224 | 0.4747 | 0.3569 | 0.0000 | 0.0000 | 79.45291 | 2.24228 | 81.37552 | 7.67129 | 22.4 | 40.380 |
| 17.000 | 12.512 | 0.4738 | 0.3562 | 0.0000 | 0.0000 | 82.89786 | 2.32550 | 84.82790 | 7.71055 | 22.2 | 43.718 |
| 17.500 | 12.798 | 0.4728 | 0.3555 | 0.0000 | 0.0000 | 86.11775 | 2.40131 | 88.05598 | 7.75129 | 22.0 | 47.023 |
| 18.000 | 13.005 | 0.4728 | 0.3555 | 0.0000 | 0.0000 | 88.10293 | 2.45674 | 90.05775 | 7.74863 | 21.9 | 49.983 |
| 18.500 | 13.284 | 0.4719 | 0.3548 | 0.0000 | 0.0000 | 91.49500 | 2.53620 | 93.43308 | 7.78720 | 21.7 | 52.643 |
| 19.000 | 13.562 | 0.4710 | 0.3541 | 0.0000 | 0.0000 | 94.96864 | 2.61667 | 96.88883 | 7.82687 | 21.6 | 55.443 |
| 19.210 | 13.648 | 0.4710 | 0.3541 | 0.0000 | 0.0000 | 95.84525 | 2.64085 | 97.80215 | 7.82831 | 21.5 | 58.649 |

$MgCl_2$; $\xi = 1.18$; $a = 0.62658$

| | | | | | | | | | | | |
|---|---|---|---|---|---|---|---|---|---|---|---|
| 0.001 | 0.001 | 0.6712 | 0.3854 | 0.1158 | 0.1709 | 0.00356 | 0.00000 | 0.00000 | 0.00000 | 78.1 | 0.887 |
| 0.002 | 0.002 | 0.6712 | 0.3854 | 0.1490 | 0.2657 | 0.00712 | 0.00000 | 0.00000 | 0.00000 | 78.1 | 0.855 |
| 0.003 | 0.003 | 0.6712 | 0.3854 | 0.1761 | 0.3375 | 0.01068 | 0.00000 | 0.00000 | 0.00000 | 78.1 | 0.829 |
| 0.004 | 0.004 | 0.6712 | 0.3853 | 0.1980 | 0.4001 | 0.01425 | 0.00000 | 0.00000 | 0.00000 | 78.1 | 0.808 |
| 0.005 | 0.005 | 0.6711 | 0.3853 | 0.2160 | 0.4582 | 0.01781 | 0.00000 | 0.00000 | 0.00000 | 78.1 | 0.791 |
| 0.006 | 0.006 | 0.6711 | 0.3853 | 0.2310 | 0.5147 | 0.02138 | 0.00000 | 0.00000 | 0.00000 | 78.1 | 0.776 |
| 0.007 | 0.007 | 0.6711 | 0.3853 | 0.2415 | 0.5741 | 0.02495 | 0.00000 | 0.00000 | 0.00000 | 78.1 | 0.765 |
| 0.008 | 0.008 | 0.6711 | 0.3853 | 0.2510 | 0.6309 | 0.02852 | 0.00000 | 0.00000 | 0.00000 | 78.1 | 0.755 |
| 0.009 | 0.009 | 0.6711 | 0.3853 | 0.2696 | 0.6616 | 0.03210 | 0.00000 | 0.00000 | 0.00000 | 78.1 | 0.739 |
| 0.010 | 0.010 | 0.6710 | 0.3853 | 0.3053 | 0.6489 | 0.03567 | 0.00000 | 0.00000 | 0.00000 | 78.1 | 0.710 |
| 0.020 | 0.020 | 0.6708 | 0.3851 | 0.3198 | 1.2406 | 0.07157 | 0.00000 | 0.00000 | 0.00000 | 78.0 | 0.675 |
| 0.030 | 0.030 | 0.6706 | 0.3850 | 0.3336 | 1.7839 | 0.10764 | 0.00000 | 0.00000 | 0.00000 | 77.9 | 0.642 |
| 0.040 | 0.040 | 0.6704 | 0.3849 | 0.3407 | 2.3286 | 0.14390 | 0.00000 | 0.00000 | 0.00000 | 77.8 | 0.614 |
| 0.050 | 0.050 | 0.6702 | 0.3848 | 0.3394 | 2.9257 | 0.18042 | 0.00000 | 0.00000 | 0.00000 | 77.7 | 0.593 |
| 0.060 | 0.060 | 0.6700 | 0.3847 | 0.3972 | 3.0000 | 0.21708 | 0.00480 | 0.06143 | 2.70572 | 77.6 | 0.576 |
| 0.070 | 0.070 | 0.6698 | 0.3845 | 0.4640 | 3.0000 | 0.25402 | 0.00561 | 0.14350 | 5.40693 | 77.5 | 0.564 |
| 0.080 | 0.080 | 0.6695 | 0.3844 | 0.5302 | 3.0000 | 0.29109 | 0.00642 | 0.22770 | 7.49440 | 77.4 | 0.553 |
| 0.090 | 0.090 | 0.6693 | 0.3843 | 0.5972 | 3.0000 | 0.32846 | 0.00724 | 0.30620 | 8.93985 | 77.3 | 0.539 |
| 0.100 | 0.100 | 0.6691 | 0.3842 | 0.6635 | 3.0000 | 0.36592 | 0.00806 | 0.37319 | 9.78936 | 77.2 | 0.520 |
| 0.200 | 0.199 | 0.6682 | 0.3836 | 1.3565 | 3.0000 | 0.74243 | 0.01628 | 1.40222 | 18.21010 | 76.3 | 0.501 |
| 0.300 | 0.298 | 0.6665 | 0.3826 | 2.1113 | 3.0000 | 1.14786 | 0.02497 | 2.53449 | 21.45428 | 74.8 | 0.488 |
| 0.400 | 0.397 | 0.6646 | 0.3816 | 2.9106 | 3.0000 | 1.58133 | 0.03411 | 3.75542 | 23.27257 | 73.4 | 0.484 |
| 0.500 | 0.496 | 0.6627 | 0.3805 | 3.7686 | 3.0000 | 2.04651 | 0.04376 | 5.08201 | 24.54909 | 71.9 | 0.488 |
| 0.600 | 0.589 | 0.6630 | 0.3807 | 4.8988 | 3.0000 | 2.47492 | 0.05300 | 6.65691 | 26.55196 | 70.4 | 0.498 |
| 0.700 | 0.687 | 0.6611 | 0.3796 | 5.9081 | 3.0000 | 2.99108 | 0.06348 | 8.20184 | 27.30979 | 69.0 | 0.511 |
| 0.800 | 0.784 | 0.6594 | 0.3786 | 7.0213 | 3.0000 | 3.53091 | 0.07438 | 9.87977 | 28.07847 | 67.5 | 0.527 |
| 0.900 | 0.882 | 0.6575 | 0.3775 | 8.1182 | 3.0000 | 4.11674 | 0.08594 | 11.60353 | 28.53895 | 66.1 | 0.552 |

*(Continued)*

**TABLE 3.9 (*Continued*)**
**Calculated Values of Thermodynamic Quantities Included in Formulae (4.84) and (4.85)**

| $m$ | $c_k$ | $R$ (nm) | $R_{H_2O}$ (nm) | $-\Psi_\gamma^0$ | $\chi$ | $-E_{sp}$ | $E_H$ | $U_{ot}$ | $B$ | $\varepsilon_S$ | $\gamma_\pm$ |
|---|---|---|---|---|---|---|---|---|---|---|---|
| 1.000 | 0.979 | 0.6558 | 0.3765 | 9.3820 | 3.0000 | 4.73758 | 0.09813 | 13.56076 | 29.21003 | 64.6 | 0.598 |
| 1.250 | 1.217 | 0.6523 | 0.3745 | 13.1060 | 3.0000 | 6.41652 | 0.13077 | 19.05889 | 30.80719 | 60.9 | 0.671 |
| 1.500 | 1.442 | 0.6505 | 0.3735 | 18.5293 | 3.0000 | 8.19894 | 0.16568 | 26.38427 | 33.66173 | 57.3 | 0.774 |
| 1.750 | 1.675 | 0.6470 | 0.3715 | 23.8033 | 3.0000 | 10.42226 | 0.20716 | 34.01437 | 34.70750 | 53.8 | 0.909 |
| 2.000 | 1.906 | 0.6436 | 0.3695 | 29.5403 | 3.0000 | 13.00343 | 0.25433 | 42.47322 | 35.30079 | 50.4 | 1.084 |
| 2.250 | 2.137 | 0.6401 | 0.3675 | 34.0847 | 3.0000 | 15.96068 | 0.30700 | 50.12084 | 34.50986 | 47.3 | 1.307 |
| 2.500 | 2.347 | 0.6384 | 0.3666 | 42.1777 | 3.0000 | 18.88938 | 0.36047 | 61.29654 | 35.94470 | 44.5 | 1.589 |
| 2.750 | 2.575 | 0.6349 | 0.3645 | 42.4205 | 3.0000 | 22.70181 | 0.42603 | 65.50079 | 32.49891 | 41.8 | 1.946 |
| 3.000 | 2.779 | 0.6333 | 0.3636 | 47.4809 | 3.0000 | 26.46990 | 0.49276 | 74.48339 | 31.95133 | 39.2 | 2.399 |
| 3.250 | 3.004 | 0.6299 | 0.3616 | 32.0620 | 3.0000 | 31.16118 | 0.57063 | 63.90387 | 23.67202 | 37.0 | 2.972 |
| 3.500 | 3.202 | 0.6282 | 0.3607 | 21.8110 | 3.0000 | 35.63029 | 0.64735 | 58.27559 | 19.02871 | 34.9 | 3.700 |
| 3.750 | 3.398 | 0.6265 | 0.3597 | 0.0000 | 0.0000 | 40.50430 | 0.72985 | 41.49025 | 12.01639 | 33.0 | 4.624 |
| 4.000 | 3.619 | 0.6232 | 0.3578 | 0.0000 | 0.0000 | 46.73727 | 0.82868 | 47.86176 | 12.20862 | 31.3 | 5.797 |
| 4.250 | 3.811 | 0.6215 | 0.3569 | 0.0000 | 0.0000 | 52.44987 | 0.92227 | 53.72103 | 12.31259 | 29.8 | 7.291 |
| 4.500 | 4.000 | 0.6199 | 0.3559 | 0.0000 | 0.0000 | 58.49450 | 1.02017 | 59.91071 | 12.41360 | 28.4 | 9.195 |
| 4.750 | 4.186 | 0.6182 | 0.3550 | 0.0000 | 0.0000 | 65.23133 | 1.02017 | 66.78466 | 12.50913 | 27.0 | 11.627 |
| 5.000 | 4.370 | 0.6166 | 0.3540 | 0.0000 | 0.0000 | 72.36751 | 1.12853 | 74.06779 | 12.60732 | 25.8 | 14.923 |
| 5.250 | 4.551 | 0.6150 | 0.3531 | 0.0000 | 0.0000 | 79.43965 | 1.24185 | 81.29244 | 12.70563 | 24.8 | 19.210 |
| 5.500 | 4.731 | 0.6134 | 0.3522 | 0.0000 | 0.0000 | 87.04407 | 1.35244 | 89.01157 | 12.80077 | 23.8 | 23.977 |
| 5.750 | 4.909 | 0.6117 | 0.3512 | 0.0000 | 0.0000 | 94.95415 | 1.46985 | 96.98746 | 12.89029 | 23.0 | 28.592 |
| *NaCl; $\xi = 0.97$; $a = 0.53835$* | | | | | | | | | | | |
| 0.001 | 0.001 | 0.5705 | 0.3854 | 0.0337 | 0.0242 | 0.00083 | 0.00000 | 0.00000 | 0.00000 | 78.1 | 0.966 |
| 0.002 | 0.002 | 0.5704 | 0.3854 | 0.0485 | 0.0337 | 0.00165 | 0.00000 | 0.00000 | 0.00000 | 78.1 | 0.951 |

| | | | | | | | | | | | |
|---|---|---|---|---|---|---|---|---|---|---|---|
| 0.003 | 0.003 | 0.5704 | 0.3854 | 0.0582 | 0.0422 | 0.00248 | 0.00000 | 0.00000 | 0.00000 | 78.1 | 0.941 |
| 0.004 | 0.004 | 0.5704 | 0.3854 | 0.0659 | 0.0497 | 0.00331 | 0.00000 | 0.00000 | 0.00000 | 78.1 | 0.933 |
| 0.005 | 0.005 | 0.5704 | 0.3854 | 0.0715 | 0.0574 | 0.00414 | 0.00000 | 0.00000 | 0.00000 | 78.0 | 0.927 |
| 0.006 | 0.006 | 0.5704 | 0.3853 | 0.0771 | 0.0638 | 0.00497 | 0.00000 | 0.00000 | 0.00000 | 78.0 | 0.921 |
| 0.007 | 0.007 | 0.5704 | 0.3853 | 0.0817 | 0.0704 | 0.00580 | 0.00000 | 0.00000 | 0.00000 | 78.0 | 0.916 |
| 0.008 | 0.008 | 0.5704 | 0.3853 | 0.0830 | 0.0792 | 0.00663 | 0.00000 | 0.00000 | 0.00000 | 78.0 | 0.914 |
| 0.009 | 0.009 | 0.5704 | 0.3853 | 0.0909 | 0.0815 | 0.00746 | 0.00000 | 0.00000 | 0.00000 | 78.0 | 0.906 |
| 0.010 | 0.010 | 0.5704 | 0.3853 | 0.1068 | 0.0771 | 0.00829 | 0.00000 | 0.00000 | 0.00000 | 78.0 | 0.891 |
| 0.020 | 0.020 | 0.5702 | 0.3852 | 0.1219 | 0.1358 | 0.01664 | 0.00000 | 0.00000 | 0.00000 | 77.8 | 0.870 |
| 0.030 | 0.030 | 0.5701 | 0.3852 | 0.1340 | 0.1867 | 0.02505 | 0.00000 | 0.00000 | 0.00000 | 77.6 | 0.852 |
| 0.040 | 0.040 | 0.5700 | 0.3851 | 0.1442 | 0.2328 | 0.03351 | 0.00000 | 0.00000 | 0.00000 | 77.4 | 0.836 |
| 0.050 | 0.050 | 0.5699 | 0.3850 | 0.1510 | 0.2802 | 0.04204 | 0.00000 | 0.00000 | 0.00000 | 77.2 | 0.823 |
| 0.060 | 0.060 | 0.5698 | 0.3849 | 0.1542 | 0.3310 | 0.05062 | 0.00000 | 0.00000 | 0.00000 | 77.0 | 0.813 |
| 0.070 | 0.070 | 0.5697 | 0.3849 | 0.1564 | 0.3839 | 0.05928 | 0.00000 | 0.00000 | 0.00000 | 76.8 | 0.804 |
| 0.080 | 0.079 | 0.5720 | 0.3864 | 0.1571 | 0.4641 | 0.06569 | 0.00000 | 0.00000 | 0.00000 | 76.6 | 0.798 |
| 0.090 | 0.090 | 0.5695 | 0.3847 | 0.1583 | 0.4931 | 0.07671 | 0.00000 | 0.00000 | 0.00000 | 76.4 | 0.788 |
| 0.100 | 0.100 | 0.5694 | 0.3846 | 0.1735 | 0.5038 | 0.08555 | 0.00000 | 0.00000 | 0.00000 | 76.2 | 0.769 |
| 0.200 | 0.199 | 0.5692 | 0.3845 | 0.1166 | 1.6361 | 0.17464 | 0.00000 | 0.00000 | 0.00000 | 74.4 | 0.742 |
| 0.300 | 0.298 | 0.5684 | 0.3840 | 0.1533 | 2.0000 | 0.26962 | 0.01434 | 0.08666 | 1.27751 | 72.7 | 0.717 |
| 0.400 | 0.395 | 0.5685 | 0.3840 | 0.2207 | 2.0000 | 0.36465 | 0.01940 | 0.22074 | 2.40565 | 71.2 | 0.698 |
| 0.500 | 0.495 | 0.5669 | 0.3830 | 0.2813 | 2.0000 | 0.47110 | 0.02485 | 0.36561 | 3.11055 | 70.0 | 0.684 |
| 0.600 | 0.590 | 0.5671 | 0.3831 | 0.3608 | 2.0000 | 0.56990 | 0.03009 | 0.52748 | 3.70559 | 68.9 | 0.674 |
| 0.700 | 0.690 | 0.5656 | 0.3821 | 0.4285 | 2.0000 | 0.68688 | 0.03596 | 0.70088 | 4.11941 | 67.8 | 0.668 |
| 0.800 | 0.784 | 0.5656 | 0.3821 | 0.5262 | 2.0000 | 0.79518 | 0.04164 | 0.89715 | 4.55384 | 66.5 | 0.663 |
| 0.900 | 0.883 | 0.5644 | 0.3813 | 0.6135 | 2.0000 | 0.92385 | 0.04805 | 1.10416 | 4.85771 | 65.2 | 0.659 |
| 1.000 | 0.983 | 0.5630 | 0.3803 | 0.7006 | 2.0000 | 1.06191 | 0.05480 | 1.32299 | 5.10310 | 64.0 | 0.657 |
| 1.250 | 1.220 | 0.5617 | 0.3795 | 0.9783 | 2.0000 | 1.38922 | 0.07120 | 1.91879 | 5.69616 | 61.4 | 0.656 |
| 1.500 | 1.455 | 0.5604 | 0.3786 | 1.2793 | 2.0000 | 1.74142 | 0.08858 | 2.56621 | 6.12402 | 59.1 | 0.658 |

*(Continued)*

**TABLE 3.9 (*Continued*)**
**Calculated Values of Thermodynamic Quantities Included in Formulae (4.84) and (4.85)**

| $m$ | $c_k$ | $R$ (nm) | $R_{H_2O}$ (nm) | $-\Psi_\gamma^0$ | $\chi$ | $-E_{sp}$ | $E_H$ | $U_{ot}$ | $B$ | $\varepsilon_S$ | $\gamma_\pm$ |
|---|---|---|---|---|---|---|---|---|---|---|---|
| 1.750 | 1.687 | 0.5590 | 0.3777 | 1.6334 | 2.0000 | 2.12472 | 0.10727 | 3.30254 | 6.50753 | 56.9 | 0.664 |
| 2.000 | 1.916 | 0.5577 | 0.3768 | 2.0335 | 2.0000 | 2.53680 | 0.12716 | 4.11520 | 6.84070 | 54.7 | 0.672 |
| 2.250 | 2.157 | 0.5552 | 0.3751 | 2.3205 | 2.0000 | 3.03285 | 0.14988 | 4.89866 | 6.90883 | 52.8 | 0.682 |
| 2.500 | 2.383 | 0.5539 | 0.3742 | 2.7498 | 2.0000 | 3.51915 | 0.17263 | 5.81437 | 7.11961 | 50.8 | 0.692 |
| 2.750 | 2.606 | 0.5526 | 0.3733 | 3.1981 | 2.0000 | 4.03467 | 0.19652 | 6.77976 | 7.29247 | 49.1 | 0.704 |
| 3.000 | 2.827 | 0.5514 | 0.3725 | 3.6701 | 2.0000 | 4.59312 | 0.22212 | 7.81249 | 7.43474 | 47.3 | 0.718 |
| 3.250 | 3.046 | 0.5501 | 0.3716 | 4.1271 | 2.0000 | 5.18883 | 0.24913 | 8.86695 | 7.52348 | 45.7 | 0.733 |
| 3.500 | 3.263 | 0.5488 | 0.3708 | 4.5460 | 2.0000 | 5.82258 | 0.27755 | 9.92075 | 7.55557 | 44.1 | 0.749 |
| 3.750 | 3.478 | 0.5476 | 0.3699 | 4.8860 | 2.0000 | 6.49114 | 0.30721 | 10.93144 | 7.52145 | 42.7 | 0.767 |
| 4.000 | 3.691 | 0.5463 | 0.3691 | 5.1494 | 2.0000 | 7.20735 | 0.33870 | 11.91188 | 7.43415 | 41.3 | 0.786 |
| 4.250 | 3.902 | 0.5451 | 0.3683 | 5.2585 | 2.0000 | 7.95982 | 0.37144 | 12.77497 | 7.26994 | 39.9 | 0.807 |
| 4.500 | 4.112 | 0.5439 | 0.3674 | 5.1437 | 2.0000 | 8.76152 | 0.40593 | 13.46203 | 7.01011 | 38.7 | 0.829 |
| 4.750 | 4.320 | 0.5426 | 0.3666 | 4.7642 | 2.0000 | 9.60307 | 0.44179 | 13.92449 | 6.66239 | 37.5 | 0.853 |
| 5.000 | 4.526 | 0.5414 | 0.3658 | 4.0718 | 2.0000 | 10.49574 | 0.47952 | 14.12360 | 6.22596 | 36.4 | 0.878 |
| 5.250 | 4.700 | 0.5414 | 0.3657 | 4.7279 | 2.0000 | 11.24031 | 0.51344 | 15.52705 | 6.39242 | 35.3 | 0.904 |
| 5.500 | 4.902 | 0.5402 | 0.3649 | 3.3597 | 2.0000 | 12.21123 | 0.55392 | 15.12309 | 5.77108 | 34.3 | 0.928 |
| 5.750 | 5.103 | 0.5390 | 0.3641 | 1.3401 | 2.0000 | 13.23419 | 0.59613 | 14.11518 | 5.00508 | 33.3 | 0.950 |
| 6.000 | 5.303 | 0.5378 | 0.3633 | 0.0000 | 0.0000 | 14.30378 | 0.63978 | 13.83440 | 4.57079 | 32.4 | 0.975 |
| 6.144 | 5.397 | 0.5377 | 0.3633 | 0.0000 | 0.0000 | 14.78888 | 0.66144 | 14.33340 | 4.58064 | 31.9 | 1.006 |
| *$NaNO_3$; $\xi = 1.03$; $a = 0.57989$* | | | | | | | | | | | |
| 0.001 | 0.001 | 0.6463 | 0.3854 | 0.0361 | 0.0377 | 0.00041 | 0.00000 | 0.00000 | 0.00000 | 78.0 | 0.964 |
| 0.002 | 0.002 | 0.6463 | 0.3854 | 0.0481 | 0.0565 | 0.00081 | 0.00000 | 0.00000 | 0.00000 | 78.0 | 0.952 |
| 0.003 | 0.003 | 0.6463 | 0.3854 | 0.0583 | 0.0699 | 0.00122 | 0.00000 | 0.00000 | 0.00000 | 78.0 | 0.942 |

| | | | | | | | | | | | |
|---|---|---|---|---|---|---|---|---|---|---|---|
| 0.004 | 0.004 | 0.6463 | 0.3853 | 0.0668 | 0.0814 | 0.00162 | 0.00000 | 0.00000 | 0.00000 | 78.0 | 0.933 |
| 0.005 | 0.005 | 0.6462 | 0.3853 | 0.0742 | 0.0915 | 0.00203 | 0.00000 | 0.00000 | 0.00000 | 78.0 | 0.926 |
| 0.006 | 0.006 | 0.6462 | 0.3853 | 0.0807 | 0.1010 | 0.00243 | 0.00000 | 0.00000 | 0.00000 | 78.0 | 0.920 |
| 0.007 | 0.007 | 0.6462 | 0.3853 | 0.0855 | 0.1112 | 0.00284 | 0.00000 | 0.00000 | 0.00000 | 78.0 | 0.915 |
| 0.008 | 0.008 | 0.6462 | 0.3853 | 0.0900 | 0.1207 | 0.00324 | 0.00000 | 0.00000 | 0.00000 | 78.0 | 0.910 |
| 0.009 | 0.009 | 0.6462 | 0.3853 | 0.0987 | 0.1238 | 0.00365 | 0.00000 | 0.00000 | 0.00000 | 78.0 | 0.902 |
| 0.010 | 0.010 | 0.6461 | 0.3853 | 0.1155 | 0.1175 | 0.00406 | 0.00000 | 0.00000 | 0.00000 | 78.0 | 0.887 |
| 0.020 | 0.020 | 0.6460 | 0.3852 | 0.1342 | 0.2022 | 0.00813 | 0.00000 | 0.00000 | 0.00000 | 78.0 | 0.866 |
| 0.030 | 0.030 | 0.6458 | 0.3851 | 0.1538 | 0.2640 | 0.01221 | 0.00000 | 0.00000 | 0.00000 | 78.0 | 0.846 |
| 0.040 | 0.040 | 0.6456 | 0.3850 | 0.1716 | 0.3155 | 0.01632 | 0.00000 | 0.00000 | 0.00000 | 77.9 | 0.827 |
| 0.050 | 0.050 | 0.6454 | 0.3848 | 0.1864 | 0.3622 | 0.02043 | 0.00000 | 0.00000 | 0.00000 | 77.9 | 0.811 |
| 0.060 | 0.060 | 0.6452 | 0.3847 | 0.1987 | 0.4076 | 0.02456 | 0.00000 | 0.00000 | 0.00000 | 77.9 | 0.798 |
| 0.070 | 0.070 | 0.6450 | 0.3846 | 0.2068 | 0.4558 | 0.02870 | 0.00000 | 0.00000 | 0.00000 | 77.9 | 0.788 |
| 0.080 | 0.080 | 0.6449 | 0.3845 | 0.2139 | 0.5035 | 0.03286 | 0.00000 | 0.00000 | 0.00000 | 77.8 | 0.779 |
| 0.090 | 0.090 | 0.6447 | 0.3844 | 0.2293 | 0.5271 | 0.03702 | 0.00000 | 0.00000 | 0.00000 | 77.8 | 0.763 |
| 0.100 | 0.100 | 0.6445 | 0.3843 | 0.2604 | 0.5156 | 0.04122 | 0.00000 | 0.00000 | 0.00000 | 77.8 | 0.737 |
| 0.200 | 0.199 | 0.6438 | 0.3839 | 0.2621 | 1.0177 | 0.08267 | 0.00000 | 0.00000 | 0.00000 | 77.6 | 0.703 |
| 0.300 | 0.296 | 0.6438 | 0.3839 | 0.2647 | 1.5521 | 0.12401 | 0.00000 | 0.00000 | 0.00000 | 77.0 | 0.670 |
| 0.400 | 0.393 | 0.6429 | 0.3834 | 0.2834 | 2.0000 | 0.16789 | 0.01552 | 0.00801 | 0.10911 | 76.0 | 0.642 |
| 0.500 | 0.491 | 0.6413 | 0.3824 | 0.3637 | 2.0000 | 0.21521 | 0.01973 | 0.09887 | 1.05907 | 75.1 | 0.619 |
| 0.600 | 0.588 | 0.6400 | 0.3816 | 0.4491 | 2.0000 | 0.26366 | 0.02402 | 0.20146 | 1.77291 | 74.2 | 0.601 |
| 0.700 | 0.685 | 0.6386 | 0.3808 | 0.5369 | 2.0000 | 0.31431 | 0.02843 | 0.31381 | 2.33300 | 73.3 | 0.586 |
| 0.800 | 0.777 | 0.6385 | 0.3807 | 0.6366 | 2.0000 | 0.36093 | 0.03263 | 0.43702 | 2.83078 | 72.5 | 0.573 |
| 0.900 | 0.873 | 0.6371 | 0.3799 | 0.7315 | 2.0000 | 0.41470 | 0.03723 | 0.56123 | 3.18651 | 71.7 | 0.560 |
| 1.000 | 0.969 | 0.6356 | 0.3790 | 0.8304 | 2.0000 | 0.47104 | 0.04198 | 0.68738 | 3.46117 | 70.9 | 0.544 |
| 1.250 | 1.204 | 0.6328 | 0.3773 | 1.0906 | 2.0000 | 0.61432 | 0.05397 | 1.05542 | 4.13359 | 69.1 | 0.527 |
| 1.500 | 1.427 | 0.6314 | 0.3765 | 1.3687 | 2.0000 | 0.75153 | 0.06556 | 1.43497 | 4.62669 | 67.8 | 0.510 |
| 1.750 | 1.657 | 0.6285 | 0.3747 | 1.6251 | 2.0000 | 0.91071 | 0.07827 | 1.81508 | 4.90199 | 66.5 | 0.494 |

*(Continued)*

**TABLE 3.9 (*Continued*)**
**Calculated Values of Thermodynamic Quantities Included in Formulae (4.84) and (4.85)**

| $m$ | $c_k$ | $R$ (nm) | $R_{H_2O}$ (nm) | $-\Psi_\gamma^0$ | $\chi$ | $-E_{sp}$ | $E_H$ | $U_{ot}$ | $B$ | $\varepsilon_S$ | $\gamma_\pm$ |
|---|---|---|---|---|---|---|---|---|---|---|---|
| 2.000 | 1.885 | 0.6256 | 0.3730 | 1.8769 | 2.0000 | 1.07947 | 0.09142 | 2.20219 | 5.09183 | 65.4 | 0.480 |
| 2.250 | 2.097 | 0.6242 | 0.3722 | 2.1655 | 2.0000 | 1.23489 | 0.10383 | 2.61609 | 5.32613 | 64.3 | 0.468 |
| 2.500 | 2.320 | 0.6214 | 0.3705 | 2.4089 | 2.0000 | 1.42072 | 0.11775 | 3.01591 | 5.41402 | 63.3 | 0.457 |
| 2.750 | 2.526 | 0.6199 | 0.3697 | 2.6892 | 2.0000 | 1.58888 | 0.13069 | 3.43775 | 5.56045 | 62.4 | 0.447 |
| 3.000 | 2.727 | 0.6186 | 0.3688 | 3.1043 | 2.0000 | 1.77843 | 0.14524 | 4.01575 | 5.84444 | 60.9 | 0.438 |
| 3.250 | 2.946 | 0.6157 | 0.3671 | 3.8959 | 2.0000 | 2.07999 | 0.16734 | 5.07601 | 6.41194 | 57.6 | 0.430 |
| 3.500 | 3.142 | 0.6143 | 0.3663 | 4.9083 | 2.0000 | 2.36168 | 0.18862 | 6.33924 | 7.10426 | 54.8 | 0.422 |
| 3.750 | 3.335 | 0.6129 | 0.3654 | 6.0415 | 2.0000 | 2.65783 | 0.21071 | 7.73800 | 7.76249 | 52.3 | 0.415 |
| 4.000 | 3.525 | 0.6115 | 0.3646 | 7.2458 | 2.0000 | 2.96360 | 0.23323 | 9.21795 | 8.35428 | 50.2 | 0.409 |
| 4.250 | 3.739 | 0.6086 | 0.3629 | 8.2169 | 2.0000 | 3.34820 | 0.25956 | 10.54049 | 8.58403 | 48.3 | 0.403 |
| 4.500 | 3.925 | 0.6072 | 0.3621 | 9.4990 | 2.0000 | 3.68631 | 0.28366 | 12.13062 | 9.03945 | 46.6 | 0.398 |
| 4.750 | 4.109 | 0.6058 | 0.3612 | 10.7971 | 2.0000 | 4.03740 | 0.30835 | 13.74967 | 9.42554 | 45.1 | 0.392 |
| 5.000 | 4.290 | 0.6044 | 0.3604 | 12.1407 | 2.0000 | 4.40159 | 0.33371 | 15.42743 | 9.77216 | 43.7 | 0.388 |
| 5.250 | 4.469 | 0.6031 | 0.3596 | 12.4464 | 2.0000 | 4.68569 | 0.35263 | 15.99407 | 9.58757 | 43.2 | 0.383 |
| 5.500 | 4.647 | 0.6016 | 0.3587 | 12.4507 | 2.0000 | 4.95968 | 0.37040 | 16.25085 | 9.27410 | 43.0 | 0.379 |
| 5.750 | 4.822 | 0.6002 | 0.3579 | 12.3888 | 2.0000 | 5.23500 | 0.38807 | 16.44348 | 8.95662 | 42.8 | 0.375 |
| 6.000 | 4.996 | 0.5988 | 0.3571 | 12.2454 | 2.0000 | 5.52028 | 0.40612 | 16.56453 | 8.62153 | 42.6 | 0.371 |
| 6.250 | 5.168 | 0.5974 | 0.3562 | 12.1734 | 2.0000 | 5.82874 | 0.42560 | 16.77917 | 8.33363 | 42.2 | 0.367 |
| 6.500 | 5.338 | 0.5961 | 0.3554 | 12.0699 | 2.0000 | 6.15085 | 0.44578 | 16.97614 | 8.04970 | 41.8 | 0.364 |
| 6.750 | 5.468 | 0.5961 | 0.3554 | 12.9411 | 2.0000 | 6.37043 | 0.46174 | 18.04889 | 8.26263 | 41.4 | 0.361 |
| 7.000 | 5.634 | 0.5947 | 0.3546 | 12.8886 | 2.0000 | 6.73019 | 0.48418 | 18.33175 | 8.00321 | 40.8 | 0.357 |
| 7.250 | 5.799 | 0.5933 | 0.3538 | 12.6958 | 2.0000 | 7.10243 | 0.50711 | 18.48696 | 7.70592 | 40.3 | 0.354 |
| 7.500 | 5.963 | 0.5919 | 0.3529 | 12.3091 | 2.0000 | 7.48141 | 0.53013 | 18.45656 | 7.35921 | 39.8 | 0.352 |

| | | | | | | | | | | | |
|---|---|---|---|---|---|---|---|---|---|---|---|
| 7.750 | 6.125 | 0.5906 | 0.3521 | 11.7760 | 2.0000 | 7.86706 | 0.55332 | 18.28615 | 6.98565 | 39.4 | 0.350 |
| 8.000 | 6.243 | 0.5905 | 0.3521 | 12.5299 | 2.0000 | 8.10839 | 0.57026 | 19.26406 | 7.14067 | 39.0 | 0.347 |
| 8.250 | 6.402 | 0.5892 | 0.3513 | 11.7413 | 2.0000 | 8.50857 | 0.59394 | 18.85134 | 6.70910 | 38.5 | 0.344 |
| 8.500 | 6.561 | 0.5878 | 0.3505 | 10.7080 | 2.0000 | 8.92200 | 0.61805 | 18.20734 | 6.22708 | 38.1 | 0.342 |
| 8.750 | 6.718 | 0.5864 | 0.3497 | 9.4745 | 2.0000 | 9.33651 | 0.64196 | 17.36454 | 5.71769 | 37.8 | 0.340 |
| 9.000 | 6.827 | 0.5864 | 0.3497 | 9.0039 | 2.0000 | 9.58241 | 0.65884 | 18.12299 | 5.81453 | 37.4 | 0.337 |
| 9.250 | 6.982 | 0.5851 | 0.3489 | 8.4272 | 2.0000 | 10.01546 | 0.68342 | 16.95501 | 5.24415 | 37.0 | 0.335 |
| 9.500 | 7.136 | 0.5837 | 0.3480 | 6.5934 | 2.0000 | 10.45221 | 0.70787 | 15.53396 | 4.63864 | 36.7 | 0.333 |
| 9.750 | 7.238 | 0.5837 | 0.3481 | 6.9554 | 2.0000 | 10.69156 | 0.72416 | 16.11940 | 4.70522 | 36.4 | 0.331 |
| 10.000 | 7.390 | 0.5823 | 0.3472 | 4.7468 | 2.0000 | 11.14811 | 0.74938 | 14.34241 | 4.04561 | 36.1 | 0.329 |
| 10.250 | 7.542 | 0.5810 | 0.3464 | 2.2126 | 2.0000 | 11.61163 | 0.77459 | 12.24680 | 3.34208 | 35.8 | 0.327 |
| 10.500 | 7.639 | 0.5810 | 0.3464 | 2.3243 | 2.0000 | 11.85411 | 0.79079 | 12.58547 | 3.36415 | 35.5 | 0.325 |
| 10.750 | 7.788 | 0.5796 | 0.3456 | 0.0000 | 0.0000 | 12.32969 | 0.81636 | 10.71308 | 2.77395 | 35.2 | 0.323 |
| 10.830 | 7.818 | 0.5796 | 0.3456 | 0.0000 | 0.0000 | 12.40809 | 0.82157 | 10.78463 | 2.77476 | 35.2 | 0.322 |
| *$NiSO_4$; ξ = 1.18; a = 0.6844* | | | | | | | | | | | |
| 0.001 | 0.001 | 0.76 69 | 0.3854 | 0.2885 | 1.0725 | 0.00184 | 0.0000 | 0.00000 | 0.00000 | 77.9 | 0.748 |
| 0.002 | 0.002 | 0.7669 | 0.3854 | 0.3416 | 1.8110 | 0.00367 | 0.0000 | 0.00000 | 0.00000 | 77.9 | 0.708 |
| 0.003 | 0.003 | 0.7669 | 0.3853 | 0.4638 | 2.0000 | 0.00551 | 0.0001 | 0.06574 | 78.50597 | 77.9 | 0.668 |
| 0.004 | 0.004 | 0.7668 | 0.3853 | 0.6181 | 2.0000 | 0.00735 | 0.0002 | 0.15858 | 142.01173 | 77.9 | 0.627 |
| 0.005 | 0.005 | 0.7668 | 0.3853 | 0.7743 | 2.0000 | 0.00919 | 0.0003 | 0.25571 | 183.07050 | 77.9 | 0.590 |
| 0.006 | 0.006 | 0.7668 | 0.3853 | 0.9288 | 2.0000 | 0.01103 | 0.0003 | 0.35268 | 210.39442 | 77.9 | 0.556 |
| 0.007 | 0.007 | 0.7667 | 0.3853 | 1.0832 | 2.0000 | 0.01288 | 0.0004 | 0.44962 | 229.88414 | 77.9 | 0.524 |
| 0.008 | 0.008 | 0.7667 | 0.3852 | 1.2375 | 2.0000 | 0.01472 | 0.0004 | 0.54474 | 243.67912 | 77.9 | 0.493 |
| 0.009 | 0.009 | 0.7667 | 0.3852 | 1.3916 | 2.0000 | 0.01656 | 0.0005 | 0.63362 | 251.91983 | 77.9 | 0.461 |
| 0.010 | 0.010 | 0.7666 | 0.3852 | 1.5457 | 2.0000 | 0.01841 | 0.0005 | 0.71522 | 255.90279 | 77.9 | 0.428 |
| 0.020 | 0.020 | 0.7663 | 0.3850 | 3.0950 | 2.0000 | 0.03695 | 0.0011 | 2.19753 | 392.28418 | 77.8 | 0.393 |
| 0.030 | 0.030 | 0.7659 | 0.3848 | 4.6366 | 2.0000 | 0.05559 | 0.0017 | 3.65864 | 434.73038 | 77.7 | 0.356 |
| 0.040 | 0.040 | 0.7655 | 0.3847 | 6.1892 | 2.0000 | 0.07439 | 0.0023 | 5.12320 | 455.58005 | 77.6 | 0.320 |

*(Continued)*

**TABLE 3.9 (*Continued*)**
**Calculated Values of Thermodynamic Quantities Included in Formulae (4.84) and (4.85)**

| $m$ | $c_k$ | $R$ (nm) | $R_{H_2O}$ (nm) | $-\Psi_\gamma^0$ | $\chi$ | $-E_{sp}$ | $E_H$ | $U_{ot}$ | $B$ | $\varepsilon_S$ | $\gamma_\pm$ |
|---|---|---|---|---|---|---|---|---|---|---|---|
| 0.050 | 0.050 | 0.7652 | 0.3845 | 7.7264 | 2.0000 | 0.09327 | 0.0029 | 6.58060 | 467.41882 | 77.6 | 0.290 |
| 0.060 | 0.060 | 0.7648 | 0.3843 | 9.2820 | 2.0000 | 0.11233 | 0.0035 | 8.07238 | 476.78784 | 77.5 | 0.267 |
| 0.070 | 0.070 | 0.7645 | 0.3841 | 10.8145 | 2.0000 | 0.13144 | 0.0041 | 9.54187 | 482.32471 | 77.4 | 0.246 |
| 0.080 | 0.080 | 0.7641 | 0.3839 | 12.3728 | 2.0000 | 0.15076 | 0.0047 | 11.06061 | 488.15680 | 77.4 | 0.232 |
| 0.090 | 0.089 | 0.7666 | 0.3852 | 14.1599 | 2.0000 | 0.16507 | 0.0053 | 12.75744 | 509.10718 | 77.3 | 0.209 |
| 0.100 | 0.100 | 0.7634 | 0.3836 | 15.4610 | 2.0000 | 0.18968 | 0.0060 | 13.90533 | 489.15903 | 77.2 | 0.175 |
| 0.200 | 0.200 | 0.7599 | 0.3818 | 30.8420 | 2.0000 | 0.39191 | 0.0122 | 29.19632 | 504.21524 | 76.5 | 0.131 |
| 0.300 | 0.298 | 0.7581 | 0.3809 | 46.6729 | 2.0000 | 0.59624 | 0.0184 | 44.94874 | 513.91956 | 75.8 | 0.099 |
| 0.400 | 0.395 | 0.7562 | 0.3799 | 62.6633 | 2.0000 | 0.80773 | 0.0248 | 60.90946 | 518.10095 | 75.2 | 0.078 |
| 0.500 | 0.491 | 0.7542 | 0.3790 | 78.7916 | 2.0000 | 1.02629 | 0.0313 | 77.07101 | 520.07660 | 74.5 | 0.065 |
| 0.600 | 0.597 | 0.7476 | 0.3757 | 91.0332 | 2.0000 | 1.31606 | 0.0391 | 89.46686 | 483.65533 | 73.9 | 0.057 |
| 0.700 | 0.688 | 0.7474 | 0.3756 | 109.8296 | 2.0000 | 1.53699 | 0.0456 | 108.38937 | 502.09875 | 73.0 | 0.052 |
| 0.800 | 0.784 | 0.7450 | 0.3743 | 126.4785 | 2.0000 | 1.80146 | 0.0529 | 125.21887 | 499.95575 | 72.1 | 0.048 |
| 0.900 | 0.876 | 0.7435 | 0.3736 | 144.7052 | 2.0000 | 2.05584 | 0.0600 | 143.63179 | 505.53061 | 71.3 | 0.045 |
| 1.000 | 0.973 | 0.7405 | 0.3721 | 160.8129 | 2.0000 | 2.35834 | 0.0680 | 159.96848 | 496.96826 | 70.5 | 0.042 |
| 1.250 | 1.241 | 0.7281 | 0.3658 | 181.8813 | 2.0000 | 3.36452 | 0.0921 | 181.95351 | 417.37628 | 68.7 | 0.039 |
| 1.500 | 1.489 | 0.7209 | 0.3622 | 204.7389 | 2.0000 | 4.35141 | 0.1156 | 205.73051 | 376.12158 | 67.0 | 0.037 |
| 1.750 | 1.721 | 0.7163 | 0.3599 | 236.8952 | 2.0000 | 5.36739 | 0.1397 | 238.83151 | 361.14005 | 64.9 | 0.035 |
| 2.250 | 2.201 | 0.7044 | 0.3540 | 273.4643 | 2.0000 | 8.12814 | 0.2011 | 278.11688 | 292.33069 | 59.6 | 0.035 |
| 2.500 | 2.428 | 0.6999 | 0.3517 | 282.9312 | 2.0000 | 9.69817 | 0.2352 | 289.15619 | 259.84247 | 57.0 | 0.036 |

| *RbCl; $\xi$ = 1.02; a = 0.61812* | | | | | | | | | | | |
|---|---|---|---|---|---|---|---|---|---|---|---|
| 0.010 | 0.010 | 0.7266 | 0.3852 | 0.0620 | 0.2908 | 0.00436 | 0.00000 | 0.00000 | 0.00000 | 78.0 | 0.935 |
| 0.020 | 0.020 | 0.7275 | 0.3857 | 0.1046 | 0.3449 | 0.00862 | 0.00000 | 0.00000 | 0.00000 | 78.0 | 0.892 |
| 0.030 | 0.030 | 0.7277 | 0.3858 | 0.1333 | 0.4058 | 0.01289 | 0.00000 | 0.00000 | 0.00000 | 78.0 | 0.863 |
| 0.040 | 0.040 | 0.7276 | 0.3857 | 0.1496 | 0.4814 | 0.01719 | 0.00000 | 0.00000 | 0.00000 | 78.0 | 0.845 |
| 0.050 | 0.050 | 0.7250 | 0.3844 | 0.1617 | 0.5515 | 0.02210 | 0.00000 | 0.00000 | 0.00000 | 78.0 | 0.830 |
| 0.060 | 0.060 | 0.7252 | 0.3845 | 0.1724 | 0.6203 | 0.02643 | 0.00000 | 0.00000 | 0.00000 | 78.0 | 0.817 |
| 0.070 | 0.070 | 0.7249 | 0.3843 | 0.1829 | 0.6809 | 0.03090 | 0.00000 | 0.00000 | 0.00000 | 78.0 | 0.805 |
| 0.080 | 0.080 | 0.7249 | 0.3843 | 0.1896 | 0.7495 | 0.03526 | 0.00000 | 0.00000 | 0.00000 | 78.0 | 0.796 |
| 0.090 | 0.090 | 0.7251 | 0.3844 | 0.2021 | 0.7923 | 0.03954 | 0.00000 | 0.00000 | 0.00000 | 78.0 | 0.782 |
| 0.100 | 0.100 | 0.7250 | 0.3844 | 0.2264 | 0.7847 | 0.04395 | 0.00000 | 0.00000 | 0.00000 | 78.0 | 0.760 |
| 0.200 | 0.199 | 0.7226 | 0.3831 | 0.2212 | 1.5929 | 0.08950 | 0.00000 | 0.00000 | 0.00000 | 77.8 | 0.727 |
| 0.300 | 0.301 | 0.7179 | 0.3806 | 0.2666 | 2.0000 | 0.14138 | 0.01502 | 0.04258 | 0.59926 | 77.1 | 0.696 |
| 0.400 | 0.400 | 0.7156 | 0.3794 | 0.3614 | 2.0000 | 0.19296 | 0.02029 | 0.14358 | 1.49578 | 76.3 | 0.667 |
| 0.500 | 0.499 | 0.7135 | 0.3783 | 0.4584 | 2.0000 | 0.24649 | 0.02567 | 0.26664 | 2.19561 | 75.7 | 0.650 |
| 0.600 | 0.597 | 0.7114 | 0.3772 | 0.5570 | 2.0000 | 0.30206 | 0.03116 | 0.39404 | 2.67301 | 75.1 | 0.634 |
| 0.700 | 0.694 | 0.7094 | 0.3761 | 0.6590 | 2.0000 | 0.35965 | 0.03677 | 0.53167 | 3.05666 | 74.4 | 0.621 |
| 0.800 | 0.791 | 0.7075 | 0.3751 | 0.7609 | 2.0000 | 0.41878 | 0.04244 | 0.67401 | 3.35708 | 73.9 | 0.611 |
| 0.900 | 0.887 | 0.7057 | 0.3741 | 0.8662 | 2.0000 | 0.47993 | 0.04823 | 0.82252 | 3.60514 | 73.3 | 0.602 |
| 1.000 | 0.983 | 0.7039 | 0.3732 | 0.9771 | 2.0000 | 0.54342 | 0.05416 | 0.97900 | 3.82111 | 72.6 | 0.592 |
| 1.200 | 1.171 | 0.7005 | 0.3714 | 1.2078 | 2.0000 | 0.67546 | 0.06629 | 1.32027 | 4.21025 | 71.4 | 0.583 |
| 1.400 | 1.347 | 0.6990 | 0.3706 | 1.4639 | 2.0000 | 0.79963 | 0.07790 | 1.68055 | 4.56005 | 70.2 | 0.574 |
| 1.600 | 1.522 | 0.6968 | 0.3694 | 1.7294 | 2.0000 | 0.93441 | 0.09011 | 2.06129 | 4.83561 | 69.0 | 0.566 |
| 1.800 | 1.707 | 0.6929 | 0.3674 | 1.9855 | 2.0000 | 1.09699 | 0.10387 | 2.46147 | 5.00905 | 67.9 | 0.559 |
| 2.000 | 1.875 | 0.6910 | 0.3663 | 2.2695 | 2.0000 | 1.24205 | 0.11654 | 2.87359 | 5.21201 | 66.9 | 0.553 |

*(Continued)*

**TABLE 3.9 (*Continued*)**
**Calculated Values of Thermodynamic Quantities Included in Formulae (4.84) and (4.85)**

| $m$ | $c_k$ | $R$ (nm) | $R_{H_2O}$ (nm) | $-\Psi_\gamma^0$ | $\chi$ | $-E_{sp}$ | $E_H$ | $U_{ot}$ | $B$ | $\varepsilon_S$ | $\gamma_\pm$ |
|---|---|---|---|---|---|---|---|---|---|---|---|
| 2.500 | 2.312 | 0.6832 | 0.3622 | 3.5169 | 2.0000 | 1.76257 | 0.15940 | 4.60630 | 6.10828 | 61.7 | 0.548 |
| 3.000 | 2.744 | 0.6755 | 0.3581 | 5.0080 | 2.0000 | 2.38833 | 0.20813 | 6.68516 | 6.78965 | 57.3 | 0.545 |
| 3.500 | 3.145 | 0.6697 | 0.3551 | 6.6538 | 2.0000 | 3.04177 | 0.25774 | 8.94909 | 7.33929 | 54.0 | 0.544 |
| 4.000 | 3.538 | 0.6640 | 0.3520 | 8.2729 | 2.0000 | 3.76867 | 0.31048 | 11.26008 | 7.66603 | 51.3 | 0.545 |
| 4.500 | 3.891 | 0.6602 | 0.3500 | 10.0600 | 2.0000 | 4.46712 | 0.36116 | 13.71513 | 8.02720 | 49.0 | 0.548 |
| 5.000 | 4.269 | 0.6545 | 0.3470 | 11.2568 | 2.0000 | 5.32859 | 0.41878 | 15.73804 | 7.94376 | 47.2 | 0.552 |
| 5.500 | 4.605 | 0.6507 | 0.3450 | 12.5412 | 2.0000 | 6.12479 | 0.47234 | 17.78873 | 7.96071 | 45.7 | 0.556 |
| 6.000 | 4.932 | 0.6470 | 0.3430 | 13.8741 | 2.0000 | 7.01378 | 0.53075 | 19.97664 | 7.95603 | 44.0 | 0.562 |
| 6.500 | 5.252 | 0.6433 | 0.3410 | 14.8208 | 2.0000 | 7.95897 | 0.59098 | 21.83327 | 7.80931 | 42.6 | 0.567 |
| 7.000 | 5.566 | 0.6396 | 0.3391 | 15.2247 | 2.0000 | 8.95914 | 0.65275 | 23.19987 | 7.51286 | 41.4 | 0.573 |
| 7.500 | 5.875 | 0.6359 | 0.3371 | 14.9779 | 2.0000 | 10.01888 | 0.71623 | 23.97261 | 7.07498 | 40.2 | 0.578 |
| 7.832 | 6.061 | 0.6341 | 0.3362 | 14.9179 | 2.0000 | 10.67717 | 0.75608 | 24.54766 | 6.86283 | 39.6 | 0.582 |

*Notes:* $m$, in mol/kg $H_2O$; $c_k$, in mol/dm$^3$.

Value $A$ can be found based on the following dependence at temperature from 0°C to 100°C:

$$A = A_0 + A_1 \cdot T^{*1.5} + A_2 \cdot T^{*3} + A_3 \cdot T^{*3.5}, \tag{3.56}$$

where

$A_0 = -0.595684$
$A_1 = -11.039345$
$A_2 = 17.449275$
$A_3 = -16.028445$

In Equation 3.56,

$$T^* = 1 - \frac{T}{T_k}. \tag{3.57}$$

Data about water activity and saturated vapor pressure over electrolyte solutions from dilute to saturated concentrations and temperatures of 0°C–100°C for a large number of aqueous electrolyte solutions are systematized and added with proper experimental researches, as shown in [58].

## REFERENCES


1. Salem R.R. 2004. *Physical Chemistry. Thermodynamics*. FIZMATLIT, Moscow, Russia, p. 352.
2. Aseyev G.G. 2001. *Electrolytes: Methods for Calculation of the Physicochemical Parameteres of Multicomponent Systems*. Begell House, Inc. New York, p. 368.
3. Aseyev G.G., Ryshchenko I.M. Savenkov A.S. 2005. *Electrolytes. Physical and Chemical Parameters of Concentrated Multicomponent Systems*. NTU "HPI", Kharkov, Ukraine p. 448.
4. Aseyev G.G. 1998. *Elektrolites: Interparticle Interactions. Theory, Calculation Methods, and Experimental Data*. Begell House, Inc. New York, p. 754.
5. Croxton K. 1978. *Physics of Liquid State*. Lane from English. Mir, Moscow, Russia, p. 356.
6. Krestov G.A., Novoselov N.P., Perlygin I.S. et al. 1987. *Ionic Solvation*. Nauka, Moscow, Russia, p. 320.
7. Drakin S.I., Shpakova S.G., Pino H.D. 1976. *Structure of Akvo Kompleks in Crystallohydrates Rich with Water and Coordination Numbers of Ions in Solutions. Physics of Molecules*, Vol. 2. Naukova Dumka, Kiev, Ukraine, pp. 75–83.
8. Hripun M.K., Lilich L.S., Yefimov A.Yu. et al. 1983. *Development of Structurally Dynamic Ideas of Strong Solutions of Electrolytes. Problems of the Modern Chemistry of Coordination Compounds*. LGU, Leningrad, Russia, pp. 58–65.
9. Under the editorship of Mishchenko K.P. and Ravdel A.A. 1965. *Short Reference Book of the Physicist of Chemical Sizes*. Chemistry, Moscow, Russia, p. 161.
10. Borisov M.V., Shvarov Yu.V. 1992. *Termodinamika of Geochemical Processes*. Moscow State University, Moscow, Russia, p. 256.
11. Kostromina N.A. (ed.). 1990. *Chemistry of Coordination Compounds: Textbook*. Higher School, Moscow, Russia, p. 432.

12. Kiselyov Yu.M., Dobrynina N.A. 2007. *Chemistry of Coordination Compounds.* Academy, Moscow, Russia, p. 325.
13. Shakhparonov M.I. 1976. *Introduction in the Modern Theory of Solutions (Intermolecular Interactions. Structure. Prime Liquids).* The highest school, Moscow, Russia, p. 296.
14. Tanganov B.B. 2009. *Interactions in Solutions of Electrolytes: Model Operation of Solvation Processes, Equilibriums in Solutions of Polyelectrolytes and Mathematical Prediction of Properties of Chemical Systems.* Akad. Natural Sciences, Moscow, Russia, p. 350.
15. Aseyev G.G., Zaytsev I.D. 1996. *Volumetric Properties of Elektrolite Solutions: Estimation Methods and Experimental Data.* Begell House, Inc. New York, p. 1572.
16. Aseyev G.G., Ryshchenko I.M. Savenkov A.S. 2007. Equations and determination of physical and chemical properties of solutions of sulfate-nitrate of an ammonium. *Zhurn. Prikl. khimii.* 80 (2): 213–220.
17. Aseyev G.G. 1992. *Properties of Aqueous Solutions of Elektrolites.* CRC Press, Boca Raton, FL, p. 1774.
18. Prezhdo V.V., Kraynov N. 1994. *Molecular Interactions and Electric Properties of Molecules.* Publishing house "Osnova", Kharkov, Ukraine. p. 240.
19. Eremin V.V., Kargov S.I., Uspenskaya I.A. et al. 2005. *Fundamentals of Physical Chemistry. Theory and Tasks.* Moscow, Russia, p. 480.
20. Lyashchenko A.K. 2002. *Structural and Molecular-Kinetic Properties of Concentrated Solutions and Phase Equilibria in Water-Salt Systems Concentrated and Saturated Solutions.* Nauka, Moscow, Russia, pp. 93–118. (A series problem of chemistry of solutions.)
21. Loginova D.V., Lileev A.S., Lyashchenko A.K., Harkin V.S. 2003. Hydrophobic hydration of anions on the example of solutions of propionates of alkali metals. *Achievements in Chemistry and Engineering Chemistry.* RHTU of D. I. Mendeleyev. 17 (7 (32)): 26–29.
22. Kirovskaya I.A. 2009. *Chemical Thermodynamics. Solutions.* Publishing House of OmGTU, Omsk, Russia, p. 236.
23. Bazhin N.M., Ivanchenko V.A., Parmon V.I. 2004. *Thermodynamics for Chemists,* 2nd edn. Chemistry, Kolos, Moscow, Russia, p. 416.
24. Stepanov N.F. 2001. *Of Quantum Mechanics and Quantum Chemistry.* Mir, Moscow, Russia, p. 519.
25. Mushrooms L.A., Mushtakov S. 1999. *Quantum Chemistry.* Gardarika, Moscow, Russia, p. 390.
26. Bader R. 2001. *Atoms in Molecules. Quantum Theory.* Mir, Moscow, Russia, p. 532.
27. Mayer I. 2006. *Selected Chapters of Quantum Chemistry: Proofs of Theorems and Formulas.* BINOM: Lab. Knowledge, Moscow, Russia, p. 384.
28. Tatevsky V.M. 1973. *Classical Theory of Molecular Structure and Quantum Mechanics.* Chemistry, Moscow, Russia, p. 320.
29. Mueller M.R. 2001. *Fundamentals of Quantum Chemistry: Molecular Spectroscopy and Modern Electronic Structure Computations.* Kluwer, New York, p. 265.
30. Baranovsky V.I. 2008. *Quantum Mechanics and Quantum Chemistry.* Academy, Moscow, Russia, p. 384.
31. Gankin V.Y., Gankin Y.V. 1998. *How Chemical Bonds Form and Chemical Reactions Proceed.* ITC, Boston, MA, p. 315.
32. Zelentsov S.V. 2006. *Introduction to Modern Quantum Chemistry.* Novosibirsk, Nizhniy Novgorod, Russia, p. 126.
33. Tsirelson V.G. 2010. *Quantum Chemistry. Molecules, Molecular Systems and Solid Bodies.* BINOM Lab. Knowledge, Moscow, Russia, p. 496.
34. Novakovskaya Yu.V. 2004. Molecular systems. *The Structure and Interaction Theory with a Radiation: Common Bases of a Quantum Mechanics and Symmetry Theory.* Editorial of URSS, Moscow, Russia, p. 104.

35. Bartyk-Pбrtay A. 2010. *The Gaussian Approximation Potential: An Interatomic Potential Derived from First Principles Quantum Mechanics.* Springer, Berlin, Germany, p. 96.
36. Gelman G. 2012. *Quantum Chemistry.* BINOM Lab. Knowledge, Moscow, Russia, p. 520.
37. Stepanov N.F., Pupyshev V.I. 1991. *Quantum Mechanics of Molecules and Quantum Chemistry.* izd-vo MGU, Moscow, Russia, p. 384.
38. Unger F.G. 2007. *Quantum Mechanics and Quantum Chemistry.* TGU, Tomsk, Russia, p. 240.
39. Hehre W.J.A. 2003. *Guide to Molecular Mechanics and Quantum Chemical Calculations.* Wavefunction, Inc., Irvine, CA, p. 796.
40. Inagaki S. 2009. *Orbitals in Chemistry.* Springer, Berlin, Germany, p. 327.
41. Koch W.A., Holthausen M.C. 2001. *Chemist's Guide to Density Functional Theory*, 2nd edn. Wiley-VCH Verlag GmbH, Weinheim, Germany, p. 293.
42. Levine I.N. 1999. *Quantum Chemistry*, 5th edn. Prentice Hall, Brooklyn, New York, p. 739.
43. Lowe J.P., Peterson K.A. 2006. *Quantum Chemistry*, 3rd edn. Elsevier Academic Press, p. 703.
44. Maruani J., Minot C. (eds.). 2002. *New Trends in Quantum Systems in Chemistry and Physics. 2. Advanced Problems and Complex Systems.* Kluwer Academic Publishers, Boston, MA, p. 313.
45. Piela L. 2007. *Ideas of Quantum Chemistry.* Elsevier Science, Amsterdam, the Netherlands, p. 1086.
46. Reiher M., Wolf A. 2009. *Relativistic Quantum Chemistry. The Fundamental Theory of Molecular Science.* Wiley-VCH, Weinheim, Germany, p. 691.
47. Kilimnik A.B., Yarmolenko V.V. 2008. *Methods of Definition and Calculation of Jet Components of an Impedance and Average Oscillation Frequencies of Hydrated Ions.* TGTU Publishing House, Tambov, Russia, p. 116.
48. Knorre D.G., Krylova L.F., Musikantov V.S. 1990. *Physical Chemistry.* The Highest School, Moscow, Russia, p. 416.
49. Gurov A.A., Badayev F.Z., Ovcharenko L.P. 2004. *Chemistry.* MGTU Publishing House of N.E. Bauman, Moscow, Russia, p. 748.
50. Politzer P., Truhlar D.G. 1981. *Chemical Applications of Atomic and Molecular Electrostatic Potentials.* Plenum, New York, p. 420.
51. Lyashchenko A.K., Karataeva I.M. 2010. The activity of the water and the dielectric constant of water solutions of electrolytes. *J. Phys. Chem.* 84 (2): 376–384.
52. Lyashchenko A.K., Zasetsky A.Yu. 1998. Complex dielectric permittivity and relaxation parameters of concentrated aqueous electrolyte solutions in millimeter and centimeter wavelength ranges. *J. Mol. Liquid* 77: 61–65.
53. Lileev A.S., Balakaeva I.V., Lyashchenko A.K. 2001. Dielectric properties, hydration and ionic association in binary and multicomponent formate water–salt systems. *J. Mol. Liquid* 87: 11–20.
54. Lileev A.S., Lyaschenko A.K., Harkin V.S. 1992. Dielectric properties of aqueous solutions of yttrium and copper. *J. Inorg. Chem.* 37 (10): 2287–2291.
55. Lyashchenko A.K. 1994. Structure and structure-sensitive properties of aqueous solutions of electrolytes and nonelectrolytes. In: Coffey W. (ed). *Relaxation Phenomena in Condensed Matter. Adv. Chem. Phys.*, Vol. LXXXVII, pp. 379–426.
56. Lyashchenko A.K. 2001. Concentration transition from water-electrolyte to electrolyte water solvents and ionic clusters in solutions. *J. Mol. Liquids.* 91: 21–31.
57. Lyashchenko A.K., Karatayeva I.M. 2007. Communication of activity of water with a statistical dielectric constant of strong solutions of electrolytes. *Dokl. Akad. Sci.* 414 (3): 357–359.

58. Aseyev G.G. 1998. *Electrolytes: Equilibria in Solutions and Phase Equilibria: Calculation of Multicomponent Systems and Experimental Data on the Activities of Water, Pressure Vapor and Osmotic Coefficients*. Begell House, Inc. New York, p. 758.
59. Akhadov Y.Y. 1981. *Dielectric Properties of Binary Solutions*. Pergamon Press, New York, p. 432.
60. Zatsepina G.M. 1974. *Structure and Properties of Water*. Moscow State University, Moscow, Russia, p. 168.
61. Eisenberg D., Kauzmann W. 1975. *The Structure and Properties of Water*. Lane from English A.K. Shemelina. Leningrad: Gidrometeoizdat, Russia, p. 280.

# 4 Supermolecular Interactions

## 4.1 VAN DER WAALS COMPONENTS OF ATTRACTION POTENTIAL

The electromagnetic interaction in the system of many particles in water appears in the form of exchange, multipole, fluctuation, and other interaction forces. Among such interaction forces, one can distinguish those that deplete at large distances as a power function. Such forces are frequently called long-range forces in contrast to short-range forces that deplete rapidly (usually exponentially) with the increase of distance. The short-range forces-distance type of dependence is not most often universal and can be determined by a certain structure of electron shells of atoms. The distinctive feature of long-range forces is, on the contrary, known versatility of their behavior at long distances. At that, if the average charge density and average dipole moment density are equal to zero in an equilibrium medium, then the main long-range forces, in general, are the forces of fluctuation origin that are frequently called the van der Waals forces. They can be calculated by the quantum mechanical approach, but meanwhile accurate values are provided by experimental measurements.

Describing the van der Waals interaction among particles, the specific nature of macroscopic condensed media can be fully estimated by using their dielectric permittivities in the solution or polarizabilities. Fruitfulness of using dielectric formalism in the theory of van der Waals forces is connected with the aforementioned versatility of their dependence from distance among particles. Results of the theory of van der Waals forces, in which only dielectric permittivities or particle polarizabilities are used in formulation, are often called macroscopic. The aggregate of such results is sometimes called the macroscopic approach in the theory of van der Waals forces or the macroscopic theory of van der Waals forces. Macroscopic scales (i.e., characteristic lengths, longer compared to atomic size $a \approx 10^{-10}$ m) actually always appear during consideration of the van der Waals interaction. They are, for example, distances between particles $R$ and length of wave $\lambda_0$ (usually $\lambda_0 \approx 10^{-7}$–$10^{-8}$ m), typical for the body absorption spectrum. At the same time, the statistical description of van der Waals interaction is possible, and even to a greater extent, by the nature of microscopic fluctuation mechanism leading to van der Waals forces among particles. van der Waals forces are directly associated with existence of fluctuation charge densities with macroscopic particles and their interaction through a long-wave fluctuating electromagnetic field. The fluctuating nature of van der Waals forces is mainly the same as in the case of macroscopic condensed bodies and for separate dipoles and molecules. Therefore, macroscopic and microscopic aspects of the theory of van der Waals forces turn to be tightly interconnected.

In this section, as an example, we will provide statistical averaging and derivation of the formula for calculation of dipole–dipole interaction. To date, Boltzmann statistics has been commonly used during statistical averaging. When commenting on Figures 2.1 and 2.3, it was concluded that Boltzmann statistics is not suitable for electrolyte solutions, and wrong theoretical results can be obtained while using it. We obtained statistics (2.9) and (2.11) that adequately describes distribution of particles in the electrolyte solution in the beginning of the concentration area and then dramatically decreases. Certainly, ionic and ion-aqueous clusters (supermolecules) are formed closer to the second concentration area, Figure 1.1, and no statistical distribution exists already. Our difference is that for statistical averaging we will use only energy values. If statistical averaging is not available when using other formulas of supermolecular forces, we will refer the reader to this example but we will present the result of the statistical averaging.

### 4.1.1 Perturbation Operator

Let us assume that $R$ is large and one can neglect overlapping of electron clouds of dipoles $A$ and $B$. In this case, wave function $\Psi$ of the system, consisting of molecules $A$ and $B$, can be presented as the product of wave functions of separate isolated dipoles.

Let us adopt that both dipoles are electrically neutral and located in their main quantum states, then

$$\Psi = \Psi_{0A}\Psi_{0B}. \tag{4.1}$$

Hamiltonian of the system:

$$\hat{H} = \hat{H}_{\infty} + \hat{H}_{*} = \hat{T}_{\infty} + \hat{U}_{\infty} + \Delta U, \tag{4.2}$$

where

$T$ is the average kinetic energy of electrons
$U$ is the potential energy

Hence, perturbation operator $\hat{H}_{*}$ can be found by a change of potential energy of the system with decrease of the distance from infinitely large to $R$:

$$\hat{H}_{*} = \Delta U = U - U_{\infty}, \tag{4.3}$$

where $U_{\infty}$ is the potential energy of two isolated dipoles $A$ and $B$. In order to calculate the system energy, it is required to find out perturbation operator $\hat{H}_{*} = \Delta U$.

Potential energy of system $U$ depends on positions of charges in dipoles $A$ and $B$ and on the distance between dipoles; $U$ is calculated according to (2.43) using classical electrostatics methods. We will consider that both dipoles are in a vacuum. Complications, occurring when dipoles $A$ and $B$ are surrounded by other particles, will be considered further.

### 4.1.2 Interaction of Two Dipoles

Let us assume that in dipole $A$, there are $m_A$ of atomic nuclei having charges $e_{A1},\ldots,e_{Am_A}$ and $m_\nu$ of electrons. In dipole $B$, there are $n_B$ of atomic nuclei having charges $e_{B1},\ldots,e_{Bm_B}$ and $n_\nu$ of electrons.

Let us assume that the electrical neutrality condition of every of the dipole is kept. Then,

$$\sum_{A=1}^{m_A} e_{m_A} = -m_\nu e; \quad \sum_{B=1}^{m_B} e_{n_B} = -n_\nu e, \tag{4.4}$$

where $e$ is the electron charge.

Let us specify that positions of all charges in dipoles $A$ and $B$ are fixed. Distances among atomic nuclei in every dipoles correspond to experimental data about the structure of these molecules. Electrons are located not far from atomic nuclei. Values

$$r_A^+ \equiv e_A^{-1}\sum_A e_{m_A} r_A; \quad r_B^- \equiv e_B^{-1}\sum_B e_{n_B} r_B \tag{4.5}$$

represent radius-vectors defining location of dipole centers of gravity in some two randomly chosen Cartesian coordinate systems, which will be indicated by indexes $A$ and $B$. In Equation 4.5, $r_A$ and $r_B$ are radius-vectors of atomic nuclei having charges $e_{m_A}$ and $e_{n_B}$, respectively.

Let us place origins of both coordinates at the points where centers of gravity of dipoles $A$ and $B$ are located. Then $r_A$ and $r_B$ will be equal to zero. We orient coordinate axes so that axes $Z_A$ and $Z_B$ coincide; axes $X_A$, $X_B$ and $Y_A$, $Y_B$ are parallel. Coordinates of centers of gravity of electrons will be found out by the following ratios:

$$\frac{X_A^- = \sum_\nu \frac{eX_{A\nu}}{m_\nu e} = \frac{X_A}{m_\nu},}{Z_B^- = \sum_\nu \frac{eZ_{B\nu}}{n_\nu e} = \frac{Z_B}{n_\nu},} \tag{4.6}$$

where $X_A \equiv \sum_{\nu=1}^{m_\nu} X_{A\nu};\ldots;Z_B \equiv \sum_{\nu=1}^{n_\nu} Z_{B\nu}$.

Dipole $A$ is a system consisting of two charges. Charge $+m_\nu e$ is located in the center of gravity of positive charges; charge $-m_\nu e$ is in the center of gravity of negative charges. Similarly, dipole $B$ is considered as a system consisting of charges $+n_\nu e$ and $-n_\nu e$ located, respectively, in the centers of gravity of positive and negative charges of this dipole. If distance $R_{H_2O}$ between centers of gravity of positive charges of dipoles $A$ and $B$ is large, such simplification will not influence the results of our calculations. When conducting calculations, it can be observed that multipliers $m_\nu$ and

$n_\nu$ are canceled. The first nonvanishing expansion term $\Delta U$ as a series in powers $R^{-1}$ takes the following form:

$$\Delta U = \hat{H}_* = \frac{e^2}{R^3_{H_2O}}\left(X_A X_B + Y_A Y_B - 2Z_A Z_B\right) + 0(R^{-4}). \tag{4.7}$$

Here, $X_A$, …, $Z_B$ are the sums of coordinates of electrons.

## 4.2 PERTURBATION METHOD

Dipoles $A$ and $B$ can be in the main quantum state or in excited quantum states:

$$0,1,2,\ldots,k,\ldots,m, \quad \text{for dipole } A,$$

$$0,1,2,\ldots,l,\ldots,n, \quad \text{for dipole } B.$$

The system, consisting of dipoles $A$ and $B$ interacting with one another, as well can be in the main or any excited state when one or both dipoles are excited. Let us denote electron states of the system by indexes

$$0,1,\ldots,i,\ldots,j,\ldots$$

If perturbation operator $\hat{H}_*$ is only a small correction to operator $\hat{H}_\infty$ of the unperturbed system, then the interaction energy of system molecules in the $i$ quantum state can be represented as

$$E_i = E_i' + E_i'', \tag{4.8}$$

where $E_i'$ and $E_i''$ are the contributions to interaction energy calculated by means of the perturbation method in the first or second approximation of this method. When proper values of Hamilton operator $\hat{H}_\infty$ are not degenerated, then [1]

$$E_i' = \left\langle \psi_i^{(0)} \middle| H_* \middle| \psi_j^{(0)} \right\rangle, \tag{4.9}$$

$$E_i'' = \sum_{i \neq j} \frac{\left|H_{ij}\right|^2}{E_i^{(0)} - E_j^{(0)}}, \tag{4.10}$$

where

$\psi_i^{(0)}$ is the proper function of operator $\hat{H}_\infty$ of the unperturbed system in the $i$ quantum state

$E_i^{(0)}$ and $E_j^{(0)}$ are the energy of the unperturbed system in quantum states $i$ and $j$

index (0) hereinafter denotes that it goes about the unperturbed state of the system

$H_{ij}$ is a perturbation matrix element:

$$H_{ij} = \left\langle \psi_i^{(0)} \middle| \hat{H}_* \middle| \psi_j^{(0)} \right\rangle. \quad (4.11)$$

Equations 4.9 and 4.10, as mentioned earlier, are true for undegenerated unperturbed systems, but they are also formally applicable in case of degenerate systems, if during calculation of matrix elements of perturbation one chooses linear combinations of functions $\psi_i^{(0)}$ as wave functions. It should also be stated that the perturbation method is applicable provided that

$$\left|H_{ij}\right| = \left|E_i^{(0)} - E_j^{(0)}\right|, \quad (4.12)$$

that is, matrix elements of perturbation must be small compared to the corresponding remainders of the unperturbed energy levels of the system.

## 4.3 INTERACTION ENERGY IN THE FIRST APPROXIMATION OF THE PERTURBATION METHOD

### 4.3.1 Interaction of Two Molecular Dipoles

The operator of the dipole moment of the molecule in the designations adopted by us takes the following form:

$$\sum e\vec{r}_\nu = eX\vec{i}_X + eY\vec{i}_Y + eZ\vec{i}_Z, \quad (4.13)$$

where

$$X = \sum_\nu X_\nu$$

$$Y = \sum_\nu Y_\nu$$

$$Z = \sum_\nu Z_\nu$$

$\vec{i}_X$, $\vec{i}_Y$, and $\vec{i}_Z$ are unit vectors directed along axes $X$, $Y$, and $Z$

In the energy presentation, that is, when wave functions of the molecule are proper functions of the system energy, values of the operator of the dipole moment form matrix $M$, where the elements take the following form:

$$\vec{M}_{ij} = e^{i(E_i - E_j/h)j} \left\langle \psi_i \middle| \sum_\nu e\vec{r}_\nu \middle| \psi_j \right\rangle, \quad (4.14)$$

where $h$ is the Planck constant.

If $i \neq j$, then amplitude of matrix element of the dipole moment is called transition dipole moment $\mu_{ij}$:

$$\mu_{ij} = \left\langle \psi_i \left| \sum_{\nu} e\vec{r}_{\nu} \right| \psi_j \right\rangle. \tag{4.15}$$

Let us assume that dipoles $A$ and $B$ are in main quantum states. If their dipole moments are not equal to zero, then

$$\mu_A = \left\langle \psi_{A0} \left| \sum_{\nu} e\vec{r}_{A\nu} \right| \psi_{A0} \right\rangle \neq 0,$$
$$\mu_B = \left\langle \psi_{B0} \left| \sum_{\nu} e\vec{r}_{B\nu} \right| \psi_{B0} \right\rangle \neq 0. \tag{4.16}$$

Interaction energy of dipoles $E_{DD}$ between one another, caused by the presence of constant dipole moments in them, can be found by means of the first approximation of the perturbation method:

$$E_{DD} = E_0' = \left\langle \psi_0^{(0)} \left| \hat{H}' \right| \psi_0^{(0)} \right\rangle = \left\langle \psi_{A0}\psi_{B0} \left| \Delta U \right| \psi_{A0}\psi_{B0} \right\rangle. \tag{4.17}$$

Substituting the value $\hat{H}_* = \Delta U$ from Equation 4.7 and introducing according to (4.13) and (4.16) the following designations of components of constant dipole moments of dipoles $A$ and $B$, available in main quantum states,

$$\frac{\mu_{A0X} = \left\langle \mu_{A0} \left| eX_A \right| \mu_{A0} \right\rangle,}{\mu_{B0Z} = \left\langle \mu_{B0} \left| eZ_B \right| \mu_{B0} \right\rangle,} \tag{4.18}$$

we will obtain

$$E_{DD} = R_{H_2O}^{-3}\left(\mu_{A0X}\mu_{B0X} + \mu_{A0Y}\mu_{B0Y} - 2\mu_{A0Z}\mu_{B0Z}\right). \tag{4.19}$$

Proceeding to spherical coordinates, we have

$$\begin{aligned}
&\mu_{A0X} = \mu_A \sin\vartheta_A \cos\varphi_A; \quad \mu_{B0X} = \mu_B \sin\vartheta_B \cos\varphi_B,\\
&\mu_{A0Y} = \mu_A \sin\vartheta_A \sin\varphi_A; \quad \mu_{B0Y} = \mu_B \sin\vartheta_B \sin\varphi_B,\\
&\mu_{A0Z} = \mu_A \cos\vartheta_A; \qquad\qquad \mu_{B0Z} = \mu_B \cos\vartheta_B.
\end{aligned} \tag{4.20}$$

We will substitute values (4.20) into (4.19) and at that we will take into account that $\mu_{A0} \equiv \mu_{B0} \equiv \mu_W$ for water dipoles, and after several transformations, we will obtain

$$E_{DD} = \frac{\mu_W^2}{R_{H_2O}^3}\left[\sin\vartheta_A \sin\vartheta_B \cos\left(\varphi_A - \varphi_B\right) - 2\cos\varphi_A \cos\varphi_B\right]. \tag{4.21}$$

Polar angles $\vartheta_A$ and $\vartheta_B$ and azimuth angles $\varphi_A$ and $\varphi_B$ define directions of dipole moments of molecules $\mu_{A0}$ and $\mu_{B0}$. Dipole moments of molecules can be determined experimentally.

Knowing the values of dipole moments, using formula (4.21), one can calculate the dipole interaction energy of molecules, if distances $R$ between dipoles and angles, determining orientation of dipoles in space, are set.

If, for example, dipole moments of dipoles $\mu_{A0}$ and $\mu_{B0}$ are equidirectional along axis $Z$, that is, $\vartheta_A = 0$, $\vartheta_B = 0$ or $\vartheta_A = 180°$, $\vartheta_B = 180°$ and $\varphi_A = \varphi_B$, then $E_{DD}$ is negative and possesses a minimum value. Then dipoles are attracted to one another and the energy of their attraction $E_{DD\max}$ is maximum. If dipole moments of molecules are contradirectional along axis $Z$, that is, $\vartheta_A = 0$, $\vartheta_B = 180°$ or $\vartheta_B = 180°$, $\vartheta_B = 0$, and $\varphi_A = \varphi_B$, then they are repulsed from one another with energy $E_{DD\max}$. When $\vartheta_A = 90°$, $\vartheta_B = 90°$, and $\varphi_A - \varphi_B = 90°$, the dipole interaction energy of dipoles $A$ and $B$ is equal to zero, etc.

### 4.3.2 Averaging over Orientations

Let us assume that dipoles $A$ and $B$ can take any orientations to one another. If dipoles are polar, that is, possess constant dipole moments, the dipole energy interaction depends on their mutual orientation and that is why different orientations, in general, possess various probabilities. In cases when $E_{DD}$ is defined by Equation 4.21, its value, averaged over any orientations at some constant value $R$ among point dipole moments, is of interest. In statistical physics, averaging over orientations is usually made using Maxwell–Boltzmann statistics, which is not correct. The author highlights this several times in this work. Maxwell–Boltzmann statistics is absolutely inappropriate for electrolyte solutions (see comments to Figures 2.1 and 2.3). That is why averaging over orientations will be made as averaging over interaction energies of dipoles $A$ and $B$:

$$\left\langle E_{DD}\right\rangle = \frac{1}{4\pi\varepsilon_0\varepsilon_S}\cdot\frac{\iint E_{DD}\exp\left(-E_{DD}\right)d\omega_A\, d\omega_B}{\iint \exp\left(-E_{DD}\right)d\omega_A\, d\omega_B}, \tag{4.22}$$

where $d\omega = \sin\vartheta\, d\vartheta\, d\vartheta$. Integration in (4.22) is made by means of $\vartheta$ from 0 to $\pi$ and by means of $\varphi$ from 0 to $2\pi$. Hereinafter, unless expressly stated, brackets $\langle\cdots\rangle$ designate the statistical averaging operation.

Exponential $\exp(-E_{DD})$ can be expanded into series being restricted to two expansion terms:

$$\exp\left(-E_{DD}\right) = 1 - \frac{E_{DD}}{1!} + \cdots, \quad E_{DD}^2 < \infty. \tag{4.23}$$

We will take into account that

$$\langle\cos\vartheta\rangle = \langle\cos\varphi\rangle = \langle\sin\vartheta\rangle = \langle\sin\vartheta\rangle = 0,$$
$$\langle\cos^2\vartheta\rangle = \frac{1}{3};\quad \langle\sin^2\vartheta\rangle = \frac{2}{3};\quad \langle\cos^2\varphi\rangle = \frac{1}{2}, \tag{4.24}$$

and having substituted (4.22) through (4.24) into (4.21) after some transformations, we will obtain

$$\langle E_{DD}\rangle = -\frac{1}{6\pi\varepsilon_0\varepsilon_S}\cdot\frac{\mu_W^4}{R_{H_2O}^6}. \tag{4.25}$$

### 4.3.3 Other Expansion Terms of Interaction Energy as a Series in Powers $R^{-1}$

Formulas (4.19) through (4.21) express only the part of a power series describing dependence $E_{DD}$ from $R$ that is connected with dipole term (4.7) at the expansion of potential interaction energy $\Delta U$ (2.43) as a series in powers $R^{-1}$. Other terms of power function $E_0 = f(R)$ can be obtained in the same way by means of the first approximation of the perturbation method.

Let us assume that we have a system of charges $e_i$ ($i$ = 1,2,3,...), $e_i = |z_i|e$. The quadrupole moment of this system is symmetric tensor, where components $Q_{\alpha\beta}$ are determined by the following equations:

$$Q_{\alpha\beta} = \sum_i e_i\left(3X_i^\alpha X_i^\beta - \delta_{\alpha\beta}r_i^2\right), \tag{4.26}$$

where

$r_i$ is a radius-vector determining position of $i$ charge
$e_i, X_i^\alpha, X_i^\beta(\alpha,\beta = 1,2,3,\ldots)$ are Cartesian components of the radius-vector $r_i$
$\delta_{\alpha\beta}$ is the Kronecker symbol

$$\delta_{\alpha\beta} = \begin{cases} 1, & \alpha = \beta, \\ 0, & \alpha \neq \beta. \end{cases}$$

It is estimated that the origin of coordinates is inside the system of charges under consideration.

For example, the quadrupole moment is determined as tensor $Q$ with elements

$$
\begin{aligned}
&Q_{xx} = \frac{1}{2}\sum_i e_i\left(3X_i^2 - r_i^2\right), \quad Q_{xy} = \frac{3}{2}\sum_i e_i X_i Y_i, \\
&Q_{yy} = \frac{1}{2}\sum_i e_i\left(3Y_i^2 - r_i^2\right), \quad Q_{xz} = \frac{3}{2}\sum_i e_i X_i Z_i, \\
&Q_{yy} = \frac{1}{2}\sum_i e_i\left(3Y_i^2 - r_i^2\right), \quad Q_{yz} = \frac{3}{2}\sum_i e_i Y_i Z_i.
\end{aligned}
\tag{4.27}
$$

From (4.26), it is easy to see that

$$
Q_{xx} + Q_{yy} + Q_{zz} = 0. \tag{4.28}
$$

The coordinate system can be chosen so that all off-diagonal tensor elements (4.27) become zero. The corresponding diagonal elements $Q_{xx}$, $Q_{yy}$, $Q_{zz}$ are called principal values of tensor. In virtue of condition (4.28), only two of them are independent. An example of the system, having zero dipole moment and nonvanishing quadrupole one, are two oppositely oriented dipoles shown in Figure 4.1 (in Figure 4.1, two positive and two negative charges of the same absolute value are located on the parallelogram apex—from here we have the name *quadrupole*).

At axial symmetry of dipole along axis $z$,

$$
Q_{xx} = Q_{yy}, \tag{4.29}
$$

which in combination with (4.27) provides

$$
Q_{zz} = -2Q_{xx} = -2Q_{yy}. \tag{4.30}
$$

$Q_{zz}$ is usually taken as an independent component; scalar value $Q = Q_{zz}$ is called quadrupole moment of dipole. Depending on location of charges, this value can be positive or negative. All linear dipoles rank among axially symmetrical.

Let us demonstrate results of calculations without a conclusion [2,3]. Let us assume that $Q_W$ is quantum-mechanical average quadrupole moment of water dipole. It is estimated in the following that quadrupole moment is cylindrically symmetrical.

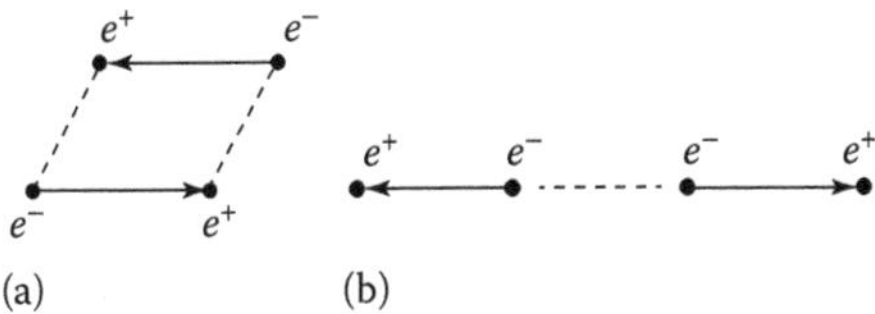


**FIGURE 4.1** Systems of charges with with (a) nonvanishing- and (b) zero dipole quadrupole moments.

Angles $\vartheta$ and $\varphi$ characterize orientation of the symmetry axis of quadrupole moment. Interaction of dipole moment $\mu_W$ and quadrupole moment $Q_W$ makes the following contribution to $E_{\mu Q}$:

$$E_{\mu Q}=-\frac{3\mu_W Q_W}{4R_{H_2O}^3}\left[\left(3\cos\vartheta_B-1\right)\cos\vartheta_A-2\cos\vartheta_A\cos\left(\varphi_A-\varphi_B\right)\right]. \tag{4.31}$$

The quadrupole–quadrupole interaction possesses the following energy:

$$E_{QQ}=-\frac{3Q_W^2}{R_{H_2O}^5}\left\{\begin{array}{l}1-5\cos^2\vartheta_A-5\cos^2\vartheta_B-15\cos^2\vartheta_A\cos^2\vartheta_B\\ +2\left[\sin\vartheta_A\sin\vartheta_B\cos\left(\vartheta_A-\vartheta_B\right)-4\cos\vartheta_A\cos\vartheta_B\right]^2\end{array}\right\}. \tag{4.32}$$

Statistical averaging over all angles for fixed values $R_{H_2O}$ provides

$$\left\langle E_{\mu Q}\right\rangle=-\frac{1}{4\pi\varepsilon_0\varepsilon_S}\cdot\frac{\mu_W^2 Q_W^2}{R_{H_2O}^8}, \tag{4.33}$$

$$\left\langle E_{QQ}\right\rangle=-\frac{7}{160\pi\varepsilon_0\varepsilon_S}\cdot\frac{Q_W^4}{R_{H_2O}^{10}}. \tag{4.34}$$

Aside from the analyzed multipole moments, let us also analyze other components of the van der Waals interactions.

Let us consider the physical aspect of the issue of van der Waals component forces with regard to electrolyte solutions (induction and dispersion effects). Solvent molecules (water) as a whole are not charged, but, as it is known, charges, included into composition of a water molecule, are distributed asymmetrically. In terms of these conditions, a water molecule possesses dipole electric moment. A salt molecule, dissolved in a solvent, dissociates into ions possessing electric charge. Forces occur, which orient water dipoles in a certain way relative to ions. Motion of molecules stops to be independent. Mutual orientation of water electric moments relative to ions becomes such that attractive forces occur and potential energy of liquid decreases.

### 4.3.4 Polarization (Induction) Interdipole Interactions

Polarization (induction) interdipole interactions are conditioned by the fact that distribution of dipole charges in the external field is changed in such a way that the energy of the dipole–external field system became minimum. Dipole and quadrupole moments are induced in dipoles as the external field increases from zero to any finite. If $\vec{\mu}$ is the dipole moment of dipole in the external field and $\vec{\mu}_0$ is its dipole moment under the absence of the field, then induced dipole moment makes $\vec{\mu}_{ind}=\vec{\mu}-\vec{\mu}_0$ (for each of the Cartesian components $\mu_{\beta\,ind}=\mu_\beta-\mu_{\beta_0}$, where $\beta$ = ($x$, $y$, $z$)). Induced dipole moment depends on the intensity of external field $E$ and on the polarizability of molecule α: $\vec{\mu}_{ind} = \alpha E$; α here means dipole polarizability. Induced quadrupole moment is associated with quadrupole polarizability. But usually polarizability means dipole one, as stated in reference books. Dipole polarizability α represents tensor with nine components (thus, $\alpha_{xx}$ characterizes polarizability in direction $x$ stipulated by the field in direction $x$, $\alpha_{xy}$ is polarizability in direction $x$ stipulated by the field in direction $y$, and so on, so that, e.g., $\mu_{x\ ind} = \alpha_{xx}E_x + \alpha_{xy}E_y + \alpha_{xz}E_z$, where $E_x$, $E_y$, and $E_z$ are Cartesian components of the intensity vector). Axes $x$, $y$, $z$ can be oriented in a molecule so that all off-diagonal components of polarizability tensor became zero. The corresponding diagonal components $\alpha_{xx}$, $\alpha_{yy}$, $\alpha_{zz}$ are called principal and designated as $\alpha_x$, $\alpha_y$, $\alpha_z$. Average polarizability of a molecule is determined as $(\alpha_x + \alpha_y + \alpha_z)/3$. For axially symmetrical molecules, polarizability is characterized by two components $\alpha_{\parallel}$ and $\alpha_{\perp}$ (parallel and perpendicular to component dipole axis); average polarizability for them is expressed as $(\alpha_{\parallel} + 2\alpha_{\perp})/3$. Polarizability is determined experimentally usually from data about the refraction index and dielectric constant. The difference between $\alpha_{\parallel}$ and $\alpha_{\perp}$, that is, polarizability anisotropy, is measured by experiments of depolarization of polarized light, dissipated by molecules, and by means of the Kerr effect [4]. Upon availability of polarizability anisotropy, the direction of the induced dipole moment may differ from the direction of the external field.

Induction interaction between dipoles $A$ and $B$ occurs if at least one of them possesses constant electric moment. In the field of molecule $A$, which possesses constant electric moment, redistribution of electron density takes place inside molecule $B$ and the dipole moment directed over the field (or almost over the field) is induced. As a result, the system energy decreases, that is, induction interaction always comes to attraction. If dipole $B$ also possesses constant electric moment (as in the case with water dipoles), it in turn polarizes dipole $A$, which also results in decrease of the system energy. Induction interaction as well as direct electrostatic one, considered earlier, can be described by means of classical electrostatics. The *constant dipole–induced dipole* interaction energy for two polar dipoles with dipole moments $\mu_A$, $\mu_B$ and polarizabilities $\alpha_A$, $\alpha_B$ is as follows:

$$E_{ind} = \frac{-\left[\alpha_A\mu_B^2\left(3\cos^2\vartheta_B + 1\right) + \alpha_B\mu_A^2\left(3\cos^2\vartheta_A + 1\right)\right]}{2R_{H_2O}^6}. \tag{4.35}$$

At averaging over disordered orientations and interaction of *constant dipole–induced dipole* and *induced dipole–induced dipole* of similar dipoles from formula (4.35), we will obtain

$$\left\langle E_{ind}\right\rangle = -\frac{1}{2\pi\varepsilon_0\varepsilon_S}\cdot\frac{\alpha_W^2\mu_W^2}{R_{H_2O}^6}. \tag{4.36}$$

Aside from interaction of *constant dipole–induced dipole*, estimated in formulas (4.35) and (4.36), interactions of constant quadrupole–induced dipole, induced dipole–induced quadrupole, etc., contribute to total energy. The full expression for the energy of induction interaction of molecules, possessing constant dipole and quadrupole moments, includes terms proportional to $R^{-7}$, $R^{-8}$, etc., where the dipole–dipole term (4.35) is fundamentally depleting most slowly ($R^{-6}$) with distance.

### 4.3.5 Dispersion (London) Interactions

Dispersion (London) interactions between water dipoles can be described in the framework of quantum mechanics. They are represented as the result of the fact that instantaneous dipoles and other multipoles, occurring at fluctuations of dipole electron density, induce multipole moments from another dipole. They are interactions among these instantaneous moments of two dipoles that determine dispersion energy, which always comes to attraction for dipole in the principal state.

The value of energy depends on polarizability of the first dipole, since the value of instantaneous dipole $\alpha_A$ depends on the amount of control over external electrons from the part of nuclear potential. It also depends on polarizability of the second dipole because value $\alpha_B$ is associated with the guidance ratio of this moment. Full estimation of attractive forces between two induced dipoles, that is, *dispersion forces*, is rather complicated; however, such expansion is an appropriate approximation.

The energy of dispersion interactions is represented in the form of a series

$$E_{disp} = -C_6R^{-6} - C_8R^{-8} - C_{10}R^{-10} - \cdots, \tag{4.37}$$

where the first term is associated with interactions of instantaneous dipoles, the second with interaction of instantaneous dipole and quadrupole, etc. The main contribution to value $E_{disp}$ is made by dipole–dipole expansion term $C_6/R^6$. Value $C_6$ is determined through dynamic polarizabilities $\alpha_A$ and $\alpha_B$ of interacting dipoles $A$ and $B$. The exact expression for $C_6$ represents the integral of the product of polarizabilities depending on imaginary argument $i\omega$:

$$C_6 = \frac{3\hbar}{\pi}\int_0^{\infty}\alpha_A(i\omega)\alpha_B(i\omega)d\omega, \tag{4.38}$$

where $\hbar$ is the reduced Planck constant.

Coefficients of series (4.37) are called dispersion constants. The London theory allowed expressing these constants through dipole polarizability and frequencies of electron oscillators, and further through the first ionization potential $I$. The following expression (London–Margenau formula) is widely used. From (4.38), we obtain

$$C_6 = \left(\frac{1}{16\pi^2}\right)\alpha_A\alpha_B\frac{3I_AI_B}{2(I_A + I_B)}. \tag{4.39}$$

For interaction of one-type dipoles, $C_6 = 3/4\alpha_W^2 I_W$, and in dipole approximation

$$E_{disp} = -\frac{3}{64\pi^2\varepsilon_S}\cdot\frac{\alpha_W^2 I_W}{R_{H_2O}^6}. \tag{4.40}$$

## 4.4 EVALUATION OF THE ENERGY OF VAN DER WAALS INTERACTIONS IN STRUCTURE FORMATION OF THE GUEST–HOST SOLUTION

As mentioned earlier, dipole, London, and polarization interactions are, generally, of electromagnetic nature. The differences among them are conditional from this point of view, although they possess as well objective content since dipole interactions are directly associated with constant dipole moments, London ones with correlation of electron motions, and polarization ones with influence of constant electric moments of water molecules on electron motion in the surrounding medium.

Let us make a numeric evaluation of energies of the aforementioned interactions. We write the sum in the right part (3.14) for dipole–dipole interaction (4.25):

$$\langle E_{DD}\rangle = -\frac{1}{6\pi\varepsilon_0\varepsilon_S}\cdot\frac{\mu_W^4}{R_{\mathrm{H_2O}}^6}\cdot\frac{N_A a c_k}{RkT}(\nu_i n_{hi}+\nu_j n_{hj}). \tag{4.41}$$

We make the same operations for dipole–quadrupole interaction (4.33):

$$\langle E_{\mu Q}\rangle = -\frac{1}{4\pi\varepsilon_0\varepsilon_S}\cdot\frac{\mu_W^2 Q_W^2}{R_{\mathrm{H_2O}}^8}\cdot\frac{N_A a c_k}{RkT}(\nu_i n_{hi}+\nu_j n_{hj}). \tag{4.42}$$

Similarly, for quadrupole–quadrupole interaction (4.34):

$$\langle E_{QQ}\rangle = -\frac{7}{160\pi\varepsilon_0\varepsilon_S}\cdot\frac{Q_W^4}{R_{\mathrm{H_2O}}^{10}}\cdot\frac{N_A a c_k}{RkT}(\nu_i n_{hi}+\nu_j n_{hj}). \tag{4.43}$$

For induction interaction (4.36):

$$\langle E_{ind}\rangle = -\frac{1}{2\pi\varepsilon_0\varepsilon_S}\cdot\frac{\alpha_W^2\mu_W^2}{R_{\mathrm{H_2O}}^6}\cdot\frac{N_A a c_k}{RkT}(\nu_i n_{hi}+\nu_j n_{hj}). \tag{4.44}$$

For dispersion interaction (4.40):

$$E_{disp} = -\frac{3}{64\pi^2\varepsilon_S}\cdot\frac{\alpha_W^2 I_W}{R_{\mathrm{H_2O}}^6}\cdot\frac{N_A a c_k}{RT}(\nu_i n_{hi}+\nu_j n_{hj}). \tag{4.45}$$

The Boltzmann constant is omitted in (4.45) because thermodynamics will not be coordinated with quantum-mechanical derivation of formula (4.40) with its availability.

The evaluation of energy contribution of formulas (4.41) through (4.45) of the van der Waals interactions will be made for solution $CaCl_2$–$H_2O$, $m$ = 1.2 mol/kg $H_2O$, $T$ = 298 K. For it, $c_k$ = 1.167 mol/dm$^3$ and $\rho$ = 1102.03 kg/m$^3$ [5]. From Table 3.2, for $Ca^{2+}$: $a_i$ = 0.242 nm and $n_{hi}$ = 6, for $Cl^-$: $a_j$ = 0.319 nm and $n_{hj}$ = 6. Molecular weight $CaCl_2$ = 110.99, $\nu_i$ = 1, $\nu_j$ = 2. From Table 3.6, for this concentration $CaCl_2$ and temperature, $\varepsilon_S = 53.5$. Let us make additional calculations. Based on formula (3.35), we calculate value $a$:

$$a = 1 \cdot (0.242 + 0.319) = 0.561 \text{ nm}.$$

Here, parameter $\xi$ is adopted equal to one.

Based on formula (3.39), we will find the distance between water dipoles:

$$R_{H_2O} = 1.47 \cdot \left( \frac{18.02}{1102.03} \right)^{1/3} = 0.373 \text{ nm}.$$

We find as well the distance between ions in the solution:

$$R = 1.47 \cdot \left( \frac{110.99}{1102.03} \right)^{1/3} = 0.684 \text{ nm}.$$

Value $R < a$, that is why electrostatic interactions are completely screened.

Water dipole moment $\mu_W = 6.172 \cdot 10^{-30}$ C/m; Avogadro's constant $N_A = 6.022 \cdot 10^{26}$ kmol$^{-1}$; electric constant $\varepsilon_0 = 8.85 \cdot 10^{-12}$ F/m; and Boltzmann constant $k = 1.38 \cdot 10^{-26}$ kJ/K. We substitute all values into formula (4.41):

$$\langle E_{DD} \rangle = -\frac{(6.172 \cdot 10^{-30})^4 \cdot 6.022 \cdot 10^{26} \cdot 0.561 \cdot 1.167 \cdot (1 \cdot 6 + 2 \cdot 6)}{6\pi \cdot 8.85 \cdot 10^{-12} \cdot 53.5 \cdot (0.373 \cdot 10^{-9})^6 \cdot 0.684 \cdot 1.38 \cdot 10^{-26} \cdot 298}$$
$$= -0.12 \text{ kJ mol}^{-1}.$$

For calculation under formula (4.42), we will take into account that quadrupole moment of water dipole $Q_W = 3.9 \cdot 10^{-10}$ D m and 1 D = $3.336 \cdot 10^{-30}$ C m. Finally, $Q_W = 1.3 \cdot 10^{-39}$ C m$^2$. We substitute all values into formula (4.42):

$$\langle E_{\mu Q} \rangle = -\frac{(6.172 \cdot 10^{-30})^2 (1.3 \cdot 10^{-39})^2 \cdot 6.022 \cdot 10^{26} \cdot 0.561 \cdot 1.167 \cdot (1 \cdot 6 + 2 \cdot 6)}{4\pi \cdot 8.85 \cdot 10^{-12} \cdot 53.5 \cdot (0.373 \cdot 10^{-9})^8 \cdot 0.684 \cdot 1.38 \cdot 10^{-26} \cdot 298}$$
$$= -5.70 \cdot 10^{-4} \text{ kJ mol}^{-1}.$$

Let us proceed to formula (4.43):

$$\langle E_{QQ} \rangle = -\frac{7 \cdot (1.3 \cdot 10^{-39})^4 \cdot 6.022 \cdot 10^{26} \cdot 0.561 \cdot 1.167 \cdot (1 \cdot 6 + 2 \cdot 6)}{160\pi \cdot 8.85 \cdot 10^{-12} \cdot 53.5 \cdot (0.373 \cdot 10^{-9})^{10} \cdot 0.684 \cdot 1.38 \cdot 10^{-26} \cdot 298}$$
$$= -9.45 \cdot 10^{-6} \text{ kJ mol}^{-1}.$$

When substituting data into formula (4.44), we will take into account that water polarizability $\alpha_W = 1.49 \cdot 10^{-30}$ C m²/V. Finally,

$$\langle E_{ind} \rangle = -\frac{(1.49\cdot 10^{-30})^2 \cdot (6.172\cdot 10^{-30})^2 \cdot 6.022\cdot 10^{26} \cdot 0.561\cdot 1.167\cdot (1\cdot 6+2\cdot 6)}{2\pi \cdot 8.85\cdot 10^{-12} \cdot 53.5\cdot (0.373\cdot 10^{-9})^6 \cdot 0.684\cdot 1.38\cdot 10^{-26} \cdot 298}$$

$$= -1.4\cdot 10^{-3} \text{ kJ mol}^{-1}.$$

For calculation of $E_{disp}$ under formula (4.45), we will take into account that the first ionization potential for water is equal to $2 \cdot 10^{-21}$ kJ. We substitute all values into (4.45) and obtain

$$E_{disp} = -\frac{3\cdot (1.49\cdot 10^{-30})^2 \cdot (2\cdot 10^{-21}) \cdot 6.022\cdot 10^{26} \cdot 0.561\cdot 1.167\cdot (6+6)}{64\pi^2 \cdot 53.5\cdot (0.373\cdot 10^{-9})^6 \cdot 0.684\cdot 298}$$

$$= -1.35\cdot 10^{-4} \text{ kJ mol}^{-1}.$$

We will make the same calculation of $E_{DD}$ for solution $KNO_3$–$H_2O$ at $m = 2.03$ mol/kg $H_2O$, $T = 298$ K. For it, $c_k = 1.845$ mol/dm³ and $\rho = 1326.1$ kg/m³ [5]. From Table 3.2, for $K^+$: $a_i = 0.271$ nm and $n_{hi} = 4$, for $NO_3^-$: $a_j = 0.327$ nm and $n_{hj} = 5$. Molecular weight $KNO_3$ = 101.105, $\nu_i = 1$, $\nu_j = 1$. From Table 3.6, for this concentration $KNO_3$ and temperature $\varepsilon_S = 70.0$.

Under formula (3.35), value $a = 0.598$ nm; under formula (3.39) $R_{H_2O} = 0.351$ nm; value $R = 0.623$ nm. Using formula (4.45), we will obtain $E_{DD} = -0.24$ kJ/mol. This interaction should be taken into account further on.

It is clear, as seen from the given calculations, that neither quadrupole nor induction nor dispersion interactions can be a reason for occurrence of somewhat stable complexes or ion and ion-aqueous clusters in the liquid phase. Even the statement of the question about occurrence of ion-aqueous clusters at the expense of quadrupole, induction, and dispersion interactions does not make any sense because the idea about such interactions in principle is inappropriate at any distances between water dipoles whatsoever. It is quite obvious that the energy of van der Waals interactions plays just a small part in the properties of electrolyte solutions. This part is often exaggerated because calculations are made for too small values of $R$, equal to sizes of water dipoles, which is unreasonable.

If one considers two polar water dipoles located at a distance $R$ from one another, apart from the fact that they possess constant moments, their electron clouds fluctuate and they can be considered as particles possessing, in addition, induced dipole moment that constantly changes its value and direction. Let us assume that one of the dipoles turns quickly into an electron configuration that provides induced dipole. This dipole polarizes another molecule and induces in it its induced dipole. Both dipoles are tied up together and that is why molecules are attracted to one another. Although the direction of the dipole in the first molecule will continue changing, the second molecule will react on that change and due to such correlation the attraction effect will not average out to zero. Dipoles tend to take a position with minimum potential energy forming at that an ion-aqueous cluster in the liquid phase, and at further increase of concentration of ions they contribute as well to forming various supramolecular assemblies. Sharp decrease of dielectric permittivity of the solution contributes to this too. Formation of ion-aqueous clusters, etc. in the liquid phase at the expense of overlapping of electron clouds will be considered a little further.

In principle, total interaction energy between dipoles is given by the sum of Equations 4.41 through 4.45. All these energies are negative meaning decrease of energy at approach of dipoles, that is, points at the attractive forces between molecules. But energies (4.42) through (4.45) turned to be of low significance in the electrolyte solution compared to (4.41).

## 4.5 ION–DIPOLE INTERACTIONS IN THE FIRST APPROXIMATION OF THE PERTURBATION THEORY

Supramolecular interactions define the difference of the real system from the ideal solution. Interactions in condensed systems, such as liquid solutions, are very intensive. The obvious demonstration of forces of interparticle attractions is the fact that ions and molecules of liquid are kept together (liquid possesses free surface on the border with gas), and considerable energy—evaporation energy is required for transfer of liquid molecules into the gas phase. Little compressibility of substances in the condensed state indicates repulsion of molecules at close distances, which allows talking about such characteristic feature of molecules as occupied volume.

Potential energy of interparticle interactions $U$ depends on the system configuration, that is, on positions of solvent ions and molecules in space. When studying dense systems, a question arises about the type of this function for the majority of big quantity of particles; however, the main data about interparticle interactions relate to isolated several particles. Dependence $U$ on the position of particles can be represented as the potential energy surface. Potential gradient over coordinates of this particle determines the force acting on the particle on the part of the other ones. Potential energy $U$ of such particles depends only on the distance between them $R$; the force of interaction $F = -\mathrm{d}U/\mathrm{d}R$ is also a function but for the distance that defines such force as central. Potential energy of the system from $N$ particles, interacting by means of central forces, is assumed by center-of-mass coordinates of molecules: $U = U(R_1, \ldots, R_N)$. If particle symmetry differs from spherical, interaction energy depends not only on intermolecular distance $R$ but also on orientation of molecules (on angle $\theta$ at interaction of a diatomic molecule with an atom in *s*-condition; on angles $\theta_A$, $\theta_B$, $\varphi_A$, $\varphi_B$ for a pair of diatomic molecules; on Euler angles in case of nonlinear molecules); on dielectric permittivity of the medium. Generally speaking, the bond lengths in molecules should be included into the number of variables that influence interaction potential, but interaction energy in a majority of cases is determined at the fixed position of nuclei. Potential energy of molecule interaction with internal rotations depends on their conformations.

The sources of information about the potential of interparticle interactions are quantum theory and experiment [6–52]. At that, main results are related to interaction potential of several particles. Experimental information about pair interaction potential is obtained mainly during research of properties of rarefied gases (the second virial coefficient and various non-equilibrium properties: viscosity, heat conductivity, self-diffusion coefficient, etc.). At present, molecular-beam scattering is apparently the most efficient method for investigating interparticle interactions. Spectral methods [53–56] provide big opportunities for studying interparticle interactions because these interactions influence molecule spectra. Investigation of interparticle interactions based on the data about the structure and thermodynamic properties of liquids and solutions is constrained by complexity of bonds between these properties and potential, though such tasks are set at the present time.

Theoretical consideration shows that supramolecular interactions are mainly of electrostatic nature and their energy depends on distribution of electron density of interacting ions and molecules. Quantum-mechanical calculations of interaction potential involve solving the Schrödinger equation at various relative positions of molecules and determining potential energy for every of the considered configurations. In typical cases, 100–150 points on the surface of potential energy are calculated. Calculation results are approximated by analytic expressions. There is voluminous literature about quantum-mechanical calculations of potential functions, which is summarized in a series of monographs and reviews [6–13]. Astonishing success has been gained in this area during the last decades. Interaction potentials between one-type molecules and interaction potentials with water have been obtained for many substances, which represents a theoretical base for studying solvation processes. Although the main volume of quantum-mechanical calculations is related to interaction of two particles, calculations for groups consisting of more than three and larger number of molecules are conducted as well, and the interest for such calculations rises more and more. The method called the supramolecular approach has gained momentum in quantum-mechanical calculations of solvation effects. The idea of the approach involves consideration of a dissolved molecule and a certain number of solvent molecules as a unified system for which the Schrödinger equation is solved (see Section 1.2 for more details). Quantum-mechanical calculations of solvation effects at part-task simulation of dissolved molecule–solvent interaction are widely conducted; macroscopic models of dielectric are often taken for estimation of solvent influence [7,57]. Quantum-mechanical potentials are usually represented as functions with a large number of parameters that limits a possibility of their usage in analytical theories. At the same time, calculations using ECM by means of Monte Carlo and molecular dynamic methods have been as well successfully implemented for such complex multiparameter potentials. Apart from the fact that quantum-mechanical calculations provide theoretical potentials directly, they help in finding simpler model empirical potentials offering a substantiated form of functional dependences.

However, further, we will discuss only interaction potentials among particles defined at the macroscopic level.

The following main contributions to total interaction between ions and water can be approximately distinguished:

- Direct electrostatic interaction conditioned by availability of constant electric moments—dipole, quadrupole, etc., with both interacting ion and water dipole particles, which is explained by difference of distribution of charges on isolated particles from spherically symmetrical.

- Interaction explained by orientational and deformation polarization of water dipoles in the ion field.
- Dispersion interaction explained by correlation of electron density in instantaneous distributions in water dipoles; it occurs among any particles, including those that do not have constant electric moments.
- Induction (polarization) interaction, related to redistribution of electron density of a water molecule in the ion field, possessing constant electric moment.
- Exchange (repulsive) interaction occurring at detectable overlapping of filled-in molecule electron shells and quickly depleting with increase of the intermolecular distance.

Having performed operations in the first approximation of the perturbation theory, as shown in Sections 1.5.1 through 1.5.3, we will obtain:

Contribution into energy explained by interaction of free charge $e_i = \nu_i |z_i| e$ of ion $i$ with dipole moment $\mu_W$, water dipole

$$E_{i\mu} = -\frac{e_i \mu_W}{R^2} \cos\vartheta_B, \tag{4.46}$$

or after statistical averaging over all orientations

$$\langle E_{i\mu} \rangle = -\frac{1}{24\pi^2 \varepsilon_0 \varepsilon_S} \cdot \frac{e_i^2 \mu_W^2}{R^4}. \tag{4.47}$$

Interaction energy of charge $e_i$ with quadrupole moment $Q_W$

$$E_{iQ} = -\frac{e_i Q_W}{4R^3} (3\cos\vartheta_B - 1), \tag{4.48}$$

or after statistical averaging

$$\langle E_{iQ} \rangle = -\frac{1}{80\pi\varepsilon_0 \varepsilon_S} \cdot \frac{e_i^2 Q_W^2}{R^6}. \tag{4.49}$$

Orientational and deformation polarization of water dipoles in the field of ions:

$$\langle E_{i\alpha} \rangle = -\frac{1}{8\pi\varepsilon_0 \varepsilon_S} \cdot \frac{e_i^2 \alpha_W^2}{R^4}. \tag{4.50}$$

A pair of ion–dipole dispersion (London) interactions looks like as follows:

$$E_{id} = -\left(\frac{3}{32\pi^2 \varepsilon_S}\right) \frac{\alpha_i \alpha_W}{R^6} \cdot \frac{e_i^2 I_i I_W}{I_i + I_W}, \tag{4.51}$$

where

$\alpha_i$ is the ion polarizability in the aqueous solution
$I_i$ is the first ion ionization potential

Similarly as in the case of dipole–dipole interactions described earlier, ion–dipole polarization (induction) interactions are explained by the fact that distribution of dipole charges in the ion field changes in such a way that the energy of the dipole–ion system is minimum:

$$\langle E_{iP}\rangle = -\frac{1}{2\pi\varepsilon_0\varepsilon_S}\cdot\frac{e_i^2\alpha_W\mu_W^2}{R^6}. \tag{4.52}$$

We will assume, as in Section 1.5.4, the sum of right part (3.14) for ion–dipole interactions (4.47), (4.49), and (4.50), which after several transformations will look as follows.

Ion–dipole interaction (4.47):

$$\langle E_{i\mu}\rangle = -\frac{1}{24\pi^2\varepsilon_0\varepsilon_S}\cdot\frac{e^2\mu_W^2}{R^5}\cdot\frac{N_A a c_k}{kT}\left(\nu_i^2 z_i^2 n_{hi} + \nu_j^2 z_j^2 n_{hj}\right)\left(\nu_i n_{hi} + \nu_j n_{hj}\right). \tag{4.53}$$

Ion–quadrupole interaction (4.49):

$$\langle E_{iQ}\rangle = -\frac{1}{80\pi\varepsilon_0\varepsilon_S}\cdot\frac{e^2 Q_W^2}{R^7}\cdot\frac{N_A a c_k}{kT}\left(\nu_i^2 z_i^2 n_{hi} + \nu_j^2 z_j^2 n_{hj}\right)\left(\nu_i n_{hi} + \nu_j n_{hj}\right). \tag{4.54}$$

Orientation and deformation of water dipoles in the field of ions:

$$\langle E_{i\alpha}\rangle = -\frac{1}{8\pi\varepsilon_0\varepsilon_S}\cdot\frac{e^2\alpha_W^2}{R^5}\cdot\frac{N_A a c_k}{kT}\left(\nu_i^2 z_i^2 n_{hi} + \nu_j^2 z_j^2 n_{hj}\right)\left(\nu_i n_{hi} + \nu_j n_{hj}\right). \tag{4.55}$$

Ion–dipole dispersion (London) interactions:

$$\langle E_{id}\rangle = -\frac{3}{32\pi^2\varepsilon_S}\cdot\frac{e^2\alpha_W}{R^7}\cdot\frac{N_A a c_k I_W}{kT}\cdot\left(\frac{\alpha_i\nu_i^2 z_i^2 n_{hi} I_i}{I_i + I_W} + \frac{\alpha_j\nu_j^2 z_j^2 n_{hj} I_j}{I_j + I_W}\right)\left(\nu_i n_{hi} + \nu_j n_{hj}\right). \tag{4.56}$$

Ion–dipole polarization (induction) interactions:

$$\langle E_{iP}\rangle = -\frac{1}{2\pi\varepsilon_0\varepsilon_S}\cdot\frac{e^2\alpha_W\mu_W^2}{R^7}\cdot\frac{N_A a c_k}{kT}\left(\nu_i^2 z_i^2 n_{hi} + \nu_j^2 z_j^2 n_{hj}\right)\left(\nu_i n_{hi} + \nu_j n_{hj}\right). \tag{4.57}$$

Let us calculate the contribution of ion–dipole electric interactions (4.53) through (4.57) in the same way as in Section 1.5.3 for solution $CaCl_2$–$H_2O$ having the same parameters.

We substitute the required values for the calculation as per formula (4.53) taking into account that electron charge is equal $e = 1.60219 \cdot 10^{-19}$ C and $\nu_i = 1$, $\nu_j = 2$, $z_i = 2$, $z_j = 1$.

Ion–dipole interaction (4.53):

$$\langle E_{i\mu}\rangle = -\frac{(1.60219\cdot 10^{-19})^2\cdot(6.172\cdot 10^{-30})^2\cdot 6.022\cdot 10^{26}\cdot 0.561\cdot 10^{-9}\cdot 1.167}{24\pi^2\cdot 8.85\cdot 10^{-12}\cdot 53.5\cdot(0.684\cdot 10^{-9})^5\cdot 1.38\cdot 10^{-26}\cdot 298}$$

$$\times (1^2\cdot 2^2\cdot 6+2^2\cdot 1^2\cdot 6)(1\cdot 6+2\cdot 6) = -4.47\ \text{kJ mol}^{-1}.$$

Ion–quadrupole interaction (4.54):

$$\langle E_{iq}\rangle = -\frac{(1.6022\cdot 10^{-19})^2\cdot(1.3\cdot 10^{-30})^2\cdot 6.022\cdot 10^{26}\cdot 0.561\cdot 10^{-9}\cdot 1.167(1^2\cdot 2^2\cdot 6+2^2\cdot 1^2\cdot 6)}{80\pi\cdot 8.85\cdot 10^{-12}\cdot 53.5\cdot(0.684\cdot 10^{-9})^7\cdot 1.38\cdot 10^{-26}\cdot 298}$$

$$\times\left(1\cdot 6+2\cdot 6\right) = 3.82\cdot 10^{-12}\ \text{kJ mol}^{-1}.$$

Orientation and deformation of water dipoles in the field of ions (4.55):

$$\langle E_{i\alpha}\rangle = -\frac{(1.6022\cdot 10^{-19})^2\cdot(1.49\cdot 10^{-30})^2\cdot 6.022\cdot 10^{26}\cdot 0.561\cdot 10^{-9}\cdot 1.167(1^2\cdot 2^2\cdot 6+2^2\cdot 1^2\cdot 6)}{8\pi\cdot 8.85\cdot 10^{-12}\cdot 53.5\cdot(0.684\cdot 10^{-9})^5\cdot 1.38\cdot 10^{-26}\cdot 298}$$

$$\times (1.6+2.6) = -2.35\cdot 10^{-11}\ \text{kJ mol}^{-1}.$$

For calculation of ion–dipole dispersion (London) interactions under (4.56), we will take into account that electron polarizability of a calcium ion in aqueous solution $\alpha_{Ca} = 0.28\cdot 10^{-30}\ \text{m}^3$, ionization potential $I_{Ca} = 9.744\cdot 10^{-22}$ kJ, accordingly, and polarizability of a chloride ion in aqueous solution $\alpha_{Cl} = 3.59\cdot 10^{-30}\ \text{m}^3$, ionization potential $I_{Cl} = 2.082\cdot 10^{-22}$ kJ, accordingly [58]. First potential of water ionization $I_W = 2\cdot 10^{-21}$ kJ.

$$\langle E_{id}\rangle = -\frac{3\cdot\left(1.6022\cdot 10^{-19}\right)^2\cdot\left(1.49\cdot 10^{-30}\right)\cdot 6.022\cdot 10^{26}\cdot 0.561\cdot 10^{-9}\cdot 1.167\cdot 2\cdot 10^{-21}}{32\pi^2\cdot 8.85\cdot 10^{-12}\cdot 53.5\cdot\left(0.684\cdot 10^{-9}\right)^7\cdot 1.38\cdot 10^{-26}\cdot 298}$$

$$\times\left(\frac{0.28\cdot 10^{-30}\cdot 1^2\cdot 2^2\cdot 6\cdot 9.744\cdot 10^{-22}}{9.744\cdot 10^{-22}+2\cdot 10^{-21}}+\frac{3.59\cdot 10^{-30}\cdot 2^2\cdot 1^2\cdot 6\cdot 2.082\cdot 10^{-22}}{2.082\cdot 10^{-22}+2\cdot 10^{-21}}\right)$$

$$\times(1\cdot 6+2\cdot 6)$$

$$= -1.05\cdot 10^{-14}\ \text{kJ mol}^{-1}.$$

Ion–dipole polarization (induction) interactions (4.57):

$$\langle E_{iP}\rangle = -\frac{\left(1.6022\cdot 10^{-19}\right)^2\cdot\left(1.49\cdot 10^{-30}\right)\cdot\left(6.172\cdot 10^{-30}\right)^2\cdot 6.022\cdot 10^{26}\cdot 0.561\cdot 10^{-9}\cdot 1.167}{2\pi\cdot 8.85\cdot 10^{-12}\cdot 53.5\cdot\left(0.684\cdot 10^{-9}\right)^7\cdot 1.38\cdot 10^{-26}\cdot 298}$$

$$\times\left(1^2\cdot 2^2\cdot 6+2^2\cdot 1^2\cdot 6\right)\left(1\cdot 6+2\cdot 6\right) = -5.1\cdot 10^{-21}\ \text{kJ mol}^{-1}.$$

Let us estimate the calculation results of dipole interactions under formulas (4.41) through (4.45) and ion–dipole ones under formulas (4.53) through (4.57). All these interactions exist objectively, though the major ones of them are dipole–dipole (4.41) and ion–dipole (4.53). The rest can be neglected due to their minor contribution, which will not influence our further investigations in any way. We sum formulas (4.41) and (4.53) up, and after some transformations we have

$$U_{sp}+U_{vv}=-\frac{1}{6\pi\varepsilon_0\varepsilon_S}\cdot\frac{\mu_W^2 N_A a c_k\left(\nu_i n_{hi}+\nu_j n_{hj}\right)}{RkT}\left[\frac{e^2}{4\pi R^4}\left(\nu_i^2 z_i^2 n_{hi}+\nu_j^2 z_j^2 n_{hj}\right)+\frac{\mu_W^2}{R_{H_2O}^6}\right]. \tag{4.58}$$

Formula (4.58) is not very convenient for practical calculations. Let us substitute into it the most important physical constants and after some transformations we will obtain the energy of supermolecular interactions in host solution, $E_{sp}$:

$$E_{sp}=-2.0376\cdot 10^{-35}\frac{a c_k(\nu_i n_{hi}+\nu_j n_{hj})}{\varepsilon_S RT}\left(\frac{\nu_i^2 z_i^2 n_{hi}+\nu_j^2 z_j^2 n_{hj}}{R^4}+\frac{1.8639\cdot 10^{-20}}{R_{H_2O}^6}\right). \tag{4.59}$$

In formula (4.59), $R$ and $R_{H_2O}$ are substituted in meter, and under dependence (3.39), if discarding $10^{-9}$ the radius can be calculated in nanometer. Let us change (4.59) so that $R$ and $R_{H_2O}$ can be substituted in nanometer:

$$E_{sp}=-20.376\frac{a c_k(\nu_i n_{hi}+\nu_j n_{hj})}{\varepsilon_S RT}\left(\frac{\nu_i^2 z_i^2 n_{hi}+\nu_j^2 z_j^2 n_{hj}}{R^4}+\frac{1.8639\cdot 10^{-2}}{R_{H_2O}^6}\right). \tag{4.60}$$

Redistribution of electron density among water dipoles is sharply intensified with the decrease of distance $R$ between ion centers. Its role cannot be neglected at small $R$. Interacting complexes of ions and dipoles form a unified system. Interaction becomes more complicated. It cannot be already divided into independent components as at large $R$. Another complex interaction becomes stronger together with ion–dipole and dipole–dipole forces, which is called weak chemical.

## 4.6 WEAK CHEMICAL INTERACTIONS IN THE HOST SOLUTION

Concentrated electrolyte solutions take a peculiar place among survey targets. The distinctive feature of such liquid-phase systems, where there is a deficit of water molecules and competition of ions for the solvent, is a subordinated role of water–water interactions. Intensification of interparticle interactions results in an increase of a number of complexes becoming the main structural component and playing an important role in structural formation of solutions.

As it was said earlier, attraction, provoked by chemical interactions, that is, those that are accompanied by collectivization of electrons, starts increasing at small distances between water dipoles. These are interactions that define peculiarities of the structure of electrolyte solutions. That is why electrolyte solutions can be considered macroscopic molecules, where their structure varies constantly resulting in minor deviations from mean under the influence of heat motion. The unified structure of the solvent and ions starts forming. Hence, it follows the notion about small molecules that some chemically isolated particles are not that rigid even for gases and even less rigid for liquids [59,60]. It keeps its meaning if chemical bonds among molecules are a lot weaker than chemical bonds of atoms in molecules. The notion about ions in the condensed phase bears approximate nature, too. It is justified to the extent that chemical interaction between ions and their surroundings can be neglected.

Every electrolyte solution, as already mentioned, can be considered a gigantic macromolecule, although in such huge macromolecules one-type small fragments occur containing small quantity of atomic nuclei, the mutual positioning of which is more or less fixed. Ordered formations, occurring as a result of chemical interaction among particles, are called complexes. Complexes, in principle, are one-type formations and differ only in their composition. Complexes consist of heterostructural particles. Then, ion-aqueous clusters are formed with increase of ion concentration in a solution; further on, various supramolecular assemblies are formed with the decrease of dielectric permittivity (see Figure 1.1).

A distinction should be made between the question of stability of one particle of a complex (e.g., in a vacuum) and the question of stability of many such particles in a solution. These two questions are different both in content and in the solving method. The answer for the first question is determined by the type of potential energy as a function of distances among atomic nuclei in one isolated particle of a complex in a vacuum. If the potential function at some coordinate values of atomic nuclei passes through the minimum, the particle of the corresponding complex is stable itself. The question about the existence of potential energy minimum of one isolated particle of a complex could be solved, for example, by means of quantum-chemical calculations.

The second question is related to the behavior of the solution macroscopic system under the physical conditions in which the solution stays. To be more exact, in such case, it goes about behavior of the corresponding thermodynamic potential. When a solution is under predefined values of temperature $T$ (in thermostat) and pressure $P$, stability of many particles of complexes in a solution is determined by the properties of Gibbs free energy $G = H - TS$ as a function of $T$, $P$, and concentration of complexes $x_1, \ldots, x_k$. Under the thermodynamic equilibrium state, Gibbs free energy is minimum at certain values of concentrations $x_1, \ldots, x_k$ (here $x_i$ is a mole fraction of ions of $i$-sort). Values $x_1, \ldots, x_k$, in principle, could be determined using $k$ conditions. Each of these conditions has a form of

$$\frac{\partial G\left(T, P, x_1, \ldots, x_k\right)}{\partial x_i} = 0.$$

Liquid phase would not contain complexes or ion-aqueous clusters only in case if Gibbs free energy had minimum at $x_1 = 1, x_2, \ldots, x_k = 0$. In such case at predefined external conditions (defined $T$ and $P$), the complexes, defined under index *1*, would be stable enough in liquid phase although in reality liquids always contain a whole series of various chemical particles, complexes, ion-aqueous clusters, etc., formed in the course of interparticle interactions with participation of particles of sort *1*. That is why the solution of complexes, composition of which is determined by values $x_i$, transferring $G_B$ minimum, is always a thermodynamically stable liquid phase. Calculation of function $G(x_1, \ldots, x_i, \ldots, x_k)$ is a task of statistical mechanics. Difficulties, related to solving this task, for liquid systems have not yet been overcome.

There is no clear dependence between the structure of complexes and ion-aqueous clusters and their stability although those complexes and ion-aqueous clusters, in which enthalpy of formation under these conditions is larger, are usually more stable relatively in liquid phase at predefined $T$ and $P$. We are going to explain this using the following qualitative arguments. Let us assume that a solution with complexes $A$ and $B$ is at temperature $T$ and pressure $P$. With the appearance of the chemical bond between $A$ and $B$, the number of external degrees of freedom of particles $A$ and $B$ decreases. Based on the laws of statistical mechanics, it leads to a decrease in entropy $S$ of a solution, so that $\Delta S < 0$. The enthalpy of such solution usually also decreases $\Delta H < 0$. If module $|\Delta H|$ is greater than module $|\Delta G|$, Gibbs free energy of a solution decreases for value $\Delta H = \Delta H - T\Delta S$. It is clear that large absolute values $\Delta H$ under other factors being equal contribute to stability of ion-aqueous cluster $AB$.

*Hydrogen bond* possesses an intermediate nature between valence and interparticle interaction. It is typical for liquids and solutions, where molecule (water, spirits, acids) composition includes a positively polarized hydrogen atom. Small sizes and the absence of inner electrons allow this hydrogen atom to enter into an additional interaction with a covalently not bound negatively polarized atom of another (or the same) molecule:

$$\overset{q-}{X}-\overset{q+}{H}\cdots\overset{q-}{X}-\overset{q+}{H}.$$

Such specific bond bears features of electrostatic and donor–acceptor interaction and results in the formation of ion-aqueous clusters and supramolecular assemblies in a solution. A hydrogen bond has no essential difference from any other chemical bond and is determined by a set of interactions of nuclei and shells participating in an atomic bond. Little energy of a hydrogen bond (compared to chemical) is explained by relatively large distance OH···O and low electron density of a hydrogen atom and, correspondingly, the effects related to it (charge transfer, polarization, etc.). Hydrogen bond and donor–acceptor interaction were studied by many researchers [6,8,11,13,61–67].

High dipole moment of water molecules ($\mu_W = 1.86$ D) and their ability to form four hydrogen bonds (two as a proton donor and two as a proton acceptor) lead to the formation of molecular associates in water. Hydrogen bonds are partially destroyed in liquid water; dynamic equilibrium among molecule associates and free molecules traveling in cavities among associates can be observed in this structure. With temperature increase, two processes are taking place simultaneously: increase of occupation density of cavities with free water molecules resulting in increase of density; increase of cavity sizes and decrease of associate sizes resulting in decrease of system density. Peculiarities of the structure (availability of inner hollows), high dielectric permittivity, ability to show proton–donor and proton–acceptor as well as electron–donor and electron–acceptor properties, low density give rise to unique properties of water as a solvent [68].

During the formation of a hydrogen bond, besides pure electrostatic effect (mentioned before) of interaction of polar bonds $A\text{–H}\cdots B\text{–}R_2$, delocalization of electron charge takes place, that is, partial transfer of charge from donor molecule $B\text{–R}_2$ to acceptor molecule $R_1\text{–H}$. For simplicity, let us consider only the bridge $A\text{–H}\cdots B$. Positive charge on atom H itself is low in bond $A$–H. But during formation of H-bond, electron charge *flows* to atom $A$ from atom H, thereby releasing s-orbital of hydrogen to accept lone-pair electron charge from atom $B$, which will tie atoms H and $B$ with a hydrogen bond. At that, release of s-orbital of atom H *strips* the proton. The proton field is large and attraction of electron charge of atom $B$ by it is rather effective, at the same time there are no other electrons near the proton, that's why repulsion of molecule $BR_2$ from $R_1AH$ within the area of atom H decreases greatly. Both results of release of s-orbital during formation of H-bond contribute to its strengthening. At the same time, they explain why hydrogen (sometimes lithium) is able to form the similar bond. The nucleus does not get stripped and inner shells provide repulsion of the second molecule from electron shells at any other atom during release of orbital. As we can see, mechanism of formation of hydrogen bond is close enough to donor–acceptor [67]. The described electron charge transfer mechanism requires bond $A$–H to differ by visible polarizability, atom $A$ by high electronegativity (high ionization potential) and atom $B$ by donor properties. Availability of lone-pair electrons with atom $B$ and much lower potential of molecule ionization contribute to donor properties.

Geometry of complexes with hydrogen bond is of utmost interest. Almost in all cases, the hydrogen atom lies on a line connecting centers of two heavy atoms, though the energy required for its displacement from that line is probably very low. Calculations show that molecules are bent very easily in the gas phase [65].

Both electrostatic attraction and charge transfer energy depend on the angle of curvature, that's why certain equilibrium should be present between them. In order to optimize charge transfer energy, a hydrogen atom should approach along the line of maximum electron density of orbital of lone pair $B$ because at that the overlap integral $S_{d\alpha}$ is maximum. However, it was shown that only that factor does not explain geometry of complexes, since such line is not obligatory optimal for electrostatic energy.

Taking into account the wide interval of hydrogen bond strength in complexes, it is difficult to find a rule to be applicable for all such complexes. Based on calculation analysis according to the method of molecular orbitals and experiment data, L. Allen offered a formula for calculation of bond energy $E$ having the following form:

$$E_{\mathrm{H}} = \frac{1.87}{e} \cdot \frac{\mu_{AH} I_{\mathrm{H}}}{R_{\mathrm{H_2O}}}, \tag{4.61}$$

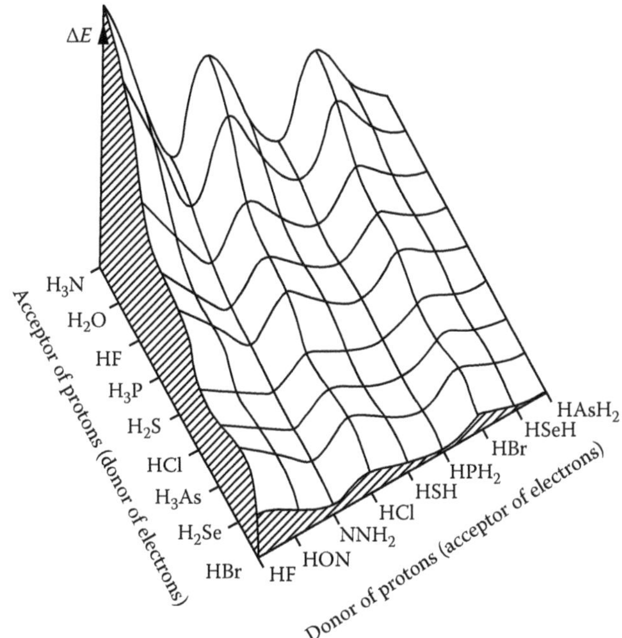


**FIGURE 4.2** Energies of complexes with hydrogen bond calculated as per formula (4.61). (From Marell D. et al., *Chemical Bond*, The Lane with English, Moscow, Russia Mir., Moscow, Russia, 1980, p. 378.)

where $\mu_{AH}$ is the dipole-binding moment *AH*. Murrell writes that this value raises certain doubts, only just total molecule dipole moment is expressly resolved. His comment is fair, he considers dipole-binding moment of *A*H complexes containing hydrogen and it can be seen from Figure 4.2. It shows the *bond energy surface* for 81 complexes calculated based on expression (4.61) and provides visual presentation about interconnection among complexes.

We consider dipole–dipole interactions in the electrolyte solution and we can calculate without any particular harm for errors of calculation under formula (4.61) that $\mu_{AH} \equiv \mu_W$. Further on, Murrell as per formula (4.61) points out that $I_H$ is ionization potential of hydrogen atom acceptor measured with regard to ionization potential of inert gas isoelectronic atom (e.g., $H_2O$ relating to Ne). It is rather complicated, but we will assume that $I_H$ is the initial ionization potential of a hydrogen atom and equal to 13.6 eV or $13.6 \cdot 1.602176 \cdot 10^{-22}$ kJ [58].

It is rather interesting that, several years before Allen, Guryanova et al. [66] published regressional dependence of calculation for heat formation in donor–acceptor bonds (in kcal/mol) based on the structure very similar to formula (4.61).

$$-\Delta H = \frac{35.3\mu_{DA}}{eR_{DA}}, \tag{4.62}$$

where the correlation coefficient is equal to 0.933.

In this dependence, there is no ionization potential of hydrogen atom $I_H$. If we multiply coefficient 1.87 under formula (4.61) by ionization potential of hydrogen atom $I_H$, we will obtain the coefficient equal to 25.43. The value is very close to coefficient 35.3 in dependence (4.62) although further on we will use formula (4.61) as much substantiated theoretically.

Guryanova investigated more than 50 various systems different from the ones shown in Figure 4.2. She made a conclusion that the established ratio between charge transfer ratio and intermolecular bond durability is common for donor–acceptor complexes.

Let us go back to Equation 4.61. We will make statistical averaging and summation as per (3.13). After some transformations, we will obtain a calculation formula for weak hydrogen bonds in the electrolyte solution:

$$\langle E_H \rangle = \frac{1.87}{e} \cdot \frac{\mu_W^2 I_H}{\varepsilon_S R_{H_2O}} \cdot \frac{N_A a c_k}{RkT} \left( \nu_i n_{hi} + \nu_j n_{hj} \right). \tag{4.63}$$

We will substitute numerical values of physical constants and data of the example for $CaCl_2$–$H_2O$ after formula (4.45).

$$\langle E_H \rangle = \frac{1.87 \cdot \left(6.172 \cdot 10^{-30}\right)^2 13.6 \cdot 1.602176 \cdot 10^{-22} \cdot 6.022 \cdot 10^{26} \cdot 0.561 \cdot 1.167}{1.602190 \cdot 10^{-19} \cdot 53.5 \cdot 0.373 \cdot 10^{-9} \cdot 0.684 \cdot 1.38 \cdot 10^{-26} \cdot 298}$$
$$\times \left(1 \cdot 6 + 2 \cdot 6\right) = 0.15 \text{ kJ mol}^{-1}.$$

In our calculations under the absolute value of energy in dipole–dipole interactions in Section 1.5.3 $\langle E_{DD} \rangle = -0.12$ kJ mol$^{-1}$, ion–dipole interactions in Section 1.5.4 $\langle E_{i\mu} \rangle = -4.47$ kJ mol$^{-1}$, weak hydrogen bonds $\langle E_H \rangle = 0.15$ kJ mol$^{-1}$. According to the classics [69], the energy of weak hydrogen bonds can be found by the module between dipole and ion–dipole interactions, which is observed in our case.

Let us substitute the values of fundamental physical constants into formula (4.63) and we will have

$$\langle E_H \rangle = \frac{4.2276 \cdot 10^{-8}}{\varepsilon_S R_{H_2O}} \cdot \frac{a c_k}{RT} \left( \nu_i n_{hi} + \nu_j n_{hj} \right). \tag{4.64}$$

If substituting values $a$, $R_{H_2O}$, and $R$ in nm into formula (4.64), then (4.64) will be as follows:

$$\langle E_H \rangle = \frac{42.276}{\varepsilon_S R_{H_2O}} \cdot \frac{a c_k}{RT} \left( \nu_i n_{hi} + \nu_j n_{hj} \right). \tag{4.65}$$

Let us proceed to the consideration of repulsive forces in the electrolyte solution.

## 4.7 SHORT-RANGE REPULSION IN THE HOST SOLUTION

Short-range repulsion always occurs at contact of particles. Particles would not have volume if short-range repulsion did not exist. They would be permeable and solutions or bodies could not exist. There would be no collisions and no exchange of kinetic and potential energies between particles. The state of thermodynamic equilibrium would be unattainable.

Empirical correlations, describing the dependence of repulsion energy $E$ or $U$ from distance $R$ among molecules, are obtained investigating properties of gases, liquids, and solid bodies as well as processes of collisions between molecules. At any calculation of $E(R)$, it is necessary to take into account that repulsion counteracts attraction of molecules. If the dependence of attraction energy from $R$ is well known, calculations of repulsion energy become much easier. Attractive forces, acting among atoms in inert elements and among ions with completely filled electron shells, such as $Na^+$, $K^+$, $Cs^+$, $Cl^-$, and $Br^-$, are studied better than others. And even for these particles the information about attraction energy is not very accurate. It is enough for empirical expressions for repulsion energy to be not accurate, either.

Basic information about repulsion potential is obtained by quantum-chemical calculations, as well as during experimental research of scattered molecular beams. Beam scattering experiments allow to achieve distances of 0.1–0.2 nm for approaching molecules, where potential energy is determined mainly by repulsion. Short-range repulsion forces, which appear at close distances between chemically saturated molecules (in the simplest case—between atoms with filled electron shells), have as a prototype the interaction between two hydrogen atoms in the ground state with parallel electron spins and are associated with manifestation of the Pauli principle. Calculations show that repulsion potential between atoms with filled electron shells represents the sum of terms of type $P(R)\exp(-aR)$, where $P(R)$ is a relatively slow-changing function $R$ (polynom type) [13].

Some researchers try to use the Lennard-Jones potential in order to describe repulsion forces in electrolyte solutions. The Lennard-Jones potential (potential 6–12) is a simple model of pair interaction of *nonpolar* molecules describing the dependence of interaction energy of two particles from the distance between them. This model produces rather realistically properties of real interaction of *spherical nonpolar molecules* [10]:

$$U(R) = 4\varepsilon\left[\left(\frac{\sigma}{R}\right)^{12} - \left(\frac{\sigma}{R}\right)^{6}\right], \qquad (4.66)$$

where

- $\varepsilon$ is the potential well depth
- $\sigma$ is the distance at which interaction energy turns to be equal to zero

Parameters $\varepsilon$ and $\sigma$ are characteristics of molecules of the corresponding substance.

At large $R$ molecules are attracted, which corresponds to term $(\sigma/R)^6$ in formula (4.66). This dependence can be substantiated theoretically and it is explained by

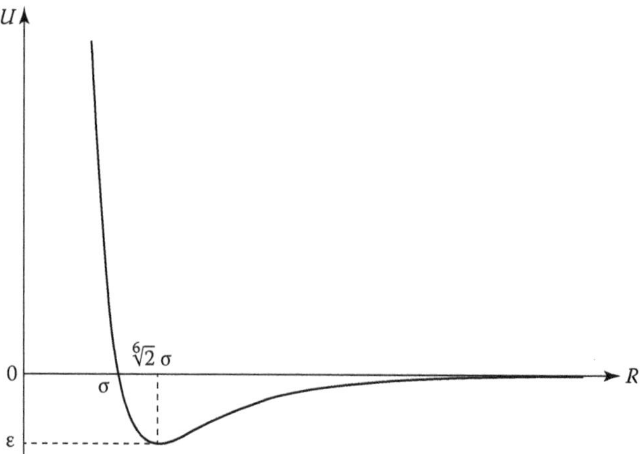


**FIGURE 4.3** Lennard-Jones potential.

van der Waals forces (dipole–dipole induced interaction) [10]. According to Section 1.5.4 for concentrated solution $CaCl_2$–$H_2O$, these interactions are extremely low $\langle E_{ind} \rangle = -1.4 \cdot 10^{-3}\,\text{kJ}\,\text{mol}^{-1}$.

At short distances, the molecules are repulsed due to exchange interaction (molecules start repulsing intensively in case of overlapping of electron clouds), which corresponds to term $(\sigma/R)^{12}$. This certain type of repulsion potential, in contrast to the type of attraction potential $(\sigma/R)^6$, is not substantiated theoretically.

From the point of view of thermodynamics, the Lennard-Jones potential can be used when describing gaseous, liquid, and solid phases of the substance. The least value of free energy for a conventional substance, for which the Lennard-Jones potential is fair, is attained in case of hexagonal close packing (see Figure 4.3).

At temperature increase, the structure with the least free energy is exchanged for cubic face-centered close packing, and then to-liquid change takes place. Transfer from cubic close packing to hexagonal close packing takes place under pressure for the structure with the least energy [10].

### 4.7.1 Other Model Potentials

Quantum mechanics allows calculating approximately potential curves of their interaction in molecules based on the known structure of atoms. However, it is much more difficult for interparticle interaction. That is why for the theoretical description one usually proceeds not from initial principles but uses some model potentials selecting parameters included into the expression for $U(R)$ so that the calculations based on these potentials conform well to experimental data.

It is clear that, except in cases of interaction of the simplest molecules, rigorous calculation of interparticle potential is beyond contemporary computational capabilities. Actually there is not much known qualitatively about the type of interparticle potential within the intermediate zone of distances (e.g., atomic diameter order).

That is why, as a rule, data about properties of transfer (and thermodynamic properties) serve as a means of obtaining information about the nature of interparticle interaction but not vice versa. One usually acts as follows: construct any kind of *model potentials* describing main properties of interparticle interaction and containing a small number of free parameters, the values of which are selected according to experiment results. Certainly, at sufficient number of parameters, it is always possible to provide concordance with experimental data, and such procedure is justified only in case if this number is small. At the same time, there should be enough parameters in order to describe all characteristics of particles depending on the type of interparticle potential (e.g., compressibility, line broadening in optical and microwave ranges, etc., belong to them besides transition coefficients [10,11]).

In this section, we will describe some model potentials that turned to be the most useful and applied the most frequently and will discuss in brief their advantages and disadvantages [10,11,13,70–75].

#### 4.7.1.1 Hard Sphere

It is assumed in the hard-sphere model that molecules are similar to billiard balls with the diameter of σ, so that interaction potential is

$$U(R) = \begin{cases} \infty, & R < \sigma, \\ 0, & R > \sigma. \end{cases} \tag{4.67}$$

Its schematic representation is shown in Figure 4.4.

It is the simplest of all models: it contains the only parameter σ. Its obvious advantages are excellent visualization and a possibility to obtain accurate results analytically. Owing to the first of these advantages, it was often used in the early (heuristic) kinetic theory; owing to the second one, it is convenient for applied calculations and for verification of approximation methods. Certainly, the hard-sphere model is rather far from reality, namely at *small* distances the potential grows too steeply and at large distances the interaction is not taken into account at all. That is why it approximately describes temperature dependence of transition coefficients (as well as other properties).

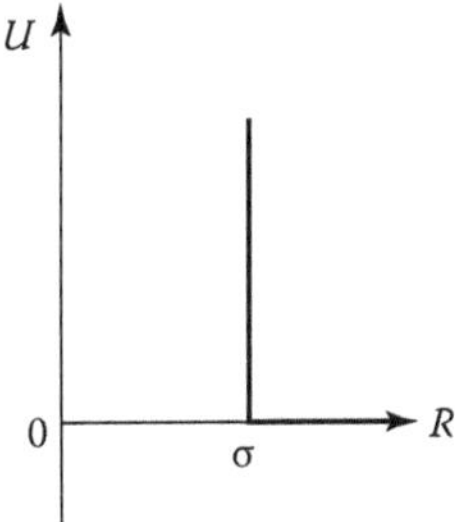


**FIGURE 4.4** Hard-sphere model.

#### 4.7.1.2 Point Center of Repulsion

This potential is purely repulsive as well; however, it grows less steeply than the hard-sphere model potential. It can be expressed as follows:

$$U(R) = \frac{A}{R^n}, \quad (4.68)$$

Where $n$ is called the repulsion index. In contrast to hard-sphere model, this model accepts compressibility of a molecule and owing to availability of two parameters it allows describing the transfer phenomenon better. Temperature dependence of kinetic coefficients is fully determined by index $n$. For the majority of real particles, $n$ has values located between $n = 9$ (*soft* particles) and $n = 15$ (*hard* particles).

Its schematic representation is shown in Figure 4.5.

The long-range part of interparticle potential that corresponds to attractive forces, that is, no potential well, is not taken into account in the considered model.

#### 4.7.1.3 Rectangular-Well Potential

The rectangular-well potential is described by the following formula [75,76]:

$$U(R_{12}) = \begin{cases} \infty, & R_{12} < a, \\ -\varepsilon, & a < R_{12} < d, \\ 0, & R_{12} > d, \end{cases} \quad (4.69)$$

where

- $d$ is the radius of action of pair potential
- $\varepsilon$ is the potential well depth
- $a$ is the effective molecular diameter

Its schematic representation is shown in Figure 4.6.

Regardless of simplicity of the rectangular-well potential, it describes adequately main peculiarities of the potential curve in interatomic interaction, including availability of the potential well and potential barrier at $R_{12} = a$.

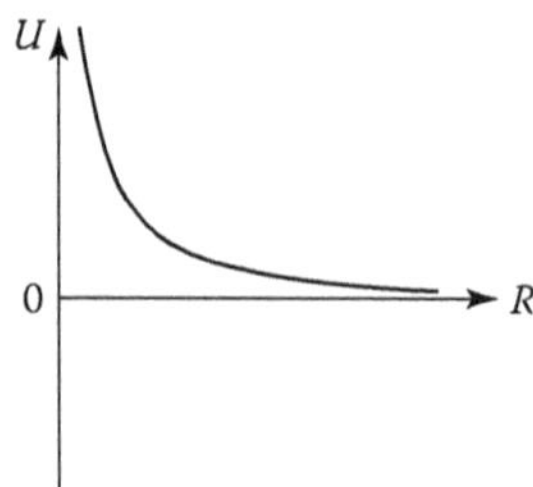


**FIGURE 4.5** Model of point center of repulsion.

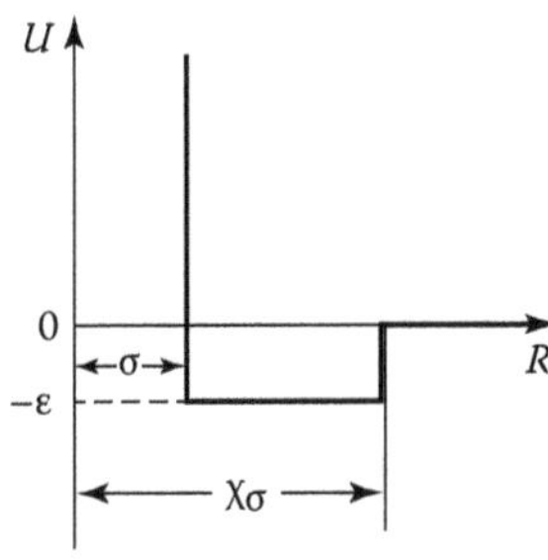


**FIGURE 4.6** Model of rectangular-well potential.

#### 4.7.1.4 Model of Triangular-Well Potential

In contrast to the previous potential, it takes into account gradual decrease of attraction energy with increase in the distance between molecules:

$$U(R)=\begin{cases}\infty, & R<\sigma,\\ \dfrac{-\varepsilon(\lambda-R/\sigma)}{(\lambda-1)}, & \sigma\le R<\lambda\sigma,\\ 0, & R\ge\lambda a.\end{cases} \tag{4.70}$$

Its schematic representation is shown in Figure 4.7.

#### 4.7.1.5 Sutherland Model

The disadvantage of previous models is that they contain only a repulsive part of real potential and do not describe attraction. The next model to be considered here was actually obtained by superposition of one of the repulsion potentials and attraction potential described earlier. Sutherland offered a model in which repulsion is described by hard-sphere potential and attraction by the power function:

$$U(R)-\begin{cases}\infty, & R<\sigma,\\ -\varepsilon\left(\dfrac{\sigma}{R}\right)^{\nu}.\end{cases} \tag{4.71}$$

Its schematic representation is shown in Figure 4.8.

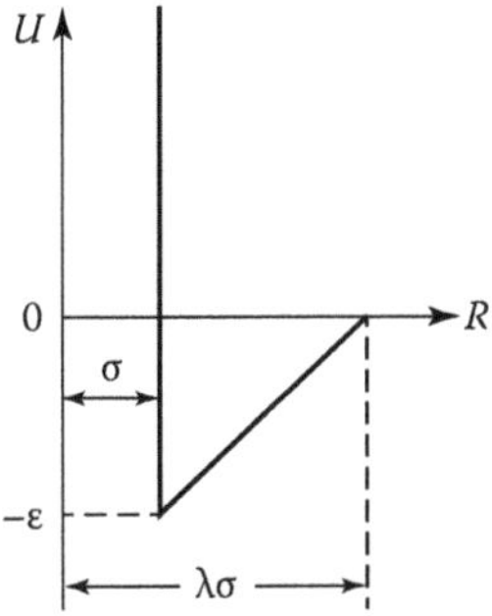


**FIGURE 4.7** Model of triangular-well potential.

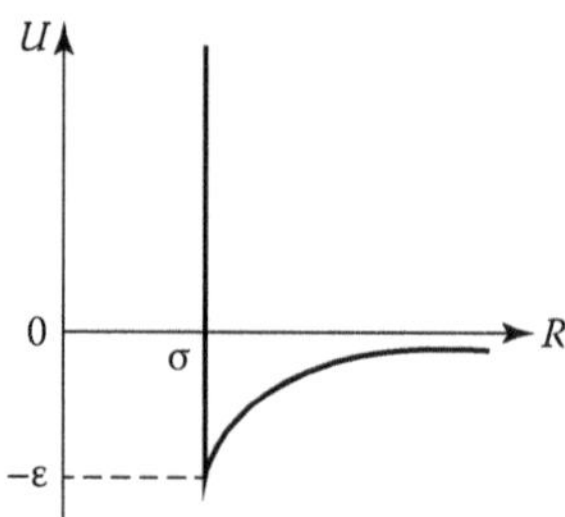


**FIGURE 4.8** Sutherland model.

It allows obtaining a good result for temperature dependence of transition coefficients and it describes with fine accuracy a number of other properties at the same values of parameters. The Sutherland potential is rather convenient for calculations too. However, its disadvantage is that it grows too steeply in the repulsion area and since kinetic coefficients are determined by repulsion at high temperatures, the Sutherland model does not provide good results.

Potentials (4.67) and (4.69) through (4.72) describe interaction of particles with a *hard core* apart from potential (4.68), which accepts its usage for *soft* or *hard* particles based on the researcher's choice. Potentials for particles with a *soft core* are given further reflecting much realistically dependence of repulsion energy from distance than the hard-sphere models. The models discussed in the following are offered in an attempt to eliminate this considerable disadvantage.

*Power potential* ($n-m$):

$$U(R) = bR^{-n} - aR^{-m}.$$

Here, the first term reproduces repulsion, the second one attraction. Parameters $b$, $a$, $n$, and $m$ are positive and $n > m$. This potential can be expressed as follows:

$$U(R) = \frac{m\varepsilon}{n-m}\left[\left(\frac{R_0}{R}\right)^n - \frac{n}{m}\left(\frac{R_0}{R}\right)^m\right], \tag{4.72}$$

or

$$U(R) = \frac{m}{n-m}\left(\frac{n}{m}\right)^{m/(n-m)} \varepsilon\left[\left(\frac{\sigma}{R}\right)^n - \left(\frac{\sigma}{R}\right)^m\right], \tag{4.73}$$

where

$\varepsilon$ is the potential well depth on curve $U(R)$
$R_0$ is the distance corresponding to minimum so that $U(R_0) = -\varepsilon$
$\sigma$ is the distance at which curve $U(R)$ crosses the axis of abscissas $U(\sigma) = 0$ (see Figure 4.9)

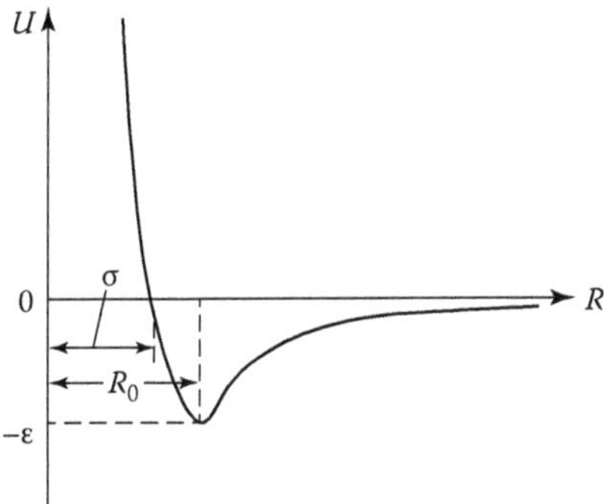


**FIGURE 4.9** Model of power potential.

At that, $\sigma = R_0(m/n)^{1/(n-m)}$. At decrease of the distance in area $R < \sigma$, repulsion grows very quickly and parameter σ is sometimes called the molecule diameter.

Values $n$ and $m$ can act as running parameters of the potential (then, potential (4.72) is four-parameter) or get specified in advance. Potential $n - 6$ is used sometimes, where parameter $n$ is a running parameter. It usually takes values from 12 to 25.

#### 4.7.1.6 Buckingham Potential

The potential includes a repulsive term approximated by the exponential and terms determined by dispersion attraction at the expense of dipole–dipole ($\approx R^{-6}$) and dipole–quadrupole ($\approx R^{-8}$) interactions:

$$U(R) = Ae^{-aR} - BR^{-6} - CR^{-8}. \tag{4.74}$$

The modified Buckingham potential appreciates only a dipole–dipole term:

$$U(R) = Ae^{-aR} - BR^{-6}, \tag{4.75}$$

(potential exp–6). It is a three-parameter potential that can be expressed as follows:

$$U(R) = \frac{\varepsilon}{1-6/s}\left\{\frac{6}{s}\exp\left[\alpha\left(1-\frac{R}{R_0}\right)\right]-\left(\frac{R_0}{R}\right)^6\right\}. \tag{4.76}$$

Parameters ε and $R_0$, like for power potential, outline the potential well depth and minimum position (see Figure 4.10); parameter $s = aR_0$ defines the steepness of the repulsion potential ($a$ is a coefficient in exponential for function (4.75)).

As mentioned already, exponential dependence for repulsion potential is theoretically more substantiated than power. However, the drawback of potentials (4.74) and (4.75) is their nonphysical behavior at small $R$: availability of maximum on curve $U(R)$ and tending $U(R)$ to $-\infty$ at $R \to 0$ (Equation 4.74). These potentials can be used for calculations for which behavior of function $U(R)$ at small distances is negligible.

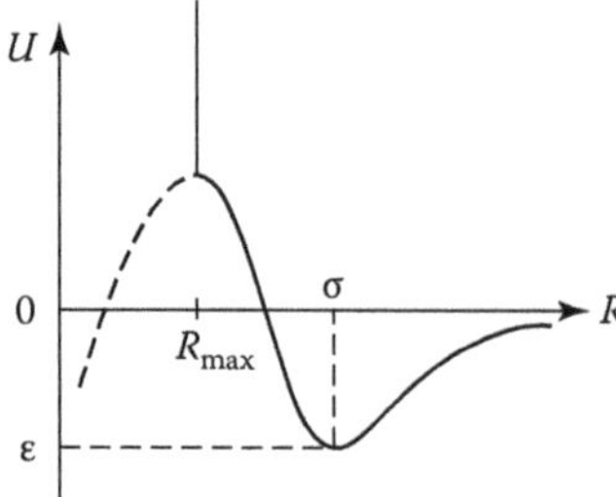


**FIGURE 4.10** The Buckingham model (4.75) (the dashed curve corresponds to unmodified potential (4.74)).

Dependence (exp −6) is widely used, in particular, in the variant of atom–atom potential to be discussed later. Potential (exp −6) is more accurate (if excluding the area of small distances) compared to potential (12 − 6), but calculations using it are more time consuming. The Buckingham potential describes experimental data well enough, but it is difficult for calculations. In order to eliminate excessive computational problems, value $\alpha$ is often get specified (value $\alpha$ is usually taken within interval of 12–16). At that, the model becomes a two-parameter one.

#### 4.7.1.7 Kihara Spherical Model (*Spherical Core* Model)

According to this model, a molecule is attributed a hard core and repulsion is subdivided into two parts, one of which is explained by the hard core and the second one by the soft repulsive part of the potential. The model is aimed at reflecting very quick growth of repulsion potential at small distances, which is impossible to do using either exponential function or power function. Let us assume that $d$ is the hard core diameter and $\rho$ is the distance between surfaces of cores for the considered pair of molecules. Obviously, $\rho = R - d$, where $R$ is the distance between molecule centers. Value $\rho$ is introduced instead of variable $R$ into potential (4.72), so that for optional $n$ and $m$:

$$U(R) = \frac{m\varepsilon}{n-m}\left[\left(\frac{\rho_0}{\rho}\right)^n - \frac{n}{m}\left(\frac{\rho_0}{\rho}\right)^m\right], \tag{4.77}$$

for $m = 6$, $n = 12$

$$U(R) = \varepsilon\left[\left(\frac{\rho_0}{\rho}\right)^{12} - 2\left(\frac{\rho_0}{\rho}\right)^6\right]. \tag{4.78}$$

In case of potential (12 − 6), $\sigma = 2^{-1/6}\rho_0 + d$. With increase of ratio $d/R_0$, the potential well becomes narrower (Figure 4.11).

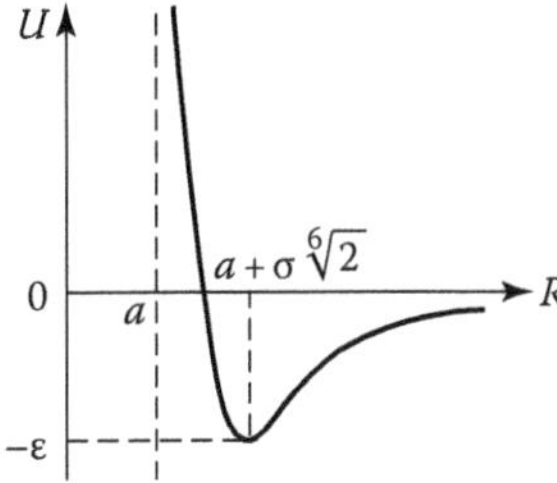


**FIGURE 4.11** Kihara model.

The hard core can be introduced in the same way for potential (exp −6). Introduction of the hard core ameliorates description not only of repulsion but also of attraction, for term $-2\varepsilon(\rho_0/\rho)^6$ corresponds to the sum of several terms in coordinates $R$:

$$\left(\frac{\rho_0}{\rho}\right)^6=\frac{\rho_0^6}{(r-d)^6}=\rho_0^6\left(R^{-6}+6R^{-7}d+21R^{-8}d^2+\cdots\right). \tag{4.79}$$

Hence, the Kihara potential allows describing dispersion contribution in a higher approximation than dipole–dipole. At the same time, the availability of the term proportional $R^{-7}$ to other odd degrees is a drawback of the potential.

There are a lot of other various two- and three-parameter potentials similar in structure to those considered [10,11,13,70,71,74]. Their consideration is not among the tasks of our narration. One can make sufficient conclusions from the considered number of potentials. Nevertheless, two- and three-parameter potentials do not allow rather adequately describing curve $U(R)$ in the whole area of distances $R$. Approximated nature of potential functions is largely settled by the potential parameter adjustment procedure based on experimental data, but values of parameters appear to be different depending on the fact in what area of distances $R$ the property under question is more sensitive. Multiparameter potential functions, found for some substances, do not have such drawbacks; however, these will not be considered here. Analytical theories, one of which is stated here, cannot use similar potentials and they are used only for numerical calculations by means of ECM.

The considered potentials of the type depicted in Figures 4.3, 4.4, 4.6 through 4.11 are absolutely unfit for the electrolyte solution. An attempt to represent forces of repulsive nature and attractive forces is made by means of a single formula. Such two mixed-nature conditions are impossible to be represented from the point of view of mathematics. Either of these conditions will always prevail and no selections of parametric coefficients can be found to avoid it. The deep potential well with a negative sign at $R \to 0$ raises many doubts.

Supramolecular interactions in electrolyte solutions in general should be described as noncentral, that is, to take into account the dependence of interaction energy from mutual orientation of polar water dipoles. The role of anisotropy of interaction forces is substantial to a variable degree for different contributions into interaction potential in electrolyte solutions. The potential of point center of repulsion, shown in

Figure 4.5, is worth attention for electrolyte solutions. The curve passes smoothly the whole concentration area without any sharp potential wells at change of electrolyte concentration in the solution. In the beginning of the curve, when $R \to 0$ ($m \to \infty$), its graphics approaches infinity, which means a sharp increase of repulsive forces. With increase of the radius, that is, at $R \to \infty$ ($m \to 0$), the curve smoothly asymptotically approaches zero. It testifies about decrease of repulsive forces at decrease of electrolyte concentration in the solution and reaching of the diluted form by it. Fulfilling the condition to accept value $n = 9$ (see the description for Figure 4.5), we will obtain one-parameter potential of point center of repulsion for real *soft* particles being supermolecular complexes and supramolecular assemblies.

Parameter $A$ is empirical. By not changing commonness of reasoning, we can consider that $A = 4.58518 \cdot 10^{-51}B$. We return to (3.14). Let us write the sum taking into account (4.68), and for repulsion potential $U_{ot}$ after several transformations, we will have

$$U_{ot} = \frac{4.58518 \cdot 10^{-51} B}{\varepsilon_S (R_{H_2O})^9} \cdot \frac{N_A a c_k}{RkT} (\nu_i n_{hi} + \nu_j n_{hj}). \tag{4.80}$$

As stated earlier, for the potential of point center of repulsion, we accept $n = 9$, and in (4.80), we will take into account that $R_{H_2O} = R_{H_2O} \cdot 10^{-9}\,\text{m}$. Then, finally

$$U_{ot} = \frac{4.58518 \cdot 10^{-51} B}{\varepsilon_S R_{H_2O}} \cdot \frac{N_A a c_k}{RkT} (\nu_i n_{hi} + \nu_j n_{hj}), \tag{4.81}$$

where $a, R_{H_2O}$, and $R$ are substituted in nanometer.

When substituting values of all fundamental physical constants into Equation 4.81, we will have

$$U_{ot} = \frac{200}{\varepsilon_S R_{H_2O}} \cdot \frac{B a c_k}{RT} (\nu_i n_{hi} + \nu_j n_{hj}). \tag{4.82}$$

From conditions of the example in Section 1.5.4 for system $CaCl_2$–$H_2O$ in Figure 4.12, the following are shown: the calculated curve of repulsive interactions under formula (4.82), curve 1; curve 2 for weak hydrogen bonds under formula (4.65); and curves 1 and 2, which are the resultant of curve 3.

It is clear from the diagrams that with increase of $R_{H_2O}$, ($m \to 0$), the curves asymptotically approach zero. On the contrary, at $R \to 0$ ($m \to \infty$), the values of repulsive forces and weak hydrogen bonds increase sharply. It confirms robustness of the chosen models for weak hydrogen bonds, formula (4.65), and repulsive forces, formula (4.82). It should be stated that repulsion energy for overlapping of electron clouds is rather sensitive to change of $R_{H_2O}$. In calculations under dependence (4.82), value $B$ was taken equal to 2. It is not very correct, as we will see in later text, that the value of parameter $B$ depends on the electrolyte concentration in the solution, but for qualitative conclusions of reasonable usage of the potential of point center of repulsion, it is quite sufficient.

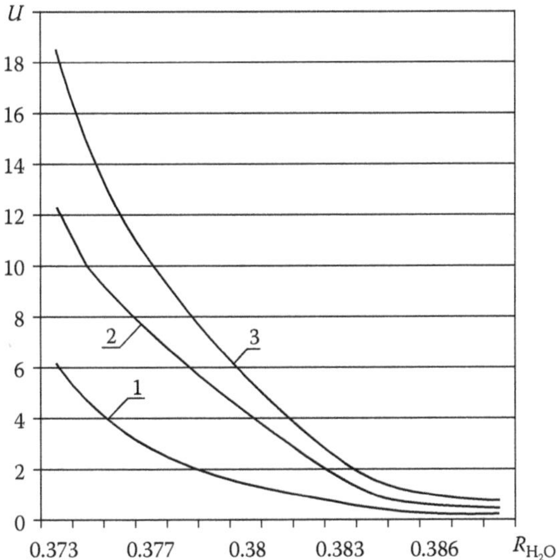


**FIGURE 4.12** Course of change of repulsive forces, hydrogen bonds, and resultant curve: 1—hydrogen bonds; 2—repulsive forces; and 3—resultant curve.

It is clear from Figure 4.12 that curves 1 and 2 behave identically, proceeding from which formulas (4.65) and (4.82) can be combined, and as a result we will obtain

$$U_x = \langle E_H \rangle + U_{ot} = \frac{42.276}{\varepsilon_S R_{H_2O}} \cdot \frac{ac_k (\nu_i n_{hi} + \nu_j n_{hj})}{RT} (1 + 4.74B). \tag{4.83}$$

Now let us proceed to the definition for a number of electrolyte solutions of parameter $B$ taking into account experimental data of activity coefficients.

## 4.8 ADJUSTMENT OF POTENTIAL OF SUPRAMOLECULAR INTERACTIONS

Activity coefficients $\gamma_\pm$ will be calculated as per formula (3.34), where $f_\pm$ will be presented as follows:

$$\ln f_\pm = E_{sp} + \langle E_H \rangle + U_{ot} + \Psi_\gamma^0, \tag{4.84}$$

where

$E_{sp}$ is calculated as per formula (4.60)
$\langle E_H \rangle$ as per (4.65)
$U_{ot}$ as per (4.82)

Electrostatic interactions $\Psi_{\gamma}^{0}$ in (4.84) will be represented as

$$\Psi_{\gamma}^{0} = \frac{\Omega_g c_k}{\chi \varepsilon_S T} \Psi_{el}^{0}, \tag{4.85}$$

where

$\Omega_g$ is determined by formula (3.15)
$\Psi_{el}^{0}$ by (2.80)
$\varepsilon_S$ by (3.50)

In formulas (2.80), (4.60), (4.65), and (4.82), values $a$ are evaluated as per dependences (3.35) or (3.36); $R_{H_2O}$ and $R$ as per (3.39); and solution density, $\rho_k$, as per (3.40).

Table 3.9 presents calculated thermodynamic values included into formulas (4.84) and (4.85) of a number of electrolytes at 25°C, corresponding to Table 3.6. In table columns from left to right, the following values are shown: electrolyte concentration, $m$, mol/kg $H_2O$; electrolyte concentration, $c$, mol/dm$^3$; distance between ions, $R$, nm; distance between water dipoles, $R_{H_2O}$, nm; electrostatic interactions, $-\Psi_{\gamma}^{0}$; value of constant $\chi$ in (4.85); energy of supermolecular interactions, $-E_{sp}$; energy of hydrogen bonds, $E_H$; energy of repulsive interactions, $U_{ot}$; values of coefficient $B$ in (4.82); values of dielectric permittivity, $\varepsilon_S$; and values of activity coefficients, $\gamma_{\pm}$. Activity coefficients $\gamma_{\pm}$ were taken from [77]. Before every electrolyte, values $\xi$ are given as per formula (3.35) and, proceeding from that, value $a$.

From Table 3.9, two 1–2 electrolytes and one 1–1 electrolyte were chosen, and all supramolecular interactions existing in the solution from very dilute to the concentrated one are shown in Figure 4.13 for $CaCl_2$, Figure 4.14 for $MgCl_2$, and Figure 4.15 for LiBr. Axis $x$ has plotted numbers of concentration points, $m$, mol/kg $H_2O$, and axis $y$ has thermodynamic values from Table 3.9: electrostatic interactions, $-\Psi_{\gamma}^{0}$; energy of supermolecular interactions, $-E_{sp}$; energy of hydrogen bonds, $E_H$; energy of repulsive interactions, $U_{ot}$; values of dielectric permittivity, $\varepsilon_S$; and values of activity coefficients, $\gamma_{\pm}$.

The axis of abscissas has plotted numbers of concentration points, $m$, mol/kg $H_2O$. Every number of the concentration point is attributed its own concentration from Table 3.9:

Figures 4.13 through 4.15 show that the behavior of interparticle forces in electrolyte solutions is of similar nature regardless of electrolyte type (1–2 or 1–1). Short-range repulsion forces, curve 2, occur at the end of the first area of the concentration structural transition scheme, Figure 1.1, where there are still residues of volumetric tetrahedral water. Hydrogen bonds, curve 3, conform to the same regularity as curve 2. In all considered cases in Table 3.9, short-range repulsion forces are considerably higher than values of hydrogen bonds.

The behavior of electrostatic forces in the first area of structural transition, Figure 1.1, is of interest. Starting from dilute solutions, the electrostatic forces grow monotonously under the absolute value till concentrations $m \approx 0.05/0.1$ mol/kg $\cdot H_2O$, and by the end of the first concentration area, they increase rapidly and decrease steeply to zero in the narrow range of concentrations. This points to the fact that the network of H–water bonds is finally disintegrated and complete shielding of

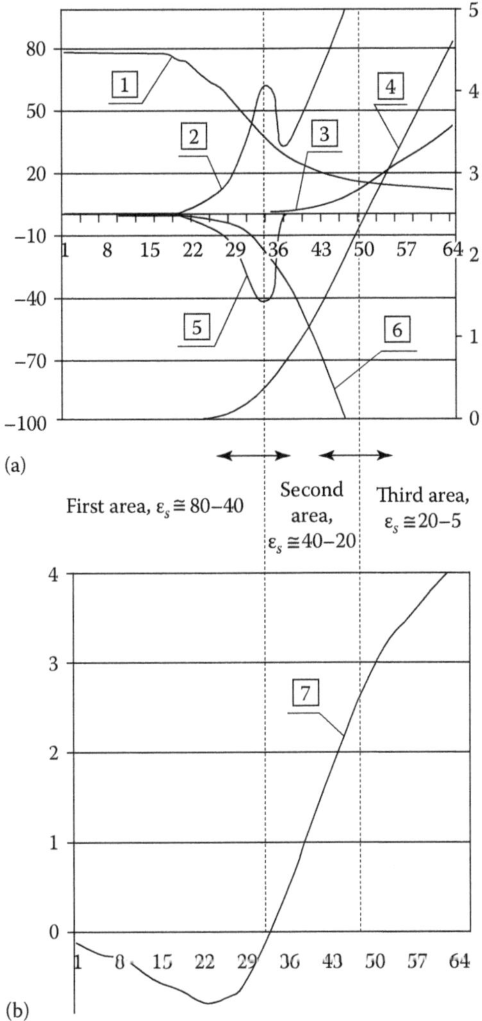


**FIGURE 4.13** Supramolecular interactions (a) (curves 1–6) and the potential of interparticle interactions (b) (curve 7) for solution of $CaCl_2$: 1—dielectric permittivity of solution, $\varepsilon_S$; 2—short-range repulsion forces, $U_{oi}$; 3—activity coefficient, $\gamma_\pm$; 4—hydrogen bonds, $E_H$; 5—electrostatic forces, $-\Psi_\gamma^0$; 6—ion–dipole and dipole–dipole forces, $-E_{sp}$; and 7—interparticle interactions potential, $\Psi^0$. The auxiliary axis on the right at (a) is for curves 4 and 5. Numerals along axis $x$ on diagrams (a) and (b) designate numbers of concentration points from Table 3.9, that is, for solution $CaCl_2$ the concentrations were taken from minimum value 0.004 till maximum 5.75$m$. The same goes to diagrams in Figures 4.14 and 4.15.

Figure 4.13 for $CaCl_2$

| **Point Number** | **1** | **8** | **15** | **22** | **29** | **36** | **43** | **50** | **57** | **64** |
|---|---|---|---|---|---|---|---|---|---|---|
| Concentration, $m$ | 0.001 | 0.008 | 0.060 | 0.400 | 1.250 | 3.000 | 4.750 | 6.500 | 8.000 | 9.750 |

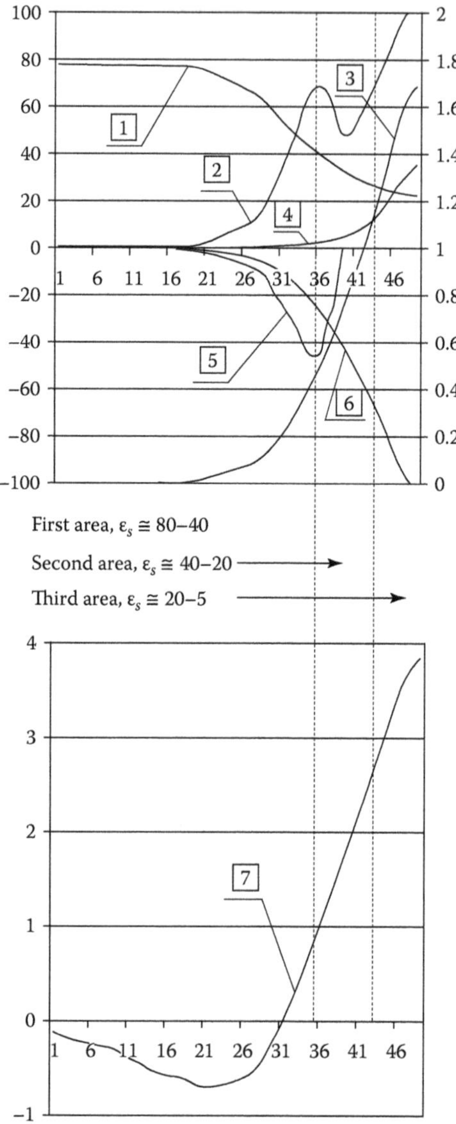


**FIGURE 4.14** Supramolecular interactions (a) (curves 1–6) and the potential of interparticle interactions (b) (curve 7) for solution of $MgCl_2$: 1—dielectric permittivity of solution, $\varepsilon_S$; 2—short-range repulsion forces, $U_{or}$; 3—hydrogen bonds, $E_H$; 4—activity coefficient, $\gamma_\pm$; 5—electrostatic forces, $-\Psi_\gamma^0$; 6—ion–dipole and dipole–dipole forces, $-E_{sp}$; and 7—interparticle interactions potential, $\Psi^0$. The auxiliary axis on the right at (a) is for curves 3 and 5.

Figure 4.14 for $MgCl_2$

| **Point Number** | **1** | **6** | **11** | **16** | **21** | **26** | **31** | **36** | **41** | **46** |
|---|---|---|---|---|---|---|---|---|---|---|
| Concentration, *m* | 0.001 | 0.008 | 0.060 | 0.400 | 1.250 | 3.000 | 4.750 | 3.000 | 4.250 | 5.500 |

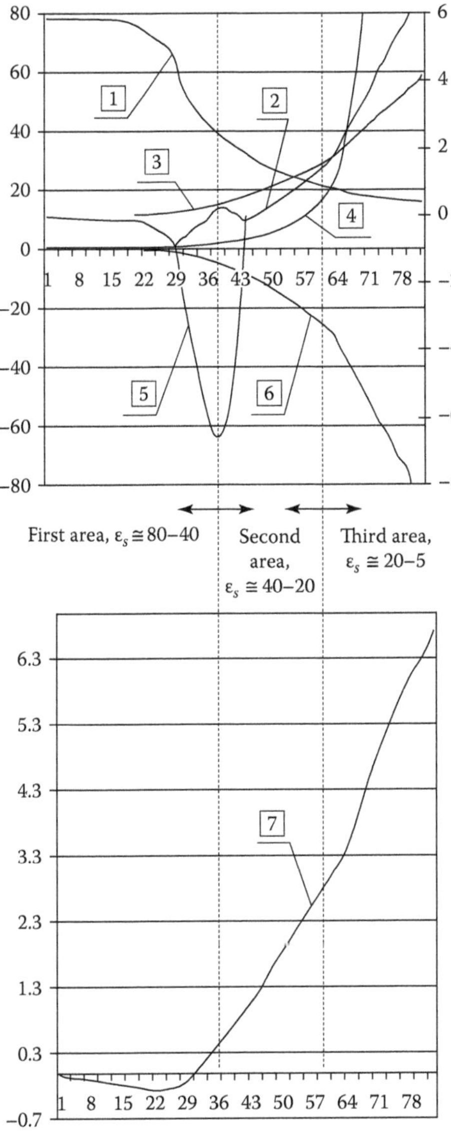


**FIGURE 4.15** Supramolecular interactions (a) (curves 1–6) and the potential of interparticle interactions (b) (curve 7) for solution of LiBr: 1—dielectric permittivity of solution, $\varepsilon_S$; 2—short-range repulsion forces, $U_{oi}$; 3—hydrogen bonds, $E_H$; 4—activity coefficient, $\gamma_{\pm}$; 5—electrostatic forces, $-\Psi_\gamma^0$; 6—ion-dipole and dipole-dipole forces, $-E_{sp}$; and 7—interparticle interactions potential, $\Psi^0$. The auxiliary axis on the right at (a) is for curves 3 and 5.

Figure 4.15 for LiBr

| **Point Number** | 1 | 8 | 15 | 22 | 29 | 36 | 43 | 50 | 57 | 64 | 71 | 78 |
|---|---|---|---|---|---|---|---|---|---|---|---|---|
| Concentration, *m* | 0.001 | 0.008 | 0.060 | 0.400 | 1.250 | 3.000 | 4.750 | 6.500 | 8.250 | 10.00 | 13.50 | 17.00 |

the electrostatic ion field by residues of volumetric tetrahedral water takes place. Transition to the second concentration area of structural transition is characterized by availability of ion-aqueous clusters and groupings. Such behavior pattern of electrostatic forces is present in all electrolyte solutions shown in Table 3.9.

The second concentration area of structural transition is characterized by further decrease of dielectric permittivity and formation of stable ion-aqueous assemblies. The electrostatic ion field is completely shielded and only ion–dipole together with dipole–dipole interactions as well as hydrogen bonds are present in the solution, which by the end of the second concentration area grow monotonously (values of ion–dipole and dipole–dipole interactions are understood as taken by the absolute value). Forces of repulsive nature as well as electrostatic interactions gain maximum at the border of the first and second concentration areas. In view of complete shielding of the strong ion field, the short-range repulsion forces decrease and then start monotonously growing with increase of solution concentration. Concentration of ion-aqueous clusters and associates grows and potential energy of all interactions tends to take the minimum value characterized by values of dielectric permittivity of solution $\varepsilon_S \approx 40$ in the beginning of the second area.

The second and third concentration areas are characterized by further decrease of dielectric permittivity. At the beginning of the third concentration area of structural transition, as Lyashchenko writes, further formation of new ion configurations, related to complex formation and formation of aggregates determined directly by regularities of coordination chemistry where water molecules act as equivalent ligands, takes place [60]. A more detailed analysis of structural transition toward highly concentrated solutions based on strong high frequency (SHF) dielectric investigations, data of electronic spectroscopy, small-angle x-ray scattering, and other radiophysical methods can be examined in other works by Lyashchenko et al. [57,59]. It must be highlighted that the theoretical *notion* of the scheme in Figure 1.1, obtained based on such precision measurements, confirms reasonably correctness of all concepts and prerequisites.

Let us go back to Figures 4.13 through 4.15. Below every figure under letter (b), there is the potential of interparticle interactions subdivided into three concentration areas. The first concentration zone finishes in the transition area from the potential well to positive values, the second zone starts in the curve break area, and the third one in the area of occurring asymptotic approach of the curve to any line with increase of electrolyte concentration or beginning of its crystallization. This assumption requires additional investigation. It is worth mentioning that the potential of interparticle interactions in electrolyte solutions bears absolutely no resemblance to the Lennard-Jones potential (see Figure 4.3). We emphasize once again on its problematic usage for electrolyte solutions.

As it is seen from Figures 4.13 through 4.15, weak hydrogen bonds (curve 3, in which values are set on the additional axis on the right) possess small values. These values are obviously not enough to form any ion-aqueous clusters and associates. The solution structure depends to a great extent on low values of dielectric permittivity of the solution. The contemporary chemical literature evidently overestimates weak hydrogen bonds for formation of any ionic, ion-aqueous clusters or associates.

## 4.9 THERMODYNAMICS OF WATER ACTIVITY INTERACTION

Let us consider the Gibbs–Duhem equation* for electrolyte solutions:

$$\frac{1000}{M_0}\,\mathrm{d}\ln a_w + \nu_k m_k\,\mathrm{d}\ln(\gamma_{\pm} m_k) = 0, \tag{4.86}$$

where

$M_0$ is the molecular weight of the solvent
$k$ is a component number of electrolyte in the solution

We write product $\ln(\gamma_{\pm} m_k)$ in (4.86), and after some transformations, we obtain

$$\ln a_w = -\frac{M_0 \nu_k}{1000}\left(\int_0^{m_k} m\,\mathrm{d}\ln\gamma_{\pm} + \int_0^{m_k} m\,\mathrm{d}\ln m\right). \tag{4.87}$$

The second integral in round brackets (4.87) will be calculated twice in parts, and after some transformations, we will obtain the thermodynamic ratio in the Gibbs–Duhem equation for electrolyte solutions between water activity and average ion activity coefficient (for water it is taken as $M_0 = 18.02$):

$$\ln a_w = -\frac{\nu_k m_k}{55.51}\left(1 + \frac{1}{m_k}\int_0^{m_k} m\,\mathrm{d}\ln\gamma_{\pm}\right). \tag{4.88}$$

Let us consider the integral on the right side (4.88). Let us take it in parts, assuming $u = m$; $\mathrm{d}u = \mathrm{d}m$; $\mathrm{d}v = \mathrm{d}\ln\gamma_{\pm}$; and $v = \ln\gamma_{\pm}$. Then, we have

$$\int_0^{m_k} m\,\mathrm{d}\ln\gamma_{\pm} = m_k \ln\gamma_{\pm} - \int_0^{m_k} \ln\gamma_{\pm}\,\mathrm{d}m. \tag{4.89}$$

Activity coefficients $\ln\gamma_{\pm}$ will be calculated as per formula (3.34), where $\ln f_{\pm}$ will be presented as follows:

$$\ln f_{\pm} = E_{sp} + U_x + \Psi_{\gamma}^{0}, \tag{4.90}$$

where

$E_{sp}$ is calculated as per formula (4.60)
$U_x$ as per (4.83)
$\Psi_{\gamma}^{0}$ as per (4.85)

* It is interesting to mention that Harned and Owen [78, p. 30] call it the Bjerrum ratio, and later Robinson and Stokes [79] and the subsequent physiochemical literature call it the Gibbs–Duhem equation. The author took this position too.

Let us depict (4.60) as follows:

$$E_{sp} = -c_k(\tau_{1sp} + \tau_{2sp}), \tag{4.91}$$

where

$$\tau_{1sp} = \frac{20.376a(\nu_i n_{hi} + \nu_j n_{hj})\left(\nu_i^2 z_i^2 n_{hi} + \nu_j^2 z_j^2 n_{hj}\right)}{\varepsilon_S T R^5}, \tag{4.92}$$

$$\tau_{2sp} = \frac{0.3798a(\nu_i n_{hi} + \nu_j n_{hj})}{\varepsilon_S T R R_{H_2O}}. \tag{4.93}$$

Values $U_x$ in (4.90) will be presented similarly to (4.91):

$$U_x = c_k \tau_x, \tag{4.94}$$

where

$$\tau_x = \frac{42.276\,a(\nu_i n_{hi} + \nu_j n_{hj})}{\varepsilon_S T R R_{H_2O}}(1 + 4.74B). \tag{4.95}$$

These designations in (4.91) and (4.103) are required for subsequent logics of narration not to be encumbered with long formulas.

Using formulas (4.91) and (4.94), we substitute into Equation 4.89, and after some transformations, the integral on the right side (4.89) will be as follows:

$$-\int_0^{m_k} \ln \gamma_\pm \,\mathrm{d}m = -\int_0^{m_k} \left[ c_k \left( \tau_x - \frac{\tau_{1sp}}{R^4} - \tau_{2sp} - \frac{\Omega_g}{\chi \varepsilon_S T} \Psi_{el}^0 \right) - \ln\left(1 + \frac{\nu_k m_k}{55.51}\right)\right] \mathrm{d}m. \tag{4.96}$$

The formulation under the integral sign on the right side (4.96) depends on two variables: $c_k$ and $m$. Let us bring them to one variable, $m$. We will use the approximate ratio for dilute solutions:

$$m \approx 10^3 \frac{c}{\rho}, \tag{4.97}$$

where $\rho$ is the solution density, kg/m$^3$, which is taken based on experimental data or determined with high accuracy as per dependence (3.40).

After differentiating (4.97):

$$\mathrm{d}m = 10^3 \frac{\mathrm{d}c}{\rho}. \tag{4.98}$$

One could try to take explicit dependence $m = m(c)$, but in the result of differentiation and subsequent integration, we have lengthy constructions bearing no physical sense.

By taking into account (4.98), we integrate the first round bracket in (4.96), and after some transformations, we obtain

$$-\int_0^{m_k} c_k\left(\tau_x - \frac{\tau_{1sp}}{R^4} - \tau_{2sp} - \frac{\Omega_g}{\chi\varepsilon_S T}\Psi_{el}^0\right)dm$$

$$= -\frac{m_k c_k}{2}\left(\tau_x - \frac{\tau_{1sp}}{R^4} - \tau_{2sp} - \frac{\Omega_g}{\chi\varepsilon_S T}\Psi_{el}^0\right) \equiv -m_k\frac{\ln f_\pm}{2}. \tag{4.99}$$

We integrate the last summand in (4.96), using tables of integrals [80], and after some transformations, we have

$$\int_0^{m_k} \ln\left(1+\frac{\nu m}{55.51}\right)dm = \frac{55.51}{\nu_k}\left(1+\frac{\nu_k m_k}{55.51}\right)\ln\left(1+\frac{\nu_k m_k}{55.51}\right) - m_k. \tag{4.100}$$

We will take into account from (3.34) that

$$\ln f_\pm = \ln\gamma_\pm + \ln\left(1+\frac{\nu m}{55.51}\right). \tag{4.101}$$

We substitute (4.101), (4.100), (4.99), and (4.89) into (4.88), and after some transformations, we obtain the thermodynamic formula for water activity calculation depending on activity coefficients:

$$\ln a_w = -\frac{\nu_k m_k}{55.51}\left(\gamma_W + S_W\right), \tag{4.102}$$

where

$$\gamma_W = \frac{\ln\gamma_\pm}{2},$$
$$S_W = \left(\frac{3}{2} + \frac{55.51}{\nu_k m_k}\right)\ln\left(1+\frac{\nu_k m_k}{55.51}\right). \tag{4.103}$$

Table 4.1 represents the comparison of experimental values of water activity $a_{w\,exp}$ with calculated values $a_{w\,calc}$ as per formula (4.102) as well as values $\gamma_W$ and $S_W$ for a number of electrolyte solutions depending on concentrations $m$, mol/kg $H_2O$ and $c$, mol/dm$^3$. Experimental values $\gamma_\pm$ were taken from [77] and water activity $a_{w\,exp}$ from [81]. As it is evident from Table 4.1, possibilities of calculation as per formula (4.102) under the concentration range for electrolytes of symmetrical type are higher than for nonsymmetrical types. Limitations are explained by condition (4.97) for dilute solutions.

**TABLE 4.1**
**Calculated Thermodynamic Components Included in Formula (4.102)**

| Electrolyte | $M$ | $c$ | $-\gamma_W$ | $S_W$ | $a_{w\,exp}$ | $a_{w\,calc}$ |
|---|---|---|---|---|---|---|
| $CaCl_2$ | 0.001 | 0.001 | 0.0600 | 1.0001 | 0.99 | 0.99 |
| | 0.002 | 0.002 | 0.0789 | 1.0001 | 0.99 | 0.99 |
| | 0.003 | 0.003 | 0.0950 | 1.0002 | 0.99 | 0.99 |
| | 0.004 | 0.004 | 0.1078 | 1.0002 | 0.99 | 0.99 |
| | 0.005 | 0.005 | 0.1191 | 1.0003 | 0.99 | 0.99 |
| | 0.006 | 0.006 | 0.1287 | 1.0003 | 0.99 | 0.99 |
| | 0.007 | 0.007 | 0.1359 | 1.0004 | 0.99 | 0.99 |
| | 0.008 | 0.008 | 0.1425 | 1.0004 | 0.99 | 0.99 |
| | 0.009 | 0.009 | 0.1546 | 1.0005 | 0.99 | 0.99 |
| | 0.010 | 0.010 | 0.1748 | 1.0005 | 0.99 | 0.99 |
| | 0.020 | 0.020 | 0.2017 | 1.0011 | 0.99 | 0.99 |
| | 0.030 | 0.030 | 0.2286 | 1.0016 | 0.99 | 0.99 |
| | 0.040 | 0.040 | 0.2521 | 1.0022 | 0.99 | 0.99 |
| | 0.050 | 0.050 | 0.2715 | 1.0027 | 0.99 | 0.99 |
| | 0.060 | 0.060 | 0.2872 | 1.0032 | 0.99 | 0.99 |
| | 0.070 | 0.070 | 0.2998 | 1.0038 | 0.99 | 0.99 |
| | 0.080 | 0.080 | 0.3099 | 1.0043 | 0.99 | 0.99 |
| | 0.090 | 0.090 | 0.3250 | 1.0049 | 0.99 | 0.99 |
| | 0.100 | 0.099 | 0.3466 | 1.0054 | 0.99 | 0.99 |
| | 0.200 | 0.198 | 0.3701 | 1.0108 | 0.99 | 0.99 |
| | 0.300 | 0.297 | 0.3894 | 1.0161 | 0.99 | 0.99 |
| | 0.400 | 0.395 | 0.4004 | 1.0214 | 0.98 | 0.99 |
| | 0.500 | 0.493 | 0.4026 | 1.0267 | 0.97 | 0.98 |
| | 0.600 | 0.590 | 0.3993 | 1.0320 | 0.97 | 0.98 |
| | 0.700 | 0.687 | 0.3915 | 1.0372 | 0.96 | 0.98 |
| | 0.800 | 0.783 | 0.3818 | 1.0425 | 0.95 | 0.97 |
| | 0.900 | 0.879 | 0.3659 | 1.0477 | 0.95 | 0.97 |
| | 1.000 | 0.975 | 0.3377 | 1.0529 | 0.94 | 0.96 |
| $Cs_2SO_4$ | 0.100 | 0.099 | 0.3839 | 1.0054 | 0.99 | 0.99 |
| | 0.200 | 0.199 | 0.4708 | 1.0108 | 0.99 | 0.99 |
| | 0.400 | 0.395 | 0.5744 | 1.0214 | 0.98 | 0.99 |
| | 0.600 | 0.585 | 0.6383 | 1.0320 | 0.98 | 0.99 |
| | 0.800 | 0.765 | 0.6813 | 1.0425 | 0.97 | 0.98 |
| | 1.000 | 0.950 | 0.7136 | 1.0529 | 0.96 | 0.97 |
| CsCl | 0.002 | 0.002 | 0.0256 | 1.0001 | 0.99 | 0.99 |
| | 0.003 | 0.003 | 0.0315 | 1.0001 | 0.99 | 0.99 |
| | 0.004 | 0.004 | 0.0357 | 1.0001 | 0.99 | 0.99 |
| | 0.005 | 0.005 | 0.0395 | 1.0002 | 0.99 | 0.99 |
| | 0.006 | 0.006 | 0.0428 | 1.0002 | 0.99 | 0.99 |
| | 0.007 | 0.007 | 0.0461 | 1.0003 | 0.99 | 0.99 |
| | 0.008 | 0.008 | 0.0472 | 1.0003 | 0.99 | 0.99 |

*(Continued)*

**TABLE 4.1 (*Continued*)**
**Calculated Thermodynamic Components Included in Formula (4.102)**

| Electrolyte | $M$ | $c$ | $-\gamma_W$ | $S_W$ | $a_{w\,exp}$ | $a_{w\,calc}$ |
|---|---|---|---|---|---|---|
| | 0.009 | 0.009 | 0.0521 | 1.0003 | 0.99 | 0.99 |
| | 0.010 | 0.010 | 0.0611 | 1.0004 | 0.99 | 0.99 |
| | 0.020 | 0.020 | 0.0754 | 1.0007 | 0.99 | 0.99 |
| | 0.030 | 0.030 | 0.0878 | 1.0011 | 0.99 | 0.99 |
| | 0.040 | 0.040 | 0.0992 | 1.0014 | 0.99 | 0.99 |
| | 0.050 | 0.050 | 0.1085 | 1.0018 | 0.99 | 0.99 |
| | 0.060 | 0.060 | 0.1166 | 1.0022 | 0.99 | 0.99 |
| | 0.070 | 0.070 | 0.1236 | 1.0025 | 0.99 | 0.99 |
| | 0.080 | 0.079 | 0.1281 | 1.0029 | 0.99 | 0.99 |
| | 0.090 | 0.089 | 0.1366 | 1.0032 | 0.99 | 0.99 |
| | 0.100 | 0.100 | 0.1526 | 1.0036 | 0.99 | 0.99 |
| | 0.200 | 0.197 | 0.1791 | 1.0072 | 0.99 | 0.99 |
| | 0.300 | 0.296 | 0.2055 | 1.0108 | 0.99 | 0.99 |
| | 0.400 | 0.391 | 0.2286 | 1.0143 | 0.98 | 0.99 |
| | 0.500 | 0.489 | 0.2471 | 1.0179 | 0.98 | 0.99 |
| | 0.600 | 0.583 | 0.2621 | 1.0214 | 0.97 | 0.98 |
| | 0.700 | 0.680 | 0.2741 | 1.0250 | 0.97 | 0.98 |
| | 0.800 | 0.772 | 0.2846 | 1.0285 | 0.96 | 0.98 |
| | 0.900 | 0.863 | 0.2944 | 1.0320 | 0.96 | 0.98 |
| | 1.000 | 0.953 | 0.3053 | 1.0355 | 0.95 | 0.97 |
| CsF | 0.100 | 0.099 | 0.1223 | 1.0036 | 0.99 | 0.99 |
| | 0.200 | 0.199 | 0.1465 | 1.0072 | 0.99 | 0.99 |
| | 0.400 | 0.398 | 0.1636 | 1.0143 | 0.99 | 0.99 |
| | 0.600 | 0.592 | 0.1670 | 1.0214 | 0.98 | 0.98 |
| | 0.800 | 0.790 | 0.1663 | 1.0285 | 0.97 | 0.98 |
| | 1.000 | 0.981 | 0.1615 | 1.0355 | 0.97 | 0.97 |
| | 2.000 | 1.916 | 0.1204 | 1.0700 | 0.93 | 0.93 |
| | 3.000 | 2.791 | 0.0651 | 1.1035 | 0.88 | 0.89 |
| | 4.000 | 3.676 | 0.0000 | 1.1361 | 0.84 | 0.85 |
| | 5.000 | 4.481 | 0.0655 | 1.1679 | 0.79 | 0.80 |
| | 6.000 | 5.288 | 0.1463 | 1.1989 | 0.75 | 0.75 |
| $CsNO_3$ | 0.100 | 0.099 | 0.1553 | 1.0036 | 0.99 | 0.99 |
| | 0.200 | 0.198 | 0.2116 | 1.0072 | 0.99 | 0.99 |
| | 0.400 | 0.390 | 0.2890 | 1.0143 | 0.99 | 0.99 |
| | 0.600 | 0.577 | 0.3456 | 1.0214 | 0.98 | 0.99 |
| | 0.800 | 0.762 | 0.3904 | 1.0285 | 0.98 | 0.98 |
| | 1.000 | 0.943 | 0.4314 | 1.0355 | 0.97 | 0.98 |
| $Cu(NO_3)_2$ | 0.003 | 0.003 | 0.0947 | 1.0002 | 0.99 | 0.99 |
| | 0.004 | 0.004 | 0.1078 | 1.0002 | 0.99 | 0.99 |
| | 0.005 | 0.005 | 0.1189 | 1.0003 | 0.99 | 0.99 |
| | 0.006 | 0.006 | 0.1285 | 1.0003 | 0.99 | 0.99 |

(*Continued*)

**TABLE 4.1 (*Continued*)**
**Calculated Thermodynamic Components Included in Formula (4.102)**

| Electrolyte | $M$ | $c$ | $-\gamma_W$ | $S_W$ | $a_{w\,exp}$ | $a_{w\,calc}$ |
|---|---|---|---|---|---|---|
| | 0.007 | 0.007 | 0.1358 | 1.0004 | 0.99 | 0.99 |
| | 0.008 | 0.008 | 0.1426 | 1.0004 | 0.99 | 0.99 |
| | 0.009 | 0.009 | 0.1543 | 1.0005 | 0.99 | 0.99 |
| | 0.010 | 0.010 | 0.1750 | 1.0005 | 0.99 | 0.99 |
| | 0.020 | 0.020 | 0.2020 | 1.0011 | 0.99 | 0.99 |
| | 0.030 | 0.030 | 0.2290 | 1.0016 | 0.99 | 0.99 |
| | 0.040 | 0.040 | 0.2528 | 1.0022 | 0.99 | 0.99 |
| | 0.050 | 0.050 | 0.2724 | 1.0027 | 0.99 | 0.99 |
| | 0.060 | 0.060 | 0.2883 | 1.0032 | 0.99 | 0.99 |
| | 0.070 | 0.069 | 0.3002 | 1.0038 | 0.99 | 0.99 |
| | 0.080 | 0.079 | 0.3107 | 1.0043 | 0.99 | 0.99 |
| | 0.090 | 0.090 | 0.3258 | 1.0049 | 0.99 | 0.99 |
| | 0.100 | 0.100 | 0.3488 | 1.0054 | 0.99 | 0.99 |
| | 0.200 | 0.199 | 0.3750 | 1.0108 | 0.99 | 0.99 |
| | 0.300 | 0.296 | 0.3966 | 1.0161 | 0.99 | 0.99 |
| | 0.400 | 0.394 | 0.4107 | 1.0214 | 0.98 | 0.99 |
| | 0.500 | 0.493 | 0.4175 | 1.0267 | 0.98 | 0.98 |
| | 0.600 | 0.587 | 0.4186 | 1.0320 | 0.97 | 0.98 |
| | 0.700 | 0.686 | 0.4163 | 1.0372 | 0.97 | 0.98 |
| | 0.800 | 0.778 | 0.4114 | 1.0425 | 0.96 | 0.97 |
| | 0.900 | 0.869 | 0.4019 | 1.0477 | 0.95 | 0.97 |
| | 1.000 | 0.968 | 0.3843 | 1.0529 | 0.95 | 0.96 |
| | 1.250 | 1.198 | 0.3575 | 1.0657 | 0.93 | 0.95 |
| $H_2SO_4$ | 0.001 | 0.001 | 0.1109 | 1.0001 | 0.99 | 0.99 |
| | 0.002 | 0.002 | 0.1466 | 1.0001 | 0.99 | 0.99 |
| | 0.003 | 0.003 | 0.1772 | 1.0002 | 0.99 | 0.99 |
| | 0.004 | 0.004 | 0.2032 | 1.0002 | 0.99 | 0.99 |
| | 0.005 | 0.005 | 0.2256 | 1.0003 | 0.99 | 0.99 |
| | 0.006 | 0.006 | 0.2450 | 1.0003 | 0.99 | 0.99 |
| | 0.007 | 0.007 | 0.2600 | 1.0004 | 0.99 | 0.99 |
| | 0.008 | 0.008 | 0.2742 | 1.0004 | 0.99 | 0.99 |
| | 0.009 | 0.009 | 0.2979 | 1.0005 | 0.99 | 0.99 |
| | 0.010 | 0.010 | 0.3403 | 1.0005 | 0.99 | 0.99 |
| | 0.020 | 0.020 | 0.3971 | 1.0011 | 0.99 | 0.99 |
| | 0.030 | 0.030 | 0.4559 | 1.0016 | 0.99 | 0.99 |
| | 0.040 | 0.040 | 0.5093 | 1.0022 | 0.99 | 0.99 |
| | 0.050 | 0.050 | 0.5540 | 1.0027 | 0.99 | 0.99 |
| | 0.060 | 0.060 | 0.5909 | 1.0032 | 0.99 | 0.99 |
| | 0.070 | 0.070 | 0.6189 | 1.0038 | 0.99 | 0.99 |
| | 0.080 | 0.080 | 0.6435 | 1.0043 | 0.99 | 0.99 |
| | 0.090 | 0.090 | 0.6780 | 1.0049 | 0.99 | 0.99 |
| | 0.100 | 0.099 | 0.7325 | 1.0054 | 0.99 | 0.99 |

*(Continued)*

**TABLE 4.1** (*Continued*)
**Calculated Thermodynamic Components Included in Formula (4.102)**

| Electrolyte | $M$ | $c$ | $-\gamma_W$ | $S_W$ | $a_{w\,exp}$ | $a_{w\,calc}$ |
|---|---|---|---|---|---|---|
| $K_2SO_4$ | 0.001 | 0.001 | 0.0624 | 1.0001 | 0.99 | 0.99 |
| | 0.002 | 0.002 | 0.0830 | 1.0001 | 0.99 | 0.99 |
| | 0.003 | 0.003 | 0.1006 | 1.0002 | 0.99 | 0.99 |
| | 0.004 | 0.004 | 0.1156 | 1.0002 | 0.99 | 0.99 |
| | 0.005 | 0.005 | 0.1284 | 1.0003 | 0.99 | 0.99 |
| | 0.006 | 0.006 | 0.1397 | 1.0003 | 0.99 | 0.99 |
| | 0.007 | 0.007 | 0.1483 | 1.0004 | 0.99 | 0.99 |
| | 0.008 | 0.008 | 0.1565 | 1.0004 | 0.99 | 0.99 |
| | 0.009 | 0.009 | 0.1710 | 1.0005 | 0.99 | 0.99 |
| | 0.010 | 0.010 | 0.1973 | 1.0005 | 0.99 | 0.99 |
| | 0.020 | 0.020 | 0.2325 | 1.0011 | 0.99 | 0.99 |
| | 0.030 | 0.030 | 0.2692 | 1.0016 | 0.99 | 0.99 |
| | 0.040 | 0.040 | 0.3029 | 1.0022 | 0.99 | 0.99 |
| | 0.050 | 0.050 | 0.3320 | 1.0027 | 0.99 | 0.99 |
| | 0.060 | 0.060 | 0.3567 | 1.0032 | 0.99 | 0.99 |
| | 0.070 | 0.069 | 0.3755 | 1.0038 | 0.99 | 0.99 |
| | 0.080 | 0.079 | 0.3926 | 1.0043 | 0.99 | 0.99 |
| | 0.090 | 0.090 | 0.4194 | 1.0049 | 0.99 | 0.99 |
| | 0.100 | 0.100 | 0.4651 | 1.0054 | 0.99 | 0.99 |
| | 0.200 | 0.197 | 0.5249 | 1.0108 | 0.99 | 0.99 |
| | 0.300 | 0.296 | 0.5853 | 1.0161 | 0.99 | 0.99 |
| | 0.400 | 0.394 | 0.6388 | 1.0214 | 0.99 | 0.99 |
| | 0.500 | 0.488 | 0.6832 | 1.0267 | 0.98 | 0.99 |
| | 0.600 | 0.581 | 0.7203 | 1.0320 | 0.98 | 0.99 |
| | 0.692 | 0.672 | 0.7489 | 1.0368 | 0.98 | 0.99 |
| KBr | 0.003 | 0.003 | 0.0304 | 1.0001 | 0.99 | 0.99 |
| | 0.004 | 0.004 | 0.0347 | 1.0001 | 0.99 | 0.99 |
| | 0.005 | 0.005 | 0.0384 | 1.0002 | 0.99 | 0.99 |
| | 0.006 | 0.006 | 0.0417 | 1.0002 | 0.99 | 0.99 |
| | 0.007 | 0.007 | 0.0444 | 1.0003 | 0.99 | 0.99 |
| | 0.008 | 0.008 | 0.0455 | 1.0003 | 0.99 | 0.99 |
| | 0.009 | 0.009 | 0.0499 | 1.0003 | 0.99 | 0.99 |
| | 0.010 | 0.010 | 0.0583 | 1.0004 | 0.99 | 0.99 |
| | 0.020 | 0.020 | 0.0714 | 1.0007 | 0.99 | 0.99 |
| | 0.030 | 0.030 | 0.0824 | 1.0011 | 0.99 | 0.99 |
| | 0.040 | 0.040 | 0.0926 | 1.0014 | 0.99 | 0.99 |
| | 0.050 | 0.050 | 0.1004 | 1.0018 | 0.99 | 0.99 |
| | 0.060 | 0.060 | 0.1072 | 1.0022 | 0.99 | 0.99 |
| | 0.070 | 0.070 | 0.1128 | 1.0025 | 0.99 | 0.99 |
| | 0.080 | 0.080 | 0.1172 | 1.0029 | 0.99 | 0.99 |
| | 0.090 | 0.089 | 0.1242 | 1.0032 | 0.99 | 0.99 |
| | 0.100 | 0.099 | 0.1372 | 1.0036 | 0.99 | 0.99 |

(*Continued*)

**TABLE 4.1 (*Continued*)**
**Calculated Thermodynamic Components Included in Formula (4.102)**

| Electrolyte | $M$ | $c$ | $-\gamma_W$ | $S_W$ | $a_{w\,exp}$ | $a_{w\,calc}$ |
|---|---|---|---|---|---|---|
| | 0.200 | 0.198 | 0.1580 | 1.0072 | 0.99 | 0.99 |
| | 0.300 | 0.295 | 0.1776 | 1.0108 | 0.99 | 0.99 |
| | 0.400 | 0.395 | 0.1943 | 1.0143 | 0.99 | 0.99 |
| | 0.500 | 0.491 | 0.2070 | 1.0179 | 0.98 | 0.99 |
| | 0.600 | 0.587 | 0.2169 | 1.0214 | 0.98 | 0.98 |
| | 0.700 | 0.683 | 0.2247 | 1.0250 | 0.98 | 0.98 |
| | 0.800 | 0.778 | 0.2310 | 1.0285 | 0.97 | 0.98 |
| | 0.900 | 0.872 | 0.2366 | 1.0320 | 0.97 | 0.97 |
| | 1.000 | 0.966 | 0.2431 | 1.0355 | 0.97 | 0.97 |
| | 1.250 | 1.194 | 0.2496 | 1.0442 | 0.96 | 0.96 |
| | 1.500 | 1.419 | 0.2546 | 1.0529 | 0.95 | 0.96 |
| | 1.750 | 1.640 | 0.2579 | 1.0615 | 0.94 | 0.95 |
| | 2.000 | 1.859 | 0.2604 | 1.0700 | 0.94 | 0.94 |
| | 2.250 | 2.075 | 0.2613 | 1.0785 | 0.93 | 0.94 |
| | 2.500 | 2.290 | 0.2613 | 1.0869 | 0.92 | 0.93 |
| | 2.750 | 2.503 | 0.2596 | 1.0952 | 0.91 | 0.92 |
| | 3.000 | 2.693 | 0.2588 | 1.1035 | 0.90 | 0.91 |
| | 3.250 | 2.902 | 0.2562 | 1.1118 | 0.89 | 0.90 |
| | 3.500 | 3.086 | 0.2537 | 1.1199 | 0.88 | 0.90 |
| | 3.750 | 3.293 | 0.2513 | 1.1281 | 0.88 | 0.89 |
| | 4.000 | 3.471 | 0.2480 | 1.1361 | 0.87 | 0.88 |
| | 4.250 | 3.676 | 0.2447 | 1.1442 | 0.86 | 0.87 |
| | 4.500 | 3.849 | 0.2414 | 1.1521 | 0.85 | 0.86 |
| | 4.750 | 4.020 | 0.2374 | 1.1601 | 0.84 | 0.85 |
| | 5.000 | 4.188 | 0.2334 | 1.1679 | 0.83 | 0.85 |
| | 5.250 | 4.354 | 0.2294 | 1.1757 | 0.82 | 0.84 |
| | 5.500 | 4.557 | 0.2255 | 1.1835 | 0.82 | 0.83 |
| | 0.001 | 0.001 | 0.0173 | 1.0000 | 0.99 | 0.99 |
| | 0.002 | 0.002 | 0.0256 | 1.0001 | 0.99 | 0.99 |
| KCl | 0.003 | 0.003 | 0.0309 | 1.0001 | 0.99 | 0.99 |
| | 0.004 | 0.004 | 0.0352 | 1.0001 | 0.99 | 0.99 |
| | 0.005 | 0.005 | 0.0384 | 1.0002 | 0.99 | 0.99 |
| | 0.006 | 0.006 | 0.0417 | 1.0002 | 0.99 | 0.99 |
| | 0.007 | 0.007 | 0.0450 | 1.0003 | 0.99 | 0.99 |
| | 0.008 | 0.008 | 0.0461 | 1.0003 | 0.99 | 0.99 |
| | 0.009 | 0.009 | 0.0505 | 1.0003 | 0.99 | 0.99 |
| | 0.010 | 0.010 | 0.0588 | 1.0004 | 0.99 | 0.99 |
| | 0.020 | 0.020 | 0.0719 | 1.0007 | 0.99 | 0.99 |
| | 0.030 | 0.030 | 0.0830 | 1.0011 | 0.99 | 0.99 |
| | 0.040 | 0.040 | 0.0932 | 1.0014 | 0.99 | 0.99 |
| | 0.050 | 0.050 | 0.1017 | 1.0018 | 0.99 | 0.99 |
| | 0.060 | 0.060 | 0.1085 | 1.0022 | 0.99 | 0.99 |

(*Continued*)

**TABLE 4.1** (*Continued*)
**Calculated Thermodynamic Components Included in Formula (4.102)**

| Electrolyte | $M$ | $c$ | $-\gamma_W$ | $S_W$ | $a_{w\,exp}$ | $a_{w\,calc}$ |
|---|---|---|---|---|---|---|
| | 0.070 | 0.070 | 0.1147 | 1.0025 | 0.99 | 0.99 |
| | 0.080 | 0.080 | 0.1185 | 1.0029 | 0.99 | 0.99 |
| | 0.090 | 0.090 | 0.1255 | 1.0032 | 0.99 | 0.99 |
| | 0.100 | 0.100 | 0.1392 | 1.0036 | 0.99 | 0.99 |
| | 0.200 | 0.198 | 0.1615 | 1.0072 | 0.99 | 0.99 |
| | 0.300 | 0.296 | 0.1819 | 1.0108 | 0.99 | 0.99 |
| | 0.400 | 0.395 | 0.2002 | 1.0143 | 0.99 | 0.99 |
| | 0.500 | 0.493 | 0.2139 | 1.0179 | 0.99 | 0.99 |
| | 0.600 | 0.587 | 0.2247 | 1.0214 | 0.98 | 0.98 |
| | 0.700 | 0.684 | 0.2334 | 1.0250 | 0.98 | 0.98 |
| | 0.800 | 0.781 | 0.2406 | 1.0285 | 0.98 | 0.98 |
| | 0.900 | 0.873 | 0.2471 | 1.0320 | 0.97 | 0.97 |
| | 1.000 | 0.969 | 0.2537 | 1.0355 | 0.97 | 0.97 |
| | 1.250 | 1.206 | 0.2621 | 1.0442 | 0.96 | 0.97 |
| | 1.500 | 1.432 | 0.2689 | 1.0529 | 0.95 | 0.96 |
| | 1.750 | 1.664 | 0.2741 | 1.0615 | 0.95 | 0.95 |
| | 2.000 | 1.883 | 0.2784 | 1.0700 | 0.94 | 0.94 |
| | 2.250 | 2.098 | 0.2802 | 1.0785 | 0.93 | 0.94 |
| | 2.500 | 2.323 | 0.2811 | 1.0869 | 0.92 | 0.93 |
| | 2.750 | 2.532 | 0.2819 | 1.0952 | 0.91 | 0.92 |
| | 3.000 | 2.738 | 0.2811 | 1.1035 | 0.90 | 0.91 |
| | 3.250 | 2.959 | 0.2802 | 1.1118 | 0.90 | 0.91 |
| | 3.500 | 3.160 | 0.2793 | 1.1199 | 0.89 | 0.90 |
| | 3.750 | 3.357 | 0.2767 | 1.1281 | 0.88 | 0.89 |
| | 4.000 | 3.552 | 0.2750 | 1.1361 | 0.87 | 0.88 |
| | 4.250 | 3.745 | 0.2724 | 1.1442 | 0.86 | 0.88 |
| | 4.500 | 3.935 | 0.2698 | 1.1521 | 0.85 | 0.87 |
| | 4.750 | 4.123 | 0.2655 | 1.1601 | 0.85 | 0.86 |
| | 5.000 | 4.308 | 0.2613 | 1.1679 | 0.84 | 0.85 |
| KF | 0.001 | 0.001 | 0.0172 | 1.0000 | 0.99 | 0.99 |
| | 0.002 | 0.002 | 0.0253 | 1.0001 | 0.99 | 0.99 |
| | 0.003 | 0.003 | 0.0305 | 1.0001 | 0.99 | 0.99 |
| | 0.004 | 0.004 | 0.0349 | 1.0001 | 0.99 | 0.99 |
| | 0.005 | 0.005 | 0.0384 | 1.0002 | 0.99 | 0.99 |
| | 0.006 | 0.006 | 0.0415 | 1.0002 | 0.99 | 0.99 |
| | 0.007 | 0.007 | 0.0444 | 1.0003 | 0.99 | 0.99 |
| | 0.008 | 0.008 | 0.0456 | 1.0003 | 0.99 | 0.99 |
| | 0.009 | 0.009 | 0.0500 | 1.0003 | 0.99 | 0.99 |
| | 0.010 | 0.010 | 0.0584 | 1.0004 | 0.99 | 0.99 |
| | 0.020 | 0.020 | 0.0712 | 1.0007 | 0.99 | 0.99 |
| | 0.030 | 0.030 | 0.0818 | 1.0011 | 0.99 | 0.99 |
| | 0.040 | 0.040 | 0.0918 | 1.0014 | 0.99 | 0.99 |

(*Continued*)

**TABLE 4.1 (*Continued*)**
**Calculated Thermodynamic Components Included in Formula (4.102)**

| Electrolyte | $M$ | $c$ | $-\gamma_W$ | $S_W$ | $a_{w\,exp}$ | $a_{w\,calc}$ |
|---|---|---|---|---|---|---|
| | 0.050 | 0.050 | 0.0998 | 1.0018 | 0.99 | 0.99 |
| | 0.060 | 0.060 | 0.1062 | 1.0022 | 0.99 | 0.99 |
| | 0.070 | 0.070 | 0.1121 | 1.0025 | 0.99 | 0.99 |
| | 0.080 | 0.079 | 0.1159 | 1.0029 | 0.99 | 0.99 |
| | 0.090 | 0.090 | 0.1227 | 1.0032 | 0.99 | 0.99 |
| | 0.100 | 0.100 | 0.1353 | 1.0036 | 0.99 | 0.99 |
| | 0.200 | 0.199 | 0.1548 | 1.0072 | 0.99 | 0.99 |
| | 0.300 | 0.299 | 0.1727 | 1.0108 | 0.99 | 0.99 |
| | 0.400 | 0.397 | 0.1874 | 1.0143 | 0.99 | 0.99 |
| | 0.500 | 0.497 | 0.1980 | 1.0179 | 0.99 | 0.99 |
| | 0.600 | 0.593 | 0.2056 | 1.0214 | 0.99 | 0.98 |
| | 0.700 | 0.694 | 0.2109 | 1.0250 | 0.99 | 0.98 |
| | 0.800 | 0.789 | 0.2147 | 1.0285 | 0.99 | 0.98 |
| | 0.900 | 0.890 | 0.2175 | 1.0320 | 0.99 | 0.97 |
| | 1.000 | 0.984 | 0.2193 | 1.0355 | 0.99 | 0.97 |
| | 1.250 | 1.234 | 0.2200 | 1.0442 | 0.99 | 0.96 |
| | 1.500 | 1.474 | 0.2180 | 1.0529 | 0.99 | 0.96 |
| | 1.750 | 1.712 | 0.2132 | 1.0615 | 0.98 | 0.95 |
| | 2.000 | 1.947 | 0.2064 | 1.0700 | 0.97 | 0.94 |
| | 2.250 | 2.200 | 0.1991 | 1.0785 | 0.96 | 0.93 |
| | 2.500 | 2.434 | 0.1911 | 1.0869 | 0.95 | 0.92 |
| | 2.750 | 2.666 | 0.1820 | 1.0952 | 0.94 | 0.92 |
| | 3.000 | 2.897 | 0.1722 | 1.1035 | 0.93 | 0.91 |
| | 3.250 | 3.127 | 0.1612 | 1.1118 | 0.92 | 0.90 |
| | 3.500 | 3.355 | 0.1496 | 1.1199 | 0.90 | 0.89 |
| | 3.750 | 3.582 | 0.1372 | 1.1281 | 0.89 | 0.88 |
| | 4.000 | 3.808 | 0.1243 | 1.1361 | 0.88 | 0.87 |
| | 4.250 | 4.033 | 0.1106 | 1.1442 | 0.88 | 0.86 |
| | 4.500 | 4.257 | 0.0967 | 1.1521 | 0.87 | 0.86 |
| $KNO_3$ | 0.001 | 0.001 | 0.0178 | 1.0000 | 0.99 | 0.99 |
| | 0.002 | 0.002 | 0.0262 | 1.0001 | 0.99 | 0.99 |
| | 0.003 | 0.003 | 0.0320 | 1.0001 | 0.99 | 0.99 |
| | 0.004 | 0.004 | 0.0368 | 1.0001 | 0.99 | 0.99 |
| | 0.005 | 0.005 | 0.0406 | 1.0002 | 0.99 | 0.99 |
| | 0.006 | 0.006 | 0.0439 | 1.0002 | 0.99 | 0.99 |
| | 0.007 | 0.007 | 0.0472 | 1.0003 | 0.99 | 0.99 |
| | 0.008 | 0.008 | 0.0483 | 1.0003 | 0.99 | 0.99 |
| | 0.009 | 0.009 | 0.0532 | 1.0003 | 0.99 | 0.99 |
| | 0.010 | 0.010 | 0.0628 | 1.0004 | 0.99 | 0.99 |
| | 0.020 | 0.020 | 0.0783 | 1.0007 | 0.99 | 0.99 |
| | 0.030 | 0.030 | 0.0914 | 1.0011 | 0.99 | 0.99 |
| | 0.040 | 0.040 | 0.1041 | 1.0014 | 0.99 | 0.99 |

(*Continued*)

**TABLE 4.1 (*Continued*)**
**Calculated Thermodynamic Components Included in Formula (4.102)**

| Electrolyte | $M$ | $c$ | $-\gamma_W$ | $S_W$ | $a_{w\,exp}$ | $a_{w\,calc}$ |
|---|---|---|---|---|---|---|
| | 0.050 | 0.050 | 0.1147 | 1.0018 | 0.99 | 0.99 |
| | 0.060 | 0.060 | 0.1236 | 1.0022 | 0.99 | 0.99 |
| | 0.070 | 0.070 | 0.1320 | 1.0025 | 0.99 | 0.99 |
| | 0.080 | 0.080 | 0.1372 | 1.0029 | 0.99 | 0.99 |
| | 0.090 | 0.089 | 0.1472 | 1.0032 | 0.99 | 0.99 |
| | 0.100 | 0.099 | 0.1670 | 1.0036 | 0.99 | 0.99 |
| | 0.200 | 0.198 | 0.2010 | 1.0072 | 0.97 | 0.99 |
| | 0.300 | 0.296 | 0.2358 | 1.0108 | 0.97 | 0.99 |
| | 0.400 | 0.393 | 0.2698 | 1.0143 | 0.97 | 0.99 |
| KOH | 0.001 | 0.001 | 0.0173 | 1.0000 | 0.99 | 0.99 |
| | 0.002 | 0.002 | 0.0251 | 1.0001 | 0.99 | 0.99 |
| | 0.003 | 0.003 | 0.0304 | 1.0001 | 0.99 | 0.99 |
| | 0.004 | 0.004 | 0.0347 | 1.0001 | 0.99 | 0.99 |
| | 0.005 | 0.005 | 0.0384 | 1.0002 | 0.99 | 0.99 |
| | 0.006 | 0.006 | 0.0417 | 1.0002 | 0.99 | 0.99 |
| | 0.007 | 0.007 | 0.0444 | 1.0003 | 0.99 | 0.99 |
| | 0.008 | 0.008 | 0.0455 | 1.0003 | 0.99 | 0.99 |
| | 0.009 | 0.009 | 0.0499 | 1.0003 | 0.99 | 0.99 |
| | 0.010 | 0.010 | 0.0583 | 1.0004 | 0.99 | 0.99 |
| | 0.020 | 0.020 | 0.0702 | 1.0007 | 0.99 | 0.99 |
| | 0.030 | 0.030 | 0.0807 | 1.0011 | 0.99 | 0.99 |
| | 0.040 | 0.040 | 0.0902 | 1.0014 | 0.99 | 0.99 |
| | 0.050 | 0.050 | 0.0974 | 1.0018 | 0.99 | 0.99 |
| | 0.060 | 0.060 | 0.1035 | 1.0022 | 0.99 | 0.99 |
| | 0.070 | 0.070 | 0.1091 | 1.0025 | 0.99 | 0.99 |
| | 0.080 | 0.079 | 0.1122 | 1.0029 | 0.99 | 0.99 |
| | 0.090 | 0.090 | 0.1185 | 1.0032 | 0.99 | 0.99 |
| | 0.100 | 0.100 | 0.1294 | 1.0036 | 0.99 | 0.99 |
| | 0.200 | 0.199 | 0.1445 | 1.0072 | 0.99 | 0.99 |
| | 0.300 | 0.300 | 0.1574 | 1.0108 | 0.99 | 0.99 |
| | 0.400 | 0.398 | 0.1656 | 1.0143 | 0.99 | 0.99 |
| | 0.500 | 0.499 | 0.1691 | 1.0179 | 0.98 | 0.98 |
| | 0.600 | 0.596 | 0.1698 | 1.0214 | 0.98 | 0.98 |
| | 0.700 | 0.698 | 0.1684 | 1.0250 | 0.97 | 0.98 |
| | 0.800 | 0.794 | 0.1656 | 1.0285 | 0.97 | 0.98 |
| | 0.900 | 0.896 | 0.1608 | 1.0320 | 0.97 | 0.97 |
| | 1.000 | 0.991 | 0.1512 | 1.0355 | 0.96 | 0.97 |
| | 1.250 | 1.245 | 0.1379 | 1.0442 | 0.95 | 0.96 |
| | 1.500 | 1.489 | 0.1185 | 1.0529 | 0.94 | 0.95 |
| | 1.750 | 1.730 | 0.0938 | 1.0615 | 0.93 | 0.94 |
| | 2.000 | 1.971 | 0.0668 | 1.0700 | 0.92 | 0.93 |
| | 2.250 | 2.209 | 0.0422 | 1.0785 | 0.91 | 0.92 |

*(Continued)*

**TABLE 4.1 (*Continued*)**
**Calculated Thermodynamic Components Included in Formula (4.102)**

| Electrolyte | $M$ | $c$ | $-\gamma_W$ | $S_W$ | $a_{w\,exp}$ | $a_{w\,calc}$ |
|---|---|---|---|---|---|---|
| | 2.500 | 2.446 | 0.0183 | 1.0869 | 0.90 | 0.91 |
| | 2.750 | 2.682 | −0.0074 | 1.0952 | 0.88 | 0.90 |
| | 3.000 | 2.917 | −0.0334 | 1.1035 | 0.87 | 0.88 |
| | 3.250 | 3.150 | −0.0611 | 1.1118 | 0.86 | 0.87 |
| | 3.500 | 3.382 | −0.0891 | 1.1199 | 0.84 | 0.86 |
| | 3.750 | 3.613 | −0.1183 | 1.1281 | 0.83 | 0.85 |
| | 4.000 | 3.842 | −0.1475 | 1.1361 | 0.81 | 0.83 |
| | 4.250 | 4.071 | −0.1778 | 1.1442 | 0.80 | 0.82 |
| | 4.500 | 4.298 | −0.2077 | 1.1521 | 0.79 | 0.80 |
| | 4.750 | 4.525 | −0.2384 | 1.1601 | 0.77 | 0.79 |
| | 5.000 | 4.751 | −0.2674 | 1.1679 | 0.75 | 0.77 |
| | 5.250 | 4.975 | −0.2991 | 1.1757 | 0.74 | 0.76 |
| | 5.500 | 5.199 | −0.3362 | 1.1835 | 0.72 | 0.74 |
| | 5.750 | 5.378 | −0.3778 | 1.1912 | 0.71 | 0.72 |
| | 6.000 | 5.599 | −0.4167 | 1.1989 | 0.70 | 0.71 |
| | 6.250 | 5.819 | −0.4507 | 1.2065 | 0.68 | 0.69 |
| | 6.500 | 6.038 | −0.4806 | 1.2141 | 0.67 | 0.67 |
| | 6.750 | 6.207 | −0.5094 | 1.2216 | 0.65 | 0.66 |
| | 7.000 | 6.423 | −0.5395 | 1.2291 | 0.63 | 0.64 |
| | 7.250 | 6.639 | −0.5720 | 1.2366 | 0.62 | 0.62 |
| | 7.500 | 6.855 | −0.6030 | 1.2440 | 0.60 | 0.61 |
| | 7.750 | 7.014 | −0.6338 | 1.2513 | 0.59 | 0.59 |
| | 8.000 | 7.227 | −0.6654 | 1.2586 | 0.57 | 0.57 |
| | 8.250 | 7.440 | −0.6987 | 1.2659 | 0.55 | 0.56 |
| | 8.500 | 7.593 | −0.7304 | 1.2731 | 0.54 | 0.54 |
| | 8.750 | 7.803 | −0.7616 | 1.2803 | 0.52 | 0.53 |
| | 9.000 | 8.013 | −0.7934 | 1.2874 | 0.51 | 0.51 |
| | 9.250 | 8.160 | −0.8266 | 1.2945 | 0.49 | 0.49 |
| | 9.500 | 8.368 | −0.8543 | 1.3016 | 0.47 | 0.48 |
| | 9.750 | 8.511 | −0.8862 | 1.3086 | 0.46 | 0.46 |
| | 10.000 | 8.717 | −0.9274 | 1.3156 | 0.44 | 0.45 |
| | 10.500 | 9.060 | −0.9841 | 1.3294 | 0.42 | 0.42 |
| | 11.000 | 9.398 | −1.0429 | 1.3431 | 0.40 | 0.39 |
| | 11.500 | 9.731 | −1.1050 | 1.3567 | 0.37 | 0.36 |
| | 12.000 | 10.059 | −1.1645 | 1.3701 | 0.35 | 0.33 |
| | 12.500 | 10.382 | −1.2251 | 1.3833 | 0.33 | 0.31 |
| | 13.000 | 10.701 | −1.2829 | 1.3964 | 0.30 | 0.29 |
| | 13.500 | 11.017 | −0.0074 | 1.4094 | 0.28 | 0.26 |
| | 14.000 | 11.328 | −0.0334 | 1.4223 | 0.26 | 0.24 |
| $Li_2SO_4$ | 0.005 | 0.005 | 0.1236 | 1.0003 | 0.99 | 0.99 |
| | 0.006 | 0.006 | 0.1340 | 1.0003 | 0.99 | 0.99 |
| | 0.007 | 0.007 | 0.1419 | 1.0004 | 0.99 | 0.99 |

(*Continued*)

**TABLE 4.1 (*Continued*)**
**Calculated Thermodynamic Components Included in Formula (4.102)**

| Electrolyte | $M$ | $c$ | $-\gamma_W$ | $S_W$ | $a_{w\,exp}$ | $a_{w\,calc}$ |
|---|---|---|---|---|---|---|
| | 0.008 | 0.008 | 0.1493 | 1.0004 | 0.99 | 0.99 |
| | 0.009 | 0.009 | 0.1623 | 1.0005 | 0.99 | 0.99 |
| | 0.010 | 0.010 | 0.1857 | 1.0005 | 0.99 | 0.99 |
| | 0.020 | 0.020 | 0.2168 | 1.0011 | 0.99 | 0.99 |
| | 0.030 | 0.030 | 0.2485 | 1.0016 | 0.99 | 0.99 |
| | 0.040 | 0.040 | 0.2771 | 1.0022 | 0.99 | 0.99 |
| | 0.050 | 0.050 | 0.3014 | 1.0027 | 0.99 | 0.99 |
| | 0.060 | 0.060 | 0.3215 | 1.0032 | 0.99 | 0.99 |
| | 0.070 | 0.070 | 0.3368 | 1.0038 | 0.99 | 0.99 |
| | 0.080 | 0.080 | 0.3505 | 1.0043 | 0.99 | 0.99 |
| | 0.090 | 0.090 | 0.3714 | 1.0049 | 0.99 | 0.99 |
| | 0.100 | 0.099 | 0.4063 | 1.0054 | 0.99 | 0.99 |
| | 0.200 | 0.199 | 0.4506 | 1.0108 | 0.99 | 0.99 |
| | 0.300 | 0.297 | 0.4932 | 1.0161 | 0.99 | 0.99 |
| | 0.400 | 0.395 | 0.5294 | 1.0214 | 0.98 | 0.99 |
| | 0.500 | 0.493 | 0.5574 | 1.0267 | 0.98 | 0.99 |
| | 0.600 | 0.591 | 0.5789 | 1.0320 | 0.97 | 0.99 |
| | 0.700 | 0.688 | 0.5950 | 1.0372 | 0.97 | 0.98 |
| | 0.800 | 0.785 | 0.6080 | 1.0425 | 0.97 | 0.98 |
| | 0.900 | 0.882 | 0.6198 | 1.0477 | 0.96 | 0.98 |
| | 1.000 | 0.978 | 0.6313 | 1.0529 | 0.96 | 0.98 |
| | 1.250 | 1.213 | 0.6419 | 1.0657 | 0.95 | 0.97 |
| LiCl | 0.001 | 0.001 | 0.0173 | 1.0000 | 0.99 | 0.99 |
| | 0.002 | 0.002 | 0.0251 | 1.0001 | 0.99 | 0.99 |
| | 0.003 | 0.003 | 0.0304 | 1.0001 | 0.99 | 0.99 |
| | 0.004 | 0.004 | 0.0347 | 1.0001 | 0.99 | 0.99 |
| | 0.005 | 0.005 | 0.0379 | 1.0002 | 0.99 | 0.99 |
| | 0.006 | 0.006 | 0.0406 | 1.0002 | 0.99 | 0.99 |
| | 0.007 | 0.007 | 0.0433 | 1.0003 | 0.99 | 0.99 |
| | 0.008 | 0.008 | 0.0450 | 1.0003 | 0.99 | 0.99 |
| | 0.009 | 0.009 | 0.0488 | 1.0003 | 0.99 | 0.99 |
| | 0.010 | 0.010 | 0.0566 | 1.0004 | 0.99 | 0.99 |
| | 0.020 | 0.020 | 0.0685 | 1.0007 | 0.99 | 0.99 |
| | 0.030 | 0.030 | 0.0777 | 1.0011 | 0.99 | 0.99 |
| | 0.040 | 0.040 | 0.0866 | 1.0014 | 0.99 | 0.99 |
| | 0.050 | 0.050 | 0.3184 | 1.0018 | 0.99 | 0.99 |
| | 0.060 | 0.060 | 0.0992 | 1.0022 | 0.99 | 0.99 |
| | 0.070 | 0.070 | 0.1035 | 1.0025 | 0.99 | 0.99 |
| | 0.080 | 0.080 | 0.1072 | 1.0029 | 0.99 | 0.99 |
| | 0.090 | 0.089 | 0.1122 | 1.0032 | 0.99 | 0.99 |
| | 0.100 | 0.099 | 0.1217 | 1.0036 | 0.99 | 0.99 |
| | 0.200 | 0.199 | 0.1339 | 1.0072 | 0.99 | 0.99 |

*(Continued)*

**TABLE 4.1 (*Continued*)**
**Calculated Thermodynamic Components Included in Formula (4.102)**

| Electrolyte | $M$ | $c$ | $-\gamma_W$ | $S_W$ | $a_{w\,exp}$ | $a_{w\,calc}$ |
|---|---|---|---|---|---|---|
| | 0.300 | 0.297 | 0.1432 | 1.0108 | 0.99 | 0.99 |
| | 0.400 | 0.397 | 0.1485 | 1.0143 | 0.99 | 0.99 |
| | 0.500 | 0.494 | 0.1499 | 1.0179 | 0.98 | 0.98 |
| | 0.600 | 0.590 | 0.1479 | 1.0214 | 0.98 | 0.98 |
| | 0.700 | 0.689 | 0.1445 | 1.0250 | 0.98 | 0.98 |
| | 0.800 | 0.784 | 0.1405 | 1.0285 | 0.98 | 0.97 |
| | 0.900 | 0.884 | 0.1339 | 1.0320 | 0.98 | 0.97 |
| | 1.000 | 0.978 | 0.1230 | 1.0355 | 0.98 | 0.97 |
| | 1.250 | 1.217 | 0.1078 | 1.0442 | 0.96 | 0.96 |
| | 1.500 | 1.454 | 0.0860 | 1.0529 | 0.95 | 0.95 |
| | 1.750 | 1.689 | 0.0588 | 1.0615 | 0.93 | 0.94 |
| | 2.000 | 1.922 | 0.0299 | 1.0700 | 0.91 | 0.93 |
| | 2.250 | 2.153 | 0.0035 | 1.0785 | 0.90 | 0.92 |
| | 2.500 | 2.383 | −0.0230 | 1.0869 | 0.89 | 0.90 |
| | 2.750 | 2.596 | −0.0508 | 1.0952 | 0.87 | 0.89 |
| | 3.000 | 2.821 | −0.0798 | 1.1035 | 0.86 | 0.88 |
| | 3.250 | 3.045 | −0.1108 | 1.1118 | 0.85 | 0.87 |
| | 3.500 | 3.267 | −0.1422 | 1.1199 | 0.83 | 0.85 |
| | 3.750 | 3.488 | −0.1757 | 1.1281 | 0.82 | 0.84 |
| | 4.000 | 3.687 | −0.2090 | 1.1361 | 0.81 | 0.82 |
| | 4.250 | 3.904 | −0.2446 | 1.1442 | 0.79 | 0.81 |
| | 4.500 | 4.121 | −0.2798 | 1.1521 | 0.77 | 0.79 |
| | 4.750 | 4.336 | −0.3167 | 1.1601 | 0.76 | 0.78 |
| | 5.000 | 4.524 | −0.3515 | 1.1679 | 0.74 | 0.76 |
| | 5.250 | 4.736 | −0.3906 | 1.1757 | 0.72 | 0.74 |
| | 5.500 | 4.947 | −0.4369 | 1.1835 | 0.71 | 0.73 |
| | 5.750 | 5.157 | −0.4894 | 1.1912 | 0.69 | 0.71 |
| | 6.000 | 5.336 | −0.5382 | 1.1989 | 0.67 | 0.69 |
| | 6.250 | 5.543 | −0.5814 | 1.2065 | 0.66 | 0.67 |
| | 6.500 | 5.750 | −0.6189 | 1.2141 | 0.64 | 0.65 |
| | 6.750 | 5.922 | −0.6548 | 1.2216 | 0.62 | 0.63 |
| | 7.000 | 6.126 | −0.6925 | 1.2291 | 0.60 | 0.62 |
| | 7.250 | 6.329 | −0.7332 | 1.2366 | 0.59 | 0.60 |
| | 7.500 | 6.495 | −0.7714 | 1.2440 | 0.57 | 0.58 |
| | 7.750 | 6.696 | −0.8092 | 1.2513 | 0.55 | 0.56 |
| | 8.000 | 6.857 | −0.8480 | 1.2586 | 0.54 | 0.54 |
| | 8.250 | 7.057 | −0.8886 | 1.2659 | 0.52 | 0.53 |
| | 8.500 | 7.214 | −0.9267 | 1.2731 | 0.50 | 0.51 |
| | 8.750 | 7.411 | −0.9639 | 1.2803 | 0.49 | 0.49 |
| | 9.000 | 7.608 | −1.0014 | 1.2874 | 0.47 | 0.48 |
| | 9.250 | 7.760 | −1.0403 | 1.2945 | 0.46 | 0.46 |
| | 9.500 | 7.955 | −1.0724 | 1.3016 | 0.44 | 0.44 |

(*Continued*)

**TABLE 4.1 (*Continued*)**
**Calculated Thermodynamic Components Included in Formula (4.102)**

| Electrolyte | $M$ | $c$ | $-\gamma_W$ | $S_W$ | $a_{w\,exp}$ | $a_{w\,calc}$ |
|---|---|---|---|---|---|---|
| | 9.750 | 8.103 | −1.1091 | 1.3086 | 0.42 | 0.43 |
| | 10.000 | 8.297 | −1.1559 | 1.3156 | 0.41 | 0.41 |
| | 10.500 | 8.634 | −1.2187 | 1.3294 | 0.38 | 0.38 |
| | 11.000 | 8.966 | −1.2826 | 1.3431 | 0.35 | 0.35 |
| | 11.500 | 9.293 | −1.3474 | 1.3567 | 0.33 | 0.33 |
| | 12.000 | 9.616 | −1.4081 | 1.3701 | 0.31 | 0.30 |
| | 12.500 | 9.936 | −1.4670 | 1.3833 | 0.29 | 0.28 |
| | 13.000 | 10.251 | −1.5220 | 1.3964 | 0.26 | 0.25 |
| | 13.500 | 10.502 | −1.5753 | 1.4094 | 0.25 | 0.23 |
| | 14.000 | 10.808 | −1.6255 | 1.4223 | 0.23 | 0.21 |
| | 14.500 | 11.112 | −1.6744 | 1.4350 | 0.21 | 0.20 |
| | 15.000 | 11.412 | −1.7206 | 1.4476 | 0.20 | 0.18 |
| | 15.500 | 11.709 | −1.7655 | 1.4600 | 0.19 | 0.17 |
| | 16.000 | 11.934 | −1.8081 | 1.4724 | 0.17 | 0.15 |
| | 16.500 | 12.224 | −1.8492 | 1.4846 | 0.16 | 0.14 |
| | 17.000 | 12.512 | −1.8889 | 1.4967 | 0.15 | 0.13 |
| $LiNO_3$ | 0.001 | 0.001 | 0.0173 | 1.0000 | 0.99 | 0.99 |
| | 0.002 | 0.002 | 0.0251 | 1.0001 | 0.99 | 0.99 |
| | 0.003 | 0.003 | 0.0304 | 1.0001 | 0.99 | 0.99 |
| | 0.004 | 0.004 | 0.0347 | 1.0001 | 0.99 | 0.99 |
| | 0.005 | 0.005 | 0.0379 | 1.0002 | 0.99 | 0.99 |
| | 0.006 | 0.006 | 0.0406 | 1.0002 | 0.99 | 0.99 |
| | 0.007 | 0.007 | 0.0433 | 1.0003 | 0.99 | 0.99 |
| | 0.008 | 0.008 | 0.0450 | 1.0003 | 0.99 | 0.99 |
| | 0.009 | 0.009 | 0.0488 | 1.0003 | 0.99 | 0.99 |
| | 0.010 | 0.010 | 0.0566 | 1.0004 | 0.99 | 0.99 |
| | 0.020 | 0.020 | 0.0685 | 1.0007 | 0.99 | 0.99 |
| | 0.030 | 0.030 | 0.0783 | 1.0011 | 0.99 | 0.99 |
| | 0.040 | 0.040 | 0.0872 | 1.0014 | 0.99 | 0.99 |
| | 0.050 | 0.050 | 0.0938 | 1.0018 | 0.99 | 0.99 |
| | 0.060 | 0.060 | 0.0992 | 1.0022 | 0.99 | 0.99 |
| | 0.070 | 0.070 | 0.1041 | 1.0025 | 0.99 | 0.99 |
| | 0.080 | 0.080 | 0.1072 | 1.0029 | 0.99 | 0.99 |
| | 0.090 | 0.089 | 0.1128 | 1.0032 | 0.99 | 0.99 |
| | 0.100 | 0.099 | 0.1230 | 1.0036 | 0.99 | 0.99 |
| | 0.200 | 0.199 | 0.1366 | 1.0072 | 0.99 | 0.99 |
| | 0.300 | 0.297 | 0.1479 | 1.0108 | 0.99 | 0.99 |
| | 0.400 | 0.397 | 0.1553 | 1.0143 | 0.99 | 0.99 |
| | 0.500 | 0.494 | 0.1587 | 1.0179 | 0.99 | 0.98 |
| | 0.600 | 0.590 | 0.1594 | 1.0214 | 0.99 | 0.98 |
| | 0.700 | 0.686 | 0.1580 | 1.0250 | 0.99 | 0.98 |
| | 0.800 | 0.779 | 0.1560 | 1.0285 | 0.99 | 0.98 |

(*Continued*)

**TABLE 4.1 (*Continued*)**
**Calculated Thermodynamic Components Included in Formula (4.102)**

| Electrolyte | $M$ | $c$ | $-\gamma_W$ | $S_W$ | $a_{w\ exp}$ | $a_{w\ calc}$ |
|---|---|---|---|---|---|---|
| | 0.900 | 0.876 | 0.1526 | 1.0320 | 0.99 | 0.97 |
| | 1.000 | 0.968 | 0.1452 | 1.0355 | 0.99 | 0.97 |
| | 1.250 | 1.205 | 0.1352 | 1.0442 | 0.97 | 0.96 |
| | 1.500 | 1.432 | 0.1204 | 1.0529 | 0.95 | 0.95 |
| | 1.750 | 1.666 | 0.1023 | 1.0615 | 0.94 | 0.94 |
| | 2.000 | 1.887 | 0.0836 | 1.0700 | 0.92 | 0.93 |
| | 2.250 | 2.104 | 0.0662 | 1.0785 | 0.91 | 0.92 |
| | 2.500 | 2.333 | 0.0494 | 1.0869 | 0.90 | 0.91 |
| | 2.750 | 2.545 | 0.0325 | 1.0952 | 0.89 | 0.90 |
| | 3.000 | 2.754 | 0.0147 | 1.1035 | 0.89 | 0.89 |
| | 3.250 | 2.960 | −0.0035 | 1.1118 | 0.88 | 0.88 |
| | 3.500 | 3.164 | −0.0215 | 1.1199 | 0.87 | 0.87 |
| | 3.750 | 3.387 | −0.0399 | 1.1281 | 0.86 | 0.85 |
| | 4.000 | 3.586 | −0.0580 | 1.1361 | 0.85 | 0.84 |
| | 4.250 | 3.784 | −0.0768 | 1.1442 | 0.84 | 0.83 |
| | 4.500 | 3.979 | −0.0949 | 1.1521 | 0.83 | 0.82 |
| | 4.750 | 4.173 | −0.1136 | 1.1601 | 0.81 | 0.80 |
| | 5.000 | 4.364 | −0.1308 | 1.1679 | 0.80 | 0.79 |
| | 5.250 | 4.554 | −0.1486 | 1.1757 | 0.79 | 0.78 |
| | 5.500 | 4.711 | −0.1693 | 1.1835 | 0.77 | 0.76 |
| | 5.750 | 4.896 | −0.1916 | 1.1912 | 0.76 | 0.75 |
| | 6.000 | 5.079 | −0.2123 | 1.1989 | 0.75 | 0.74 |
| | 6.250 | 5.261 | −0.2303 | 1.2065 | 0.73 | 0.72 |
| | 6.500 | 5.442 | −0.2461 | 1.2141 | 0.72 | 0.71 |
| | 6.750 | 5.621 | −0.2615 | 1.2216 | 0.71 | 0.70 |
| | 7.000 | 5.799 | −0.2772 | 1.2291 | 0.70 | 0.68 |
| | 7.250 | 5.937 | −0.2931 | 1.2366 | 0.69 | 0.67 |
| | 7.500 | 6.112 | −0.3089 | 1.2440 | 0.68 | 0.66 |
| | 7.750 | 6.285 | −0.3254 | 1.2513 | 0.67 | 0.64 |
| | 8.000 | 6.458 | −0.3400 | 1.2586 | 0.65 | 0.63 |
| | 8.250 | 6.587 | −0.3557 | 1.2659 | 0.64 | 0.62 |
| $MgCl_2$ | 0.001 | 0.001 | 0.0597 | 1.0001 | 0.99 | 0.99 |
| | 0.002 | 0.002 | 0.0781 | 1.0001 | 0.99 | 0.99 |
| | 0.003 | 0.003 | 0.0935 | 1.0002 | 0.99 | 0.99 |
| | 0.004 | 0.004 | 0.1062 | 1.0002 | 0.99 | 0.99 |
| | 0.005 | 0.005 | 0.1170 | 1.0003 | 0.99 | 0.99 |
| | 0.006 | 0.006 | 0.1264 | 1.0003 | 0.99 | 0.99 |
| | 0.007 | 0.007 | 0.1334 | 1.0004 | 0.99 | 0.99 |
| | 0.008 | 0.008 | 0.1400 | 1.0004 | 0.99 | 0.99 |
| | 0.009 | 0.009 | 0.1511 | 1.0005 | 0.99 | 0.99 |
| | 0.010 | 0.010 | 0.1708 | 1.0005 | 0.99 | 0.99 |
| | 0.020 | 0.020 | 0.1962 | 1.0011 | 0.99 | 0.99 |

(*Continued*)

**TABLE 4.1** (*Continued*)
**Calculated Thermodynamic Components Included in Formula (4.102)**

| Electrolyte | $M$ | $c$ | $-\gamma_W$ | $S_W$ | $a_{w\,exp}$ | $a_{w\,calc}$ |
|---|---|---|---|---|---|---|
| | 0.030 | 0.030 | 0.2214 | 1.0016 | 0.99 | 0.99 |
| | 0.040 | 0.040 | 0.2434 | 1.0022 | 0.99 | 0.99 |
| | 0.050 | 0.050 | 0.2613 | 1.0027 | 0.99 | 0.99 |
| | 0.060 | 0.060 | 0.2757 | 1.0032 | 0.99 | 0.99 |
| | 0.070 | 0.070 | 0.2864 | 1.0038 | 0.99 | 0.99 |
| | 0.080 | 0.080 | 0.2957 | 1.0043 | 0.99 | 0.99 |
| | 0.090 | 0.090 | 0.3086 | 1.0049 | 0.99 | 0.99 |
| | 0.100 | 0.100 | 0.3268 | 1.0054 | 0.99 | 0.99 |
| | 0.200 | 0.199 | 0.3456 | 1.0108 | 0.99 | 0.99 |
| | 0.300 | 0.298 | 0.3579 | 1.0161 | 0.99 | 0.99 |
| | 0.400 | 0.397 | 0.3619 | 1.0214 | 0.98 | 0.99 |
| | 0.500 | 0.496 | 0.3580 | 1.0267 | 0.97 | 0.98 |
| | 0.600 | 0.589 | 0.3479 | 1.0320 | 0.97 | 0.98 |
| | 0.700 | 0.687 | 0.3355 | 1.0372 | 0.96 | 0.97 |
| | 0.800 | 0.784 | 0.3202 | 1.0425 | 0.96 | 0.97 |
| | 0.900 | 0.882 | 0.2965 | 1.0477 | 0.95 | 0.96 |
| | 1.000 | 0.979 | 0.2567 | 1.0529 | 0.94 | 0.96 |
| | 1.250 | 1.217 | 0.1991 | 1.0657 | 0.92 | 0.94 |
| $Na_2SO_4$ | 0.001 | 0.001 | 0.0613 | 1.0001 | 0.99 | 0.99 |
| | 0.002 | 0.002 | 0.0810 | 1.0001 | 0.99 | 0.99 |
| | 0.003 | 0.003 | 0.0976 | 1.0002 | 0.99 | 0.99 |
| | 0.004 | 0.004 | 0.1117 | 1.0002 | 0.99 | 0.99 |
| | 0.005 | 0.005 | 0.1237 | 1.0003 | 0.99 | 0.99 |
| | 0.006 | 0.006 | 0.1341 | 1.0003 | 0.99 | 0.99 |
| | 0.007 | 0.007 | 0.1422 | 1.0004 | 0.99 | 0.99 |
| | 0.008 | 0.008 | 0.1496 | 1.0004 | 0.99 | 0.99 |
| | 0.009 | 0.009 | 0.1627 | 1.0005 | 0.99 | 0.99 |
| | 0.010 | 0.010 | 0.1863 | 1.0005 | 0.99 | 0.99 |
| | 0.020 | 0.020 | 0.2179 | 1.0011 | 0.99 | 0.99 |
| | 0.030 | 0.030 | 0.2501 | 1.0016 | 0.99 | 0.99 |
| | 0.040 | 0.040 | 0.2795 | 1.0022 | 0.99 | 0.99 |
| | 0.050 | 0.050 | 0.3045 | 1.0027 | 0.99 | 0.99 |
| | 0.060 | 0.060 | 0.3255 | 1.0032 | 0.99 | 0.99 |
| | 0.070 | 0.070 | 0.3414 | 1.0038 | 0.99 | 0.99 |
| | 0.080 | 0.080 | 0.3558 | 1.0043 | 0.99 | 0.99 |
| | 0.090 | 0.089 | 0.3784 | 1.0049 | 0.99 | 0.99 |
| | 0.100 | 0.099 | 0.4171 | 1.0054 | 0.99 | 0.99 |
| | 0.200 | 0.199 | 0.4679 | 1.0108 | 0.99 | 0.99 |
| | 0.300 | 0.297 | 0.5189 | 1.0161 | 0.99 | 0.99 |
| | 0.400 | 0.394 | 0.5647 | 1.0214 | 0.98 | 0.99 |
| | 0.500 | 0.495 | 0.6027 | 1.0267 | 0.98 | 0.99 |
| | 0.600 | 0.592 | 0.6338 | 1.0320 | 0.97 | 0.99 |

(*Continued*)

**TABLE 4.1** (*Continued*)
**Calculated Thermodynamic Components Included in Formula (4.102)**

| Electrolyte | $M$ | $c$ | $-\gamma_W$ | $S_W$ | $a_{w\,exp}$ | $a_{w\,calc}$ |
|---|---|---|---|---|---|---|
| | 0.700 | 0.687 | 0.6591 | 1.0372 | 0.97 | 0.99 |
| | 0.800 | 0.782 | 0.6809 | 1.0425 | 0.97 | 0.98 |
| | 0.900 | 0.877 | 0.7028 | 1.0477 | 0.96 | 0.98 |
| | 1.000 | 0.971 | 0.7273 | 1.0529 | 0.96 | 0.98 |
| NaCl | 0.001 | 0.001 | 0.0173 | 1.0000 | 0.99 | 0.99 |
| | 0.002 | 0.002 | 0.0251 | 1.0001 | 0.99 | 0.99 |
| | 0.003 | 0.003 | 0.0304 | 1.0001 | 0.99 | 0.99 |
| | 0.004 | 0.004 | 0.0347 | 1.0001 | 0.99 | 0.99 |
| | 0.005 | 0.005 | 0.0379 | 1.0002 | 0.99 | 0.99 |
| | 0.006 | 0.006 | 0.0411 | 1.0002 | 0.99 | 0.99 |
| | 0.007 | 0.007 | 0.0439 | 1.0003 | 0.99 | 0.99 |
| | 0.008 | 0.008 | 0.0450 | 1.0003 | 0.99 | 0.99 |
| | 0.009 | 0.009 | 0.0494 | 1.0003 | 0.99 | 0.99 |
| | 0.010 | 0.010 | 0.0577 | 1.0004 | 0.99 | 0.99 |
| | 0.020 | 0.020 | 0.0696 | 1.0007 | 0.99 | 0.99 |
| | 0.030 | 0.030 | 0.0801 | 1.0011 | 0.99 | 0.99 |
| | 0.040 | 0.040 | 0.0896 | 1.0014 | 0.99 | 0.99 |
| | 0.050 | 0.050 | 0.0974 | 1.0018 | 0.99 | 0.99 |
| | 0.060 | 0.060 | 0.1035 | 1.0022 | 0.99 | 0.99 |
| | 0.070 | 0.070 | 0.1091 | 1.0025 | 0.99 | 0.99 |
| | 0.080 | 0.079 | 0.1128 | 1.0029 | 0.99 | 0.99 |
| | 0.090 | 0.090 | 0.1191 | 1.0032 | 0.99 | 0.99 |
| | 0.100 | 0.100 | 0.1313 | 1.0036 | 0.99 | 0.99 |
| | 0.200 | 0.199 | 0.1492 | 1.0072 | 0.99 | 0.99 |
| | 0.300 | 0.298 | 0.1663 | 1.0108 | 0.99 | 0.99 |
| | 0.400 | 0.395 | 0.1798 | 1.0143 | 0.98 | 0.99 |
| | 0.500 | 0.495 | 0.1899 | 1.0179 | 0.98 | 0.99 |
| | 0.600 | 0.590 | 0.1973 | 1.0214 | 0.98 | 0.98 |
| | 0.700 | 0.690 | 0.2017 | 1.0250 | 0.97 | 0.98 |
| | 0.800 | 0.784 | 0.2055 | 1.0285 | 0.97 | 0.98 |
| | 0.900 | 0.883 | 0.2085 | 1.0320 | 0.96 | 0.97 |
| | 1.000 | 0.983 | 0.2100 | 1.0355 | 0.96 | 0.97 |
| | 1.250 | 1.220 | 0.2108 | 1.0442 | 0.95 | 0.96 |
| | 1.500 | 1.455 | 0.2093 | 1.0529 | 0.94 | 0.96 |
| | 1.750 | 1.687 | 0.2047 | 1.0615 | 0.93 | 0.95 |
| | 2.000 | 1.916 | 0.1987 | 1.0700 | 0.92 | 0.94 |
| | 2.250 | 2.157 | 0.1914 | 1.0785 | 0.91 | 0.93 |
| | 2.500 | 2.383 | 0.1841 | 1.0869 | 0.90 | 0.92 |
| | 2.750 | 2.606 | 0.1755 | 1.0952 | 0.90 | 0.91 |
| | 3.000 | 2.827 | 0.1656 | 1.1035 | 0.88 | 0.90 |
| | 3.250 | 3.046 | 0.1553 | 1.1118 | 0.87 | 0.89 |
| | 3.500 | 3.263 | 0.1445 | 1.1199 | 0.86 | 0.88 |

(*Continued*)

**TABLE 4.1 (*Continued*)**
**Calculated Thermodynamic Components Included in Formula (4.102)**

| Electrolyte | $M$ | $c$ | $-\gamma_W$ | $S_W$ | $a_{w\,exp}$ | $a_{w\,calc}$ |
|---|---|---|---|---|---|---|
| | 3.750 | 3.478 | 0.1326 | 1.1281 | 0.85 | 0.87 |
| | 4.000 | 3.691 | 0.1204 | 1.1361 | 0.84 | 0.86 |
| | 4.250 | 3.902 | 0.1072 | 1.1442 | 0.83 | 0.85 |
| | 4.500 | 4.112 | 0.0938 | 1.1521 | 0.82 | 0.84 |
| | 4.750 | 4.320 | 0.0795 | 1.1601 | 0.81 | 0.83 |
| | 5.000 | 4.526 | 0.0651 | 1.1679 | 0.80 | 0.82 |
| | 5.250 | 4.700 | 0.0505 | 1.1757 | 0.79 | 0.81 |
| $NaNO_3$ | 0.001 | 0.001 | 0.0183 | 1.0000 | 0.99 | 0.99 |
| | 0.002 | 0.002 | 0.0245 | 1.0001 | 0.99 | 0.99 |
| | 0.003 | 0.003 | 0.0298 | 1.0001 | 0.99 | 0.99 |
| | 0.004 | 0.004 | 0.0343 | 1.0001 | 0.99 | 0.99 |
| | 0.005 | 0.005 | 0.0382 | 1.0002 | 0.99 | 0.99 |
| | 0.006 | 0.006 | 0.0417 | 1.0002 | 0.99 | 0.99 |
| | 0.007 | 0.007 | 0.0443 | 1.0003 | 0.99 | 0.99 |
| | 0.008 | 0.008 | 0.0468 | 1.0003 | 0.99 | 0.99 |
| | 0.009 | 0.009 | 0.0513 | 1.0003 | 0.99 | 0.99 |
| | 0.010 | 0.010 | 0.0600 | 1.0004 | 0.99 | 0.99 |
| | 0.020 | 0.020 | 0.0715 | 1.0007 | 0.99 | 0.99 |
| | 0.030 | 0.030 | 0.0836 | 1.0011 | 0.99 | 0.99 |
| | 0.040 | 0.040 | 0.0947 | 1.0014 | 0.99 | 0.99 |
| | 0.050 | 0.050 | 0.1043 | 1.0018 | 0.99 | 0.99 |
| | 0.060 | 0.060 | 0.1127 | 1.0022 | 0.99 | 0.99 |
| | 0.070 | 0.070 | 0.1190 | 1.0025 | 0.99 | 0.99 |
| | 0.080 | 0.080 | 0.1248 | 1.0029 | 0.99 | 0.99 |
| | 0.090 | 0.090 | 0.1348 | 1.0032 | 0.99 | 0.99 |
| | 0.100 | 0.100 | 0.1526 | 1.0036 | 0.99 | 0.99 |
| | 0.200 | 0.199 | 0.1760 | 1.0072 | 0.99 | 0.99 |
| | 0.300 | 0.296 | 0.1997 | 1.0108 | 0.99 | 0.99 |
| | 0.400 | 0.393 | 0.2210 | 1.0143 | 0.99 | 0.99 |
| | 0.500 | 0.491 | 0.2391 | 1.0179 | 0.99 | 0.99 |
| | 0.600 | 0.588 | 0.2543 | 1.0214 | 0.99 | 0.98 |
| | 0.700 | 0.685 | 0.2670 | 1.0250 | 0.99 | 0.98 |
| | 0.800 | 0.777 | 0.2782 | 1.0285 | 0.98 | 0.98 |
| | 0.900 | 0.873 | 0.2898 | 1.0320 | 0.98 | 0.98 |
| | 1.000 | 0.969 | 0.3038 | 1.0355 | 0.98 | 0.97 |
| | 1.250 | 1.204 | 0.3198 | 1.0442 | 0.97 | 0.97 |
| | 1.500 | 1.427 | 0.3362 | 1.0529 | 0.97 | 0.96 |
| | 1.750 | 1.657 | 0.3518 | 1.0615 | 0.97 | 0.96 |
| | 2.000 | 1.885 | 0.3662 | 1.0700 | 0.96 | 0.95 |
| | 2.250 | 2.097 | 0.3792 | 1.0785 | 0.96 | 0.94 |
| | 2.500 | 2.320 | 0.3911 | 1.0869 | 0.95 | 0.94 |
| | 2.750 | 2.526 | 0.4020 | 1.0952 | 0.95 | 0.93 |

(*Continued*)

**TABLE 4.1 (*Continued*)**
**Calculated Thermodynamic Components Included in Formula (4.102)**

| Electrolyte | $M$ | $c$ | $-\gamma_W$ | $S_W$ | $a_{w\ exp}$ | $a_{w\ calc}$ |
|---|---|---|---|---|---|---|
| | 3.000 | 2.727 | 0.4122 | 1.1035 | 0.94 | 0.93 |
| | 3.250 | 2.946 | 0.4216 | 1.1118 | 0.93 | 0.92 |
| | 3.500 | 3.142 | 0.4304 | 1.1199 | 0.92 | 0.92 |
| | 3.750 | 3.335 | 0.4387 | 1.1281 | 0.90 | 0.91 |
| | 4.000 | 3.525 | 0.4464 | 1.1361 | 0.89 | 0.91 |
| | 4.250 | 3.739 | 0.4538 | 1.1442 | 0.88 | 0.90 |
| | 4.500 | 3.925 | 0.4607 | 1.1521 | 0.87 | 0.89 |
| | 4.750 | 4.109 | 0.4672 | 1.1601 | 0.86 | 0.89 |
| | 5.000 | 4.290 | 0.4734 | 1.1679 | 0.85 | 0.88 |
| | 5.250 | 4.469 | 0.4793 | 1.1757 | 0.84 | 0.88 |
| | 5.500 | 4.647 | 0.4850 | 1.1835 | 0.84 | 0.87 |
| | 5.750 | 4.822 | 0.4903 | 1.1912 | 0.84 | 0.86 |
| | 6.000 | 4.996 | 0.4954 | 1.1989 | 0.84 | 0.86 |
| | 6.250 | 5.168 | 0.5002 | 1.2065 | 0.83 | 0.85 |
| | 6.500 | 5.338 | 0.5046 | 1.2141 | 0.83 | 0.85 |
| | 6.750 | 5.468 | 0.5093 | 1.2216 | 0.83 | 0.84 |
| | 7.000 | 5.634 | 0.5139 | 1.2291 | 0.82 | 0.83 |
| | 7.250 | 5.799 | 0.5181 | 1.2366 | 0.82 | 0.83 |
| | 7.500 | 5.963 | 0.5215 | 1.2440 | 0.81 | 0.82 |
| | 7.750 | 6.125 | 0.5249 | 1.2513 | 0.81 | 0.82 |
| | 8.000 | 6.243 | 0.5286 | 1.2586 | 0.80 | 0.81 |
| | 8.250 | 6.402 | 0.5324 | 1.2659 | 0.80 | 0.80 |
| | 8.500 | 6.561 | 0.5359 | 1.2731 | 0.79 | 0.80 |
| | 8.750 | 6.718 | 0.5393 | 1.2803 | 0.79 | 0.79 |
| | 9.000 | 6.827 | 0.5427 | 1.2874 | 0.79 | 0.79 |
| | 9.250 | 6.982 | 0.5459 | 1.2945 | 0.78 | 0.78 |
| | 9.500 | 7.136 | 0.5491 | 1.3016 | 0.78 | 0.77 |
| | 9.750 | 7.238 | 0.5522 | 1.3086 | 0.78 | 0.77 |
| | 10.000 | 7.390 | 0.5554 | 1.315 | 0.77 | 0.76 |
| | 10.250 | 7.542 | 0.5586 | 1.322 | 0.77 | 0.75 |
| | 10.500 | 7.639 | 0.5615 | 1.329 | 0.76 | 0.75 |
| | 10.750 | 7.788 | 0.5638 | 1.336 | 0.76 | 0.74 |
| | 10.830 | 7.818 | 0.5657 | 1.338 | 0.76 | 0.74 |
| RbCl | 0.010 | 0.010 | 0.0333 | 1.0004 | 0.99 | 0.99 |
| | 0.020 | 0.020 | 0.0570 | 1.0007 | 0.99 | 0.99 |
| | 0.030 | 0.030 | 0.0736 | 1.0011 | 0.99 | 0.99 |
| | 0.040 | 0.040 | 0.0841 | 1.0014 | 0.99 | 0.99 |
| | 0.050 | 0.050 | 0.0928 | 1.0018 | 0.99 | 0.99 |
| | 0.060 | 0.060 | 0.1005 | 1.0022 | 0.99 | 0.99 |
| | 0.070 | 0.070 | 0.1081 | 1.0025 | 0.99 | 0.99 |
| | 0.080 | 0.080 | 0.1139 | 1.0029 | 0.99 | 0.99 |
| | 0.090 | 0.090 | 0.1224 | 1.0032 | 0.99 | 0.99 |

(*Continued*)

**TABLE 4.1 (*Continued*)**
**Calculated Thermodynamic Components Included in Formula (4.102)**

| Electrolyte | $M$ | $c$ | $-\gamma_W$ | $S_W$ | $a_{w\,exp}$ | $a_{w\,calc}$ |
|---|---|---|---|---|---|---|
| | 0.100 | 0.100 | 0.1370 | 1.0036 | 0.99 | 0.99 |
| | 0.200 | 0.199 | 0.1589 | 1.0072 | 0.99 | 0.99 |
| | 0.300 | 0.301 | 0.1806 | 1.0108 | 0.99 | 0.99 |
| | 0.400 | 0.400 | 0.2024 | 1.0143 | 0.99 | 0.99 |
| | 0.500 | 0.499 | 0.2152 | 1.0179 | 0.99 | 0.99 |
| | 0.600 | 0.597 | 0.2276 | 1.0214 | 0.99 | 0.98 |
| | 0.700 | 0.694 | 0.2376 | 1.0250 | 0.99 | 0.98 |
| | 0.800 | 0.791 | 0.2458 | 1.0285 | 0.99 | 0.98 |
| | 0.900 | 0.887 | 0.2537 | 1.0320 | 0.99 | 0.98 |
| | 1.000 | 0.983 | 0.2614 | 1.0355 | 0.98 | 0.97 |
| | 1.200 | 1.171 | 0.2695 | 1.0425 | 0.98 | 0.97 |
| | 1.400 | 1.347 | 0.2771 | 1.0494 | 0.98 | 0.96 |
| | 1.600 | 1.522 | 0.2842 | 1.0563 | 0.97 | 0.96 |
| | 1.800 | 1.707 | 0.2900 | 1.0632 | 0.97 | 0.95 |
| | 2.000 | 1.875 | 0.2955 | 1.0700 | 0.96 | 0.95 |
| | 2.500 | 2.312 | 0.3000 | 1.0869 | 0.94 | 0.93 |
| | 3.000 | 2.744 | 0.3028 | 1.1035 | 0.92 | 0.92 |
| | 3.500 | 3.145 | 0.3038 | 1.1199 | 0.90 | 0.90 |
| | 4.000 | 3.538 | 0.3028 | 1.1361 | 0.89 | 0.89 |
| | 4.500 | 3.891 | 0.3006 | 1.1521 | 0.87 | 0.87 |
| | 5.000 | 4.269 | 0.2971 | 1.1679 | 0.86 | 0.85 |
| | 5.500 | 4.605 | 0.2929 | 1.1835 | 0.85 | 0.84 |
| | 6.000 | 4.932 | 0.2881 | 1.1989 | 0.83 | 0.82 |
| | 6.500 | 5.252 | 0.2830 | 1.2141 | 0.82 | 0.80 |
| | 7.000 | 5.566 | 0.2781 | 1.2291 | 0.81 | 0.79 |

*Notes:* $m$ in mol/kg $H_2O$; $c_k$ in mol/dm$^3$, $a_{w\,exp}$ and $a_{w\,calc}$, experimental and calculated values of water activity.

## REFERENCES

1. Stepanov N.F. 2001. *Of Quantum Mechanics and Quantum Chemistry*. Mir, Moscow, Russia, p. 519.
2. Shakhparonov M.I. 1976. *Introduction in the Modern Theory of Solutions (Intermolecular Interactions. Structure. Prime Liquids)*. The Highest School, Moscow, Russia, p. 296.
3. Tatevsky V.M. 1973. *Classical Theory of Molecular Structure and Quantum Mechanics*. Chemistry, Moscow, Russia, p. 320.
4. Vereshchagin A.N. 1980. *Polarizability of the Molecules*. Science, Moscow, Russia, p. 178.
5. Aseyev G.G. 2001. *Electrolytes: Methods for Calculation of the Physicochemical Parameteres of Multicomponent Systems*. Begell House, Inc. New York, p. 368.
6. Prezhdo V.V., Kraynov N. 1994. *Molecular Interactions and Electric Properties of Molecules*. Osnova, Kharkov, Ukraine, p. 240.
7. Simkin B.Yu., Sheykhet I.I. 1989. *Quantum Chemical and Statistical Theory of Solutions. Computing Methods and Their Application*. Chemistry, Moscow, Russia, p. 256.

8. Shakhparonov M.I. 1976. *Introduction in the Modern Theory of Solutions (Intermolecular Interactions. Structure. Prime Liquids).* The Highest School, Moscow, Russia, p. 296.
9. Eremin V.V., Kargov S.I., Uspenskaya I.A. et al. 2005. *Fundamentals of Physical Chemistry. Theory and Tasks.* Examination, Moscow, Russia, p. 480.
10. Girshfelder Dzh. Curtice Ch., Byrd R. 1961. *Molecular Theory of Gases and Liquids.* Publishing House of the Foreign Lit., Moscow, Russia, p. 930.
11. Kaplan I.G. 1982. *Introduction to the Theory of Intermolecular Interactions.* Nauka, Moscow, Russia, p. 312.
12. Barash Yu. S. 1988. *Van-der-Waals Forces.* Nauka, Moscow, Russia, p. 344.
13. Smirnova N.A. 1987. *Molecular Theories of Solutions.* L.: Chemistry, p. 336.
14. Abarenkov I.V., Brothers V.F., Tulub A.V. 1989. *Beginning of the Beginning of a Quantum Chemistry.* The Highest School, Moscow, Russia, p. 303.
15. Stepanov N.F. 2001. *Of Quantum Mechanics and Quantum Chemistry.* Mir, Moscow, Russia, p. 519.
16. Zagradnik R., Polak R. 1979. *Bases of a Quantum Chemistry.* Mir, Moscow, Russia, p. 504.
17. Mushrooms L.A., Mushtakov S. 1999. *Quantum Chemistry.* Gardarika, Moscow, Russia, p. 390.
18. Zhidomirov G.M., Bagaturyants A.A., Abronin I.A. 1979. *Applied Quantum Chemistry.* Chemistry, Moscow, Russia, p. 296.
19. Bader R. 2001. *Atoms in Molecules. Quantum Theory.* Mir, Moscow, Russia, p. 532.
20. Mayer I. 2006. *Selected Chapters of Quantum Chemistry: Proofs of Theorems and Formulas.* BINOM: Lab. Knowledge, Moscow, Russia, p. 384.
21. Dmitriyev I.S., Semenov S.G. 1980. *Quantum Chemistry—Its Past and the Present. Development of Electronic Ideas of the Nature of a Chemical Bond.* Atomizdat, Moscow, Russia, p. 160.
22. Tatevsky V.M. 1973. *Classical Theory of Molecular Structure and Quantum Mechanics.* Chemistry, Moscow, Russia, p. 320.
23. Kozman U. 1960. *Introduction to Quantum Chemistry.* Publishing House of the Aliens. Lit., Moscow, Russia, p. 558.
24. Davtyan O.K. 1962. *Quantum Chemistry.* The Highest School, Moscow, Russia, p. 784.
25. Nagakura S., Nakajima T. (eds.). 1982. *Introduction to Quantum Chemistry.* The Lane with the Jap. Mir, Moscow, Russia, p. 264.
26. Sinanoglu O. (ed.). 1968. *Modern Quantum Chemistry: In 2t.* The Lane from English. Mir, Moscow, Russia.
27. Mueller M.R. 2001. *Fundamentals of Quantum Chemistry: Molecular Spectroscopy and Modern Electronic Structure Computations.* Kluwer, New York, p. 265.
28. Baranovsky V.I. 2008. *Quantum Mechanics and Quantum Chemistry.* Academy, Moscow, Russia, p. 384.
29. Brodsky A.M. 1968. *Modern Quantum Chemistry. 2.* Mir, Moscow, Russia, p. 319.
30. Gankin V.Y., Gankin Y.V. 1998. *How Chemical Bonds Form and Chemical Reactions Proceed.* ITC, Boston, MA, p. 315.
31. Gubanov V.A., Zhukov V.P., Litinskii A.O. 1976. *Semi-Empirical Molecular Orbital Methods in Quantum Chemistry.* Nauka, Moscow, Russia, p. 219.
32. Zelentsov S.V. 2006. *Introduction to Modern Quantum Chemistry.* Novosibirsk, Nizhniy Novgorod, Russia, p. 126.
33. Tsirelson V.G. 2010. *Quantum Chemistry. Molecules, Molecular Systems and Solid Bodies.* BINOM Lab. Knowledge, Moscow, Russia, p. 496.
34. Novakovskaya Yu.V. 2004. Molecular Systems. *The Structure and Interaction Theory with a Radiation: Common Bases of a Quantum Mechanics and Symmetry Theory.* Editorial of URSS, Moscow, Russia, p. 104.
35. Segal J. 1980. *Semi-Empirical Methods of Calculation of the Electronic Structure.* Mir, Moscow, Russia, p. 327.

36. Bartók-Pártay A. 2010. *The Gaussian Approximation Potential: An Interatomic Potential Derived from First Principles Quantum Mechanics*. Springer, Berlin, Heidelberg, p. 96.
37. Sinanoglu O. 1966. *Multielectronic Theory of Atoms, Molecules and Their Interactions*. The Lane with English. Mir, Moscow, Russia, p. 152.
38. Gelman G. 2012. *Quantum Chemistry*. BINOM Lab. Knowledge, Moscow, Russia, p. 520.
39. Stepanov N.F., Pupyshev V.I. 1991. *Quantum Mechanics of Molecules and Quantum Chemistry*. izd-vo MGU, Moscow, Russia, p. 384.
40. Unger F.G. 2007. *Quantum Mechanics and Quantum Chemistry*. TGU, Tomsk, Russia, p. 240.
41. Flarri R. 1985. *Quantum Chemistry*. Mir, Moscow, Russia, p. 472.
42. Fudzinaga C. 1983. *Method of Molecular Orbitals*. Mir, Moscow, Russia, p. 461.
43. Hedwig P. 1977. *Applied Quantum Chemistry*. Mir, Moscow, Russia, p. 596.
44. Shalva O., Dodel R., Dean S., Malryyo Zh.-P. 1978. *Localization and Delocalization in a Quantum Chemistry. Atoms and Molecules in a Ground State*. Mir, Moscow, Russia, p. 416.
45. Fitts D.D. 2002. *Principles of Quantum Mechanics: As Applied to Chemistry and Chemical Physics*. Cambridge University Press, New York, p. 352.
46. Hehre W.J.A. 2003. *Guide to Molecular Mechanics and Quantum Chemical Calculations*. Wavefunction, Inc., Irvine, CA, p. 796.
47. Inagaki S. 2009. *Orbitals in Chemistry*. Springer, Berlin, Germany, p. 327.
48. Koch W.A., Holthausen M.C. 2001. *Chemist's Guide to Density Functional Theory*, 2nd edn. Wiley-VCH Verlag GmbH, Weinheim, Germany, p. 293.
49. Levine I.N. 1999. *Quantum Chemistry*, 5th edn. Prentice Hall, Brooklyn, NY, p. 739.
50. Lowe J.P., Peterson K.A. 2006. *Quantum Chemistry*, 3rd edn. Elsevier Academic Press, London, p. 703.
51. Maruani J., Minot C. (eds.). 2002. *New Trends in Quantum Systems in Chemistry and Physics. 2. Advanced Problems and Complex Systems*. Kluwer Academic Publishers, Dordrecht, the Netherlands, p. 313.
52. Piela L. 2007. *Ideas of Quantum Chemistry*. Elsevier Science, Amsterdam, the Netherlands, p. 1086.
53. Atkins. 1980. *Physical Chemistry*. 2. Mir, Mosow, Russia, p. 584.
54. Günter H. 1984. *Introduction in a Course of a Spectroscopy of a Nuclear Magnetic Resonance*. The Lane with English. Mir, Moscow, Russia, p. 478.
55. Wainstein L.A. 1991. *Atomic Spectrum Analysis (Ranges of Atoms and Ions)*. MFTI, Moscow, Russia, p. 76.
56. Deroum E. 1992. *The Nuclear Magnetic Resonances Modern Methods for Chemical Researches*. The Lane with English. Mir, Moscow, Russia, p. 403.
57. Lyashchenko A.K. 2002. *Structural and Molecular-Kinetic Properties Of Concentrated Solutions and Phase Equilibria in Water-Salt Systems Concentrated and Saturated Solutions*. Nauka, Moscow, Russia, pp. 93–118. (A series problem of chemistry of solutions).
58. Lyashchenko A.K., Zasetsky A. Yu. 1998. Change of a structural state, dynamics of molecules of water and properties of solutions upon transition to electrolytic water solvent. *Zhurn. Struct. Chem.* 39 (5): 851–863.
59. Lyashchenko A.K. 2010. Communication of activity of water with a statistical dielectric constant of strong solutions of electrolytes. *Zhurn. Inorganic Chem.* 55 (11): 1930–1936.
60. Knorre D.G., Krylova L.F., Musikantov V.S. 1990. *Physical Chemistry*. The Highest School, Moscow, Russia, p. 416.
61. Kaplan I.G. 1982. *Introduction to the Theory of Intermolecular Interactions*. Nauka, Moscow, Russia, p. 312.
62. Emisli D. 1993. *Elements*. Mir, Moscow, Russia, p. 256.
63. Sechkarev B.A., Titov F.V. 2006. *Chemical Bond*. Kuzbassvuzizdat, Kemerovo, Russia, p. 156.

64. Under Edition of Sokolova N.D. 1989. *Hydrogen Bridge*. Nauka, Moscow, Russia, p. 326.
65. Marell D., Kettl S., Tedder D. 1980. *Chemical Bond*. The Lane with English. Mir, Moscow, Russia, p. 378.
66. Guryanova E.N., Goldstein I.P., Romm I.P. 1973. *Donor–Acceptor Bond*. Chemistry, Moscow, Russia, p. 406.
67. Under Edition of Krasnov K.S. 2001. *Physical Chemistry. Substance Structure. Thermodynamics*. The Highest School, Moscow, Russia, p. 512.
68. Gurova N.Ya. 1997. *Inorganic Chemistry in Tables*. Nauka, Moscow, Russia, p. 111.
69. Stid Dzh. B., Atwood J.L. 2007. *Supramolecular Chemistry*. In 2 t. Akademkniga Moscow, Russia.
70. Gray C.G., Gubbins K.E. 1984. *Theory of Molecular Fluids*. Clarendon Press, Oxford, U.K., p. 626.
71. Reed M., Gubbins K.E. 1973. *Applied Statistical Mechanics*. McGraw-Hill Kogakusha Ltd., Tokyo, Japan, p. 506.
72. Barker J.A. In: Klein M.L., Venables J.A. (eds.). 1976. *Rare Gas Solids*. Academic Press, London, U.K., pp. 212–264.
73. Mason E.A., Sperling T.H. Under the editorship of Sychev. 1972. *Virial Equation of State*. The Lane with English. Mir, Moscow, Russia, p. 420.
74. Fertsiger Dzh., Kaperr G. 1976. *The Mathematical Theory of Processes in Gases*. Mir, Moscow, Russia, p. 555.
75. Ueyles C. 1989. *Phase Equilibria in Chemical Engineering*. Mir, Moscow, Russia, p. 304.
76. Hashin V.A., Samsonov V.M. 2010. Calculation of structural thermodynamic characteristics of nanodrops on the basis of a method of a self-consistent field with use of potential of a rectangular hole. *Vestn. TvGU.* 11: 89–95.
77. Aseyev G.G. Under. edition of I.D. Zaytsev. 1992. *Properties of Aqueous Solutions of Elektrolites*. CRC Press, Boca Raton, FL, p. 1774.
78. Harned G., Owen B. 1952. *Physical Chemistry of Solutions of Electrolytes*. Publishing House of the Foreign Lit., Moscow, Russia, p. 628.
79. Robinson R., Stokes R. 1963. *Solutions of Electrolytes*. 1, 2. Publishing House of the Foreign Lit., Moscow, Russia.
80. Dvait G.B. 1977. *Tables of Integrals and Other Mathematical Equations*. Nauka, Moscow, Russia, p. 228.
81. Aseyev G.G. 1998. *Electrolytes: Equilibria in Solutions and Phase Equilibria: Calculation of Multicomponent Systems and Experimental Data on the Activities of Water, Pressure Vapor and Osmotic Coefficients*. Begell House, Inc., New York, p. 758.

# *Section II*

## *Non-Equilibrium Phenomena*

# 5 Electrical Conductivity

## 5.1 HISTORIC PREAMBLE

In the first half of the nineteenth century, the following scientists made significant contributions in the area of electricity concepts: J. Berzelius, T. Grotthuss, M. Faraday, J. Dalton, A. Avogadro, H. Davy, J. Gay-Lussac, H. Kopp, V. Regnault, H. Hess, and others. The most important line of that period is studying of physical properties, weight, and volume of substances, which provided convincing material for experimental substantiation of the atomic and, subsequently, molecular theory in the first decades of the nineteenth century.

Electricity studies were developing absolutely independently from chemistry for a long time. But gradually facts and observations were accumulated, which were taking chemistry more and more into contact with electricity studies. The invention of the *voltaic pile* attracted close attention of both chemists and physical scientists. Electrochemical investigations allowed for the first time studying in detail the question of binding and separating of *electric fluid* from a substance at electrolysis and the role of chemical processes in the formation of galvanic current. The idea, that a substance atom can carry an electric charge and because of that its (ion) properties fundamentally differ from the properties of a neutral atom, enriched physics and chemistry by the provision of fundamental importance.

The works of scientists in that period are marked with the main thesis: the motion of matter is explained by electrochemical interaction of microparticles; the physicochemical process is a process of interaction of differently charged particles. Such approach allowed considering the processes of electrolysis, acid–base properties of substances, displacement reactions, and so on from the concurrent view. The idea of duality of substance behavior and relativity of their properties, depending on the conditions and nature of the reagent, appeared in the electrochemical system for the first time. However, chemists were not yet ready for systematic investigations of chemical processes during the first decades of the nineteenth century. They had still to study in detail the properties and composition of inorganic and organic compounds.

J. Berzelius is known for his investigations in the electrochemistry area. He made his work on electrolysis (together with W. Hisinger) in 1803 and on electrochemical classification of elements in 1812. Berzelius developed the electrochemical affinity theory based on this classification in 1812–1819, according to which electric polarity of atoms is the reason for the connection of elements in certain ratios. In his theory, Berzelius considered electronegativity as the main characteristic of the element; the chemical affinity was considered by him as a tendency to equalization of electric polarities of an atom or groups of atoms.

The first hypothesis of electrolysis was proposed by T. Grotthuss. In 1805, he proposed the first correct theoretical explanation of water decomposition by electric current and formulated the first electrolysis theory, which main postulate was the idea about polarity of corpuscles initiated by electric current or occurring in the result of mutual electrization of atoms. The hydrogen ion transport principle with support by water molecules described by him is called the *Grotthuss mechanism.* In 1818, he made an assumption about spontaneous decomposition of the electrolyte without participation of external electricity. He developed electrochemical notions about acidity and basicity, according to which molecules of the solute consist of two parts—one of which is charged positively and the other one negatively. In the solution to which the electric field is applied, some of the molecules of the solute decompose by some reason into these component elements. The latter elements are transferred into the solution in the form of charged particles, which later obtained the name of *ions.* Positive ions under the influence of the electric field move to the negative electrode (cathode) and negative ions to the positive one (anode). Despite such movement, there will always be the same quantity of positive and negative ions in the electrodes in every volume element, and in case of turning the electric field off, positive ions will get connected with the negative ones into neutral molecules and no chemical decomposition of the solute will take place. Otherwise, it will take place near the electrodes. Having reached the electrodes, ions will transfer their charges to them and get evolved in the form of neutral atoms forming decomposition products of the solute.

The notions of Grotthuss prepared the ground for creation of ionic electrical conductivity theories. In 1833–1834, M. Faraday was studying the passage of electric currents through acid, salt, and alkali solutions, which led to the disclosure of electrolysis laws. These laws (Faraday laws) further on played an important role in establishing notions about discrete carriers of electric charge.

In the works of Faraday in 1833, an idea was advanced about the appearance of charges in electrolytes at the expense of their dissociation at electric field application. Faraday began studying the laws of electrochemical phenomena. The first law, established by Faraday, states that the quantity of electrochemical action depends neither on the size of electrodes, nor on current intensity, nor on strength of the decomposed solution, but only on the quantity of electricity passing in the network; in other words, the quantity of electricity is required to be proportional to the quantity of chemical action. Faraday derived this law from countless numbers of experiments in which conditions were varied endlessly by him.

The second and still more important law of electrochemical action, established by Faraday, states that the quantity of electricity required for decomposition of various substances is always inversely proportional to the atomic weight of a substance or, in other words, the same quantity of electricity is always required for decomposition of a molecule (particle) of any substance.

A closer approach between physics and chemistry was outlined in the 1850s. Atomic notions contributed to this. Studying of processes and methods on thermal dissociation of compounds advanced to the forefront in theoretical chemistry after the classical works by A. Sainte-Claire Deville.

The introduction of a new notion about dissociation in theoretical chemistry makes the biggest scholar merit of Sainte-Claire Deville. In 1857, Sainte-Claire Deville proved that aqueous vapor decomposes into oxygen and hydrogen under temperature influence. Having imagined that water, hydrogen chloride, carbon dioxide, and a number of other compounds dissociate at high temperatures, Sainte-Claire

Deville made a great contribution to developing chemical equilibrium studies. The development of this area of researches led to the creation of chemical statistics and penetration of the first and then second law of thermodynamics into chemistry. The consideration of equilibrium states as a certain aspect of the chemical process turned out to be the basis on which physics and chemistry began approaching, which deepened progressively over the years.

At the same time, it was not clear how neutral molecules are decomposed into ions in the solution. Before R. Clausius, it was considered that it takes place under the action of the applied electric field. Hence, it was believed that the electrostatic field decomposes the solute into component elements. In 1857, Clausius was the first to point out that if it were like this, some minimum electric field intensity would be required for every chemical compound in order to overcome the chemical affinity. The more affinity there is, the greater the field required. At electric field intensity less than some minimum, the solution would not contain ions and it would not be able to conduct electric current. In fact, it all happens otherwise. Experiments demonstrated that electrolytes follow Ohm's law and that is why any indefinitely weak electric field provokes electrolysis in any electrolyte. Later in 1857, Clausius demonstrated that observance of Ohm's law for electrolyte solutions even at the lowest intensities at measuring electrodes indicates constantly existing charge carriers in the electrolyte solution. If these charged particles appeared as a result of electrolysis, the voltage threshold should be lowered such that the particles would not exist in the solution.

Additional energy consumption is required for decomposition of the electrolyte solute. It is not small at all. If chemical decomposition of the electrolyte were performed by means of the electric field, the energy of electric current would be spent not only for liberation of Joule heat but also for chemical decomposition. Meanwhile, experiments demonstrated that the Joule–Lenz law is fulfilled for electrolytes according to which all current energy is spent for Joule heat.

Taking into account the stated ideas, Clausius offered another explanation for occurrence of ions in solutions. According to the kinetic theory of gases, where Clausius himself was one of the main founders, atoms inside a molecule execute chaotic heat motion. The energy of that heat motion is not usually enough in order to overcome chemical attractive forces holding oppositely charged parts of a molecule together. In solutions, however, forces of chemical affinity are weakened under the influence of the solvent, and a molecule, at least, for a short period of time *dissociates*, that is, decomposes into oppositely charged ions. In case a positive ion meets with a negative one, ions may *recombine*, that is, get connected into a neutral molecule. Other neutral molecules, vice versa, may dissociate into ions. As a result of continuous processes of dissociation and recombination, *statistical equilibrium* is set where a fraction of dissociated molecules stays in average over time. By applying the electric field, positive and negative ions start rushing in opposite directions according to the views of Grotthuss. The electric current appears and liberation of decomposition products takes place on electrodes. Thus, the electric field has nothing to do with chemical decomposition of solute molecules into component elements. The role of the electric field only comes down to separation of already existing positive and negative ions and collecting them on different electrodes. This is the essence of the electrolysis phenomenon.

In the history of chemistry during the 1870s–1880s, it is the period when the struggle for the theory of chemical composition, for specifying the notions of *atom*, *molecule*, and *ion*, was still on track, which interfered with the understanding of the electrolyte electrical conductivity mechanism.

F.G. Kohlrausch performed a series of works in the field of physicochemistry of solutions and promoted understanding of the *ion* nature. In 1879, he discovered independence of ion motion in electrolytes at infinite dilution and formulated the electrical conductivity additivity law (Kohlrausch's law). He investigated electrical conductivity of electrolytes and established that it increases with temperature rise. He was dealing with defining electrochemical equivalents of various metals. In 1885, he proposed an empirical equation expressing dependence of electrical conductivity of strong electrolyte solutions from concentration and proposed the method of determining electric resistance of electrolytes.

Then, van't Hoff came to a conclusion that Avogadro's law is valid for dilute solutions too. His invention was very important since all chemical reactions and exchange reactions inside living beings take place in solutions. He also experimentally established that osmotic pressure, representing the measure of tending of two different solutions from both sides of a membrane toward concentration equalization, in weak solutions depends on concentration and temperature and, consequently, is governed by gaseous laws of thermodynamics. Investigations of dilute solutions, conducted by van't Hoff, substantiated the theory of electrolytic dissociation by S. Arrhenius.

S. Arrhenius investigated the passage of electric current through many types of solutions. He put forward an assumption that molecules of some substances at dissipation in liquid dissociate or decompose into two and more particles, which he called ions. Despite the fact that every whole molecule is electrically neutral, its particles carry a small electric charge—either positive or negative depending on the nature of the particle. For example, molecules of sodium chloride (salt) at dissolution in water are dissolved into positively charged sodium atoms and negatively charged chlorine atoms. These charged atoms, active component elements of a molecule, are formed only in the solution and create a possibility for passage of electric current. The electric current in turn directs active component elements to oppositely charged electrodes.

He introduced the notion of the dissociation degree being the measure of dissociation. He called this measure the *activity coefficient* $\alpha$ and determined it as "the ratio of a number of ions actually existing in the electrolyte solution to a number of ions which would appear at complete decomposition of the electrolyte into simple electrostatic molecules." Arrhenius correlated the values $\alpha$ obtained from osmotic pressure and from electrical conductivity and demonstrated that they coincide (for some electrolytes).

This dependence in general can be obtained from the following simple concepts [1]. Let $n$ be the number of molecules of the solute in volume unit of the solution. From this, the number ($n\alpha$) of molecules are dissociated and the rest $n(1 - \alpha)$ of the molecules are not dissociated. Dissociated molecules may recombine. The average number of recombination processes per unit volume per unit time is proportional to $(n\alpha)^2$ and can be presented as $A(n\alpha)^2$. The average number of reverse processes, that is, new acts of dissociation, will be proportional to the number of available nondissociated molecules and is presented by the expression $Bn(1 - \alpha)$. In the set state, the number of direct processes should be equal, in average, to the number of reverse processes, that is, $A(n\alpha)^2 = Bn(1 - \alpha)$. Hence,

$$\frac{\alpha^2}{1-\alpha} = \frac{B}{An} = \frac{K}{n}. \tag{5.1}$$

Coefficients $A$ and $B$ and, consequently, their ratio $K = B/A$ depend on temperature and pressure of the solution. If concentration of molecules in the solute and concentration of ions in the solution are small, ions can be considered as *independent noninteracting particles*, similar to molecules of the ideal gas. In these conditions, coefficient $K$ does not depend on concentration $n$ and ratio $\alpha^2/(1 - \alpha)$ is inversely proportional. This provision is called the *Ostwald dilution law* (1853–1932). When $n \to 0$, then $\alpha^2/(1 - \alpha) \to \infty$ and subsequently $\alpha \to 1$. Hence, *in infinite dilute solutions all molecules of the solute are dissociated.*

The Ostwald dilution law acts well for *weak electrolytes*, that is, electrolytes whose dissociation degree is small ($\alpha \ll 1$). For *strong electrolytes*, that is, electrolytes whose value $\alpha$ is of the order of unity ($\alpha \sim 1$), the Ostwald law does not correspond to experimental data. The case with strongly dilute solutions, for which this law brings to a correct result, is certainly an exception ($\alpha = 1$). The contemporary theory of solutions states that molecules of the strong electrolyte are completely dissociated and all peculiarities of solution behavior are tried to be explained by interaction of ions between one another and with solvent molecules.

These assumptions are considered as the basis of the contemporary theory of electrolytic dissociation, which is defined actually by two provisions: (1) Arrhenius hypothesis of electrolytic dissociation and (2) relation of dissociation degree and electrolyte concentration in a solution through the law of mass action (Ostwald dilution law (5.1)) [2,3].

## 5.2 THEORIES OF ELECTRICAL CONDUCTIVITY

The notions of the classical Debye–Hückel theory about the ion atmosphere and electrochemical mobilities of ions make the basis for the contemporary theory of electrical conductivity of electrolyte solutions and its different variants. The development of this theory at different stages was always attributed to multiple attempts to liquidate its separate drawbacks involving, in particular, estimation of only long-range Coulomb forces and description of ion properties using only their charge and that of solvent properties—by its macroscopic dielectric permittivity. Such attempts were made, primarily, in the works of Falkenhagen, Onsager, Fuoss, and Harned. Concentration dependence of electrical conductivity by the Debye–Hückel–Onsager theory is explained by the existence of electrophoretic and relaxation retarding effects in ions [4–6]. Having calculated the values of retarding forces, Onsager established a calculation equation for electrical conductivity corresponding to approximation of the Debye–Hückel theory named the Onsager limiting law.

$$\Lambda = \Lambda^0 - \left(B_1\Lambda^0 + B_2\right)\frac{\sqrt{I}}{1 + B\alpha\sqrt{I}}, \tag{5.2}$$

where

$$B = \left(\frac{2e^2 N_A}{\varepsilon kT}\right), \tag{5.3}$$

$$B_1 = \frac{|z_+ z_-| F^3 q}{3\pi N_A}, \tag{5.4}$$

$$B_2 = \frac{(z_+ + |z_-|) F^3}{6\pi N_A \eta}\left(\frac{2}{\varepsilon RT}\right)^{1/2}, \tag{5.5}$$

$$q = \frac{|z_+ z_-|}{z_+ + |z_-|} \cdot \frac{\lambda_+^0 + \lambda_-^0}{|z_-|\lambda_+^0 + z_+\lambda_-^0}, \tag{5.6}$$

where
$\Lambda^0$ is the electrical conductivity of the electrolyte solution at infinite dilution
$I$ is the ionic strength of the solution, mol/dm$^3$
$e$ is the electron charge
$N_A$ is the Avogadro constant
$\varepsilon$ is the dielectric permittivity of the solvent
$k$ is the Boltzmann constant
$T$ is the temperature, K
$F$ is the Faraday number
$R$ is the universal gas constant
$\eta$ is the dynamic viscosity of the solvent
$\lambda_+^0$ and $\lambda_-^0$ are the electrical conductivities of the anion and cation at infinite dilution

When substituting constant for 1–1 valence electrolytes, the Onsager limiting law for electrical conductivity will be rearranged as follows:

$$\Lambda = \Lambda^0 - \left[\frac{8.20 \cdot 10^5}{(\varepsilon T)^{3/2}}\Lambda^0 + \frac{8.24 \cdot 10^{-4}}{(\varepsilon T)^{1/2}\eta}\right]\sqrt{c}, \tag{5.7}$$

where $c$ is the electrolyte concentration, mol/dm$^3$.

The Onsager law is valid for narrow concentration area of solutions having ionic strength not more than 0.01. The boundary of applicability for nonaqueous solutions with low dielectric permittivity is even lower.

For more concentrated solutions in the Debye–Hückel–Onsager model, the following assumptions are inapplicable:

1. One cannot neglect proper sizes of ions with decrease of ionic sphere sizes and decrease of the distance between ions.
2. Boltzmann statistical distribution with additional condition $ze\Psi \ll kT$ and the corresponding mathematical simplification are valid only for dilute

solutions. Usage of Boltzmann statistical distribution for electrolyte solutions is inapplicable, as shown in Section 1.3.3.

3. Usage of the dielectric constant of the solvent when the dielectric constant of the solution decreases considerably even at low concentrations of the electrolyte.

Similar works appeared much later too. The method of *quasi-ions*, possessing interaction potential equal to the Debye–Hückel potential, is used in a series of works by Mirtskhulava [7,8]. The work of Ebeling and colleagues [9] and other works should be attributed to this area, too. By using various distribution functions of ions in the solution, the authors [9] obtained a calculation equation justified in a rather wide range of electrolyte concentrations in the solution.

Onsager, Fuoss, Eyring, Pitts, Karman, Wien, Falkenhagen, Robinson, Stokes, Smedley, and other researchers developed new notions about electrolyte conductivity [10–13] based on the Debye–Hückel theory. However, these ideas do not explain any of the peculiarities in change of electrical conductivity of solutions depending on various factors.

Fuoss [14] provides the equation binding conductivity with concentration of a solution containing ion pairs. In work of Monica et al [15], the Onsager–Falkenhagen equation, describing electrical conductivity of dilute solutions of 11 electrolytes, is modified by means of estimating the influence of change in medium viscosity on the electrophoretic term. The modified equation describes satisfactorily the concentration dependence of electrical conductivity of electrolyte solutions in a wider weight content. The error of the description of experimentally measured dependence for transfer numbers of ions from weight content of the electrolyte in the Fuoss and Onsager [16,17] and Pitts et al. [18] equation is estimated in the work of Spiro using computer [19].

Fuoss proposed the thermodynamic theory of electrical conductivity postulating the existence of *ion–ion pair* equilibrium, telling about a possibility of existence of contact pairs, based on the conductometric determination of thermodynamic constants of formation of ion pairs for symmetric electrolytes [20].

The work of Evans and colleagues [21] is dedicated to the issues related to the mechanism of electrical conductivity of solutions, criticizing the theories of Onsager and Hubbard such as the dielectric continuum model and ignoring the molecular nature of *ion–water* interaction. The authors believe that the Wolynes theory, which takes into account hydration and considers the equilibrium structure of hydration shells and transport properties of a solvent, is more real.

Based on a primitive model of solutions, the author of work [22] developed the kinetic theory of electrical conductivity of strong electrolyte solutions. The diffusivity equation of $n$ particle in the configuration space is used as an initial equation in this theory.

Turq and Lantelme try to explain electrical conductivity of electrolytes engaging Brownian motion laws [23]. They offer the model of electrical conductivity of strong electrolytes forming a quasicrystalline lattice in concentrated solutions in sites where there are ions not participating in charge transfer.

Elkind proposed the method of calculating specific electrical conductivity of strong electrolytes, taking into account the energy of Coulomb interaction based on the assumption about availability of the structure with ion atmosphere surrounding an ion in the solution similar to the electrolyte structure in the crystalline state [24]. Calculations under the offered model provide good convergence for concentrated solutions, too. A great number of works by Semenchenko, Sukhotin, Mishchenko, Poltoratsky, Fialkov, and a few other researchers [25] are dedicated to the questions of electrical conductivity of electrolytes in aqueous and nonaqueous solutions. When explaining the behavior of electrolyte solutions, one uses ideas about association of ions, about negligibility of specific interactions among particles in a solution, about assumption of small differences in solvation energy of ion pairs and separate ions, and so on.

The use of nuclear magnetic resonance (NMR), electron paramagnetic resonance (EPR), and relaxation radiospectroscopy methods allowed the authors [26] to confirm the idea about existence of different types of relaxations in electrolyte solutions: *ion–ion atmosphere*, *ion–dipoles of the solvent*, dipoles of ion pairs in concentrated solutions, and dipoles of solvent molecules, having estimated the role of every relaxation effects. Having called their theory statico-dynamic, the authors [26] tried to take into account the discreteness and continuity of the solution structure, as well as its spatial structure and time factors. The main provisions of this theory are that instantaneous the distribution of particles depends on their structure and interaction; change in the solution composition and concentration is reflected on positioning of particles; the nature of short-range forces is anisotropic and the structure of the solution is microanisotropic; hydrogen bonds and charge transfer to ions contribute to long-range action of forces and forming of multilayered solvates; the basis for the spatial lattice in solutions, depending on concentrations, can be ions, ion pairs, or ion complexes.

The statement is put forward in works [12,27] about a possibility of using the theory of semiconductors in order to explain the mechanism of electrical conductivity and structure of electrolyte solutions; a conclusion was made about the existence of fermions in solutions being current carriers. The review of the theories about electrical conductivity of electrolyte solutions is presented in works [12,27,28]. Corrections to the electrical conductivity theory of Onsager–Fuoss–Falkenhagen, enlarging the range of its application, are proposed by authors of works [29,30].

There are theoretical and experimental works coming out dedicated to defining limiting equivalent electrical conductivities of ions and concentration dependence of equivalent electrical conductivity of aqueous solutions of strong electrolytes [31,32] using the Walden–Pisarzhevsky rule in one form or another [33]: $\kappa\eta$ = constant, where $\kappa$ is the equivalent electrical conductivity. At the same time, there are considerable contradictions in estimation of practicability of this rule by a number of authors. On the one hand, Kuznetsova [32] proposes an equation describing concentration dependence of equivalent electrical conductivity of aqueous solutions of strong electrolytes corrected on viscosity over the range of weight content from 0.5 mol to saturation. On the other hand, Fialkov and Zhytomyrskiy [31] affirm that the electrical conductivity permanence rule, both of electrolytes and ions in various solvents and under different temperatures cannot be considered as fair even at first approximation. Authors [34,35] also write that permanence of product (5.8) for various solvents is more a seldom phenomenon than a rule.

After appearance of contemporary computer machines, a possibility occurred of rather simple solving of sets of equations with a great number of unknown quantities. It predetermined the appearance of more or less substantiated theoretical equations of electrical conductivity of electrolyte solutions containing a considerable number of summands with unknown coefficients. A detailed review of these equations is given in [35,36]. One can classify any smooth surface using the power polynom with a large number of summands to which all these equations are reduced. Such correlations are gaining more in popularity [36].

Things are not better with electrical conductivity–temperature dependence, which is still classified by approximation empirical equations [36–38]. Researches in temperature dependence of electrical conductivity of aqueous electrolyte solutions at high pressures, conducted by Maksimova and colleagues [39–41], demonstrated availability of electrical conductivity maximum. The availability of extremum on dependence $k = f(T)$ cannot be predicted by any of the contemporary electrolyte solution electrical conductivity theories. Investigating empirical conductivity and kinetic salt effect, Karelson and Palm [42] came to a conclusion about a necessity of reviewing the main provisions of the strong electrolyte solution theory. According to their perceptions, already moderately concentrated solutions of strong electrolytes ($c > 2$ mol) are practically completely associated. In order to explain regularities of investigated phenomena, the authors [42] assume that ion pairs may participate in electromigration. Besides, ions and ion pairs may be in two states, one of which contributes to passing of current through the solution and the other one doesn't. In this regard, the notion of the organized and disorganized solvent is introduced corresponding to conducting and nonconducting states of ions and ion pairs.

The availability of proton energy spectrum, similar to electron spectrum, allowed representing such solutions as proton semiconductors. Krogh-Moc [43] and Eigen and Maeyer [44,45] were the first ones to pay attention to it. Chomutov [46] and Shakhparonov [47] contributed considerably to developing such issue. In works [48], electric properties of solutions are explained based on the ideas about proton excitons—exciprotons. The discovery of limiting high-frequency electrical conductivity [49] makes an interesting and important stage in developing perceptions about the mechanism of electrical conductivity of electrolyte solutions. Works [49,50] demonstrate that limiting high-frequency electrical conductivity of electrolyte solutions is explained by the properties of the solvent not electrolyte.

Attempts at estimating the influence of various factors, for example, ion associations and solvation, made by authors [16,20,24,48–59], led to the appearance of new more complicated and cumbersome equations allowing somewhat increasing accuracy of calculation of $\Lambda_0$, constant of association $K$, and other parameters.

It is worth mentioning that theoretical and experimental works dedicated to determining limiting equivalent electrical conductivities of ions and concentration dependence of electrical conductivity of aqueous and nonaqueous electrolyte solutions are emerging these days.

The work of Valyashko and Ivanov [60] demonstrates that the maximum on isotherms of specific electrical conductivity is the main property of aqueous electrolytic systems, and its position is not occasional but is under direct dependence from the nature of electrolyte and water interaction. Dependence of concentration of electrical conductivity maximum from hydration, charge, and association degree of electrolytes is traced in these works. In solutions where concentrations correspond to electrical conductivity maximum (or exceed those), the main types of intermolecular interactions are ion–water and ion–ion interactions. According to his point of view, increase of electrolyte content in such concentrated solutions should result in redistribution of water molecules among hydration shells of ions, decrease of a number of water molecules in the coordination sphere of every ion, and decrease of time of water molecule stay at one ion. As a result, prompt intensification of interionic interactions takes place, and a system of ionic bonds should appear in the solution. Reconstruction of the aqueous-like structure takes place under concentrations differing for various types of electrolytes. This concept is in line with the results of estimation of concentration limit of water structure existence in solutions, based on the Lyashchenko model [61], but there is no interionic interaction criterion.

Extreme concentration dependence is another property where specific conductivity can be explained in the general scheme of concentration structural transition. Maximums on these dependences are possessed by all salts in which solubility reaches concentrations corresponding to the aqueous-like structure zone boundary. At the same time, they are not available if saturated salt solutions correspond to weaker concentrations. The boundary of the first structural zone is close to concentrations of specific electrical conductivity maximum. As it was demonstrated, it is characteristic for quite a large number of systems [62,63]. Disappearance of tetrahedral volumetric water defines appearance of complex ionic forms in concentrated solutions, which do not obligatory correspond to groupings conforming to the pure salting component as it was assumed in [64,65]. Ionic or ion-aqueous polycondensation reaction bears its own specific nature compared to other ionic reactions in solutions. It determines the appearance of a large number of associated ions than those added into the solution [66,67]. Hence, it defines a small change in water condition with increase of electrolyte concentration. Mobility of ionic groups (complexes) being formed is substantially lower than that of separate ions. The authors [68,69] believe that due to this, the number of effective current carriers in the solution decreases simultaneously with increase of electrolyte concentration and electrical conductivity minimum appears.

In work [70], the Debye–Hückel–Onsager electrical conductivity theory, describing concentration dependence of equivalent conductivity of electrolyte solutions, is synthesized with the Debye–Falkenhagen electrical conductivity dispersion theory. Equations for dependence of equivalent electrical conductivity from concentration and electric field frequency are obtained by means of the mode conjugation theory, which is used for describing transport properties near a critical point. The mode conjugation theory allows defining a bond between transport characteristics of an individual ion and collective dynamic properties of the electrolyte, in particular its ion atmosphere. The authors [70] obtained an equation for dependence $\lambda = f(c)$ based on diffusion coefficient frequency dependence. The drawback of this work is the absence of correct estimation of electrical conductivity dispersion of the solvent itself in which contribution into ionic friction is considered in this work as constant and, according to the point of view of authors [70], will produce influence at higher frequencies. As a result, considerable deviation from the theory at high concentrations and frequencies can be observed. These are actually numerical methods with usage of adjustable parameters.

Using the Lee–Wheaton equation [72,73], Kaluginh and Panchenko [74] proposed the method for processing conductometric data for electrolyte 1–1 solutions taking into account the formation of ion pairs and three-junctions in the low dielectric permittivity media. The authors eventually succeeded in widening the concentration range up to $3 \cdot 10^{-2}$ mol/dm$^3$ using the Lee–Wheaton equation. Unfortunately, the obtained results can be applied only for the electrolytes in solvents having dielectric permittivity of not more than $\varepsilon = 10$. Besides, the necessity of using a considerable number of experimental values of electrical conductivity (at least 50) can also be attributed to limitations of the method.

The number of works dedicated to analytical ab initio researches of multicomponent electrolyte solutions is extremely low. In works [75,76], the methods for calculating electrical conductivity of aqueous solutions of strong electrolytes and of aqueous solutions of acid–salt systems are proposed. The information, given by the authors, is either provided for a small number of electrolytes at room temperatures and low concentrations, or unavailable in reference literature.

The state of ions in the framework of the *ion plasma* concept is considered in the works of Baldanov and Mokhosoev [71,77]. According to the contemporary theory of plasma processes, Debye and Hückel restricted themselves only to spatial dispersion and did not place emphasis on frequency dispersion, that is, availability of longitudinal acoustic oscillations of the aggregate of ions in electrolyte solutions with the frequency of $w_L$, approximated as ion sound. The proposed interpretation of the state of electrolyte ions allowed from uniform positions explaining such factors as Debye–Falkenhagen electrical conductivity dispersion, Wien effect, and maximum electrical conductivity value $\lambda_{i,j}^0$. In subsequent works, Baldanov and colleagues propose a concept of electrical conductivity of electrolyte solutions on the basis of the plasma-hydrodynamic theory [78–80]. The obtained results allowed describing the behavior of electrical conductivity $\Lambda$ of concentrated electrolyte solutions with satisfactory accuracy:

$$\Lambda = \frac{1.11 \cdot 10^{-12} N_A e^2 \exp(-(\hbar\omega/2kT))}{6\pi\eta R_S(1+(R_S/r_D))}, \tag{5.8}$$

where

$\hbar\omega$ is the total energy of plasma oscillations
$R_S$ is the radius of inertia of hydrated ions
$r_D$ is the Debye shielding length of charges

However, initial provisions of the plasma-hydrodynamic theory trigger a few questions. In addressing the problem, Baldanov uses the electrostatic theory of Debye–Huckel, but at the same time he also uses the Debye shielding length of charges for the whole range of concentrations in the formula whose dielectric constant of the solvent (water) is used, which cannot be used in the concentrated area. Maxwell–Boltzmann statistics is used, but as it was proved in Section 1.3.3, it is inapplicable for electrolyte solutions. The radius of inertia of hydrated ions is used for the whole concentration area of the electrolyte, which is not very clear. The network of water H-bonds is practically destroyed in the second area of concentrations and the solution contains ionic and ion-aqueous clusters and fragments of ion hydration spheres (see Figure 1.1 and description thereto), not saying anything about the third concentration area, which is described already by the laws of coordination chemistry. Provisions of the plasma-hydrodynamic theory do not touch upon the issues of structure formation in the solution. Usage of the CGS system of units of measurement, and not the SI system, is not understandable at the present time.

Hence, it can be concluded that despite certain successes in development of theoretical concepts, the contemporary theory of electrolyte solutions is underdeveloped and, to that end, the problem of theoretical consideration of electrical conductivity of electrolyte solutions is still topical.

Various theoretical approaches are used for explaining electrical conductivity of electrolyte solutions till the present time. Electrical conductivity of electrolyte solutions, being one of the most common properties, is the property studied almost in the best way and most comprehensively. It has been a long time that comparatively high accuracy and easiness of its experimental research attracted attention of multiple researchers who accumulated huge theoretical and experimental materials; these publications are systemized in [25]. However, in spite of this, many phenomena related to electrical conductivity are still incomprehensible and unexplained by the theory [1–3]. The problem of electrical conductivity in solutions was never fully cleared. It is not managed to classify the concentration dependence of electrical conductivity in a wide range of concentrations from unified positions, the transfer mechanism of electric energy in the solution is not very clear, and there are various approaches for interpreting conductivity of strong and weak electrolytes [3].

The notions of the Debye–Hückel theory about the ion atmosphere and electrochemical mobilities of ions make the basis for the majority of theories of electrical conductivity of electrolyte solutions. This theory is based only on electrostatic interactions of particles in the solvent, which does not allow describing electrical conductivity of concentrated electrolyte solutions. Many researchers point out a tight connection between electrical conductivity of solutions and their structure. Neglect of the solvent structure and, consequently, failure to estimate ion, interparticle, and other interactions make many theories semiempirical, with a large number of parameters affecting consistency—considerable decrease of dielectric permittivity of the solvent $\varepsilon_S$, characterizing its structure with an increase in electrolyte concentration in the solution, is ignored.

Further, we will consider the concentration dependence of electrical conductivity in a wide range of concentrations from a perspective of supermolecular interactions outlined in the first part.

## 5.3 GENERAL CONTINUITY EQUATION FOR ELECTROLYTE SOLUTIONS IN PERTURBED STATE

If the electrolyte experiences external disturbing forces, the distribution functions and potentials take a nonsymmetrical form [81]:

$$\begin{aligned} f_{ij}\left(\vec{R}_i,\vec{R}_{ij},U_{sp},U_x\right) &= f_{ij}^0\left(\vec{R}_i,\vec{R}_{ij},U_{sp},U_x\right)+f'_{ij}\left(\vec{R}_i,\vec{R}_{ij},U_{sp},U_x\right), \\ \Psi_{ij}\left(\vec{R}_i,\vec{R}_{ij},U_{sp},U_x\right) &= \Psi_{ij}^0\left(\vec{R}_i,\vec{R}_{ij},U_{sp},U_x\right)+\Psi'_{ij}\left(\vec{R}_i,\vec{R}_{ij},U_{sp},U_x\right), \end{aligned} \quad (5.9)$$

where

$f_{ij}^0$ is the distribution function representing concentration of complexes and supermolecular assemblies of $j$-ion in the unit volume interacting with complexes or supermolecular assemblies of $i$-ions and, vice versa, for $f_{ji}$

$\Psi_{ij}$ is the potential of interparticle (supramolecular) interactions

The terms with “0” sign designate the steady state of a medium, and those with a dash designate perturbations. Similar expressions can be obtained as well for $f_{ji}$ and $\Psi_{ji}$.

In order to express distribution functions $f_{ij}^0$ and $f_{ji}^0$ in terms of potential, let us consider the Poisson equation between charge density and potential as shown in equation 43 in [82]:

$$\nabla\cdot\nabla\Psi^0(R) = -\frac{\rho_{sr}^0}{\varepsilon_S\varepsilon_0}, \quad (5.10)$$

from where

$$\rho_{sr}^0 = -\varepsilon_S\varepsilon_0\nabla\cdot\nabla\Psi^0(R), \quad (5.11)$$

where

$\varepsilon_S$ is the dielectric constant of the solution

$\varepsilon_0$ is the electric constant equal to $8.85\cdot10^{-12}$ F/m

Let us proceed to (2.18) and we will obtain

$$|z_j|\,ef_{ij}^0 = n_i\rho_{j\,sr} = n_i\rho_{i\,sr} = -n_j\varepsilon_S\varepsilon_0\nabla_i\cdot\nabla_i\Psi_{ij}^0 = -n_i\varepsilon_S\varepsilon_0\nabla_j\cdot\nabla_j\,\Psi_{ji}^0 = |z_j|\,ef_{ji}^0, \quad (5.12)$$

where $\nabla_i\cdot\nabla_i\Psi_{ij}^0(R_{ij}) = \nabla_j\cdot\nabla_j\Psi_{ji}^0(R_{ji}) = -\nabla_i\cdot\nabla_i\Psi_{ij}^0(R_{ji})$ and is equal to (2.59).

From (5.12), we have

$$f_{ij}^0 = -\varepsilon_0\varepsilon_S\frac{n_i\nabla_j\cdot\nabla_j\,\Psi_{ji}^0}{|z_j|\,e} = -\varepsilon_0\varepsilon_S\frac{n_j\nabla_i\cdot\nabla_i\,\Psi_{ij}^0}{|z_i|\,e}. \quad (5.13)$$

The similar expression is obtained as well for $f_{ij}^0$. Taking into account (5.13) for $f'_{ij}$, it can be obtained as follows:

$$f'_{ij} = -\varepsilon_0\varepsilon_S \frac{n_i \nabla_j \cdot \nabla_j \Psi'_{ji}}{|z_j|e} = -\varepsilon_0\varepsilon_S \frac{n_j \nabla_i \cdot \nabla_i \Psi'_{ji}}{|z_i|e}. \tag{5.14}$$

Using Equations 5.13 and 5.14 from Equation 5.9, we obtain the distribution function for the electrolyte in the perturbed state in the following form:

$$f_{ij}^0 = -\frac{n_i \varepsilon_0 \varepsilon_S}{|z_j|e}\left(\nabla_j \cdot \nabla_j \Psi_{ji}^0 + \nabla_j \cdot \nabla_j \Psi'_{ji}\right). \tag{5.15}$$

The similar expression can be obtained as well for $f_{ji}$.

For the total force acting upon the complex of *i*-ion, we will obtain the following ratio:

$$\vec{F}_{ij} = \vec{F}_i - |z_i| e \nabla_i \Psi_{ij}\left(\vec{R}_i, \vec{R}_{ij}, U_{sp}, U_x\right), \tag{5.16}$$

where

$\vec{F}_i$ is the disturbing force

$\nabla_i$ is the Hamiltonian operator for *i*-ion

The term $|z_i| e \nabla_i \Psi_{ij}(\vec{R}_i, \vec{R}_{ij}, U_{sp}, U_x)$ designates force acting upon the complex of *i*-ion explained by the electrostatic field of the complex of *j*-ion and its shell. The similar expression can be obtained as well for $\vec{F}_{ji}$.

Let us consider Equation 1.32 based on 2.29, multiplying for $f_{ij}$ and $f_{ji}$ and taking divergence in the following form:

$$\nabla_i \cdot \left(f_{ij} \vec{v}_{ij}\right) = \nabla_i \cdot \left[ f_{ij} \vec{V}\left(R_i\right) + \omega_i \left(\vec{F}_{ij} f_{ij} - kT \nabla_i f_{ij}\right)\right]$$
$$+ \nabla_j \cdot \left[ f_{ji} \vec{V}\left(R_j\right) + \omega_j \left(\vec{F}_{ji} f_{ji} - kT\, \nabla_j f_{ji}\right)\right] = \nabla_j \cdot \left(f_{ji} \vec{v}_{ji}\right). \tag{5.17}$$

Substituting into (5.17) the value $F_{ij}$ and $F_{ij}$ from (5.16), we will obtain

$$\nabla_i \cdot \left(f_{ij} \vec{v}_{ij}\right) = \nabla_i \cdot \left\{ f_{ij} \vec{V}\left(R_i\right) + \omega_i \left[ f_{ij} \vec{F}_i - |z_i| e f_{ij} \nabla_i \Psi_{ij} - kT \nabla_i f_{ij}\right]\right\}$$
$$+ \nabla_j \cdot \left\{ f_{ji} \vec{V}\left(R_j\right) + \omega_j \left[ f_{ji} \vec{F}_j - |z_j| e f_{ji} \nabla_j \Psi_{ji} - kT \nabla_i f_{ij}\right]\right\} = \nabla_j \cdot \left(f_{ji} \vec{v}_{ji}\right). \tag{5.18}$$

Let us estimate what terms come to expression (5.18) at calculating divergence. The terms $\Delta_i \cdot \vec{V}(r_i)$ and $\Delta_i \cdot \vec{F}_i$ and, accordingly, those symmetric to them can be equated to zero, for usually in all studied cases, the solution rate and external electric field stay constant. Then, after several transformations, Equation 5.18 takes the following form:

$$\nabla_i \cdot \left(f_{ij}\vec{v}_{ij}\right) = \vec{V}\left(R_i\right)\nabla_i f_{ij} + \omega_i \vec{F}_i \nabla_i f_{ij} - \omega_i \left|z_i\right| e\left(\nabla_i f_{ij}\, \nabla_i \Psi_{ij} + f_{ij}\nabla_i \cdot \nabla_i \Psi_{ij}\right)$$
$$-\omega_i kT\nabla_i \cdot \nabla_i f_{ij} + \vec{V}\left(R_j\right)\nabla_j f_{ji} + \omega_j \vec{F}_j \nabla_j f_{ji} - \omega_j \left|z_j\right| e$$
$$\times\left(\nabla_j f_{ji}\, \nabla_j \Psi_{ji} + f_{ji}\nabla_j \cdot \nabla_j \Psi_{ji}\right) - \omega_j kT\nabla_j \cdot \nabla_j f_{ji} = \nabla_j \cdot \left(f_{ji}\vec{v}_{ji}\right). \quad (5.19)$$

Let us substitute values $\nabla_i \cdot (f_{ij}\vec{v}_{ij})$ and $\nabla_j \cdot (f_{ji}\vec{v}_{ji})$ from (5.19) into (2.29) and we will obtain

$$-\frac{\partial f_{ij}}{\partial t} = \vec{V}\left(R_i\right)\nabla_i f_{ij} + \omega_i \vec{F}_i \nabla_i f_{ij} - \omega_i \left|z_i\right| e\left(\nabla_i f_{ij}\nabla_i \Psi_{ij} + f_{ij}\nabla_i \cdot \nabla_i \Psi_{ij}\right)$$
$$-\omega_i kT\nabla_i \cdot \nabla_i f_{ij} + \vec{V}\left(R_j\right)\nabla_j f_{ji} + \omega_j \vec{F}_j \nabla_j f_{ji} - \omega_j \left|z_j\right| e$$
$$\times\left(\nabla_j f_{ji}\nabla_j \Psi_{ji} + f_{ji}\nabla_j \cdot \nabla_j \Psi_{ji}\right) - \omega_j kT\,\nabla_j \cdot \nabla_j f_{ji} = -\frac{\partial f_{ji}}{\partial t}. \quad (5.20)$$

Further, we substitute (5.9) into (5.20) and we will obtain the following expression:

$$-\frac{\partial f_{ij}}{\partial t} = \vec{V}\left(R_i\right)\nabla_i f_{ij}^0 + \vec{V}\left(R_j\right)\nabla_j f_{ji}^0 + \omega_i \vec{F}_i \nabla_i f_{ij}^0 + \omega_j \vec{F}_j \nabla_j f_{ji}^0$$
$$-\omega_i \left|z_i\right| e\left[\nabla_i\left(f_{ij}^0 + f_{ij}'\right)\nabla_i\left(\psi_{ij}^0 + \psi_{ij}'\right) + \left(f_{ij}^0 + f_{ij}'\right)\nabla_i \cdot \nabla_i\left(\psi_{ij}^0 + \psi_{ij}'\right)\right]$$
$$-\omega_j \left|z_j\right| e\left[\nabla_j\left(f_{ji}^0 + f_{ji}'\right)\nabla_j\left(\psi_{ji}^0 + \psi_{ji}'\right) + \left(f_{ji}^0 + f_{ji}'\right)\nabla_j \cdot \nabla_j\left(\psi_{ji}^0 + \psi_{ji}'\right)\right]$$
$$-\omega_i kT\nabla_i \cdot \nabla_i\left(f_{ij}^0 + f_{ij}'\right) - \omega_j kT\nabla_j \cdot \nabla_j\left(f_{ji}^0 + f_{ji}'\right) = -\frac{\partial f_{ji}}{\partial t}. \quad (5.21)$$

We write the products in square brackets of the fifth summand (5.21):

$$\omega_i \left|z_i\right| e\left[\nabla_i\left(f_{ij}^0 + f_{ij}'\right)\nabla_i\left(\psi_{ij}^0 + \psi_{ij}'\right) + \left(f_{ij}^0 + f_{ij}'\right)\nabla_i \cdot \nabla_i\left(\psi_{ij}^0 + \psi_{ij}'\right)\right]$$
$$= \omega_i \left|z_i\right| e\begin{pmatrix} \nabla_i f_{ij}^0 \nabla_i \psi_{ij}^0 + \nabla_i f_{ij}^0 \nabla_i \psi_{ij}' + \nabla_i f_{ij}' \nabla_i \psi_{ij}^0 + \nabla_i f_{ij}' \nabla_i \psi_{ij}' \\ + f_{ij}^0 \nabla_i \cdot \nabla_i \psi_{ij}^0 + f_{ij}^0 \nabla_i \cdot \nabla_i \psi_{ij}' + f_{ij}' \nabla_i \cdot \nabla_i \psi_{ij}^0 + f_{ij}' \nabla_i \cdot \nabla_i \psi_{ij}' \end{pmatrix}. \quad (5.22)$$

The sixth summand in square brackets will be the same as well (5.21).

The terms $\nabla_i f'_{ij}\nabla_i\psi'_{ij}$ and $f'_{ij}\nabla_i\cdot\nabla_i\psi'_{ij}$ in (5.22) as well as $\nabla_j f'_{ji}\nabla_j\psi'_{ji}$ and $f'_{ji}\nabla_j\cdot\nabla_j\psi'_{ji}$ of the sixth summand in square brackets (5.21) due to its small value compared to the other ones can be neglected at once.

Let us consider the rest of the terms in round brackets on the right side (5.22):

$$\begin{aligned}
&\nabla_i f^0_{ij}\nabla_i\psi^0_{ij} + \nabla_i f^0_{ij}\nabla_i\psi'_{ij} + \nabla_i f'_{ij}\nabla_i\psi^0_{ij} + f^0_{ij}\nabla_i\cdot\nabla_i\psi^0_{ij} + f^0_{ij}\nabla_i\cdot\nabla_i\psi'_{ij} + f'_{ij}\nabla_i\cdot\nabla_i\psi^0_{ij}\\
&= \nabla_i f^0_{ij}\nabla_i\psi^0_{ij} + f^0_{ij}\nabla_i\cdot\nabla_i\psi^0_{ij} + \nabla_i\left(f^0_{ij}\nabla_i\psi'_{ij}\right) + \nabla_i\left(f'_{ij}\nabla_i\psi^0_{ij}\right)\\
&= \nabla_i f^0_{ij}\nabla_i\psi^0_{ij} + f^0_{ij}\nabla_i\cdot\nabla_i\psi^0_{ij} + \nabla_i\left(f^0_{ij}\nabla_i\psi'_{ij} + f'_{ij}\nabla_i\psi^0_{ij}\right).
\end{aligned} \tag{5.23}$$

The same expression can be obtained for the complex of $j$-ion.

Neglecting the terms for the complex of $i$-ion $f'_{ij}\nabla_i\psi^0_{ij}$ in (5.23) and, accordingly, for $j$-ion $f'_{ji}\nabla_j\,\psi^0_{ji}$ due to its small value compared to the other ones in (5.23), we substitute (5.23) into (5.21) and we will obtain the following ratio:

$$\begin{aligned}
-\frac{\partial f_{ij}}{\partial t} &= \vec{V}(R_i)\nabla_i f^0_{ij} + \vec{V}(R_j)\nabla_j f^0_{ji} + \omega_i\vec{F}_i\nabla_i f^0_{ij} + \omega_j\vec{F}_j\nabla_j f^0_{ji} - \omega_i|z_i|e\\
&\times\left[\nabla_i f^0_{ij}\nabla_i\psi^0_{ij} + f^0_{ij}\nabla_i\cdot\nabla_i\psi^0_{ij} + \nabla_i\left(f^0_{ij}\nabla_i\psi'_{ij}\right)\right]\\
&-\omega_j|z_j|e\left[\nabla_j f^0_{ji}\nabla_j\psi^0_{ji} + f^0_{ij}\nabla_j\cdot\nabla_j\psi^0_{ji} + \nabla_j\left(f^0_{ji}\nabla_i\psi'_{ji}\right)\right]\\
&-\omega_i kT\nabla_i\cdot\nabla_i\left(f^0_{ij} + f'_{ij}\right) - \omega_j kT\nabla_j\cdot\nabla_j\left(f^0_{ji} + f'_{ji}\right) = -\frac{\partial f_{ji}}{\partial t}.
\end{aligned} \tag{5.24}$$

Then, we will analyze the summands $-\omega_i kT\nabla_i\cdot\nabla_i(f^0_{ij} + f'_{ij})$ and $-\omega_j kT\nabla_j\cdot\nabla_j(f^0_{ji} + f'_{ji})$ on the right side of (5.24). Let us consider (5.19) and take into account for the equilibrium state that $\vec{v}_{ij} = \vec{V}(r_i) = \vec{F} = 0$. At such assumptions, we can obtain $f \approx f^0$ and $\Psi \approx \Psi^0$. For the complex of $i$-ion, we will obtain

$$\omega_i|z_i|e\left(\nabla_i f^0_{ij}\nabla_i\psi^0_{ij} + f^0_{ij}\nabla_i\cdot\nabla_i\psi^0_{ij}\right) = -\omega_i kT\nabla_i\cdot\nabla_i f^0_{ij}. \tag{5.25}$$

Similarly, the same expression can be obtained for $j$-ion.

Let us substitute expression (5.25) for $i$-ion and analogous for $j$-ion into (5.24) and we will obtain continuity equation (5.21) for nonstationary fields in the final form:

$$\begin{aligned}
-\frac{\partial f_{ij}}{\partial t} &= \vec{V}(R_i)\nabla_i f^0_{ij} + \vec{V}(R_j)\nabla_j f^0_{ji} + \omega_i\vec{F}_i\nabla_i f^0_{ij} + \omega_j\vec{F}_j\nabla_j f^0_{ji}\\
&-\omega_i|z_i|e\left[\nabla_i\left(f^0_{ij}\nabla_i\psi'_{ij}\right)\right] - \omega_j|z_j|e\left[\nabla_j\left(f^0_{ji}\nabla_j\psi'_{ij}\right)\right]\\
&-\omega_i kT\nabla_i\cdot\nabla_i f'_{ij} - \omega_j kT\nabla_j\cdot\nabla_j f'_{ji} = -\frac{\partial f_{ji}}{\partial t}.
\end{aligned} \tag{5.26}$$

For the case of stationary fields,

$$\frac{\partial f_{ij}\left(\vec{r}_i,\vec{r}_{ij},U_{ij},U_x\right)}{\partial t}=\frac{\partial f_{ji}\left(\vec{r}_j,\vec{r}_{ji},U_{ji},U_x\right)}{\partial t}\equiv 0. \tag{5.27}$$

In Equation 5.26, the first four terms contain perturbing factors and the rest of the six terms asymmetric components of the potential. In order to solve these or those specific problems, Equation 5.26 can be attributed to a different form. Thus, in case of applying viscosity phenomena, the terms of speed rate of the whole solution $\vec{V}(\vec{r}_i)$ and $\vec{V}(\vec{r}_j)$ stay, and the terms containing external disturbing forces $\vec{F}_i$ and $\vec{F}_j$ disappear.

Studying such issues as electrical conductivity and diffusion, the first two terms of Equation 5.26 containing speed rates of the whole solution can be omitted. Let us proceed to electrical conductivity.

## 5.4 ELECTRICAL CONDUCTIVITY: GENERAL THEORY OF SUPERMOLECULAR FORCES

Let us consider the general equation of motion (5.26). For electrical conductivity, $\partial f_{ij}/\partial t$, $\partial f_{ji}/\partial t$ and $\vec{V}(\vec{R}_i),\vec{V}(\vec{R}_j)$ are equal to zero because we will study stationary fields and these are the members related to motion of the solution as a whole. Equation of motion (5.26) takes the form of

$$\omega_i\left[\vec{F}_i\nabla_i f_{ij}^0\left(\vec{R}\right)\right]+\omega_j\left[\vec{F}_j\nabla_j f_{ji}^0\left(\vec{R}\right)\right]-\omega_i\left|z_i\right|e\nabla_i\left[f_{ij}^0\left(\vec{R}\right)\nabla_i\psi'_{ij}\left(R_{ij}\right)\right]$$
$$-\omega_j\left|z_j\right|e\nabla_j\left[f_{ji}^0\left(\vec{R}\right)\nabla_j\psi'_{ji}\left(R_{ji}\right)\right]-\omega_i kT\,\nabla_i\cdot\nabla_i f'_{ij}\left(R_{ij}\right)-\omega_j kT\,\nabla_j\cdot\nabla_j f'_{ji}\left(R_{ji}\right)=0. \tag{5.28}$$

Let us substitute $\vec{F}_i$ and $\vec{F}_j$ into

$$\vec{F}_i=X\left|z_i\right|e,\quad \vec{F}_j=X\left|z_j\right|e, \tag{5.29}$$

where $X$ is the force (upon availability of electric field) acting in the direction of axis $x$. Then, $\nabla_i=-\nabla_j=\nabla$ and assume that $\nabla f\cong\partial f/\partial x$. Proceeding from the condition of solution for electrical neutrality, we have that $\Psi'_{ij}=\Psi'_{ji}$. Functions $f$ and $f'$ are symmetric, that is, $f'_{ij}(R_{ij})=f'_{ji}(R_{ji})$ [83,84]. After some transformations, we will obtain

$$X\left(\omega_i\left|z_i\right|e-\omega_j\left|z_j\right|e\right)\frac{\partial f_{ij}^0}{\partial x}-\left(\omega_i\left|z_i\right|e+\omega_j\left|z_j\right|e\right)\nabla\left(f_{ij}^0\nabla\psi'_{ij}\right)$$
$$-\left(\omega_i+\omega_j\right)kT\,\nabla\cdot\nabla f'_{ij}=0. \tag{5.30}$$

Let us exclude (5.30) using the function $f'$ by means of the Poisson equation (5.14) and substitute it into (5.30), and we will obtain

$$(\nabla\cdot\nabla)^2\Psi'_{ij}-\frac{1}{\varepsilon_0\varepsilon_S kT}\sum_{i,j}\frac{\left(\omega_i|z_i|e+\omega_j|z_j|e\right)|z_i|e}{(\omega_i+\omega_j)n_j}\nabla\left(f_{ij}^0\nabla\Psi'_{ij}\right)$$
$$+\frac{X}{\varepsilon_0\varepsilon_S kT}\sum_{i,j}\frac{\left(\omega_i|z_i|e-\omega_j|z_j|e\right)|z_i|e}{(\omega_i+\omega_j)n_j}\frac{\partial f_{ij}^0}{\partial x}=0. \tag{5.31}$$

For further simplifications of (5.31), we will determine the electrostatic part of distribution function $f_{ij}^0$. We will estimate expressions (2.56) in (5.13) for κ and the expression for potential of interparticle interactions (2.59). Then, distribution function $f_{ij}^0$ takes the form

$$f_{ij}^0=n_i n_j\varphi(r), \tag{5.32}$$

where

$$\varphi^0(R)=-2\left[\operatorname{csch}\left(\frac{\beta}{R_{ij}}\right)\cosh\left(\frac{\beta}{R_{ij}}\right)-2\sinh\left(\frac{\beta}{R_{ij}}\right)\right]. \tag{5.33}$$

Let us substitute (5.32) into (5.31) and we will obtain

$$(\nabla\cdot\nabla)^2\Psi'_{ij}-\frac{1}{\varepsilon_0\varepsilon_S kT}\sum_{i,j}\frac{\left(\omega_i|z_i|e+\omega_j|z_j|e\right)n_i|z_i|e}{\omega_i+\omega_j}\nabla\left(\varphi_{ij}^0\nabla\Psi'_{ij}\right)$$
$$=\frac{X}{\varepsilon_0\varepsilon_S kT}\sum_{i,j}\frac{\left(\omega_i|z_i|e-\omega_j|z_j|e\right)n_i|z_i|e}{\omega_i+\omega_j}\frac{\partial\varphi_{ij}^0}{\partial x}. \tag{5.34}$$

For the binary solution, Equation 5.34 will have the following form:

$$(\nabla\cdot\nabla)^2\Psi'_{ij}-\frac{1}{\varepsilon_0\varepsilon_S kT}\cdot\frac{\left(\omega_i|z_i|e+\omega_j|z_j|e\right)\left(n_i|z_i|e+n_j|z_j|e\right)}{\omega_i+\omega_j}\nabla\left(\varphi_{ij}^0\nabla\Psi'_{ij}\right)$$
$$=\frac{X}{\varepsilon_0\varepsilon_S kT}\frac{\left(\omega_i|z_i|e-\omega_j|z_j|e\right)\left(n_i|z_i|e+n_j|z_j|e\right)}{\omega_i+\omega_j}\frac{\partial\varphi_{ij}^0}{\partial x}. \tag{5.35}$$

Solving of differential equation (5.35) will be demonstrated in the next section.

## 5.5 RELAXATION FORCE

Application of low-frequency, weak electric field on electrodes submerged into the solution provokes the force tending to move ion complexes or ion-aqueous clusters and fragments of hydration spheres of $i$-ions to one side, and the same supermolecular formations of $n_j$ ions of type $j$ to the other side. In this regard, the regular structure of the solution as well as the spherical symmetry of ion complexes is disturbed, that is, the ions are displaced from the center of their complexes and tend to be destroyed. When the deformed ion complex comes to a new place, an attempt of restoration of supermolecular formation and spherical symmetry takes place. When the ion moves under the influence of an electric field, there will be a part of the complex behind it still not destroyed, and a newly formed spherical structure of the solution in front tends to hold the deformed ion complex in its nodes and, due to this, the force opposite to the direction of movement occurs. The time for formation of the new spherical structure and the shell of ion complexes will be certainly called the relaxation time, and the force impedes movement of the ion-relaxation.

Let us proceed to developing its mathematic expression. We introduce the following designations:

$$Q = \frac{1}{\varepsilon_0 \varepsilon_S kT} \cdot \frac{\left(\omega_i |z_i| e + \omega_j |z_j| e\right)\left(n_i |z_i| e + n_j |z_j| e\right)}{\omega_i + \omega_j}, \tag{5.36}$$

$$K = Qh, \tag{5.37}$$

where

$$h = \frac{\left(\omega_i |z_i| e - \omega_j |z_j| e\right)}{\left(\omega_i |z_i| e + \omega_j |z_j| e\right)}. \tag{5.38}$$

Taking into account designations (5.36) through (5.38), Equation 5.35 will be as follows:

$$(\nabla \cdot \nabla)^2 \Psi'_{ij} - Q\nabla\left(\varphi^0_{ij} \nabla \Psi'_{ij}\right) = XK \frac{\partial \varphi^0_{ij}}{\partial x}. \tag{5.39}$$

Equation 5.39 will be expressed as follows:

$$\nabla Y = XK \frac{\partial \varphi^0_{ij}}{\partial x}, \tag{5.40}$$

where we will bring the gradient under brackets in (5.39) and after simple transformations

$$Y = \nabla^2 \cdot \nabla \Psi'_{ij} - 2Q\varphi^0_{ij} \nabla \Psi'_{ij}. \tag{5.41}$$

We return to (5.40). The force acting on the solvent is directed along axis $x$. In the first approximation, we can assume that $\nabla \cong \nabla_x = \partial/\partial x$. We integrate both parts of Equation 5.40:

$$Y = XK\left(\varphi_{ij}^0 + C_1\right), \tag{5.42}$$

where $C_1$ is the integration constant.

We substitute (5.42) into (5.39) and we will obtain

$$\Delta\cdot\nabla\Psi'_{ij} - 2Q\varphi_{ij}^0\nabla\Psi'_{ij} = XK\left(\varphi_{ij}^0 + C_1\right). \tag{5.43}$$

Let us denote in (5.43) $\nabla\Psi'_{ij} = U$. Indices $ij$ are omitted when reproducing the following formulas. We substitute designation $U$ into (5.43) and we will obtain

$$\Delta U - 2Q\varphi^0 U = XK\left(\varphi^0 + C_1\right). \tag{5.44}$$

We write Laplacian operator (2.61) in (5.44), and after some transformations, it will be as follows:

$$\frac{1}{R}\frac{\partial^2}{\partial R^2}\left(RU\right) = \frac{1}{R}\frac{\partial}{\partial R}\left(U + R\frac{\partial U}{\partial R}\right) = \frac{\partial^2 U}{\partial R^2} + \frac{2}{R}\frac{\partial U}{\partial R}. \tag{5.45}$$

We substitute (5.45) into (5.44):

$$\frac{\partial^2 U}{\partial R^2} + \frac{2}{R}\frac{\partial U}{\partial R} - 2Q\varphi^0 U = XK\left(\varphi^0 + C_1\right). \tag{5.46}$$

We find the integral of differential equation (5.46) in the form of power series:

$$U = a_0 + a_1\frac{\beta}{R} + a_2\left(\frac{\beta}{R}\right)^2 + a_3\left(\frac{\beta}{R}\right)^3 + \cdots, \tag{5.47}$$

$$\frac{\partial U}{\partial R} = -a_1\left(\frac{\beta}{R}\right)\frac{1}{R} - 2a_2\left(\frac{\beta}{R}\right)^2\frac{1}{R} - 3a_3\left(\frac{\beta}{R}\right)^3\frac{1}{R} - \cdots, \tag{5.48}$$

$$\frac{\partial^2 U}{\partial R^2} = 2a_1\left(\frac{\beta}{R}\right)\frac{1}{R^2} + 6a_2\left(\frac{\beta}{R}\right)^2\frac{1}{R^2} + 12a_3\left(\frac{\beta}{R}\right)^3\frac{1}{R^2} + \cdots, \tag{5.49}$$

where $a_i$ are the desired coefficients.

Let us find value $\varphi^0$ in (5.33) in the form of power series. The first product $\operatorname{csch}(\beta/R)\cosh(\beta/R)$ in square brackets is equal to (2.68), where $x$ means $\beta/R$:

$$\operatorname{csch}\left(\frac{\beta}{R}\right)\cosh\left(\frac{\beta}{R}\right)=\frac{R}{\beta}+\frac{\beta}{3R}-\frac{1}{45}\left(\frac{\beta}{R}\right)^3-\cdots\quad\left[\left(\frac{\beta}{R}\right)^2<\pi^2\right]. \tag{5.50}$$

We will use expansion in series for the hyperbolic sine

$$\sinh\left(\frac{\beta}{R}\right)=\frac{\beta}{R}+\frac{1}{6}\left(\frac{\beta}{R}\right)^3+\frac{1}{120}\left(\frac{\beta}{R}\right)^5+\cdots\quad\left[\left(\frac{\beta}{R}\right)^2<\infty\right]. \tag{5.51}$$

Then, taking into account (5.50) and (5.51), after some transformations, expression (5.33) will be as follows:

$$\begin{aligned}\varphi^0&=-2\left[\operatorname{csch}\left(\frac{\beta}{R}\right)\cosh\left(\frac{\beta}{R}\right)-2\sinh\left(\frac{\beta}{R}\right)\right]\\&=-2\left[\left(\frac{R}{\beta}\right)-\frac{5}{3}\left(\frac{\beta}{R}\right)-\frac{16}{45}\left(\frac{\beta}{R}\right)^3-\frac{17}{360}\left(\frac{\beta}{R}\right)^5\right].\end{aligned} \tag{5.52}$$

We return to differential equation (5.46). As it is seen from its structure, it is necessary to calculate product $\varphi^0 U$. Taking into account (5.47) and (5.52), after some transformations, we will obtain

$$\begin{aligned}\varphi^0 U=&-a_0\left(\frac{R}{\beta}\right)-a_1+\left(\frac{5}{3}a_0-a_2\right)\left(\frac{\beta}{R}\right)+\left(\frac{5}{3}a_1-a_3\right)\left(\frac{\beta}{R}\right)^2\\&+\left(\frac{16}{45}a_0+\frac{5}{3}a_2-a_4\right)\left(\frac{\beta}{R}\right)^3+\frac{1}{3}\left(\frac{16}{45}a_1+5a_3\right)\left(\frac{\beta}{R}\right)^4.\end{aligned} \tag{5.53}$$

To avoid redundancy of a number of equations, we neglect powers $>R^{-3}$ in (5.53) because, for our purposes, it will be enough to substitute Equation 5.52 with power not $>R^{-3}$ into the right part of (5.46). We substitute expressions (5.47) through (5.49), (5.52), and (5.53) into differential equation (5.46) and we will obtain

$$\begin{aligned}&a_2\left(\frac{\beta}{R}\right)^2\frac{1}{R^2}-2Q\left[\begin{aligned}&-a_0\left(\frac{R}{\beta}\right)-a_1+\left(\frac{5}{3}a_0-a_2\right)\left(\frac{\beta}{R}\right)+\left(\frac{5}{3}a_1-a_3\right)\left(\frac{\beta}{R}\right)^2\\&+\left(\frac{16}{45}a_0+\frac{5}{3}a_2-a_4\right)\left(\frac{\beta}{R}\right)^3+\frac{1}{3}\left(\frac{16}{15}a_1+5a_3\right)\left(\frac{\beta}{R}\right)^4\end{aligned}\right]\\&=-XK\left[\left(\frac{R}{\beta}\right)-\frac{5}{3}\left(\frac{\beta}{R}\right)-\frac{16}{45}\left(\frac{\beta}{R}\right)^3+C_1\right].\end{aligned} \tag{5.54}$$

We equate the left and right parts of expression (5.54) at equal powers $R^{-j}$, and after some transformations, taking into account (5.37), we have the following system of equations:

| No. | | $a_0$ | $a_1$ | $a_2$ | $a_3$ | $a_4$ | Free Term |
|---|---|---|---|---|---|---|---|
| 1 | $R^{-4}$ | 0 | $-\frac{32}{45}Q\beta^2$ | 1 | $-\frac{10}{3}Q\beta^2$ | 0 | 0 |
| 2 | $R^{-3}$ | $-\frac{16}{45}$ | 0 | $-\frac{5}{3}$ | 0 | 1 | $\frac{8}{45}Xh$ |
| 3 | $R^{-2}$ | 0 | $-\frac{5}{3}$ | 0 | 1 | 0 | 0 |
| 4 | $R^{-1}$ | $-\frac{10}{3}$ | 0 | 2 | 0 | 0 | $\frac{5}{3}Xh$ |
| 5 | $R^{0}$ | 0 | 2 | 0 | 0 | 0 | $XhC_1$ |

In order to solve the given system of equations, one can use the theory of determinants and find coefficients $a_i$, $i = 1/5$. For that, one would have to calculate six determinants with dimension of Dim = 5 × 5, which is rather complicated. It is easier to solve this system by the substitution method.

From equation No. 5, we find coefficient $a_1$:

$$a_1 = \frac{XhC_1}{2}, \tag{5.55}$$

and substitute it into equation Nos. 1 and 3. From equation No. 3, we determine coefficient $a_3$:

$$a_3 = \frac{5XhC_1}{6}, \tag{5.56}$$

and substitute it into equation No. 1 from which we find coefficient $a_2$:

$$a_2 = \frac{47Q\beta^2 XhC_1}{15}, \tag{5.57}$$

and substitute it into equation No. 4 from which we have

$$a_0 = \frac{-5Xh}{3} + \frac{47Q\beta^2 XhC_1}{25}. \tag{5.58}$$

Let us study these two summands in (5.58). Value $Q$ possesses order $\approx 10^{22}$, value $\beta \approx 10^{-9}$, and product $Q\beta^2 \approx 10^4$. Thus, the first summand in (5.58) is approximately four orders less than the second one. The first summand in (5.58) can be neglected due to its small value, and finally (5.58) will be as follows:

$$a_0 = \frac{47Q\beta^2 XhC_1}{25}. \tag{5.59}$$

Further, values $a_0$ and $a_2$ are substituted into equation No. 2 and define coefficient $a_4$:

$$a_4 = 5.9Q\beta^2 XhC_1 - \frac{8Xh}{45}. \tag{5.60}$$

By the same reasoning as in (5.58), we neglect the second summand in (5.60) and finally

$$a_4 = 5.9Q\beta^2 XhC_1. \tag{5.61}$$

And due to small value, we can neglect coefficients $a_1$ and $a_3$. Taking into account (5.37) and coefficients $a_0$ (5.59), $a_2$ (5.57), and $a_4$ (5.61), we go back to (5.47) and we will obtain

$$U = XK\beta^2 C_1\left[\frac{47}{25} + \frac{47}{15}\left(\frac{\beta}{R}\right)^2 + 5.9\left(\frac{\beta}{R}\right)^4\right]. \tag{5.62}$$

In (5.62), we will take into account that $U = \nabla\Psi'$. Finally,

$$\nabla\Psi' = XK\beta^2 C_1\left[\frac{47}{25} + \frac{47}{15}\left(\frac{\beta}{R}\right)^2 + 5.9\left(\frac{\beta}{R}\right)^4\right]. \tag{5.63}$$

We integrate the left and right parts of (5.63). We obtain

$$\Psi' = XK\beta^2 C_1\left[\frac{47}{25}R - \frac{47}{15}\beta^2\left(\frac{1}{R}\right) - 1.96\beta^4\left(\frac{1}{R}\right)^3 + C_2\right], \tag{5.64}$$

where integration constant $C_2$ at $R \to a$ is found from the second boundary condition of (2.60):

$$C_2 = XK\beta^2 C_1\left[-\frac{47}{25}a + \frac{47}{15}\beta^2\left(\frac{1}{a}\right) + 1.96\beta^4\left(\frac{1}{a}\right)^3 + \left(E'_{sp} + U'_x\right)\right], \tag{5.65}$$

where $E'_{sp}$ is determined by (4.60) and $U'_x$ by (4.65) and (4.82).

Let us substitute (5.65) into (5.64) and we obtain

$$\Psi' = XK\beta^2 C_1\left[\frac{47}{25}(R-a)+\frac{47}{15}\beta^2\left(\frac{1}{a}-\frac{1}{R}\right)+1.96\beta^4\left(\frac{1}{a^3}-\frac{1}{R^3}\right)+\left(E'_{sp}+U'_x\right)\right]. \quad (5.66)$$

The value of field $\delta X$ provoked by distortion of shells of ion complexes or ion-aqueous clusters and fragments of hydration spheres of $i$- and $j$-ions in relaxation force is equal to

$$\delta X = -\mathrm{grad}\,\Psi'. \quad (5.67)$$

We calculate the gradient in (5.66) taking into account (4.60), (4.65), and (4.82), and after some transformations, we have

$$-\mathrm{grad}\,\Psi' = -XK\beta_R^2 C_1\left\{\begin{aligned}&\frac{47}{25}R^2+\frac{47}{15}\beta^2+5.9\beta^2\left(\frac{\beta}{R}\right)+\frac{101.88ac_k\left(\nu_i n_{hi}+\nu_j n_{hj}\right)}{\varepsilon_S T}\\ &\times\left[\frac{\nu_i^2 z_i^2 n_{hi}+\nu_j^2 z_j^2 n_{hj}}{R^4}+\frac{3.73\cdot 10^{-3}}{R_{H_2O}^6}-\frac{1.96B}{R_{H_2O}}-\frac{0.415}{R_{H_2O}}\right]\end{aligned}\right\}, \quad (5.68)$$

where $\beta_R = \beta/R$.

Let us estimate the numerical value of the values of summation in (5.68). The first three summands in parentheses (5.68) can be neglected at once, for $R \cong \beta \approx 10^{-9}$ m. We emphasize that these are the summands responsible for direct electrostatic interactions between ions. At application of electric field and motion of ion complexes or other supermolecular formations to electrodes, the electrostatic interactions actually occur between ions but they are extremely low compared to the action force of electric field on complexes.

Further, we will consider the summands in square brackets of (5.68). Let us estimate their numerical values proceeding from conditions of the example for solution $CaCl_2$–$H_2O$ having the same parameters as demonstrated in Section 1.5.4. With regard to the data from Table 3.9, we will take into account that for solution $CaCl_2$–$H_2O$ $B \approx 32$,

$$-\mathrm{grad}\,\Psi' = -XK\beta_R^2 C_1\left(\frac{1^2\cdot 2^2\cdot 6+2^2\cdot 1^2\cdot 6}{(0.684)^4}+\frac{3.73\cdot 10^{-3}}{(0.373)^6}-\frac{1.96\cdot 32}{0.373}-\frac{0.415}{0.373}\right)$$

$$= -XK\beta_R^2 C\left(219.29+1.39-1.11-168.15\right).$$

As it is evident from the given example, the second and third summands are of low significance compared to the first and fourth ones, in particular their remainder. The second summand corresponds to dipole–dipole interactions among ion complexes and the third summand is responsible for weak hydrogen bonds among them and formation of various associates or any kind of supermolecular formations.

At application of electric field and motion of ion complexes or other supermolecular formations to electrodes, the stated interactions have no time for formation. All kinds of associates and supermolecular formations are destroyed and ion–dipole as well short-range repulsion forces even increase at motion of ion complexes as seen from the coefficient in front of the square bracket of (5.68). Neglecting the stated summands of low significance in (5.68), we will obtain an equation for relaxation force:

$$\delta X = -\text{grad}\Psi' = -\frac{101.88XK\beta_R^2 C_1 ac_k\left(\nu_i n_{hi} + \nu_j n_{hj}\right)}{\varepsilon_S T}\left[\frac{\nu_i^2 z_i^2 n_{hi} + \nu_j^2 z_j^2 n_{hj}}{R^4} - \frac{1.96B}{R_{H_2O}}\right]. \tag{5.69}$$

Let us proceed to electrophoretic retardation.

## 5.6 ELECTROPHORETIC RETARDATION

At application of electric field on unit volume of a solution, opposite flows of ion complexes or ion-aqueous clusters and fragments of hydration spheres of $i$-ions and supermolecular formations alike of $n_j$ ions of type $j$ occur, which impede mutual motion. Ion complexes of $i$-ions at their motion draw apart complexes of $j$-ions heading to one another and friction force increases as well in the aggregate of all these supermolecular motions. This force is opposite to the liquid resistance force and it is called the electrophoretic force or electrophoretic retardation.

In volume unit of the solution, the sum of all forces ($F$), acting upon solvent particles and ion complexes, is equal to zero:

$$\left(n_0 F_0 + \sum_j n_j F_j\right) = 0. \tag{5.70}$$

Complexes of ions in volume element d$V$, near $i$-ion, are under the influence of force $\sum_j n_{jsr} F_j \mathrm{d}V$, where $n_{jsr}$ corresponds to concentration of ion complexes or ion-aqueous clusters and fragments of hydration spheres of $j$-ions in unit volume at distance $\vec{R}_{ij}$ from similar supermolecular formations of $i$-ion. If in volume element we add force acting upon solvent particles contained in it, we will obtain total force acting upon volume element:

$$\left(n_0 F_0 + \sum_j n_j F_j\right)\mathrm{d}V = \frac{1}{4\pi\varepsilon_0\varepsilon_S}\sum_j (n_{jsr} - n_j) F_j \mathrm{d}V. \tag{5.71}$$

We proceed to spherical coordinates. The force, acting upon the spherical layer of complex located at distance $R$ from the central ion in complex, is equal to

$$F\mathrm{d}R = \frac{1}{4\pi\varepsilon_0\varepsilon_S} 4\pi R^2 \sum_j \left(n_{jsr} - n_j\right) F_j \mathrm{d}R. \tag{5.72}$$

This force acts in the direction of applied external force and is equally distributed along the complex sphere surface. These conditions are similar to those for which Stokes' law can be applied for motion of the ion complex in viscous fluid:

$$F = 6\pi\eta v R_{ji}, \tag{5.73}$$

where
- $\eta$ is the solution viscosity, Pa s
- $v$ is the speed of complex

We take into account from the determination of distribution function (2.19) that

$$n_{j\,sr} = \frac{|z_j| e f_{ij}^0}{n_i}. \tag{5.74}$$

We substitute (5.74) into (5.72):

$$F\mathrm{d}R = \frac{1}{\varepsilon_0\varepsilon_S} R^2 \sum_j \left( \frac{|z_j| e f_{ij}^0}{n_i} - n_j \right) F_j \mathrm{d}R_{ji}. \tag{5.75}$$

According to (5.73) and (5.75), the force acting upon the $j$-ion complex will entail change in speed equal to

$$dv_j = \frac{FdR}{6\pi\eta R} = \frac{1}{6\pi\varepsilon_0\varepsilon_S} \cdot \frac{R}{\eta} \sum_j \left( \frac{|z_j| e f_{ij}^0}{n_i} - n_j \right) F_j dR_{ji}. \tag{5.76}$$

We substitute (5.32) into (5.76) and we have

$$\mathrm{d}v_j = -\frac{1}{6\pi\varepsilon_0\varepsilon_S} \cdot \frac{R}{\eta} \left( \varphi^0 + 1 \right) \sum_j |z_j| e n_j F_j \mathrm{d}R_{ji}, \tag{5.77}$$

where $\varphi^0$ is equal to (5.33).

Change in speed of the ion complex will be found by integration of the left and right parts of (5.77):

$$v_j = -\frac{1}{6\pi\varepsilon_0\varepsilon_S} \cdot \frac{1}{\eta} \sum_j |z_j| e n_j F_j \int R \left( \varphi^0 + 1 \right) \mathrm{d}R. \tag{5.78}$$

We substitute value $\varphi^0$ into (5.78) from (5.52) restricted to three expansion terms and calculate the integral:

$$\int R(\varphi^0 + 1)\mathrm{d}R = \int R \left[ \frac{2R}{\beta} - \frac{10}{3}\left( \frac{\beta}{R} \right) - \frac{32}{45}\left( \frac{\beta}{R} \right)^3 + 1 \right] \mathrm{d}R$$

$$= \left[ \frac{2R^3}{3\beta} - \frac{10}{3}\beta R + \frac{32\beta^2}{45}\left( \frac{\beta}{R} \right) + \frac{R^2}{2} \right] + C_2. \tag{5.79}$$

Constant $C_2$ will be determined from boundary condition $v = 0$ at $R \to \infty$. Using this condition, we will obtain that

$$C_2 = -\frac{2R^3}{3\beta} + \frac{10}{3}\beta R - \frac{R^2}{2} + C_3,$$

where constant $C_3$ for electrical conductivity problems $C_3 \equiv 0$ but we will need it for diffusion problems.

We substitute value $C_2$ into the right part of (5.79) and the resulting value in the right part of (5.79) in (5.78) and we have

$$v_j = -\frac{1}{\pi\varepsilon_0\varepsilon_S}\cdot\frac{16\beta^2}{135\eta}\left(\frac{\beta}{R}\right)\sum_j |z_j| e n_j F_j + C_3. \tag{5.80}$$

For electrostatic forces, we take into account in (5.80) that $F \equiv X$, $K_j = |z_j| eX_j$, $C_3 \equiv 0$ (constant $C_3$ is of significance for diffusion) and $n_j = N_A c_j$. And we will obtain the final expression for the electrophoretic force:

$$v_j = -\frac{1}{\pi\varepsilon_0\varepsilon_S}\cdot\frac{16\beta^2 \nu_k c_k N_A \beta_R}{135\eta}\sum_j K_j, \tag{5.81}$$

where $\beta_R = \beta/R$.

This expression as well as the expression for the relaxation force (5.69) will be required for determining the final formula of electrical conductivity.

## 5.7 ELECTRICAL CONDUCTIVITY OF ELECTROLYTES AT WEAK ELECTRIC FIELDS AND LOW FREQUENCIES

Due to electrophoretic effect, the speed of ion complexes or ion-aqueous clusters and fragments of hydration spheres of $i$-ions and supermolecular formations alike of $n_j$ ions of type $j$ is lower than value $|z_j| e\omega_j(X + \Delta X_j)$, where $(X + \Delta X_j)$ is the total field acting upon all complexes and various supermolecular formations.

Then, the resultant speed in the direction of axis $x$ will be

$$v_j = X\left[|z_j|\ e\omega_j + \frac{\delta X_j}{X}|z_j| e\omega_j - \frac{1}{\pi\varepsilon_0\varepsilon_S}\frac{16\beta^2 \nu_k c_k N_A e\beta_R}{135\eta}\left(|z_i| + |z_j|\right)\right], \tag{5.82}$$

where $\omega_j$ is the mobility of supermolecular formation of $j$-ion.

We take into account that mobility of supermolecular formation is speed at difference of potentials equal to one, that is, $U_j = |v_j|/X$. Then for the mobility value of supermolecular formation, we have

$$\bar{U}_j = |z_j|\ e\omega_j + \frac{\delta X_j}{X}|z_j| e\omega_j - \frac{1}{\pi\varepsilon_0\varepsilon_S}\frac{16\beta^2 \nu_k c_k N_A e\beta_R}{135\eta}\left(|z_i| + |z_j|\right), \tag{5.83}$$

where $\bar{U}_j$ is the limiting mobility of supermolecular formation. Value $|z_j| e\omega_j$ in (5.83) is the limiting mobility $\bar{U}_j^0$ at infinite dilution. We take into account that $\lambda = F\bar{U}_j$, where $F$ is the Faraday constant equal to 96,500 CL/mol. After substitution of these values into (5.83), we have

$$\lambda_j = \lambda_j^0 + \frac{\delta X_j}{X}\lambda_j^0 - \frac{1}{\pi\varepsilon_0\varepsilon_S}\frac{11{,}437\beta^2\nu_k c_k N_A e}{\eta}\left(|z_i| + |z_j|\right). \tag{5.84}$$

We substitute value $\Delta X_j/X$ into (5.84) from (5.69):

$$\lambda_j = \lambda_j^0 - \Re_\lambda, \tag{5.85}$$

where

$$\Re_\lambda = \left\{\frac{101.88K\beta_R^2 C_1 a c_k\left(\nu_i n_{hi} + \nu_j n_{hj}\right)}{\varepsilon_S T}\left[\frac{\nu_i^2 z_i^2 n_{hi} + \nu_j^2 z_j^2 n_{hj}}{R^4} - \frac{1.96B}{R_{H_2O}}\right]\right\}\lambda_j^0$$
$$+ \frac{1}{\pi\varepsilon_0\varepsilon_S}\frac{11{,}437\beta^2\nu_k c_k N_A e\beta_R}{\eta}\left(|z_i| + |z_j|\right). \tag{5.86}$$

Expressing Equation 5.85 for both types of ion complexes of this electrolyte and summing them, we obtain the following equation:

$$\Lambda = \Lambda^0 - \Re_\Lambda^*, \tag{5.87}$$

where

$$\Re_\Lambda^* = \left\{\frac{101.88K\beta_R^2 C_1 a c_k\left(\nu_i n_{hi} + \nu_j n_{hj}\right)}{\varepsilon_S T}\left[\frac{\nu_i^2 z_i^2 n_{hi} + \nu_j^2 z_j^2 n_{hj}}{R^4} - \frac{1.96B}{R_{H_2O}}\right]\right\}\Lambda^0$$
$$+ \frac{1}{\pi\varepsilon_0\varepsilon_S}\frac{11{,}437\beta^2\nu_k c_k N_A e\beta_R}{\eta}\left(|z_i| + |z_j|\right), \tag{5.88}$$

because as per Kohlrausch's law,

$$\Lambda^0 = \lambda_i^0 + \lambda_j^0. \tag{5.89}$$

We return to Equation 5.37. Value $K$ from expressions (5.37), (5.36), and (5.38) is equal to

$$K = \frac{1}{\varepsilon_0\varepsilon_S kT}\cdot\frac{\left(\omega_i|z_i|e - \omega_j|z_j|e\right)\left(n_i|z_i|e + n_j|z_j|e\right)}{\omega_i + \omega_j}. \tag{5.90}$$

Expressing mobilities of ions in Equation 5.90 in terms of corresponding electrical conductivities and using (3.11) and (3.12), we have

$$K = \frac{1}{\varepsilon_0 \varepsilon_S kT} \cdot \frac{e^2 N_A |z_i z_j| \left(|z_i| + |z_j|\right)\left(\lambda_i^0 + \lambda_j^0\right) c_k}{|z_j| \lambda_i^0 + |z_i| \lambda_j^0}. \tag{5.91}$$

Let us denote in (5.91)

$$q^* = \frac{|z_i z_j| \left(|z_i| + |z_j|\right)\left(\lambda_i^0 + \lambda_j^0\right)}{|z_j| \lambda_i^0 + |z_i| \lambda_j^0}. \tag{5.92}$$

Then (5.91) will be as follows:

$$K = \frac{1}{\varepsilon_0 \varepsilon_S kT} \cdot e^2 N_A q^* c_k. \tag{5.93}$$

For 1–1 electrolytes, expression (5.92) is reduced and equal to 2.

With regard to (5.93), Equation 5.88 acquires the following form:

$$\Re_\Lambda^* = \frac{101.88}{\varepsilon_0 \varepsilon_S^2 kT^2} \cdot e^2 N_A q^* \beta_R^2 C_1 a c_k^2 \left\{ \left(\nu_i n_{hi} + \nu_j n_{hj}\right) \left[ \frac{\nu_i^2 z_i^2 n_{hi} + \nu_j^2 z_j^2 n_{hj}}{R^4} - \frac{1.96B}{R_{H_2O}} \right] \right\} \Lambda^0$$

$$+ \frac{1}{\pi \varepsilon_0 \varepsilon_S} \frac{11{,}437 \beta^2 \nu_k c_k N_A e \beta_R}{\eta} \left(|z_i| + |z_j|\right). \tag{5.94}$$

Equation 5.94 in quadratic power depends on solution dielectric permittivity and it is evident how this value is important for investigating electrical conductivity. We denote the parenthesis in (5.94) in terms of $P_\Lambda$. We substitute values of physical constants and obtain

$$\Re_\Lambda^* = \frac{1.29 \cdot 10^{25}}{\varepsilon_S^2 T^2} \cdot q^* \beta_R^2 C_1 a c_k^2 P_\Lambda \Lambda^0 + \frac{3.97 \cdot 10^{22} \beta^2 \nu_k c_k \beta_R}{\varepsilon_S \eta} \left(|z_i| + |z_j|\right), \tag{5.95}$$

where

$$P_\Lambda = \left(\nu_i n_{hi} + \nu_j n_{hj}\right) \left( \frac{\nu_i^2 z_i^2 n_{hi} + \nu_j^2 z_j^2 n_{hj}}{R^4} - \frac{1.96B}{R_{H_2O}} \right), \tag{5.96}$$

where values $a$ in Equation 5.95 as well as $R$ and $R_{H_2O}$ in Equation 5.96 stay in nm.

In (5.89), values $\lambda_i^0$ and $\lambda_j^0$ in S cm$^2$/mol at water vapor saturation pressure with sufficient accuracy for practical calculations can be calculated as per dependence given in [25]:

$$\lambda_{i,j}^0 = \sum_{k=0}^{6} b_{k,i,j} t^k, \tag{5.97}$$

where coefficients $b_{k,i,j}$, obtained based on statistical processing of experimental data [25,36], are given in Table 5.1: $t$ is the temperature, °C.

In (5.89), values $\lambda_i^0$ and $\lambda_j^0$ for various pressure rates can be calculated as per the following equation:

$$\lambda_{i,j}^0 = A_0 + A_1 t + A_2 t^{0.7} + A_3 t^{0.5}, \tag{5.98}$$

where coefficients $A_i$ for a number of cations and anions, obtained based on statistical processing of experimental data at various pressure rates from [25], are given in Table 5.2.

We return to Equation 5.95. We rename the integration constant in (5.95) and assume that $C_1 \equiv C_3 + C_4 E^* / \aleph^*$ and represent this equation as follows:

$$\Re_\Lambda = \aleph^* \left( C_4 + \frac{C_5 E^*}{\aleph^*} \right), \tag{5.99}$$

where

$$\aleph^* = 1.29 \cdot 10^{25} \left( \frac{\beta_R c_k}{\varepsilon_S T} \right)^2 q^* a P_\Lambda \Lambda^0, \tag{5.100}$$

and

$$E^* = \frac{3.97 \cdot 10^{22} \beta^2 \nu_k c_k \beta_R}{\varepsilon_S \eta} \left( |z_i| + |z_j| \right). \tag{5.101}$$

We remove parentheses in (5.99) and we have

$$\Re_\Lambda = \aleph^* C_4 + E^* C_5. \tag{5.102}$$

In (5.102), we rename constant $C_4 = A_\Lambda \cdot 10^{-22}$, where $A_\Lambda$ is the theoretical constant individual for every electrolyte. Constant $C_5$ is assumed equal to $10^{-2}$ and we take into

## TABLE 5.1
## Coefficients of Equation 5.97 to Calculate Limiting Molar Electrical Conductivity, S cm²/mol

| Ion | Range $t$ (°C) from $t_{min}$ to $t_{max}$ | $b_0$ | $b_1$ | $b_2 \cdot 10^2$ | $b_3 \cdot 10^5$ | $b_4 \cdot 10^7$ | $b_5 \cdot 10^9$ | $b_6 \cdot 10^{12}$ | $S_\lambda$ |
|---|---|---|---|---|---|---|---|---|---|
| $Ag^+$ | 0–110 | 32.9703 | 1.0634 | 4.0694 | — | — | — | — | $8.37 \cdot 10^{-2}$ |
| $Ba^{2+}$ | 0–200 | 31.9803 | 1.2390 | –0.4792 | 6.8346 | — | — | — | $5.72 \cdot 10^{-2}$ |
| $Be^{2+}$ | 25 | 45 | — | — | — | — | — | — | 0 |
| $Ca^{2+}$ | 0–200 | 30.4966 | 0.9005 | 12.9197 | –8.7588 | 2.1945 | — | — | $2.23 \cdot 10^{-3}$ |
| $Cd^{2+}$ | 0–30 | 28.0000 | 0.7186 | 12.8569 | — | — | — | — | $1.36 \cdot 10^{-5}$ |
| $Ce^{3+}$ | 25 | 67 | — | — | — | — | — | — | 0 |
| $Co^{2+}$ | 0–30 | 28.0000 | 0.7186 | 12.8569 | — | — | — | — | $1.36 \cdot 10^{-5}$ |
| $Cr^{3+}$ | 25 | 67 | — | — | — | — | — | — | 0 |
| $Cs^+$ | 0–200 | 43.8117 | 1.2257 | 6.4220 | –7.3083 | 6.5117 | –2.9780 | 4.9572 | 0.37 |
| $Cu^{2+}$ | 0–30 | 28.0000 | 0.6064 | 19.8414 | — | — | — | — | $1.15 \cdot 10^{-5}$ |
| $Eu^{3+}$ | 25 | 67.8 | — | — | — | — | — | — | 0 |
| $Fe^{2+}$ | 0–30 | 28.0000 | 0.6510 | 14.7619 | — | — | — | — | $7.79 \cdot 10^{-7}$ |
| $Fe^{3+}$ | 25 | 68 | — | — | — | — | — | — | 0 |
| $H^+$ | 0–310 | 224.0046 | 5.3072 | 10.9686 | –1.6856 | 0.6421 | — | — | $5.39 \cdot 10^{-3}$ |
| $Hg^{2+}$ | 25 | 63.6 | — | — | — | | — | — | 0 |
| $K^+$ | 0–310 | 40.3021 | 1.2139 | 5.1356 | –2.0650 | 4.0589 | — | — | $2.59 \cdot 10^{-3}$ |
| $La^{3+}$ | 0–110 | 34.9989 | 1.2339 | 6.1616 | — | — | — | — | $3.03 \cdot 10^{-3}$ |
| $Li^+$ | 0–310 | 19.1967 | 0.7183 | 2.1053 | 3.2501 | –2.6699 | 0.5856 | — | 0.25 |
| $Mg^{2+}$ | 0–200 | 31.0311 | 0.2588 | 31.9091 | –28.9096 | 8.2089 | — | — | 0.10 |
| $Mn^{2+}$ | 0–30 | 23.5000 | 1.4410 | –15.2380 | — | — | — | — | $1.65 \cdot 10^{-6}$ |
| $Na^+$ | 0–310 | 25.7066 | 0.8501 | 5.5764 | –1.7986 | 0.2000 | — | — | $2.02 \cdot 10^{-2}$ |
| $NH_4^+$ | 0–310 | 39.9964 | 1.4167 | –3.3337 | 4.5838 | –1.0168 | — | — | $3.61 \cdot 10^{-3}$ |
| | 0–30 | 11.5000 | 0.5411 | –6.6033 | — | — | — | — | $5.46 \cdot 10^{-6}$ |
| $Nd^{3+}$ | 25 | 64.3 | — | — | | | | | 0 |
| $Ni^+$ | 0–30 | 23.0000 | 1.6908 | –26.0317 | — | — | — | — | $3.30 \cdot 10^{-6}$ |
| $Pb^{2+}$ | 0–30 | 37.5000 | 1.6357 | –21.4284 | — | — | — | — | $9.38 \cdot 10^{-6}$ |
| $Pr^{3+}$ | 25 | 65.4 | — | — | — | — | — | — | 0 |
| $Ra^{2+}$ | 0–30 | 33.0000 | 1.1861 | 6.6352 | — | — | — | — | $1.58 \cdot 10^{-5}$ |
| $Rb^+$ | 0–200 | 43.5132 | 1.2817 | 4.1039 | –1.1927 | — | — | — | $3.87 \cdot 10^{-2}$ |
| $Sc^{3+}$ | 25 | 64.7 | — | — | — | — | — | — | 0 |
| $Sr^{2+}$ | 0–30 | 30.0000 | 1.0016 | 7.9361 | — | — | — | — | $2.13 \cdot 10^{-5}$ |
| $Tl^+$ | 0–30 | 43.3000 | 1.0081 | 4.7944 | — | — | — | — | $4.64 \cdot 10^{-5}$ |
| $UO_2^+$ | 25 | 32 | — | — | — | — | — | — | 0 |
| $Zn^{2+}$ | 0–30 | 27.0000 | 0.7943 | 11.4286 | — | — | — | — | $3 \cdot 10^{-6}$ |
| $Br^-$ | 0–200 | 43.0729 | 1.4765 | –5.7738 | 12.0097 | –4.2582 | — | — | $7.96 \cdot 10^{-2}$ |
| $BrO_3^-$ | 0–110 | 30.7000 | 1.0493 | –2.6335 | 4.5709 | — | — | — | $7.07 \cdot 10^{-5}$ |
| $Cl^-$ | 0–310 | 41.3804 | 1.2330 | 7.2914 | –3.2018 | — | — | — | $2.17 \cdot 10^{-2}$ |
| $ClO_2^-$ | 25 | 52 | — | — | — | — | — | — | 0 |

*(Continued)*

**TABLE 5.1 (*Continued*)**
**Coefficients of Equation 5.97 to Calculate Limiting Molar Electrical Conductivity, S cm²/mol**

| Ion | Range $t$ (°C) from $t_{min}$ to $t_{max}$ | $b_0$ | $b_1$ | $b_2 \cdot 10^2$ | $b_3 \cdot 10^5$ | $b_4 \cdot 10^7$ | $b_5 \cdot 10^9$ | $b_6 \cdot 10^{12}$ | $S_\lambda$ |
|---|---|---|---|---|---|---|---|---|---|
| $ClO_3^-$ | 0–110 | 36.0003 | 0.8717 | 14.3880 | −9.5054 | — | — | — | $7.47 \cdot 10^{-4}$ |
| $ClO_4^-$ | 0–200 | 37.2868 | 1.3162 | −7.9320 | 14.0038 | 4.7822 | — | — | $4.4610^{-2}$ |
| $CN^-$ | 25 | 82 | — | — | — | — | — | — | 0 |
| $CNS^-$ | 0–200 | 36.9834 | 1.0402 | 5.7626 | −1.6329 | — | — | — | $4.63 \cdot 10^{-2}$ |
| $CO_3^-$ | 0–30 | 36.0000 | 0.9525 | 22.6983 | — | — | — | — | $5 \cdot 10^{-6}$ |
| $CrO_4^{2-}$ | 0–30 | 43.9860 | 1.5591 | — | — | — | — | — | $3 \cdot 10^{-2}$ |
| $HCO_3^-$ | 25 | 44.5 | — | — | — | — | — | — | 0 |
| $HPO_4^{2-}$ | 25 | 57 | — | — | — | — | — | — | 0 |
| $HS^-$ | 0–30 | 40.0000 | 0.8016 | 7.9365 | — | — | — | — | $8.71 \cdot 10^{-6}$ |
| $HSO_4^-$ | 0–310 | 24.5464 | 1.0577 | −0.0046 | 1.3330 | −0.3180 | — | — | $2.57 \cdot 10^{-2}$ |
| $I^-$ | 0–60 | 41.4082 | 1.3257 | 3.6617 | — | — | — | — | $2.93 \cdot 10^{-2}$ |
| $IO_3^-$ | 0–110 | 20.7003 | 0.6228 | 9.9069 | −5.5045 | — | — | — | $8.07 \cdot 10^{-4}$ |
| $MnO_4^-$ | 0–30 | 36.0000 | 0.8016 | 7.9365 | — | — | — | — | $8.71 \cdot 10^{-7}$ |
| $NO_2^-$ | 0–30 | 44.0000 | 0.0962 | 40.9523 | — | — | — | — | $4.83 \cdot 10^{-6}$ |
| $NO_3^-$ | 0–200 | 40.1948 | 1.2727 | −2.7081 | 7.4194 | −2.7179 | — | — | $1.83 \cdot 10^{-2}$ |
| $OH^-$ | 0–310 | 105.0064 | 3.7364 | −2.0450 | −96.9833 | 0.0506 | — | — | $6.32 \cdot 10^{-3}$ |
| $PO_4^{3-}$ | 25 | 69 | — | — | — | — | — | — | 0 |
| $S^{2-}$ | 18 | 53.5 | — | — | — | — | — | — | 0 |
| $SO_3^-$ | 25 | 72 | — | — | — | — | — | — | 0 |
| $O^{2-}$ | 0–310 | 40.6963 | 1.2019 | 14.6689 | −6.0308 | 0.8726 | — | — | $2.80 \cdot 10^{-3}$ |

*Note:* $S_\lambda$, a mean squared error of calculation, S cm²/mol.

account that in the right part (5.95) $\beta = \beta \cdot 10^{-9}$. Then taking into account constants $C_4$ and $C_5$, Equation 5.102 is as follows:

$$\Re_\Lambda = \aleph + E, \tag{5.103}$$

where the relaxation force is determined by the equation

$$\aleph = 1290\left(\frac{\beta_R c_k}{\varepsilon_S T}\right)^2 q^* A_\Lambda a P_\Lambda \Lambda^0, \tag{5.104}$$

and the electrophoretic retardation

$$E = \frac{397\beta^2 \nu_k c_k \beta_R}{\varepsilon_S \eta}\left(|z_i| + |z_j|\right), \tag{5.105}$$

where $\beta$ is already substituted in nm and from (5.87) $\Lambda = \Lambda^0 - \Re_\Lambda$.

**TABLE 5.2**
**Coefficients of Equation 5.98 to Calculate Limiting Molar Electrical Conductivity, S cm²/mol, Depending on Pressure**

| Ion: | Range $t$ (°C) from $t_{min}$ to $t_{max}$ | $A_0$ | $A_1$ | $A_2$ | $A_3$ | $S_\lambda$ | Δ (%) |
|---|---|---|---|---|---|---|---|
| $P = P_{sat}$ | | | | | | | |
| $H^+$ | 25–350 | 369.5426 | −10.7097 | −134.5211 | −204.3840 | 0.683 | 0.071 |
| | 25–150 | 331.4308 | −9.1704 | 115.5066 | −168.2700 | 0.070 | 0.013 |
| | 150–350 | −327.4930 | −3.3302 | 0.6626 | 126.8397 | 0.138 | 0.013 |
| $Li^+$ | 25–350 | 50.9253 | 1.2292 | 6.8729 | −22.8186 | 2.990 | 0.767 |
| | 25–250 | 69.1234 | 1.0725 | 10.0004 | −30.3856 | 0.911 | 0.751 |
| | 250–350 | 940.2106 | 9.0947 | −59.7945 | −2.1743 | 0.046 | 0.010 |
| $Na^+$ | 25–200 | 23.1147 | 3.1840 | −14.7717 | 17.7742 | 0.283 | 0.207 |
| | 200–350 | 4680.5342 | −45.2609 | 860.3721 | −2151.5604 | 0.492 | 0.105 |
| $K^+$ | 25–200 | 38.7388 | 2.9185 | −11.2621 | 12.4694 | 0.265 | 0.138 |
| | 200–350 | 12386.9522 | −115.2627 | 183.5713 | −5522.2886 | 1.212 | 0.250 |
| $Rb^+$ | 25–200 | 43.9527 | 2.9975 | −11.5449 | 13.0126 | 0.304 | 0.183 |
| | 200–3501 | 12542.8516 | −115.9324 | 2201.4324 | −5574.1841 | 1.249 | 0.237 |
| $Cs^+$ | 25–200 | 45.1644 | 3.2190 | −13.4640 | 16.2723 | 0.245 | 0.110 |
| | 200–350 | 10883.6407 | −99.6383 | 1902.0896 | −4822.6075 | 1.096 | 0.204 |
| $Cu^+$ | 25–200 | 18.4950 | 3.0241 | −13.7322 | 16.1124 | 0.268 | 0.196 |
| | 200–350 | 3942.3453 | −38.7995 | 735.9382 | −1832.9753 | 0.414 | 0.092 |
| $Ag^+$ | 25–200 | 26.0019 | 3.1652 | −14.4983 | 17.8522 | 0.240 | 0.139 |
| | 200–350 | 2321.2598 | −114.5626 | 2171.4881 | −5494.0499 | 1.247 | 0.263 |
| $Au^+$ | 25–200 | 37.6363 | 3.0366 | −12.4402 | 14.4380 | 0.296 | 0.176 |
| | 200–350 | 2915.3545 | −119.6770 | 269.8274 | −5745.9644 | 1.300 | 0.250 |
| $NH_4^+$ | 25–200 | 28.8589 | 3.8104 | −19.0355 | 25.5541 | 0.233 | 0.107 |
| | 200–350 | 2715.5948 | −116.5910 | 2223.4400 | −5639.2144 | 1.296 | 0.227 |
| $Mg^{2+}$ | 25–350 | 125.8727 | 0.1118 | 32.4908 | −76.1246 | 2.393 | 0.800 |
| | 25–250 | 93.3501 | 1.1794 | 18.4117 | −48.2456 | 1.107 | 0.533 |
| | 250–350 | 609.3931 | 7.0585 | −41.1156 | 5.6895 | 0.010 | 0.001 |
| $Ca^{2+}$ | 25–200 | 40.6842 | 3.1646 | −7.1964 | 2.6025 | 0.359 | 0.159 |
| | 25–200 | 40.6842 | 3.1646 | −7.1964 | 2.6025 | 0.359 | 0.159 |
| | 200–350 | 5956.6265 | −58.9225 | 1111.0758 | −2764.3343 | 0.616 | 0.101 |
| $Sr^{2+}$ | 25–200 | 42.1107 | 3.2162 | −7.6259 | 3.3889 | 0.375 | 0.190 |
| | 200–350 | 2122.0489 | −115.4948 | 2177.3238 | −5476.3794 | 1.226 | 0.197 |
| $Ba^{2+}$ | 25–250 | 99.8950 | 1.4292 | 16.4228 | −44.2332 | 1.034 | 0.447 |
| | 250–350 | 1548.7903 | −11.1227 | 97.8436 | 12.7902 | 0.020 | 0.005 |
| $Sn^{2+}$ | 25–250 | 86.1521 | 2.2517 | 7.3296 | −27.2485 | 0.860 | 0.389 |
| | 250–350 | −2949.6910 | −20.3752 | 156.7734 | 71.6094 | 0.025 | 0.003 |
| $Pb^{2+}$ | 25–150 | 12.8183 | 4.5599 | −23.7709 | 34.1378 | 0.035 | 0.025 |
| | 150–350 | 4224.9297 | −48.1850 | 879.2500 | −2123.6709 | 1.159 | 0.177 |

*(Continued)*

**TABLE 5.2 (*Continued*)**
**Coefficients of Equation 5.98 to Calculate Limiting Molar Electrical Conductivity, S cm²/mol, Depending on Pressure**

| Ion: | Range $t$ (°C) from $t_{min}$ to $t_{max}$ | $A_0$ | $A_1$ | $A_2$ | $A_3$ | $S_\lambda$ | Δ (%) |
|---|---|---|---|---|---|---|---|
| $Mn^{2+}$ | 25–350 | 207.9192 | −1.9228 | 62.0024 | −137.1410 | 2.238 | 1.104 |
| | 25–100 | 12.3765 | 4.3133 | −21.5102 | 29.5700 | 0.587 | 0.514 |
| | 100–350 | 603.2986 | −7.1843 | 151.9440 | −350.1849 | 0.862 | 0.148 |
| $Fe^{2+}$ | 25–150 | 4.4445 | 4.4035 | −23.0721 | 32.9053 | 0.987 | 0.074 |
| | 150–350 | −877.5295 | 8.4250 | −131.0945 | 349.8674 | 1.166 | 0.202 |
| $Co^{2+}$ | 25–350 | 160.1689 | −0.8199 | 45.5215 | −102.6276 | 2.039 | 0.980 |
| | 25–250 | 90.8845 | 1.2455 | 17.3106 | −45.5508 | 1.114 | 0.569 |
| | 250–350 | 405.6764 | 3.9136 | −9.0399 | −28.1978 | 0.022 | 0.002 |
| $Ni^{2+}$ | 25–350 | 146.0242 | −0.4680 | 40.3013 | −91.9250 | 2.150 | 0.868 |
| | 25–250 | 94.3187 | 1.1322 | 18.6465 | −48.4102 | 1.081 | 0.577 |
| | 250–350 | 539.0965 | 5.3500 | −22.1082 | −20.2794 | 0.022 | 0.003 |
| $Cu^{2+}$ | 25–350 | 162.9302 | −0.7683 | 45.4816 | −103.0060 | 3.013 | 0.923 |
| | 25–250 | −47.9150 | 1.2296 | 18.2358 | −47.9150 | 1.096 | 0.516 |
| | 250–350 | 329.3955 | 4.6972 | −22.8432 | 6.3687 | 0.022 | 0.003 |
| $Zn^{2+}$ | 25–150 | 4.5571 | 4.4066 | 23.1177 | 32.9516 | 0.103 | 0.080 |
| | 150–350 | −890.4237 | 8.5714 | −133.4879 | 355.5975 | 1.296 | 0.244 |
| $Cd^{2+}$ | 25–150 | 6.0254 | 4.5109 | −23.8710 | 34.2610 | 0.096 | 0.070 |
| | 150–350 | 2050.6001 | −24.1358 | 449.6535 | −1071.7354 | 0.205 | 0.031 |
| $Hg^{2+}$ | 25–150 | 7.4553 | 4.6043 | −24.4500 | 35.1758 | 0.096 | 0.075 |
| | 150–350 | 3301.4261 | −37.8943 | 694.1982 | −1676.2679 | 0.747 | 0.113 |
| $F^-$ | 25–250 | 113.3205 | 0.5594 | 23.6957 | −54.6762 | 0.897 | 0.388 |
| | 150–350 | −2580.2437 | −19.5972 | 156.6582 | 33.1857 | 0.051 | 0.007 |
| $Cl^-$ | 25–150 | 33.3301 | 3.3805 | −13.9659 | 18.4370 | 0.094 | 0.061 |
| | 150–350 | 1966.3221 | −23.5013 | 431.1163 | −1022.5805 | 0.220 | 0.035 |
| $Br^-$ | 25–150 | 35.7774 | 3.2241 | −12.7580 | 16.2673 | 0.082 | 0.054 |
| | 150–350 | 1333.2393 | −16.7656 | 308.4768 | −719.9077 | 0.072 | 0.014 |
| $I^-$ | 25–150 | 36.4363 | 3.1758 | −12.7507 | 16.2421 | 0.188 | 0.130 |
| | 150–350 | 1005.6022 | −13.3600 | 245.7728 | −564.5851 | 0.266 | 0.049 |
| $OH^-$ | 25–150 | 205.1762 | −5.1050 | 81.3865 | −129.5329 | 0.061 | 0.014 |
| | 150–350 | 560.1902 | −13.4002 | 200.4958 | −381.3855 | 0.067 | 0.008 |
| $NO_3^-$ | 25–150 | 43.6090 | 2.5423 | −7.4032 | 7.5053 | 0.013 | 0.008 |
| | 150–350 | 1811.0683 | −23.6578 | 418.3792 | −975.7835 | 0.078 | 0.015 |
| *P = 1 kbar* | | | | | | | |
| $H^+$ | 25–400 | 322.3821 | −8.8946 | 108.1199 | −149.8503 | 1.009 | 0.099 |
| | 400–1000 | −1352.5003 | 6.3848 | −184.2093 | 597.2852 | 0.888 | 0.096 |
| $Li^+$ | 25–500 | 140.9410 | −1.7252 | 43.4539 | −95.1429 | 0.907 | 0.192 |
| | 500–1000 | 2109.8536 | −88.8906 | 2180.9705 | −6581.2466 | 0.963 | 0.146 |

*(Continued)*

**TABLE 5.2 (*Continued*)**
**Coefficients of Equation 5.98 to Calculate Limiting Molar Electrical Conductivity, S cm²/mol, Depending on Pressure**

| Ion: | Range $t$ (°C) from $t_{min}$ to $t_{max}$ | $A_0$ | $A_1$ | $A_2$ | $A_3$ | $S_\lambda$ | $\Delta$ (%) |
|---|---|---|---|---|---|---|---|
| $Na^+$ | 25–400 | 191.6486 | −2.4778 | 57.6646 | −125.5061 | 1.278 | 0.454 |
| | 400–1000 | −1346.6617 | 0.4251 | −45.0325 | 233.7070 | 0.799 | 0.099 |
| $K^+$ | 25–300 | 148.5564 | −1.2485 | 39.1220 | −84.4278 | 0.014 | 0.008 |
| | 300–600 | −1289.3669 | 1.1518 | −60.1375 | 267.6196 | 0.471 | 0.070 |
| | 600–1000 | 1181.9157 | 0.6516 | 6.7118 | −61.2967 | 0.053 | 0.006 |
| $Rb^+$ | 25–300 | 155.7531 | −1.2041 | 39.2331 | −84.7095 | 0.010 | 0.004 |
| | 300–600 | −1593.3221 | 2.9638 | −98.8744 | 376.2517 | 0.738 | 0.106 |
| | 600–1000 | 1755.3370 | 1.0244 | 14.0101 | −118.6957 | 0.056 | 0.006 |
| $Cs^+$ | 25–300 | 160.5681 | −0.9776 | 37.3200 | −81.7845 | 0.131 | 0.073 |
| | 300–600 | −1979.4356 | 5.6568 | −153.0381 | 522.5436 | 1.087 | 0.151 |
| | 600–1000 | −1337.3122 | 0.6009 | −52.6158 | 258.7806 | 0.104 | 0.015 |
| $Cu^+$ | 25–400 | 190.6251 | −2.6279 | 58.7227 | −127.6058 | 1.263 | 0.418 |
| | 400–1000 | −426.7231 | −3.4036 | 45.5355 | −37.5629 | 0.187 | 0.028 |
| $Ag^+$ | 25–300 | 122.8106 | −0.7942 | 32.5477 | −71.3979 | 0.012 | 0.007 |
| | 300–600 | −1076.8072 | 0.3406 | −38.3375 | 200.0213 | 0.325 | 0.049 |
| | 600–1000 | 940.4867 | 0.5291 | 5.6392 | −45.0177 | 0.013 | 0.002 |
| $NH_4^+$ | 25–300 | 134.4420 | −0.1962 | 29.1384 | −66.9485 | 0.072 | 0.045 |
| | 300–600 | −1583.8186 | 3.9732 | −107.5977 | 387.9150 | 0.781 | 0.100 |
| | 600–1000 | 2030.0313 | 1.4610 | 22.5076 | −165.7232 | 0.010 | 0.001 |
| $Mg^{2+}$ | 25–500 | 222.6355 | −3.3933 | 74.4517 | −157.9881 | 0.968 | 0.278 |
| | 500–1000 | 12918.2295 | −51.9856 | 1263.1956 | −3759.0410 | 0.537 | 0.055 |
| $Ca^{2+}$ | 25–400 | 244.7720 | −3.8561 | 81.3917 | −171.8570 | 1.503 | 0.370 |
| | 400–1000 | −956.7893 | −2.1518 | 8.5303 | 95.5850 | 0.268 | 0.028 |
| $Sr^{2+}$ | 25 400 | 253.5636 | 4.0036 | 83.8494 | 177.0890 | 1.649 | 0.392 |
| | 400–1000 | −2651.0328 | 4.7694 | −159.9703 | 600.7720 | 1.302 | 0.123 |
| $Ba^{2+}$ | 25–400 | 262.9045 | −4.1443 | 86.3315 | −182.2490 | 1.819 | 0.465 |
| | 400–1000 | −4874.4170 | 13.9861 | −382.9714 | 1267.4880 | 2.770 | 0.258 |
| $Sn^{2+}$ | 25–300 | 122.4267 | −0.7647 | 37.8239 | −79.0770 | 0.126 | 0.077 |
| | 300–600 | −2468.7059 | 3.9752 | −146.7560 | 567.0680 | 1.093 | 0.123 |
| | 600–1000 | 4202.1792 | 2.6061 | 46.8030 | −368.1870 | 0.003 | 0.006 |
| $Pb^{2+}$ | 25–300 | 507.8505 | −10.5769 | 179.5585 | −376.3330 | 0.073 | 0.040 |
| | 300–600 | −528.8854 | −4.1541 | 52.9443 | −31.5440 | 0.293 | 0.039 |
| | 600–1000 | −207.5890 | −0.1457 | −10.8535 | 86.4640 | 0.052 | 0.005 |
| $Mn^{2+}$ | 25–400 | 231.9678 | −3.5063 | 76.5487 | −162.0670 | 1.262 | 0.340 |
| | 400–1000 | 2100.8169 | −14.3178 | 308.5051 | −808.1550 | 1.799 | 0.159 |
| $Fe^{2+}$ | 25–500 | 249.3576 | −3.9871 | 83.2882 | −176.5430 | 1.358 | 0.437 |
| | 500–1000 | 8808.0235 | −37.4520 | 895.0452 | −2624.3270 | 0.419 | 0.046 |

(*Continued*)

**TABLE 5.2 (*Continued*)**
**Coefficients of Equation 5.98 to Calculate Limiting Molar Electrical Conductivity, S cm²/mol, Depending on Pressure**

| Ion: | Range $t$ (°C) from $t_{min}$ to $t_{max}$ | $A_0$ | $A_1$ | $A_2$ | $A_3$ | $S_\lambda$ | $\Delta$ (%) |
|---|---|---|---|---|---|---|---|
| $Co^{2+}$ | 25–500 | 245.1577 | −3.8726 | 81.5842 | −173.0360 | 1.247 | 0.384 |
| | 500–1000 | 9782.0830 | −40.9289 | 982.7121 | −2894.0090 | 0.421 | 0.047 |
| $Ni^{2+}$ | 25–500 | 213.9632 | −3.5222 | 75.0850 | −157.4110 | 1.199 | 0.341 |
| | 500–1000 | 1289.5254 | −46.3605 | 1118.8026 | −3312.0710 | 0.484 | 0.054 |
| $Cu^{2+}$ | 25–500 | 242.7577 | −3.7571 | 80.4419 | −170.7930 | 1.239 | 0.361 |
| | 500–1000 | 9809.2344 | −40.8949 | 983.8165 | −2899.0401 | 0.462 | 0.050 |
| $Zn^{2+}$ | 25–400 | 225.5193 | −3.4459 | 75.1296 | −159.0299 | 1.165 | 0.333 |
| | 400–1000 | 3255.2300 | −18.9274 | 4212865 | −1148.2781 | 2.530 | 0.223 |
| $Cd^{2+}$ | 25–400 | 243.5370 | −3.7551 | 80.3541 | −170.2812 | 1.482 | 0.341 |
| | 400–1000 | −684.7483 | −3.1447 | 34.3550 | 16.3330 | 0.189 | 0.021 |
| $Hg^{2+}$ | 25–400 | 249.0252 | −3.7936 | 81.6015 | −172.9793 | 1.612 | 0.377 |
| | 400–1000 | 2414.0762 | 4.0042 | −138.3715 | 533.2743 | 1.171 | 0.104 |
| $F^-$ | 25–300 | 185.2225 | −2.4238 | 57.6657 | −118.8755 | 0.032 | 0.018 |
| | 300–600 | −1363.1746 | −3.0427 | −3.2176 | 171.7471 | 0.152 | 0.020 |
| | 600–1000 | 3620.1360 | 1.6875 | 40.3488 | −304.1622 | 0.094 | 0.010 |
| $Cl^-$ | 25–400 | 247.1344 | 394.0493 | 79.8993 | −165.8701 | 1.236 | 0.301 |
| | 400–1000 | 941.6629 | −1.2424 | −8.8810 | 131.6495 | 0.228 | 0.025 |
| $Br^-$ | 25–400 | 246.6457 | −4.0493 | 79.7423 | −165.2881 | 1.165 | 0.325 |
| | 400–1000 | 459.2093 | −7.2103 | 132.9190 | −290.1118 | 0.917 | 0.103 |
| $I^-$ | 25–400 | 243.1897 | −4.0421 | 78.2331 | −162.1883 | 1.074 | 0.295 |
| | 400–1000 | 920.2733 | 13.2760 | 278.7327 | 725.9596 | 2.249 | 0.263 |
| $OH^-$ | 25–400 | 345.5415 | −9.6851 | 137.8574 | −238.2171 | 0.206 | 0.022 |
| | 400–1000 | 1730.5433 | 5.1201 | 163.5489 | 568.5209 | 0.804 | 0.090 |
| $NO_3^-$ | 25–500 | 84.4609 | 5.5069 | 95.7673 | −196.6848 | 1.324 | 0.427 |
| | 500–100 | 1553.5850 | 59.9370 | 1455.2073 | −4373.8912 | 0.701 | 0.126 |
| *P = 2 kbar* | | | | | | | |
| $H^+$ | 25–400 | 303.1875 | −8.0246 | 95.3108 | −123.4034 | 0.962 | 0.094 |
| | 400–1000 | −1178.9650 | 5.6608 | −165.8492 | 542.6580 | 0.793 | 0.083 |
| $Li^+$ | 25–1000 | 291.7807 | −5.5200 | 94.0903 | −203.2345 | 3.690 | 1.719 |
| | 25–500 | 215.7242 | −3.8476 | 71.1311 | −151.5693 | 1.457 | 0.735 |
| | 500–1000 | −6008.1866 | 15.3862 | −437.0754 | 1457.9545 | 0.227 | 0.039 |
| $Na^+$ | 25–1000 | 282.8069 | −4.8180 | 89.8428 | −193.1024 | 2.770 | 1.105 |
| | 25–400 | 177.7495 | −2.6486 | 55.8081 | −118.3626 | 1.032 | 0.357 |
| | 400–1000 | −663.5933 | −1.3647 | 3.4825 | 71.4270 | 0.173 | 0.026 |
| $K^+$ | 25–1000 | 309.7980 | −5.4157 | 97.4654 | −206.5428 | 2.826 | 0.967 |
| | 25–400 | 201.0736 | −3.1335 | 61.8904 | −128.7088 | 1.052 | 0.312 |
| | 400–1000 | 201.6289 | −4.9406 | 86.0235 | −172.6091 | 1.363 | 0.197 |

(*Continued*)

**TABLE 5.2 (*Continued*)**
**Coefficients of Equation 5.98 to Calculate Limiting Molar Electrical Conductivity, S cm²/mol, Depending on Pressure**

| Ion: | Range $t$ (°C) from $t_{min}$ to $t_{max}$ | $A_0$ | $A_1$ | $A_2$ | $A_3$ | $S_\lambda$ | Δ (%) |
|---|---|---|---|---|---|---|---|
| $Rb^+$ | 25–1000 | 317.7858 | −5.3844 | 97.6713 | −206.9846 | 2.937 | 0.960 |
| | 25–400 | 210.2461 | −3.1172 | 62.3661 | −129.7978 | 1.038 | 0.308 |
| | 400–1000 | 492.5638 | −5.9327 | 111.8417 | −251.6020 | 1.773 | 0.241 |
| $Cs^+$ | 25–1000 | 324.7777 | −5.1436 | 95.5322 | −203.7600 | 3.145 | 0.891 |
| | 25–400 | 220.4266 | −2.9376 | 61.2209 | −128.7946 | 1.006 | 0.281 |
| | 400–1000 | 918.2599 | −7.2264 | 147.9673 | −365.4399 | 2.301 | 0.291 |
| $Cu^+$ | 25–1000 | 287.3997 | −5.0106 | 91.6810 | −197.2157 | 2.783 | 1.195 |
| | 25–400 | 182.2689 | −2.8402 | 57.6207 | −122.4106 | 1.026 | 0.395 |
| | 400–1000 | −683.5796 | −1.4696 | 3.1133 | 74.1017 | 0.229 | 0.038 |
| $Ag^{2+}$ | 25–1000 | 293.7347 | −5.0220 | 92.2225 | −197.1687 | 2.691 | 0.989 |
| | 25–400 | 189.3920 | −2.8272 | 58.0267 | −122.3819 | 1.036 | 0.331 |
| | 400–1000 | 21.7585 | −3.9284 | 65.4977 | −116.7641 | 1.115 | 0.164 |
| $Au^{2+}$ | 25–1000 | 307.2580 | −5.2906 | 96.0039 | −203.7819 | 2.840 | 0.962 |
| | 25–400 | 199.7083 | −3.0240 | 60.6815 | −126.5367 | 1.037 | 0.346 |
| | 400–1000 | 295.7402 | −5.1578 | 93.1882 | −196.4173 | 1.492 | 0.211 |
| $NH_4^+$ | 25–1000 | 284.7990 | −4.1277 | 83.6684 | −180.6528 | 2.762 | 0.795 |
| | 25–400 | 186.8538 | −2.0403 | 51.2727 | −109.9844 | 0.985 | 0.285 |
| | 400–1000 | 479.3461 | −4.6867 | 98.5658 | −228.4434 | 1.771 | 0.196 |
| $Mg^{2+}$ | 25–1000 | 302.3826 | −5.6878 | 104.0757 | −218.7347 | 2.770 | 0.836 |
| | 25–400 | 213.9415 | −3.8558 | 75.3520 | −155.6935 | 1.119 | 0.324 |
| | 400–1000 | −1025.6093 | −0.8290 | −17.2400 | 152.6130 | 1.092 | 0.135 |
| $Ca^{2+}$ | 25–1000 | 311.7070 | −5.8680 | 106.7483 | −223.4083 | 2.820 | 0.767 |
| | 25–400 | 219.4797 | −3.8851 | 76.0483 | −156.5707 | 1.166 | 0.311 |
| | 400–1000 | −919.5218 | −1.1192 | −10.0870 | 130.5243 | 0.208 | 0.023 |
| $Sr^{2+}$ | 25–1000 | 314.3118 | −5.8695 | 106.9690 | −223.8241 | 2.876 | 0.737 |
| | 25–400 | 222.4028 | −3.8697 | 76.1066 | −156.7894 | 1.180 | 0.336 |
| | 400–1000 | −879.8767 | −1.1848 | −7.7528 | 122.5543 | 0.117 | 0.013 |
| $Ba^{2+}$ | 25–1000 | 313.7369 | −5.8261 | 106.6080 | −222.7131 | 2.979 | 0.717 |
| | 25–400 | 221.3311 | −3.7874 | 75.2962 | −154.9063 | 1.188 | 0.323 |
| | 400–1000 | −774.2464 | −1.4383 | −0.0573 | 97.6282 | 0.471 | 0.054 |
| $Sn^{2+}$ | 25–1000 | 275.9626 | −5.1581 | 96.9108 | −200.2729 | 3.849 | 0.738 |
| | 25–400 | 178.4890 | −2.9140 | 62.9236 | −127.3408 | 1.301 | 0.343 |
| | 400–1000 | 34.5270 | −3.5656 | 62.1259 | −104.4593 | 2.470 | 0.262 |
| $Pb^{2+}$ | 25–1000 | 317.7988 | −5.8002 | 106.6449 | −223.2884 | 2.901 | 0.697 |
| | 25–400 | 226.3612 | −3.8016 | 75.8731 | −156.5329 | 1.182 | 0.296 |
| | 400–1000 | −844.4080 | −1.1950 | −5.8282 | 115.6402 | 0.239 | 0.025 |
| $Mn^{2+}$ | 25–1000 | 309.0584 | −5.7340 | 105.2131 | −220.8064 | 2.760 | 0.781 |
| | 25–400 | 218.8654 | −3.8293 | 75.5392 | −155.9457 | 1.143 | 0.319 |
| | 400–1000 | −991.6075 | −0.8540 | −15.7690 | 147.6727 | 0.680 | 0.080 |

(*Continued*)

**TABLE 5.2 (*Continued*)**
**Coefficients of Equation 5.98 to Calculate Limiting Molar Electrical Conductivity, S cm²/mol, Depending on Pressure**

| Ion: | Range $t$ (°C) from $t_{min}$ to $t_{max}$ | $A_0$ | $A_1$ | $A_2$ | $A_3$ | $S_\lambda$ | Δ (%) |
|---|---|---|---|---|---|---|---|
| $Fe^{2+}$ | 25–1000 | 308.0439 | –5.7978 | 105.6143 | –221.6372 | 2.754 | 0.816 |
| | 25–400 | 218.1228 | –3.9162 | 76.2183 | –157.2636 | 1.137 | 0.318 |
| | 400–1000 | –992.2372 | –0.9747 | –14.3460 | 144.5506 | 0.853 | 0.105 |
| $Co^{2+}$ | 25–1000 | 310.3585 | –5.8049 | 105.7828 | –222.2625 | 2.751 | 0.814 |
| | 25–400 | 221.1332 | –3.9421 | 76.6477 | –158.4223 | 1.138 | 0.330 |
| | 400–1000 | –1003.0589 | –0.9522 | –15.0510 | 146.8821 | 0.890 | 0.110 |
| $Ni^{2+}$ | 25–1000 | 307.8082 | –5.8474 | 105.9611 | –222.3640 | 2.758 | 0.841 |
| | 25–400 | 217.9488 | –3.9813 | 76.7326 | –158.2533 | 1.135 | 0.324 |
| | 400–1000 | –991.7588 | –1.0755 | –13.0680 | 141.7129 | 0.975 | 0.123 |
| $Cu^{2+}$ | 25–1000 | 306.0218 | –5.6718 | 104.3168 | –219.1925 | 2.751 | 0.796 |
| | 25–500 | 256.9778 | –4.7114 | 88.9187 | –184.9720 | 1.790 | 0.564 |
| | 500–1000 | 3123.2740 | 6.0032 | –192.7988 | 702.9172 | 0.124 | 0.013 |
| $Zn^{2+}$ | 25–1000 | 307.5124 | –5.7792 | 105.4080 | –221.2676 | 2.752 | 0.815 |
| | 25–400 | 217.7840 | –3.9018 | 76.0710 | –157.0183 | 1.142 | 0.326 |
| | 400–1000 | 1004.1683 | –0.9159 | –15.5613 | 148.0288 | 0.831 | 0.102 |
| $Cd^{2+}$ | 25–1000 | 312.0692 | –5.7618 | 105.6795 | –221.9151 | 2.795 | 0.755 |
| | 25–400 | 221.9357 | –3.8211 | 75.6350 | –156.5170 | 1.157 | 0.324 |
| | 400–1000 | –943.8011 | –0.9216 | –13.4273 | 138.9440 | 0.267 | 0.030 |
| $Hg^{2+}$ | 25–1000 | 309.6784 | –5.6570 | 104.6835 | –219.6336 | 2.849 | 0.717 |
| | 25–400 | 219.6196 | –3.6970 | 74.4564 | –153.9950 | 1.164 | 0.307 |
| | 400–1000 | –883.3480 | –0.9749 | –9.9622 | 126.4861 | 0.108 | 0.011 |
| $F^-$ | 25–1000 | 274.3359 | –5.2886 | 94.3110 | –192.1691 | 3.453 | 0.576 |
| | 25–400 | 222.8963 | –3.9656 | 74.9232 | –151.5199 | 1.004 | 0.265 |
| | 400–1000 | –519.0372 | –1.5495 | 6.9611 | 62.5179 | 1.560 | 0.192 |
| $Cl^-$ | 25–1000 | 272.6273 | –5.1327 | 91.3664 | –187.3371 | 2.410 | 0.457 |
| | 25–400 | 223.6429 | –3.9818 | 74.0250 | –150.2775 | 0.962 | 0.273 |
| | 400–1000 | –895.5541 | –0.4193 | –23.1205 | 156.3954 | 0.244 | 0.027 |
| $Br^-$ | 25–1000 | 276.8774 | –5.2903 | 92.8279 | –190.3123 | 2.446 | 0.504 |
| | 25–400 | 226.5960 | –4.1325 | 75.2885 | –152.6820 | 0.975 | 0.265 |
| | 400–1000 | 1040.6591 | –0.0845 | –34.3335 | 193.0287 | 0.520 | 0.069 |
| $I^-$ | 25–1000 | 279.0323 | –5.3350 | 93.0487 | –191.0398 | 2.640 | 0.566 |
| | 25–400 | 227.5617 | –4.1654 | 75.2247 | –152.6703 | 0.981 | 0.301 |
| | 400–1000 | –1266.7536 | 0.6642 | –54.2007 | 254.4026 | 0.966 | 0.137 |
| $OH^-$ | 25–400 | 292.5380 | –8.6269 | 119.1510 | –195.6152 | 0.274 | 0.031 |
| | 400–1000 | –1527.6600 | 4.7283 | –150.6344 | 522.5216 | 0.738 | 0.084 |
| $NO_3^-$ | 25–1000 | 307.0667 | 6.4541 | 105.6048 | –215.1129 | 3.432 | 0.882 |
| | 25–400 | 245.3860 | –5.0879 | 84.6222 | –169.7083 | 1.051 | 0.333 |
| | 400–1000 | –1862.8501 | 1.8131 | –98.3413 | 404.0698 | 1.706 | 0.324 |

(*Continued*)

**TABLE 5.2 (*Continued*)**
**Coefficients of Equation 5.98 to Calculate Limiting Molar Electrical Conductivity, S cm²/mol, Depending on Pressure**

| Ion: | Range $t$ (°C) from $t_{min}$ to $t_{max}$ | $A_0$ | $A_1$ | $A_2$ | $A_3$ | $S_\lambda$ | Δ (%) |
|---|---|---|---|---|---|---|---|
| *P = 3 kbar* | | | | | | | |
| $H^+$ | 25–400 | 293.5739 | −7.5298 | 88.0406 | −108.4265 | 0.920 | 0.089 |
| | 400–1000 | −4601.7261 | 19.0017 | 492.4008 | 1529.6272 | 3.576 | 0.338 |
| $Li^+$ | 25–1000 | 251.6900 | −4.6707 | 81.9642 | −174.6246 | 2.270 | 1.044 |
| | 25–400 | 186.7716 | −3.3356 | 61.0010 | −128.5496 | 1.019 | 0.474 |
| | 400–1000 | −343.5292 | −2.5997 | 29.4605 | −12.2830 | 1.913 | 0.422 |
| $Na^+$ | 25–1000 | 241.0775 | −4.2556 | 78.0873 | −165.2759 | 2.132 | 0.832 |
| | 25–400 | 167.0665 | −2.6862 | 53.6819 | −111.9872 | 0.875 | 0.319 |
| | 400–1000 | −641.5325 | −0.8992 | −4.8303 | 86.6129 | 0.137 | 0.021 |
| $K^+$ | 25–1000 | 274.1748 | −4.9483 | 87.2377 | −182.2315 | 2.714 | 0.853 |
| | 25–400 | 188.9691 | −3.1216 | 58.9160 | −120.5283 | 0.767 | 0.257 |
| | 400–1000 | −758.0992 | −0.9100 | −11.7498 | 116.8185 | 1.220 | 0.199 |
| $Rb^+$ | 25–1000 | 285.0983 | −4.9603 | 88.1202 | −184.2151 | 2.882 | 0.820 |
| | 25–400 | 199.0072 | −3.1072 | 59.4330 | −121.7737 | 0.726 | 0.221 |
| | 400–1000 | −751.8840 | −0.8644 | −12.0197 | 117.7513 | 1.595 | 0.245 |
| $Cs^+$ | 25–1000 | 298.0333 | −4.7755 | 86.9018 | −183.2473 | 3.084 | 0.775 |
| | 25–500 | 246.3796 | −3.6771 | 69.7823 | −145.8716 | 1.315 | 0.401 |
| | 500–1000 | 3510.8826 | −14.1401 | 336.4116 | −982.0034 | 0.141 | 0.023 |
| $Cu^+$ | 25–1000 | 249.5559 | −4.4613 | 80.2496 | −170.3871 | 2.113 | 0.887 |
| | 25–400 | 176.5706 | −2.9134 | 56.1748 | −117.8172 | 0.875 | 0.353 |
| | 400–1000 | −640.8237 | −1.0826 | −3.2652 | 83.4216 | 0.210 | 0.039 |
| $Ag^+$ | 25–1000 | 259.0132 | −4.5449 | 81.9132 | −172.8094 | 2.551 | 0.860 |
| | 25–400 | 179.0844 | −2.8265 | 55.3134 | −114.9050 | 0.770 | 0.264 |
| | 400–1000 | −746.6404 | −0.6051 | −14.6180 | 118.7362 | 0.994 | 0.166 |
| $Au^{2+}$ | 25–1000 | 271.8245 | −4.8301 | 85.8676 | −179.6521 | 2.740 | 0.843 |
| | 25–400 | 187.4911 | −3.0156 | 57.7578 | −118.4493 | 0.749 | 0.260 |
| | 400–1000 | −743.3250 | −0.8373 | −11.8636 | 115.2952 | 1.331 | 0.213 |
| $NH_4^+$ | 25–1000 | 252.4183 | −3.6998 | 74.0686 | −157.8150 | 2.831 | 0.688 |
| | 25–400 | 176.7496 | −2.0422 | 48.5449 | −102.4621 | 0.678 | 0.200 |
| | 400–1000 | −797.1242 | 0.5123 | −28.4524 | 150.3622 | 1.628 | 0.203 |
| $Mg^{2+}$ | 25–1000 | 250.3795 | −4.9977 | 89.4334 | −18.0410 | 2.053 | 0.505 |
| *P = 4 kbar* | | | | | | | |
| $Ag^+$ | 25–1000 | 234.9543 | −4.1755 | 74.3231 | −155.1813 | 1.932 | 0.661 |
| | 25–400 | 176.7206 | −2.9038 | 54.7411 | −112.6976 | 0.689 | 0.241 |
| | 400–1000 | −670.6685 | −0.6396 | −12.3666 | 106.7895 | 0.309 | 0.055 |
| $Au^+$ | 25–1000 | 246.4314 | −4.4590 | 78.1836 | −161.7046 | 2.001 | 0.646 |
| | 25–400 | 184.3677 | −3.1115 | 57.3825 | −116.5080 | 0.682 | 0.234 |
| | 400–1000 | −697.3897 | −0.7976 | −11.7433 | 110.3943 | 0.351 | 0.060 |

(*Continued*)

**TABLE 5.2 (*Continued*)**
**Coefficients of Equation 5.98 to Calculate Limiting Molar Electrical Conductivity, S cm²/mol, Depending on Pressure**

| Ion: | Range $t$ (°C) from $t_{min}$ to $t_{max}$ | $A_0$ | $A_1$ | $A_2$ | $A_3$ | $S_\lambda$ | Δ (%) |
|---|---|---|---|---|---|---|---|
| $NH_4^+$ | 25–1000 | 229.4291 | −3.3427 | 66.6533 | −140.6076 | 1.941 | 0.477 |
| | 25–400 | 176.4722 | −2.1632 | 48.5940 | −101.5825 | 0.607 | 0.183 |
| | 400–1000 | −688.5593 | 0.2941 | −22.1491 | 126.9827 | 0.374 | 0.046 |
| $Mg^{2+}$ | 25–1000 | 221.3651 | −4.5944 | 80.9571 | −164.1694 | 2.139 | 0.436 |
| | 25–400 | 188.8231 | −3.7822 | 68.9352 | −138.8000 | 0.891 | 0.296 |
| | 400–1000 | −786.0761 | −0.4665 | −18.8853 | 134.6915 | 0.122 | 0.019 |
| $Ca^{2+}$ | 25–1000 | 227.1427 | −4.5945 | 80.9571 | −164.1694 | 2.139 | 0.436 |
| | 25–400 | 188.8231 | −3.7822 | 68.9351 | −138.8000 | 0.891 | 0.296 |
| | 400–1000 | −786.0761 | −0.4665 | −18.8853 | 134.6916 | 0.122 | 0.019 |
| | 25–1000 | 227.1427 | −4.6527 | 82.0174 | −165.7438 | 2.199 | 0.410 |
| | 25–400 | 190.2374 | −3.7546 | 68.6368 | −137.3657 | 0.854 | 0.255 |
| | 400–1000 | −846.7917 | −0.2867 | −23.8074 | 151.5103 | 0.219 | 0.029 |
| $Sr^{2+}$ | 25–1000 | 230.5336 | −4.6374 | 82.0945 | −166.0561 | 2.237 | 0.405 |
| | 25–400 | 193.1337 | −3.7273 | 68.5204 | −137.2567 | 0.836 | 0.260 |
| | 400–1000 | −881.5080 | −0.1281 | −27.2772 | 161.9936 | 0.249 | 0.031 |
| $Ba^{2+}$ | 25–1000 | 226.9308 | −4.6374 | 82.0945 | −166.0561 | 2.237 | 0.405 |
| | 25–400 | 187.6962 | −3.6074 | 66.9793 | −133.5258 | 0.809 | 0.245 |
| | 400–1000 | −900.4161 | 0.0124 | −29.6773 | 168.8330 | 0.306 | 0.037 |
| $Sn^{2+}$ | 25–1000 | 200.6755 | −3.9860 | 73.4119 | −146.3851 | 2.504 | 0.460 |
| | 25–400 | 148.9668 | −2.7832 | 55.2608 | −107.5360 | 0.755 | 0.203 |
| | 400–1000 | −1026.7564 | 0.9541 | −46.6662 | 214.3132 | 0.591 | 0.068 |
| $Pb^{2+}$ | 25–1000 | 234.1230 | −4.5562 | 81.6315 | −165.2990 | 2.248 | 0.387 |
| | 25–400 | 196.7693 | −3.6472 | 68.0798 | −136.5531 | 0.824 | 0.238 |
| | 400–1000 | −891.2179 | 0.0028 | −28.9702 | 166.4960 | 0.274 | 0.031 |
| $Mn^{2+}$ | 25–1000 | 227.1775 | −4.5831 | 81.3783 | −164.9598 | 2.157 | 0.410 |
| | 25–400 | 192.4479 | −3.7270 | 68.6660 | −138.0678 | 0.866 | 0.266 |
| | 400–1000 | −801.4356 | −0.3785 | −20.3858 | 139.7958 | 0.146 | 0.019 |
| $Fe^{2+}$ | 25–1000 | 226.7786 | −4.6700 | 82.0752 | −166.3462 | 2.146 | 0.427 |
| | 25–400 | 192.4316 | −3.8230 | 69.5086 | −139.7708 | 0.880 | 0.269 |
| | 400–1000 | −791.6220 | −0.5041 | −18.7332 | 135.5074 | 0.120 | 0.019 |
| $Co^{2+}$ | 25–1000 | 232.2694 | −4.6985 | 82.6525 | −168.0533 | 2.132 | 0.427 |
| | 25–400 | 199.1129 | −3.8756 | 70.4582 | −142.2933 | 0.880 | 0.280 |
| | 400–1000 | −778.8425 | −0.5608 | −17.4569 | 131.6792 | 0.129 | 0.019 |
| $Ni^{2+}$ | 25–1000 | 227.5104 | −4.7446 | 82.7680 | −167.7721 | 2.144 | 0.441 |
| | 25–400 | 193.5264 | −3.9042 | 70.2953 | −141.3986 | 0.890 | 0.300 |
| | 400–1000 | −786.8494 | −0.5934 | −17.6699 | 132.9447 | 0.114 | 0.019 |
| $Cu^{2+}$ | 25–1000 | 225.2101 | −4.5544 | 80.9255 | −164.2097 | 2.129 | 0.415 |
| | 25–400 | 192.4157 | −3.7373 | 68.8261 | −138.6669 | 0.882 | 0.283 |
| | 400–1000 | −781.4724 | −0.4314 | −18.8075 | 134.3504 | 0.118 | 0.017 |

(*Continued*)

**TABLE 5.2 (*Continued*)**
**Coefficients of Equation 5.98 to Calculate Limiting Molar Electrical Conductivity, S cm²/mol, Depending on Pressure**

| Ion: | Range $t$ (°C) from $t_{min}$ to $t_{max}$ | $A_0$ | $A_1$ | $A_2$ | $A_3$ | $S_\lambda$ | Δ (%) |
|---|---|---|---|---|---|---|---|
| $Zn^{2+}$ | 25–1000 | 226.3608 | −4.6511 | 81.8696 | −165.9903 | 2.153 | 0.426 |
| | 25–400 | 192.4091 | −3.8133 | 69.4409 | −139.7101 | 0.880 | 0.269 |
| | 400–1000 | −809.8594 | −0.4228 | −20.5099 | 140.7108 | 0.141 | 0.021 |
| $Cd^{2+}$ | 25–1000 | 232.2537 | −4.5705 | 81.4580 | −165.6877 | 2.223 | 0.413 |
| | 25–400 | 197.9240 | −3.7196 | 68.8496 | −139.0520 | 0.840 | 0.248 |
| | 400–1000 | −849.8128 | −0.1546 | −25.4597 | 154.5972 | 0.206 | 0.023 |
| $Hg^{2+}$ | 25–1000 | 219.4685 | −4.3867 | 78.9461 | −159.4422 | 2.242 | 0.408 |
| | 25–400 | 179.4506 | −3.4251 | 64.5575 | −128.8401 | 0.857 | 0.258 |
| | 400–1000 | −872.1222 | 0.0417 | −28.4538 | 162.6690 | 0.251 | 0.030 |
| $F^-$ | 25–1000 | 198.3685 | −4.0272 | 70.1360 | −137.3226 | 2.476 | 0.460 |
| | 25–400 | 190.3777 | −3.6805 | 65.6388 | −128.7876 | 0.565 | 0.163 |
| | 400–1000 | −957.9237 | 0.7955 | −45.9208 | 208.8546 | 0.484 | 0.063 |
| $Cl^-$ | 25–1000 | 199.1936 | −4.0180 | 69.0544 | −135.8500 | 2.305 | 0.499 |
| | 25–400 | 194.2412 | −3.7387 | 65.5914 | −129.5529 | 0.668 | 0.205 |
| | 400–1000 | −792.6563 | 0.1806 | −31.5930 | 163.5082 | 0.222 | 0.032 |
| $Br^-$ | 25–1000 | 202.9723 | −4.1701 | 70.4141 | −138.5548 | 2.257 | 0.512 |
| | 25–400 | 197.1023 | −3.8765 | 66.7033 | −131.6780 | 0.699 | 0.221 |
| | 400–1000 | −782.4351 | −0.0118 | −29.3493 | 158.3530 | 0.185 | 0.032 |
| $I^-$ | 25–1000 | 203.7215 | −4.1852 | 70.1768 | −138.2498 | 2.202 | 0.517 |
| | 25–400 | 197.1846 | −3.8821 | 66.2944 | −130.9659 | 0.730 | 0.236 |
| | 400–1000 | −743.3222 | −0.1815 | −25.8327 | 147.3853 | 0.157 | 0.030 |
| $OH^-$ | 25–400 | 260.0699 | −7.7582 | 104.1126 | −162.2982 | 0.388 | 0.045 |
| | 400–1000 | −1282.0071 | 4.1370 | −133.0673 | 463.0600 | 0.659 | 0.076 |
| $NO_3$ | 25–1000 | 228.4350 | −5.1970 | 81.2707 | 159.3330 | 2.015 | 0.598 |
| | 25–400 | 213.6640 | −4.7453 | 74.9361 | −146.5099 | 0.834 | 0.295 |
| | 400–1000 | −694.8750 | −1.3556 | −11.2577 | 116.7918 | 0.093 | 0.023 |
| *P = 5 kbar* | | | | | | | |
| $H^+$ | 25–400 | 288.4577 | −7.1178 | 82.4065 | −97.0880 | 0.853 | 0.082 |
| | 400–1000 | −1038.2925 | 4.9390 | −148.4473 | 493.3215 | 0.719 | 0.075 |
| $Li^+$ | 25–1000 | 224.9882 | −4.1572 | 71.5583 | −151.0413 | 1.546 | 0.715 |
| | 25–400 | 183.9350 | −3.2664 | 57.7918 | −121.1208 | 0.802 | 0.412 |
| | 400–1000 | −634.7484 | −0.8804 | −9.2926 | 94.4230 | 0.216 | 0.055 |
| $Na^+$ | 25–1000 | 197.1143 | −3.6377 | 65.3699 | −135.4407 | 1.693 | 0.560 |
| | 25–400 | 182.7712 | −3.2314 | 57.6202 | −116.0013 | 0.633 | 0.228 |
| | 400–1000 | −620.1340 | −0.3992 | −13.6931 | 102.7792 | 0.158 | 0.260 |
| $K^+$ | 25–1000 | 232.0396 | −4.3534 | 74.6791 | −152.6930 | 1.858 | 0.543 |
| | 25–400 | 182.7712 | −3.2314 | 57.6202 | −116.0013 | 0.633 | 0.228 |
| | 400–1000 | −543.5813 | −1.1891 | −1.9845 | 77.0219 | 0.080 | 0.014 |

*(Continued)*

**TABLE 5.2 (*Continued*)**
**Coefficients of Equation 5.98 to Calculate Limiting Molar Electrical Conductivity, S cm²/mol, Depending on Pressure**

| Ion: | Range $t$ (°C) from $t_{min}$ to $t_{max}$ | $A_0$ | $A_1$ | $A_2$ | $A_3$ | $S_\lambda$ | Δ (%) |
|---|---|---|---|---|---|---|---|
| $Rb^+$ | 25–1000 | 246.1567 | −4.3952 | 76.0219 | −155.8035 | 1.934 | 0.511 |
| | 25–400 | 196.9860 | −3.2314 | 57.6202 | −116.0013 | 0.633 | 0.228 |
| | 400–1000 | −543.5813 | −1.1891 | −1.9845 | 77.0220 | 0.080 | 0.014 |
| $Cs^+$ | 25–1000 | 272.9125 | −4.3270 | 76.8975 | −160.1575 | 1.844 | 0.456 |
| | 25–400 | 224.8219 | −3.2131 | 60.0561 | −124.0655 | 0.551 | 0.172 |
| | 400–1000 | −372.3527 | −1.5952 | 11.3452 | 34.8709 | 0.075 | 0.014 |
| $Cu^+$ | 25–1000 | 213.1856 | −3.8679 | 68.1557 | −142.4861 | 1.669 | 0.588 |
| | 25–400 | 172.3959 | −2.9362 | 53.9852 | −112.0135 | 0.695 | 0.288 |
| | 400–1000 | −602.5491 | −0.6315 | −10.8244 | 95.4271 | 0.168 | 0.033 |
| $Ag^+$ | 25–1000 | 219.6806 | −3.9530 | 69.5389 | −143.9291 | 1.832 | 0.539 |
| | 25–400 | 175.5613 | −2.9345 | 54.1170 | −110.8525 | 0.629 | 0.245 |
| | 400–1000 | −574.7735 | −0.7070 | −9.0646 | 91.5144 | 0.085 | 0.016 |
| $Au^+$ | 25–1000 | 230.6701 | −4.2544 | 73.6052 | −150.7702 | 1.827 | 0.531 |
| | 25–400 | 181.5497 | −3.1376 | 56.6331 | −114.2654 | 0.628 | 0.211 |
| | 400–1000 | −545.7029 | −1.0659 | −3.4774 | 79.8782 | 0.066 | 0.009 |
| $NH_4^+$ | 25–1000 | 214.9856 | −3.1408 | 62.1135 | −129.8271 | 1.964 | 0.380 |
| | 25–400 | 176.4035 | −2.2069 | 48.1512 | −100.1590 | 0.564 | 0.196 |
| | 400–1000 | −572.1394 | 0.1690 | −17.4181 | 107.0757 | 0.117 | 0.015 |
| $Mg^{2+}$ | 400–1000 | 204.3872 | −4.3242 | 75.5721 | −151.8623 | 2.074 | 0.422 |
| | 25–400 | 181.0836 | −3.7093 | 66.6300 | −133.2214 | 0.812 | 0.249 |
| | 400–1000 | −850.6086 | −0.0297 | −28.4623 | 159.9273 | 0.242 | 0.033 |
| $Ca^{2+}$ | 25–1000 | 207.4160 | −4.3584 | 76.1090 | −152.0814 | 2.170 | 0.419 |
| | 25–400 | 182.3233 | −3.6900 | 66.3904 | −131.8397 | 0.772 | 0.258 |
| | 400–1000 | −827.8453 | −0.0925 | −26.9050 | 155.9298 | 0.206 | 0.027 |
| $Sr^{2+}$ | 25–1000 | 210.4521 | −4.3377 | 76.0691 | −152.1053 | 2.203 | 0.417 |
| | 25–400 | 185.4418 | −3.6640 | 66.3032 | −131.8128 | 0.752 | 0.244 |
| | 400–1000 | 809.9216 | −0.1066 | −25.9405 | 152.5362 | 0.197 | 0.027 |
| $Ba^{2+}$ | 25–1000 | 204.8575 | −4.2481 | 74.8971 | −149.0094 | 2.236 | 0.409 |
| | 25–400 | 178.9750 | −3.5542 | 64.8346 | −128.0850 | 0.741 | 0.227 |
| | 400–1000 | −811.1630 | −0.0209 | −26.9252 | 154.8763 | 0.186 | 0.025 |
| $Sn^{2+}$ | 25–1000 | 178.7893 | −3.6956 | 67.2134 | −131.7032 | 2.579 | 0.434 |
| | 25–400 | 144.4096 | −2.7896 | 54.0434 | −104.2420 | 0.683 | 0.181 |
| | 400–1000 | −779.0781 | 0.4421 | −31.5069 | 160.8351 | 0.191 | 0.025 |
| $Pb^{2+}$ | 25–1000 | 214.5796 | −4.2589 | 75.6506 | −151.4844 | 2.217 | 0.397 |
| | 25–400 | 190.2318 | −3.5979 | 66.0974 | −131.6695 | 0.743 | 0.225 |

*Note:* $S_\lambda$, a mean squared error of calculation, S cm²/mol and Δ, the average relative error of calculation, %.

In (5.105), for calculating solution viscosity within the interval of temperatures, one can use a semiempirical formula for calculation of concentrated multicomponent solution in Pa s [85]:

$$\eta = \eta_0 \exp V, \quad (5.106)$$

where

$$V = \sum_{i=1}^{k} c_i \left( A_{0i} + A_{1i}t + A_{2i}c_i + A_{3i}t^2 \right). \quad (5.107)$$

where

- $k$ is a component number of the solution
- $c$ is the mass concentration of electrolyte in the solution, %
- $t$ is the temperature, °C
- $A_{ji}$ are the coefficients obtained based on processing experimental data on viscosimetry of binary solutions of about 150 electrolytes, using regression analysis methods, given in Table 5.3

Binary solution viscosity data were summarized from the world scientific literature and substantially supplemented with proper experimental measurements and added into [85]. The correlation coefficient in Equation 5.106 is equal to 0.942 obtained on the basis of studying viscosity of multicomponent solutions. Coefficients given in this chapter are much more accurate than those in [86].

In (5.106), water viscosity $\eta_0$ in Pa s in the temperature interval of 0°C–200°C with mean relative approximation error of 0.07% is calculated by the following equation:

$$\eta_0 = S_0 + S_1 t + \frac{S_2}{t} + S_3 \exp(-0.1t) + \frac{S_4}{t^2} + S_5\sqrt{t} + S_6 \ln t, \quad (5.108)$$

where coefficients $S_i$ are equal to

$$S_0 = 4{,}496{,}839{,}782; \quad S_1 = -4{,}644{,}252; \quad S_2 = -7{,}025{,}494{,}506; \quad S_3 = -159{,}419{,}139;$$

$$S_4 = 4{,}104{,}442{,}382; \quad S_5 = 316{,}705{,}387; \quad S_6 = -1{,}486{,}934{,}219.$$

## 5.8 RESULTS AND THEIR DISCUSSION

1. As it was shown in Equation 5.68, at application of weak electric field and motion of ion complexes or other supermolecular formations to electrodes, direct electrostatic interactions between ions of type $n_i$ and $n_j$ occur but they are extremely low compared to the action force of electric field on complexes. The relaxation force does not depend on direct electrostatic interactions between ions.

**TABLE 5.3**
**Coefficients of Equation 5.106 to Calculate Dynamic Viscosity, Pa s, of Multicomponent Solutions**

| Electrolyte | $A_0 \cdot 10^3$ | $A_1 \cdot 10^4$ | $A_2 \cdot 10^5$ | $A_3 \cdot 10^6$ | Range $c$ (%) from $c_{min}$ to $c_{max}$ | Range $t$ (°C) from $t_{min}$ to $t_{max}$ | $S_\eta \cdot 10^3$ | Δ (%) |
|---|---|---|---|---|---|---|---|---|
| $Al(NO_3)_3$ | 14.558 | 2.770 | 102.71 | −2.0 | 5–30 | 30–90 | 5.28 | 2.18 |
| | 55.819 | −2.599 | 34.20 | 1.0 | 30–40 | 30–90 | 12.81 | 2.34 |
| $Al_2(SO_4)_3$ | 10.851 | 18.475 | 99.99 | −17.1 | 5–25 | 25–90 | 6.79 | 2.38 |
| $BaBr_2$ | −0.278 | 1.459 | 18.01 | −0.9 | 5–45 | 25–90 | 1.67 | 1.01 |
| $BaCl_2$ | 8.282 | 2.045 | 3.24 | −1.1 | 0–20 | 10–90 | 0.77 | 0.96 |
| | 5.049 | 1.657 | 24.50 | −0.6 | 20–30 | 10–90 | 0.87 | 0.92 |
| $Ba(ClO_4)_2$ | 16.692 | −7.754 | 39.37 | 7.5 | 5–15 | 25–75 | 0.73 | 0.83 |
| | 2.9330 | −0.766 | 25.52 | 0.7 | 15–50 | 25–75 | 1.60 | 0.99 |
| | −13.270 | −0.624 | 59.53 | 0.4 | 50–60 | 25–75 | 2.31 | 0.90 |
| $BaI_2$ | 1.038 | 0.902 | 11.49 | −0.4 | 5–45 | 25–75 | 0.88 | 0.76 |
| | −7.265 | 1.061 | 30.92 | −0.7 | 45–60 | 25–75 | 1.82 | 0.93 |
| $Ba(NO_2)_2$ | 8.149 | 0.922 | 16.55 | −1.2 | 5–25 | 15–75 | 0.91 | 0.74 |
| | 3.430 | 1.275 | 39.11 | −1.9 | 25–40 | 15–75 | 1.91 | 1.01 |
| $Ba(NO_3)_2$ | 17.972 | −2.517 | 27.92 | 1.6 | 0–22 | 50–100 | 0.54 | 0.84 |
| $Ba(OH)_2$ | −19.468 | 9.574 | 65.26 | −7.4 | 0–30 | 40–90 | 1.59 | 1.56 |
| | 59.803 | −8.136 | 10.83 | 4.6 | 30–40 | 40–90 | 0.56 | 0.50 |
| BaS | 51.202 | 3.007 | −204.15 | −2.6 | 0–15 | 25–90 | 1.87 | 1.92 |
| | 37.603 | −0.584 | −59.56 | 1.2 | 15–30 | 25–90 | 2.11 | 1.27 |
| $CaCl_2$ | 17.282 | 1.063 | 70.95 | −1.4 | 0–40 | 1–100 | 8.97 | 2.60 |
| $Ca(ClO_4)_2$ | 10.243 | −1.049 | 4.55 | 2.7 | 5–25 | 5–50 | 1.36 | 0.80 |
| | 0.959 | 0.220 | 38.02 | 0.5 | 25–45 | 5–50 | 2.71 | 0.97 |
| | −17.856 | 0.177 | 81.99 | 0.1 | 45–60 | 5–50 | 7.13 | 0.99 |
| $CaCr_2O_7$ | 29.123 | 4.772 | −8.95 | 4.6 | 5–20 | 25–90 | 1.78 | 1.83 |
| | 14.749 | −2.018 | 38.02 | 1.7 | 20–45 | 25–90 | 5.57 | 2.55 |
| $CaI_2$ | 0.497 | 1.877 | 15.83 | −1.4 | 5–45 | 25–90 | 1.63 | 1.33 |
| | −19.782 | 0.693 | 69.16 | −0.5 | 45–60 | 25–90 | 4.97 | 2.05 |
| $Ca(NO_2)_2$ | 18.079 | −0.709 | 60.98 | −0.3 | 5–35 | 25–90 | 3.84 | 1.66 |
| | 10.196 | −1.131 | 98.15 | −1.0 | 35–50 | 25–90 | 13.24 | 2.74 |
| $Ca(NO_3)_2$ | 7.824 | 1.614 | 45.30 | −1.2 | 10–30 | 10–100 | 2.41 | 1.06 |
| | 7.689 | 0.769 | 56.00 | −0.9 | 30–45 | 10–100 | 13.09 | 2.36 |
| | 0.596 | −2.250 | 94.82 | 1.0 | 45–60 | 10–100 | 50.89 | 3.40 |
| $CdBr_2$ | 11.111 | 0.258 | −2.47 | −0.1 | 5–30 | 25–75 | 1.61 | 1.18 |
| | −3.810 | 1.963 | 41.75 | −2.2 | 30–50 | 25–75 | 2.22 | 1.17 |
| $CdCl_2$ | 16.447 | 0.732 | 13.64 | −1.4 | 5–30 | 25–75 | 2.22 | 1.25 |
| | 1.888 | −0.337 | 72.96 | −0.5 | 30–45 | 25–75 | 2.66 | 0.88 |
| $CdI_2$ | 9.146 | −0.439 | 9.31 | 0.2 | 5–45 | 15–75 | 1.60 | 0.81 |
| $Cd(NO_3)_2$ | 12.972 | 2.126 | −14.80 | −0.9 | 5–25 | 30–95 | 1.97 | 1.64 |
| | 10.472 | 0.271 | 32.59 | −0.2 | 25–50 | 30–95 | 4.32 | 1.60 |
| | 10.929 | −1.191 | 49.64 | 0.4 | 50–60 | 30–95 | 7.41 | 1.85 |

*(Continued)*

**TABLE 5.3 (*Continued*)**
**Coefficients of Equation 5.106 to Calculate Dynamic Viscosity, Pa s, of Multicomponent Solutions**

| Electrolyte | $A_0 \cdot 10^3$ | $A_1 \cdot 10^4$ | $A_2 \cdot 10^5$ | $A_3 \cdot 10^6$ | Range $c$ (%) from $c_{min}$ to $c_{max}$ | Range $t$ (°C) from $t_{min}$ to $t_{max}$ | $S_\eta \cdot 10^3$ | Δ (%) |
|---|---|---|---|---|---|---|---|---|
| $CdSO_4$ | 26.622 | 1.739 | 51.91 | −3.9 | 5–40 | 15–75 | 9.21 | 1.62 |
| $CoBr_2$ | 10.096 | 2.414 | 18.42 | −2.8 | 5–35 | 25–90 | 1.87 | 1.47 |
| | 2.295 | 0.553 | 59.13 | −1.4 | 35–55 | 25–90 | 6.81 | 2.20 |
| $CoCl_2$ | 21.220 | 10.096 | 15.45 | −15.0 | 5–30 | 20–60 | 4.62 | 1.88 |
| | −34.000 | 16.024 | 33.15 | −10.1 | 5–30 | 60–100 | 1.51 | 1.21 |
| $CoI_2$ | 0.414 | 2.331 | 27.78 | 2.3 | 5–55 | 25–90 | 5.07 | 2.39 |
| | 15.320 | 0.078 | 71.24 | −0.7 | 55–70 | 25–90 | 11.14 | 2.98 |
| $Co(NO_3)_2$ | 9.930 | 4.255 | 41.56 | −2.9 | 5–25 | 20–80 | 3.01 | 1.31 |
| | 29.675 | 2.363 | 1.49 | −2.2 | 25–40 | 20–80 | 4.06 | 1.51 |
| | −11.372 | −1.288 | 133.30 | 0.4 | 40–50 | 20–80 | 24.39 | 3.62 |
| $CoSO_4$ | 34.870 | 1.084 | 78.29 | −3.8 | 5–30 | 15–75 | 4.35 | 1.54 |
| $CrCl_3$ | 42.444 | −0.607 | 85.92 | −2.4 | 5–30 | 25–90 | 5.27 | 1.98 |
| $Cr(NO_3)_3$ | 3.511 | 8.837 | 49.33 | −9.6 | 5–25 | 25–75 | 2.14 | 1.05 |
| | −3.624 | 6.722 | 104.50 | −8.2 | 25–50 | 25–75 | 5.70 | 1.78 |
| $Cr_2(SO_4)_3$ | −16.096 | 24.329 | 106.09 | −27.4 | 5–20 | 25–75 | 4.30 | 1.62 |
| | 4.801 | 14.963 | 146.57 | 20.220 | 20–45 | 25–75 | 51.13 | .44 |
| CsBr | −8.958 | 2.086 | 4.92 | −0.9 | 10–50 | 20–90 | 0.58 | 0.86 |
| $Cs_2CO_3$ | 0.793 | 5.135 | −0.03 | −4.1 | 10–40 | 20–90 | 2.62 | 2.09 |
| | −5.229 | 2.520 | 32.60 | −2.0 | 40–60 | 20–90 | 3.67 | 1.77 |
| CsCl | −8.253 | 2.693 | −0.45 | −1.1 | 5–35 | 1–100 | 1.28 | 1.20 |
| | −11.520 | 2.302 | 13.86 | −1.0 | 35–60 | 1–100 | 1.89 | 1.49 |
| CsI | 4.263 | 0.416 | −29.14 | 1.2 | 5–25 | 5–75 | 1.75 | 1.43 |
| | −9.625 | 1.633 | 11.33 | −0.7 | 25–55 | 5–75 | 1.03 | 0.93 |
| $CsNO_3$ | −7.478 | 1.845 | 2.81 | −0.6 | 2–20 | 20–90 | 0.73 | 0.85 |
| $Cs_2SO_4$ | 7.756 | 0.608 | −21.54 | 0.1 | 0–25 | 20–90 | 1.33 | 1.57 |
| | −2.531 | 1.047 | 21.68 | −0.4 | 25–65 | 20–90 | 2.00 | 1.30 |
| $CuCl_2$ | 26.519 | 1.408 | 2.04 | −3.1 | 5–25 | 20–90 | 1.74 | 1.00 |
| | 24.377 | −0.623 | 33.19 | −1.2 | 25–45 | 20–90 | 4.92 | 1.55 |
| $Cu(NO_3)_2$ | 2.238 | 4.127 | 53.12 | −3.2 | 5–35 | 25–75 | 2.33 | 1.68 |
| | 13.831 | 1.052 | 48.21 | −1.6 | 35–50 | 25–75 | 12.50 | 2.74 |
| $CuSO_4$ | 49.180 | −4.610 | 29.97 | 3.0 | 5–20 | 20–75 | 1.93 | 1.01 |
| | 35.674 | −6.625 | 148.24 | 3.5 | 20–32 | 20–75 | 1.56 | 0.88 |
| $FeCl_2$ | −8.463 | 14.395 | −16.82 | 10.3 | 5–18 | 40–80 | 1.41 | 0.99 |
| | 31.382 | 0.094 | 23.49 | 0.0 | 18–30 | 40–80 | 1.97 | 1.07 |
| $FeCl_3$ | 49.609 | 2.048 | −95.52 | −2.9 | 0–15 | 25–90 | 2.26 | 1.12 |
| | 50.964 | −3.808 | 28.68 | 0.9 | 15–50 | 25–90 | 21.67 | 2.39 |

(*Continued*)

**TABLE 5.3 (*Continued*)**
**Coefficients of Equation 5.106 to Calculate Dynamic Viscosity, Pa s, of Multicomponent Solutions**

| Electrolyte | $A_0 \cdot 10^3$ | $A_1 \cdot 10^4$ | $A_2 \cdot 10^5$ | $A_3 \cdot 10^6$ | Range $c$ (%) from $c_{min}$ to $c_{max}$ | Range $t$ (°C) from $t_{min}$ to $t_{max}$ | $S_\eta \cdot 10^3$ | Δ (%) |
|---|---|---|---|---|---|---|---|---|
| $Fe(NO_3)_3$ | 16.128 | 3.455 | 83.76 | −4.5 | 5–15 | 25–90 | 1.20 | 1.49 |
| | 23.398 | 0.975 | 56.20 | −1.0 | 15–35 | 25–90 | 4.83 | 2.39 |
| | 32.037 | 1.818 | 59.16 | 0.4 | 35–55 | 25–90 | 22.61 | 3.01 |
| $FeSO_4$ | 34.319 | −2.011 | 106.83 | 0.5 | 5–30 | 25–75 | 5.83 | 1.89 |
| HCl | 11.024 | 0.348 | 15.20 | 4.7 | 0–40 | 10–30 | 2.92 | 1.03 |
| $HClO_4$ | −5.040 | 1.468 | 18.36 | −0.7 | 5–35 | 1–90 | 1.75 | 1.21 |
| | −12.344 | 0.755 | 45.94 | −0.3 | 35–70 | 1–90 | 7.32 | 1.66 |
| HI | −12.105 | 2.673 | 3.02 | −2.4 | 5–55 | 10–40 | 1.81 | 1.64 |
| $HNO_3$ | 2.111 | 1.616 | 12.09 | −0.8 | 5–35 | 15–90 | 1.81 | 1.54 |
| | 11.289 | 0.426 | −0.62 | −0.2 | 35–57 | 15–90 | 3.27 | 1.95 |
| $H_3PO_4$ | 29.166 | −0.444 | 15.18 | −0.7 | 5–35 | 25–90 | 2.57 | 1.68 |
| | 28.189 | −0.720 | 17.50 | −0.1 | 35–65 | 25–90 | 11.85 | 2.74 |
| | 31.351 | 1.832 | 2.14 | −2.3 | 65–80 | 25–90 | 29.01 | 2.37 |
| $H_4P_2O_7$ | 43.853 | −2.000 | 193.66 | 0.5 | 8–18 | 20–90 | 4.39 | 1.98 |
| | 68.963 | −1.588 | 44.30 | 0.1 | 18–38 | 20–90 | 24.11 | 2.74 |
| $H_2SO_4$ | 28.559 | 0.615 | −41.60 | −1.2 | 0–20 | 1–85 | 3.43 | 1.92 |
| | 17.150 | 0.852 | 19.09 | −1.0 | 20–60 | 1–85 | 8.21 | 2.19 |
| | 14.701 | 0.126 | 26.29 | −0.5 | 60–70 | 1–85 | 35.33 | 3.04 |
| $H_2SiF_6$ | 8.682 | 2.802 | 27.00 | −2.0 | 5–30 | 1–90 | 3.03 | 1.70 |
| | 6.323 | 1.737 | 42.84 | −1.4 | 30–45 | 1–90 | 5.04 | 1.38 |
| $Hg(NO)_3$ | −6.284 | 7.101 | 24.51 | −15.4 | 18–35 | 15–45 | 2.01 | 1.58 |
| | −9.710 | 4.381 | 36.42 | −8.6 | 35–55 | 15–45 | 2.16 | 1.36 |
| | −29.748 | 2.747 | 79.07 | −5.9 | 55–65 | 15–45 | 8.44 | 2.41 |
| KBr | −8.457 | 2.062 | 8.53 | −0.6 | 5–40 | 20–90 | 9.03 | 0.97 |
| $K_2CO_3$ | 16.280 | 3.351 | 21.83 | −2.4 | 0–45 | 25–100 | 10.64 | 1.71 |
| KCl | −9.372 | 3.473 | 12.18 | −1.3 | 0–25 | 15–90 | 0.89 | 1.12 |
| $K_2CrO_4$ | 6.170 | 1.681 | 15.15 | −1.5 | 5–40 | 1–90 | 2.67 | 1.22 |
| $K_2Cr_2O_7$ | 2.593 | 1.093 | 4.03 | −1.0 | 5–30 | 30–60 | 0.19 | 0.24 |
| KF | 20.006 | 0.769 | 30.45 | −0.9 | 0–44 | 20–90 | 6.55 | 1.27 |
| $KHCO_3$ | 18.054 | 0.486 | −0.95 | 0.5 | 0–30 | 20–80 | 1.61 | 1.00 |
| $KHSO_4$ | 7.733 | 0.433 | 23.46 | −1.0 | 5–35 | 25–90 | 1.82 | 1.52 |
| $KH_2PO_4$ | 12.392 | 1.588 | 60.23 | −2.0 | 5–35 | 18–75 | 2.33 | 1.97 |
| $K_2HPO_4$ | 15.232 | 6.850 | −15.50 | −8.5 | 5–20 | 25–75 | 1.26 | 1.09 |
| | −0.121 | 1.958 | 113.60 | −3.9 | 20–40 | 25–75 | 6.35 | 2.42 |
| | 37.972 | 2.389 | 21.65 | −4.8 | 40–55 | 25–75 | 22.71 | 2.62 |
| KI | −11.212 | 2.799 | 8.36 | −1.6 | 5–55 | 1–90 | 1.82 | 1.46 |
| $KNO_3$ | −9.306 | 3.912 | 9.72 | −2.3 | 0–30 | 10–90 | 0.89 | 0.81 |

(*Continued*)

**TABLE 5.3 (*Continued*)**
**Coefficients of Equation 5.106 to Calculate Dynamic Viscosity, Pa s, of Multicomponent Solutions**

| Electrolyte | $A_0 \cdot 10^3$ | $A_1 \cdot 10^4$ | $A_2 \cdot 10^5$ | $A_3 \cdot 10^6$ | Range $c$ (%) from $c_{min}$ to $c_{max}$ | Range $t$ (°C) from $t_{min}$ to $t_{max}$ | $S_\eta \cdot 10^3$ | Δ (%) |
|---|---|---|---|---|---|---|---|---|
| KOH | 20.601 | 2.328 | −8.95 | −1.5 | 0–20 | 1–100 | 2.05 | 0.97 |
| | 17.720 | 0.584 | 36.05 | −0.4 | 20–40 | 1–100 | 14.69 | 2.64 |
| | 18.104 | −1.153 | 49.74 | 0.4 | 40–50 | 1–100 | 12.94 | 1.75 |
| $K_3PO_4$ | 27.366 | 1.750 | 28.46 | −2.9 | 5–35 | 25–90 | 3.18 | 1.54 |
| | 7.902 | 3.027 | 75.42 | −4.0 | 35–55 | 25–90 | 24.33 | 2.89 |
| $K_4P_2O_7$ | 32.204 | −4.698 | 28.49 | 4.3 | 0–25 | 20–90 | 3.07 | 2.16 |
| KSCN | −10.055 | 2.954 | 17.40 | −1.8 | 4–44 | 20–90 | 1.09 | 1.00 |
| | −13.496 | 1.259 | 37.19 | −0.6 | 44–64 | 20–90 | 2.72 | 1.65 |
| $K_2SO_4$ | 10.255 | 2.791 | −29.48 | −1.4 | 0–16 | 1–100 | 1.52 | 0.80 |
| | 13.671 | 1.375 | −39.16 | −0.4 | 3–10 | 100–200 | 0.09 | 0.29 |
| $LaCl_3$ | 23.461 | 2.082 | 18.65 | −2.0 | 1–25 | 15–55 | 2.96 | 1.14 |
| | 14.972 | 0.029 | 75.66 | −0.3 | 25–40 | 15–55 | 7.96 | 1.24 |
| LiBr | 8.805 | 1.595 | 8.59 | −1.1 | 5–30 | 20–90 | 1.07 | 0.92 |
| | 4.374 | 1.436 | 30.42 | −1.3 | 30–45 | 20–90 | 1.95 | 1.36 |
| | 10.947 | 1.385 | 10.42 | −0.3 | 25–35 | 1–130 | 3.56 | 1.88 |
| | 3.414 | 0.864 | 39.37 | −0.2 | 35–50 | 1–130 | 7.60 | 2.33 |
| LiCNS | 7.209 | 1.452 | 32.13 | −1.9 | 0–23 | 20–90 | 1.64 | 1.54 |
| LiCl | 30.387 | 2.107 | 23.06 | −1.4 | 0–30 | 20–90 | 5.18 | 1.60 |
| LiI | 8.220 | 0.341 | −2.61 | −0.3 | 5–30 | 1–90 | 2.87 | 1.51 |
| | −3.035 | 0.764 | 28.52 | −0.3 | 30–55 | 1–90 | 4.58 | 1.78 |
| $LiNO_3$ | 11.451 | 0.867 | 30.26 | −0.4 | 5–35 | 20–90 | 2.26 | 1.41 |
| | 7.328 | 0.550 | 47.77 | −0.9 | 35–56 | 20–90 | 6.89 | 1.89 |
| LiOH | 69.098 | −9.377 | −93.15 | 2.9 | 0–10 | 1–90 | 13.29 | 1.89 |
| $Li_2SO_4$ | 57.632 | −1.278 | 30.02 | 0.2 | 5–20 | 25–275 | 1.44 | 1.02 |
| $MgBr_2$ | 7.442 | 3.779 | 24.32 | −3.9 | 5–30 | 25–75 | 2.34 | 1.65 |
| | 2.306 | 0.544 | 74.66 | −1.4 | 30–50 | 25–75 | 9.84 | 2.34 |
| $MgCl_2$ | 38.205 | 1.239 | 86.44 | −0.1 | 0–28 | 20–100 | 5.56 | 1.41 |
| $Mg(ClO_4)_2$ | 13.233 | 1.198 | 8.63 | −0.8 | 5–30 | 25–75 | 1.58 | 1.01 |
| $MgCrO_4$ | 46.170 | −0.729 | 48.02 | −1.4 | 5–20 | 25–80 | 1.30 | 0.92 |
| | 42.740 | −3.020 | 107.67 | 0.2 | 20–35 | 25–80 | 9.47 | 1.88 |
| $MgCr_2O_7$ | 25.440 | −0.848 | 5.39 | 0.3 | 0–30 | 25–80 | 1.95 | 1.48 |
| | 8.480 | −0.626 | 68.86 | −0.4 | 30–45 | 25–80 | 3.98 | 1.31 |
| | −39.610 | −2.122 | 187.72 | 0.7 | 45–55 | 25–80 | 27.28 | 2.71 |
| $MgI_2$ | 6.495 | 1.119 | 17.67 | −1.4 | 5–40 | 25–90 | 1.71 | 1.47 |
| | 0.410 | −0.225 | 44.19 | −0.4 | 40–55 | 25–90 | 4.16 | 1.77 |
| $Mg(NO_3)_2$ | 20.223 | 1.483 | 21.11 | −0.5 | 5–25 | 25–80 | 1.48 | 0.95 |
| | 17.027 | 0.302 | 55.48 | −0.2 | 25–40 | 25–80 | 3.98 | 1.04 |
| | 28.570 | −3.709 | 121.09 | 0.3 | 60–75 | 100–150 | 48.07 | 4.14 |

(*Continued*)

**TABLE 5.3** (*Continued*)
**Coefficients of Equation 5.106 to Calculate Dynamic Viscosity, Pa s, of Multicomponent Solutions**

| Electrolyte | $A_0 \cdot 10^3$ | $A_1 \cdot 10^4$ | $A_2 \cdot 10^5$ | $A_3 \cdot 10^6$ | Range $c$ (%) from $c_{min}$ to $c_{max}$ | Range $t$ (°C) from $t_{min}$ to $t_{max}$ | $S_\eta \cdot 10^3$ | Δ (%) |
|---|---|---|---|---|---|---|---|---|
| $MgSO_4$ | 66.760 | −4.182 | 122.86 | 1.6 | 0–10 | 25–150 | 1.89 | 1.16 |
| $MnCl_2$ | 22.098 | 5.320 | 8.50 | −5.5 | 5–25 | 25–75 | 2.48 | 1.29 |
| | 7.0480 | 6.521 | 69.46 | −7.2 | 25–40 | 25–75 | 8.08 | 1.82 |
| $Mn(NO_3)_2$ | 24.590 | −1.037 | 6.24 | 1.6 | 10–20 | 30–90 | 0.84 | 0.87 |
| | 18.910 | 0.379 | 30.73 | −0.3 | 20–40 | 30–90 | 2.96 | 1.64 |
| | 0.580 | −0.698 | 88.30 | 0.1 | 40–50 | 30–90 | 8.36 | 2.43 |
| $MnSO_4$ | 41.830 | −2.129 | 86.89 | 0.8 | 5–35 | 20–80 | 14.80 | 1.38 |
| $NH_3$ | 39.710 | −3.673 | −124.01 | 3.2 | 2–16 | 10–100 | 1.74 | 1.89 |
| $NH_4Br$ | −11.356 | 2.240 | 13.31 | −1.0 | 5–50 | 20–75 | 1.07 | 1.18 |
| $NH_4Cl$ | −5.938 | 3.901 | −8.07 | −2.2 | 0–30 | 1–100 | 1.92 | 1.21 |
| $NH_4ClO_4$ | −8.993 | 1.967 | 16.50 | −1.2 | 5–35 | 25–90 | 0.75 | 0.82 |
| $(NH_4)_2CrO_4$ | 5.976 | 1.781 | 10.7 | −1.5 | 5–40 | 10–90 | 1.07 | 1.09 |
| $(NH_4)_2Cr_2O_7$ | −4.757 | 2.196 | 16.24 | −1.7 | 5–45 | 25–90 | 0.80 | 0.92 |
| $NH_4F$ | 31.99 | −1.035 | 12.84 | −0.7 | 5–45 | 25–75 | 3.38 | 1.51 |
| $NH_4HCO_3$ | 15.027 | 3.392 | 28.32 | −0.5 | 2–16 | 15–100 | 1.01 | 0.61 |
| $NH_4H_2PO_4$ | 14.993 | 2.967 | 49.24 | −3.3 | 5–25 | 20–75 | 1.66 | 1.00 |
| | 21.494 | 2.573 | 17.40 | −2.4 | 25–50 | 20–75 | 2.36 | 0.89 |
| $NH_4HSO_4$ | 12.190 | −0.411 | 15.95 | −0.2 | 5–45 | 20–75 | 1.50 | 1.10 |
| | 0.266 | −0.175 | 40.47 | −0.4 | 45–65 | 20–75 | 2.36 | 0.89 |
| $NH_4I$ | −15.379 | 4.497 | 8.98 | −3.6 | 5–40 | 10–75 | 0.85 | 1.05 |
| | −15.466 | 2.793 | 17.41 | −1.8 | 40–60 | 10–75 | 0.96 | 0.90 |
| $NH_4NO_3$ | −13.765 | 3.956 | 13.88 | −2.1 | 5–40 | 25–95 | 1.50 | 1.32 |
| | −11.476 | 2.000 | 26.34 | −1.0 | 40–70 | 25–95 | 2.41 | 1.56 |
| $(NH_4)_2SO_4$ | 15.433 | 0.744 | 12.29 | 0.0 | 0–50 | 20–90 | 3.54 | 1.72 |
| $NaBr$ | −0.002 | 1.156 | 27.49 | −0.7 | 10–40 | 20–90 | 2.44 | 1.61 |
| $Na_2CO_3$ | 48.490 | 3.803 | 49.64 | −5.8 | 0–26 | 20–10 | 6.50 | 2.84 |
| | 56.170 | −1.205 | −8.85 | 0.2 | 0–15 | 100–270 | 0.22 | 0.67 |
| $NaCl$ | 13.953 | 0.769 | 27.44 | −0.2 | 2–26 | 20–150 | 1.58 | 1.02 |
| $NaClO_4$ | −1.376 | 1.427 | 25.19 | −1.3 | 5–50 | 25–90 | 3.91 | 1.44 |
| | −7.25 | −0.403 | 52.06 | −0.3 | 50–60 | 25–90 | 3.62 | 1.35 |
| | −25.392 | 0.868 | 78.65 | −1.6 | 60–70 | 25–90 | 14.91 | 2.52 |
| $Na_2CrO_4$ | 39.280 | −1.256 | 5.72 | 0.4 | 5–25 | 20–200 | 2.20 | 1.64 |
| | 37.850 | −1.735 | 23.28 | 0.5 | 25–40 | 20–200 | 12.24 | 2.59 |
| $Na_2Cr_2O_7$ | 3.349 | 2.538 | 18.78 | −2.8 | 5–35 | 25–90 | 1.80 | 1.56 |
| | −0.580 | −0.897 | 56.37 | 0.1 | 35–55 | 25–90 | 5.50 | 1.96 |
| $NaF$ | 77.330 | −10.252 | −233.26 | 15.2 | 0–34 | 5–55 | 1.05 | 0.67 |
| $NaHCO_3$ | 28.439 | 5.141 | 107.23 | −1.4 | 0–10 | 20–50 | 0.76 | 0.61 |

(*Continued*)

**TABLE 5.3 (*Continued*)**
**Coefficients of Equation 5.106 to Calculate Dynamic Viscosity, Pa s, of Multicomponent Solutions**

| Electrolyte | $A_0 \cdot 10^3$ | $A_1 \cdot 10^4$ | $A_2 \cdot 10^5$ | $A_3 \cdot 10^6$ | Range $c$ (%) from $c_{min}$ to $c_{max}$ | Range $t$ (°C) from $t_{min}$ to $t_{max}$ | $S_\eta \cdot 10^3$ | Δ (%) |
|---|---|---|---|---|---|---|---|---|
| $NaH_2PO_4$ | 38.732 | −0.487 | 23.82 | −2.3 | 5–25 | 25–75 | 2.76 | 1.59 |
| | 33.370 | −2.159 | 66.75 | −1.1 | 25–50 | 25–75 | 20.48 | 2.61 |
| NaHS | 24.173 | 4.015 | −59.75 | −2.3 | 5–25 | 25–75 | 2.76 | 1.59 |
| | 29.410 | −0.808 | 31.71 | 0.0 | 15–30 | 25–90 | 4.15 | 1.84 |
| $NaHSO_4$ | −4.193 | 6.593 | 49.35 | −6.8 | 5–20 | 25–90 | 1.32 | 1.35 |
| | 19.220 | −1.506 | 38.04 | 0.0 | 20–55 | 25–90 | 6.66 | 1.36 |
| NaI | −1.448 | 1.254 | 11.37 | −0.6 | 10–40 | 10–90 | 1.40 | 1.08 |
| | −7.396 | 0.189 | 35.76 | 0.1 | 40–55 | 10–90 | 4.10 | 1.79 |
| $Na_2MoO_4$ | 35.235 | −1.736 | 13.13 | 0.5 | 0–40 | 15–200 | 6.75 | 2.46 |
| $NaNO_2$ | 10.377 | −0.052 | 29.16 | −0.5 | 5–30 | 25–90 | 1.47 | 1.05 |
| | 8.890 | −0.374 | 42.79 | −0.7 | 30–50 | 25–90 | 2.88 | 1.45 |
| $NaNO_3$ | 5.063 | 0.990 | 31.52 | −0.6 | 5–50 | 10–90 | 6.16 | 1.75 |
| NaOH | 71.137 | −7.223 | 82.71 | 4.3 | 0–25 | 25–100 | 7.94 | 2.50 |
| | 91.390 | −5.538 | 2.71 | 1.8 | 25–40 | 25–100 | 45.07 | 3.00 |
| $Na_3PO_4$ | 41.674 | 5.000 | 90.48 | −7.4 | 5–20 | 50–90 | 1.29 | 1.04 |
| $Na_2S$ | 48.360 | −0.226 | 64.48 | −1.5 | 5–27 | 25–75 | 2.70 | 1.03 |
| NaSCN | 2.523 | 1.565 | 35.39 | −1.4 | 4–40 | 20–90 | 3.84 | 1.59 |
| | 4.150 | −1.177 | 54.98 | 0.2 | 40–55 | 20–90 | 11.64 | 2.27 |
| $Na_2SO_3$ | 48.889 | −0.670 | 27.05 | −2.4 | 5–20 | 25–90 | 1.90 | 0.99 |
| $Na_2SO_4$ | 37.068 | −0.708 | 20.26 | 0.1 | 6–20 | 25–200 | 1.24 | 0.77 |
| $Na_2S_2O_3$ | 18.793 | 0.320 | 40.30 | −1.7 | 5–30 | 25–90 | 2.52 | 1.30 |
| | 20.230 | −2.210 | 69.38 | −0.5 | 30–50 | 25–90 | 11.25 | 2.20 |
| $Na_2SiO_3$ | 22.370 | 11.481 | 112.23 | 12.7 | 5–20 | 25–90 | 4.86 | 2.76 |
| $Na_2WO_4$ | 21.590 | −0.671 | 2.00 | 0.2 | 5–40 | 20–200 | 2.74 | 1.31 |
| $NiCl_2$ | 20.085 | 8.212 | −5.97 | −7.2 | 5–15 | 20–100 | 1.50 | 1.36 |
| | 28.000 | 1.843 | 40.56 | −1.8 | 15–30 | 20–100 | 2.74 | 1.19 |
| | 15.578 | 0.257 | 50.30 | −0.4 | 25–45 | 20–95 | 7.07 | 1.80 |
| $Ni(NO_3)_2$ | 13.170 | 3.488 | 18.99 | −2.5 | 5–25 | 20–95 | 2.17 | 1.31 |
| $NiSO_4$ | 62.792 | −9.820 | 57.03 | 11.9 | 0–26 | 25–55 | 5.02 | 1.54 |
| | −3.318 | 17.942 | 2.28 | 14.1 | 0–26 | 55–100 | 4.70 | 2.25 |
| $Pb(NO_3)_2$ | 0.440 | 0.788 | 24.98 | −0.4 | 4–30 | 25–100 | 0.81 | 0.75 |
| $Rb_2CO_3$ | 6.406 | 1.063 | 22.55 | −0.4 | 10–25 | 20–80 | 1.14 | 1.00 |
| | 5.375 | 0.636 | 29.58 | −0.4 | 25–55 | 20–80 | 7.38 | 2.29 |
| RbCl | −7.376 | 1.823 | 4.99 | 0.1 | 0–41 | 15–90 | 1.27 | 1.23 |
| $RbNO_3$ | −10.135 | 1.634 | 16.75 | −0.4 | 5–35 | 20–90 | 0.92 | 1.00 |
| RbOH | 2.245 | 0.667 | 3.39 | −0.2 | 20–48 | 20–80 | 2.79 | 1.20 |
| $Rb_2SO_4$ | 5.4220 | 2.053 | −1.93 | −1.2 | 5–30 | 20–90 | 0.78 | 0.77 |

*(Continued)*

**TABLE 5.3 (*Continued*)**
**Coefficients of Equation 5.106 to Calculate Dynamic Viscosity, Pa s, of Multicomponent Solutions**

| Electrolyte | $A_0 \cdot 10^3$ | $A_1 \cdot 10^4$ | $A_2 \cdot 10^5$ | $A_3 \cdot 10^6$ | Range c (%) from $c_{min}$ to $c_{max}$ | Range t (°C) from $t_{min}$ to $t_{max}$ | $S_\eta \cdot 10^3$ | Δ (%) |
|---|---|---|---|---|---|---|---|---|
| $SrBr_2$ | 9.333 | 1.481 | 9.60 | −1.5 | 5–30 | 25–75 | 1.14 | 1.01 |
| | −4.565 | 0.901 | 63.03 | −1.0 | 30–45 | 25–75 | 1.99 | 1.12 |
| $SrCl_2$ | 15.542 | 0.168 | 29.70 | 0.5 | 5–35 | 20–70 | 2.48 | 1.36 |
| $Sr(ClO_4)_2$ | 7.753 | −0.026 | 1.47 | 1.8 | 5–30 | 5–50 | 2.25 | 1.42 |
| | −5.144 | 0.364 | 46.49 | 0.3 | 30–55 | 5–50 | 8.88 | 2.36 |
| | −33.980 | −1.056 | 105.07 | 1.5 | 55–65 | 5–50 | 22.07 | 2.49 |
| $Sr(NO_3)_2$ | 12.286 | −0.436 | 37.07 | 0.2 | 5–40 | 20–75 | 4.46 | 1.83 |
| $ZnCl_2$ | 28.334 | 0.539 | −9.37 | −1.5 | 5–40 | 25–75 | 3.28 | 1.88 |
| | 17.340 | −2.613 | 37.99 | 1.8 | 40–55 | 25–75 | 14.03 | 2.83 |
| | −0.120 | −9.067 | 105.06 | 6.5 | 55–70 | 25–75 | 99.26 | 6.60 |
| $ZnI_2$ | 8.537 | −0.588 | 17.52 | 0.1 | 5–55 | 25–90 | 2.28 | 1.24 |
| | 2.200 | −0.846 | 33.53 | −0.1 | 55–70 | 25–90 | 4.76 | 1.68 |
| $Zn(NO_3)_2$ | 14.865 | 2.768 | 6.89 | −1.7 | 10–25 | 25–90 | 1.75 | 1.38 |
| | 16.989 | 0.027 | 36.08 | 0.0 | 25–45 | 25–90 | 5.63 | 2.39 |
| $ZnSO_4$ | 76.444 | −19.768 | 65.84 | 17.0 | 0–30 | 25–60 | 8.03 | 2.39 |

*Note:* $S_\lambda$, a mean squared error of calculation, S cm²/mol and Δ, the average relative error of calculation, %.

2. With regard to numerical evaluation of summands in Equation 5.68, it can be seen that dipole–dipole interactions among ion complexes practically do not occur. This points to the fact that in the electrolyte concentration area at $\varepsilon_S \approx$ 80–40, electrical conductivity depends only on ions and complexes in the solution, and at $\varepsilon_S \approx$ 40–20, it depends on ion and ion-aqueous clusters and fragments of hydration spheres of ions. The solution structure is completely absent.
3. Numerical evaluation of summands in Equation 5.68 also demonstrates that formation of various associates or any kind of supermolecular formations neither influences nor determines the electrical conductivity mechanism. All stated formations get simply destroyed at application of weak electrical field.
4. Let us consider the nature of dependence of specific electrical conductivity, κ, $\Omega^{-1}$ $cm^{-1}$, from solution concentration, *C*, mol/dm³, expressed by a peak curve, given in Figure 5.1.

In the area of low concentrations κ increases with concentration growth (for the number of ions in 1 m³ of the solution increases with concentration growth), reaches the maximum value and decreases after that. According to the contemporary views, this decrease in strong electrolyte solutions is explained by retardation of motion of ions

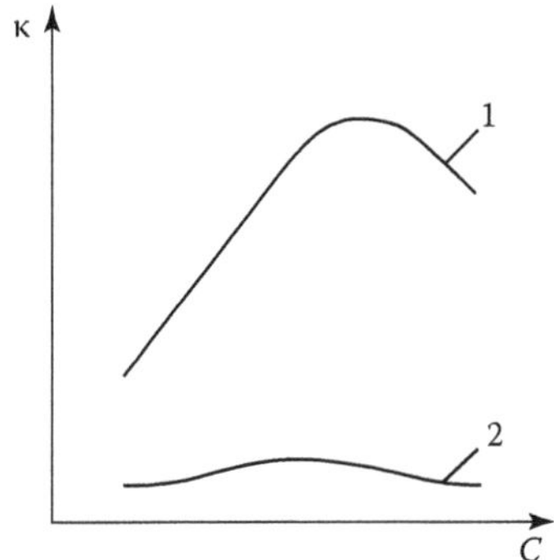


**FIGURE 5.1** Specific electrical conductivity and electrolyte solution concentration dependence: 1—strong electrolyte and 2—weak electrolyte.

due to the relaxation effect and the electrophoretic retardation effect, and in solutions of weak electrolytes by decrease of the dissociation degree [87].

We will explain inflections on curves 1 and 2 of the specific electrical conductivity and electrolyte solution concentration dependence, Figure 5.1, from the positions of principles stated earlier. Dielectric permittivity of the solution decreases to values of $\varepsilon_S \approx 20$–5 at increase of electrolyte concentration in the solution and its approach to the third area of the concentration structural transition, Figure 1.1. Formation of supramolecular assemblies in a host solution occurs to an appreciable extent with formation of polymer aqueous melts of crystalline hydrates or salts, hydroxides, and acids because water molecules are already not enough for their entire intraspheric hydration. Water structure is completely destroyed. From our point of view, supramolecular host–guest structure interchanges its places, and zone of coordination chemistry regularities appears, where water-host acts as a ligand-guest. Mobility of supramolecular formations decreases substantially, and relaxation and electrophoretic forces increase at further decrease of $\varepsilon_S$. Accordingly, decrease of specific electrical conductivity values takes place, which is characterized by the break in Figure 5.1.

Table 5.4 demonstrates components of formulas (5.104) and (5.105) for calculation of electrical conductivity as per formula (5.87) for a number of electrolyte solutions. The first column contains value $m$, mol/kg $H_2O$; the second column $c_k$, mol/dm$^3$; the third column $\Lambda$, S cm$^2$/mol; the fourth column viscosity value $\eta$, Pa s calculated as per dependence (5.106); the fifth column values of relaxation force $\aleph$; the sixth column $P_\Lambda$; the seventh column values of electrophoretic retardation $E$; and the eighth column values of dielectric permittivity of the solution taken from Table 3.9. Auxiliary values required for calculation are given below the electrolyte name: values $a$, nm, of minimum approach of electrostatic forces among ion complexes taken from Table 3.9; $\lambda_i^0$, S cm$^2$/mol, electrical conductivity of the ion of sort $i$ at infinite dilution; $\lambda_j^0$, S cm$^2$/mol, electrical conductivity of the ion of sort $j$ at infinite dilution; and $q^*$ determined by formula (5.92) and $B$, the averaged values of which were obtained for each electrolyte from Table 3.9. Values $\lambda_i^0$ and $\lambda_j^0$ are calculated as per dependence (5.97). Electrical conductivity $\Lambda$ of electrolytes for each $m$ was interpolated based on data of [25]. For calculation of $P_\Lambda$ as per (5.96), values $n_{hi}$ and $n_{hj}$ were taken from Tables 3.2 and 3.4, and $R$ and $R_{H_2O}$ from Table 3.9.

## TABLE 5.4
## Components of Formula (5.103) to Calculate Electrical Conductivity, S cm²/mol, According to Formula (5.87)

| Electrolyte | $m$ | $c_k$ | $\Lambda$ | $\eta$ | $\aleph$ | $P_\Lambda$ | $E$ | $\varepsilon_S$ |
|---|---|---|---|---|---|---|---|---|
| $CaCl_2$: $a = 0.66198$ nm; $\lambda_i^0 = 59.80\,\mathrm{S\,cm^2\,mol^{-1}}$; $\lambda_j^0 = 76.26\,\mathrm{S\,cm^2\,mol^{-1}}$; $q^* = 4.17$; $B = 19.87$ | | | | | | | | |
| | 0.200 | 0.198 | 0.198 | 928 | 90.69 | 1699.38 | 0.045 | 76.60 |
| | 0.300 | 0.297 | 87.14 | 949 | 48.84 | 1743.25 | 0.078 | 73.30 |
| | 0.400 | 0.395 | 83.65 | 972 | 52.29 | 1777.00 | 0.120 | 70.30 |
| | 0.500 | 0.493 | 80.34 | 997 | 55.55 | 1815.44 | 0.172 | 67.60 |
| | 0.600 | 0.590 | 77.12 | 1024 | 58.71 | 1848.61 | 0.232 | 65.20 |
| | 0.700 | 0.687 | 74.16 | 1054 | 61.60 | 1884.41 | 0.302 | 63.00 |
| | 0.800 | 0.783 | 71.28 | 1085 | 64.40 | 1918.56 | 0.383 | 60.90 |
| | 0.900 | 0.879 | 68.86 | 1120 | 66.73 | 1953.20 | 0.471 | 59.10 |
| | 1.000 | 0.975 | 66.44 | 1157 | 69.05 | 1990.57 | 0.574 | 57.30 |
| | 1.250 | 1.208 | 61.93 | 1240 | 73.19 | 2064.71 | 0.940 | 52.60 |
| | 1.500 | 1.437 | 58.07 | 1336 | 76.57 | 2134.45 | 1.419 | 48.70 |
| | 1.750 | 1.663 | 54.51 | 1449 | 79.46 | 2208.53 | 2.089 | 45.00 |
| | 2.000 | 1.886 | 51.06 | 1579 | 82.01 | 2284.69 | 2.991 | 41.60 |
| | 2.250 | 2.106 | 47.66 | 1732 | 84.36 | 2358.44 | 4.045 | 38.80 |
| | 2.500 | 2.323 | 44.38 | 1909 | 85.94 | 2434.17 | 5.737 | 35.60 |
| | 2.750 | 2.520 | 41.14 | 2009 | 86.89 | 2474.12 | 8.030 | 33.00 |
| | 3.000 | 2.731 | 37.88 | 2235 | 87.68 | 2550.25 | 10.498 | 30.70 |
| | 3.250 | 2.940 | 34.63 | 2499 | 87.95 | 2626.14 | 13.479 | 28.60 |
| | 3.500 | 3.125 | 31.46 | 2649 | 86.76 | 2666.49 | 17.841 | 26.70 |
| | 3.750 | 3.305 | 28.67 | 2811 | 84.57 | 2703.97 | 22.816 | 25.10 |
| | 4.000 | 3.506 | 26.19 | 3180 | 82.83 | 2781.44 | 27.042 | 23.70 |
| | 4.250 | 3.681 | 23.97 | 3389 | 78.02 | 2823.28 | 34.067 | 22.30 |
| | 4.500 | 3.851 | 22.12 | 3617 | 72.19 | 2860.83 | 41.750 | 21.10 |
| | 4.750 | 4.046 | 20.36 | 4139 | 68.93 | 2941.64 | 46.773 | 20.10 |
| | 5.000 | 4.211 | 18.73 | 4436 | 62.66 | 2980.57 | 54.667 | 19.20 |
| | 5.250 | 4.373 | 17.20 | 4762 | 54.64 | 3019.38 | 64.222 | 18.30 |
| | 5.500 | 4.532 | 15.70 | 5119 | 47.83 | 3058.65 | 72.528 | 17.60 |
| | 5.750 | 4.689 | 14.25 | 5510 | 39.62 | 3101.56 | 82.188 | 16.90 |
| | 6.000 | 4.843 | 12.81 | 5940 | 32.06 | 3142.36 | 91.194 | 16.30 |
| $CsCl$: $a = 0.66865$ nm; $\lambda_i^0 = 77.55\,\mathrm{S\,cm^2\,mol^{-1}}$; $\lambda_j^0 = 76.26\,\mathrm{S\,cm^2\,mol^{-1}}$; $q^* = 2.00$; $B = 5.08$ | | | | | | | | |
| | 0.100 | 0.100 | 130.48 | 887 | 23.33 | 98.00 | 0.000 | 76.90 |
| | 0.200 | 0.197 | 128.61 | 885 | 25.20 | 98.72 | 0.000 | 75.60 |
| | 0.300 | 0.296 | 126.78 | 881 | 27.03 | 107.47 | 0.000 | 74.30 |
| | 0.400 | 0.391 | 125.00 | 879 | 28.81 | 110.97 | 0.001 | 73.10 |
| | 0.500 | 0.489 | 123.29 | 875 | 30.52 | 119.34 | 0.001 | 71.90 |
| | 0.600 | 0.583 | 121.59 | 873 | 32.22 | 124.19 | 0.001 | 70.80 |
| | 0.700 | 0.680 | 119.94 | 869 | 33.87 | 132.15 | 0.001 | 69.70 |
| | 0.800 | 0.772 | 118.37 | 867 | 35.44 | 136.81 | 0.001 | 68.60 |

(Continued)

**TABLE 5.4 (*Continued*)**
**Components of Formula (5.103) to Calculate Electrical Conductivity, S cm²/mol, According to Formula (5.87)**

| Electrolyte | $m$ | $c_k$ | $\Lambda$ | $\eta$ | $\aleph$ | $P_\Lambda$ | $E$ | $\varepsilon_S$ |
|---|---|---|---|---|---|---|---|---|
| | 0.900 | 0.863 | 117.02 | 865 | 36.79 | 141.55 | 0.002 | 67.60 |
| | 1.000 | 0.953 | 115.80 | 863 | 38.01 | 146.26 | 0.002 | 66.60 |
| | 1.250 | 1.180 | 113.50 | 857 | 40.31 | 161.23 | 0.003 | 64.20 |
| | 1.500 | 1.403 | 111.65 | 850 | 42.16 | 176.28 | 0.004 | 62.00 |
| | 1.750 | 1.626 | 109.98 | 844 | 43.82 | 193.06 | 0.005 | 60.20 |
| | 2.000 | 1.832 | 108.38 | 840 | 45.42 | 204.53 | 0.007 | 58.50 |
| | 2.250 | 2.037 | 106.97 | 836 | 46.83 | 217.26 | 0.009 | 57.10 |
| | 2.500 | 2.262 | 105.68 | 830 | 48.12 | 237.83 | 0.010 | 55.90 |
| | 2.750 | 2.465 | 104.44 | 826 | 49.36 | 252.11 | 0.012 | 55.10 |
| | 3.000 | 2.667 | 103.18 | 822 | 50.62 | 267.16 | 0.014 | 54.40 |
| | 3.250 | 2.840 | 101.90 | 820 | 51.89 | 274.91 | 0.016 | 53.70 |
| | 3.500 | 3.039 | 100.63 | 2306 | 53.17 | 290.95 | 0.006 | 53.10 |
| | 3.750 | 3.239 | 99.39 | 2464 | 54.41 | 307.89 | 0.007 | 52.30 |
| | 4.000 | 3.403 | 98.18 | 2549 | 55.62 | 316.41 | 0.007 | 51.50 |
| | 4.250 | 3.602 | 97.02 | 2730 | 56.78 | 333.94 | 0.008 | 50.70 |
| | 4.500 | 3.762 | 95.90 | 2827 | 57.90 | 342.96 | 0.008 | 49.90 |
| | 4.750 | 3.919 | 94.87 | 2928 | 58.93 | 352.16 | 0.009 | 49.10 |
| | 5.000 | 4.118 | 93.95 | 3146 | 59.85 | 371.53 | 0.010 | 48.30 |
| | 5.250 | 4.272 | 93.00 | 3262 | 60.80 | 380.83 | 0.010 | 47.50 |
| | 5.500 | 4.425 | 92.03 | 3385 | 61.77 | 391.22 | 0.011 | 46.80 |
| | 5.750 | 4.576 | 90.85 | 3513 | 62.95 | 401.49 | 0.012 | 46.00 |
| | 6.000 | 4.725 | 89.59 | 3647 | 64.21 | 411.34 | 0.012 | 45.30 |
| | 6.250 | 4.874 | 88.10 | 3788 | 65.70 | 422.36 | 0.013 | 44.60 |
| | 6.500 | 5.020 | 86.59 | 3936 | 67.21 | 432.61 | 0.014 | 43.80 |
| | 6.750 | 5.166 | 85.03 | 4092 | 68.76 | 443.56 | 0.015 | 42.70 |
| | 7.000 | 5.311 | 83.46 | 4255 | 70.33 | 454.87 | 0.017 | 41.70 |
| | 7.250 | 5.478 | 81.84 | 4426 | 71.95 | 470.32 | 0.019 | 40.60 |
| | 7.500 | 5.630 | 80.23 | 4606 | 73.56 | 483.51 | 0.021 | 39.60 |
| *CsNO₃*: $a = 0.630$ nm; $\lambda_i^0 = 77.55\,\text{S}\,\text{cm}^2\,\text{mol}^{-1}$; $\lambda_j^0 = 71.37\,\text{S}\,\text{cm}^2\,\text{mol}^{-1}$; $q^* = 2.00$; $B = 3.54$ | | | | | | | | |
| | 0.200 | 0.198 | 115.43 | 879 | 33.49 | 96.34 | 0.000 | 76.70 |
| | 0.400 | 0.390 | 107.97 | 872 | 40.95 | 104.28 | 0.000 | 75.20 |
| | 0.600 | 0.577 | 101.84 | 864 | 47.08 | 112.65 | 0.001 | 73.90 |
| | 0.800 | 0.762 | 100.93 | 858 | 47.99 | 122.42 | 0.001 | 72.70 |
| | 1.000 | 0.943 | 100.03 | 852 | 48.89 | 131.92 | 0.001 | 71.60 |
| *Cu(NO₃)₂*: $a = 0.73575$ nm; $\lambda_i^0 = 55.56\,\text{S}\,\text{cm}^2\,\text{mol}^{-1}$; $\lambda_j^0 = 71.37\,\text{S}\,\text{cm}^2\,\text{mol}^{-1}$; $q^* = 4.17$; $B = 17.69$ | | | | | | | | |
| | 0.003 | 0.003 | 48.60 | 890 | 78.33 | 38.60 | 0.000 | 78.40 |
| | 0.004 | 0.004 | 48.58 | 890 | 78.35 | 38.60 | 0.000 | 78.30 |
| | 0.005 | 0.005 | 48.57 | 890 | 78.36 | 39.32 | 0.000 | 78.30 |
| | 0.006 | 0.006 | 48.56 | 890 | 78.37 | 39.32 | 0.000 | 78.30 |

(*Continued*)

**TABLE 5.4 (*Continued*)**
**Components of Formula (5.103) to Calculate Electrical Conductivity, S cm²/mol, According to Formula (5.87)**

| Electrolyte | $m$ | $c_k$ | $\Lambda$ | $\eta$ | $\aleph$ | $P_\Lambda$ | $E$ | $\varepsilon_S$ |
|---|---|---|---|---|---|---|---|---|
| | 0.007 | 0.007 | 48.55 | 890 | 78.38 | 39.65 | 0.000 | 78.20 |
| | 0.008 | 0.008 | 48.54 | 890 | 78.39 | 39.65 | 0.000 | 78.20 |
| | 0.009 | 0.009 | 48.54 | 890 | 78.39 | 40.37 | 0.000 | 78.20 |
| | 0.010 | 0.010 | 48.53 | 890 | 78.40 | 40.37 | 0.000 | 78.20 |
| | 0.020 | 0.020 | 48.46 | 890 | 78.47 | 42.82 | 0.000 | 78.00 |
| | 0.030 | 0.030 | 48.24 | 900 | 78.69 | 45.65 | 0.001 | 77.90 |
| | 0.040 | 0.040 | 47.66 | 900 | 79.27 | 48.49 | 0.001 | 77.70 |
| | 0.050 | 0.050 | 47.09 | 900 | 79.84 | 50.97 | 0.001 | 77.50 |
| | 0.060 | 0.060 | 46.48 | 900 | 80.45 | 53.83 | 0.001 | 77.30 |
| | 0.070 | 0.069 | 45.86 | 900 | 81.07 | 34.75 | 0.001 | 77.20 |
| | 0.080 | 0.079 | 45.23 | 900 | 81.70 | 40.37 | 0.002 | 77.00 |
| | 0.090 | 0.090 | 44.60 | 911 | 82.33 | 62.10 | 0.002 | 76.80 |
| | 0.100 | 0.100 | 43.97 | 911 | 82.96 | 65.01 | 0.002 | 76.70 |
| | 0.200 | 0.199 | 40.84 | 937 | 86.09 | 84.11 | 0.004 | 75.00 |
| | 0.300 | 0.296 | 38.91 | 951 | 88.01 | 98.75 | 0.007 | 73.30 |
| | 0.400 | 0.394 | 37.47 | 984 | 89.45 | 122.57 | 0.009 | 71.70 |
| | 0.500 | 0.493 | 36.21 | 1022 | 90.71 | 151.94 | 0.012 | 70.10 |
| $H_2SO_4$: $a = 0.63879$ nm; $\lambda_i^0 = 363.30\,\mathrm{S\,cm^2\,mol^{-1}}$; $\lambda_i^0 = 51.18\,\mathrm{S\,cm^2\,mol^{-1}}$; $q^* = 5.34$; $B = 17.06$ | | | | | | | | |
| | 1.000 | 0.961 | 218.09 | 1356 | 196.35 | 3929.31 | 0.045 | 61.70 |
| | 1.500 | 1.417 | 196.61 | 1513 | 217.77 | 4120.20 | 0.102 | 54.40 |
| | 2.000 | 1.848 | 175.94 | 1646 | 238.34 | 4264.60 | 0.203 | 48.40 |
| | 2.500 | 2.280 | 157.75 | 1744 | 256.38 | 4471.64 | 0.352 | 43.90 |
| | 3.000 | 2.686 | 142.90 | 1907 | 270.95 | 4632.38 | 0.626 | 38.60 |
| | 3.500 | 3.081 | 129.62 | 2029 | 283.83 | 4795.31 | 1.030 | 34.60 |
| | 4.000 | 3.444 | 117.52 | 2161 | 295.39 | 4908.05 | 1.569 | 31.60 |
| | 4.500 | 3.821 | 107.09 | 2306 | 305.13 | 5084.69 | 2.258 | 29.00 |
| | 5.000 | 4.163 | 98.01 | 2464 | 313.03 | 5204.39 | 3.435 | 26.30 |
| | 5.500 | 4.495 | 89.63 | 2637 | 319.97 | 5327.52 | 4.879 | 24.20 |
| | 6.000 | 4.819 | 81.84 | 2730 | 326.16 | 5457.81 | 6.477 | 22.60 |
| | 6.500 | 5.135 | 74.93 | 2928 | 331.17 | 5582.32 | 8.379 | 21.20 |
| | 7.000 | 5.445 | 69.04 | 3034 | 334.91 | 5714.15 | 10.530 | 20.00 |
| | 7.500 | 5.709 | 64.01 | 3262 | 337.06 | 5784.04 | 13.407 | 18.90 |
| | 8.000 | 6.006 | 59.44 | 3385 | 338.98 | 5921.39 | 16.058 | 18.00 |
| | 8.500 | 6.254 | 55.33 | 3647 | 339.75 | 5988.76 | 19.397 | 17.20 |
| | 9.000 | 6.541 | 51.43 | 3788 | 340.66 | 6130.64 | 22.387 | 16.50 |
| | 9.500 | 6.776 | 47.68 | 3936 | 340.82 | 6200.51 | 25.984 | 15.90 |
| | 10.000 | 7.054 | 44.07 | 4092 | 341.00 | 6342.78 | 29.413 | 15.30 |
| | 10.500 | 7.279 | 41.31 | 4255 | 340.61 | 6415.20 | 32.563 | 14.90 |
| | 11.000 | 7.499 | 39.09 | 4426 | 339.30 | 6494.09 | 36.088 | 14.50 |
| | 11.500 | 7.713 | 37.05 | 4606 | 338.52 | 6568.34 | 38.913 | 14.20 |

(*Continued*)

**TABLE 5.4 (*Continued*)**
**Components of Formula (5.103) to Calculate Electrical Conductivity, S cm²/mol, According to Formula (5.87)**

| Electrolyte | $m$ | $c_k$ | $\Lambda$ | $\eta$ | $\aleph$ | $P_\Lambda$ | $E$ | $\varepsilon_S$ |
|---|---|---|---|---|---|---|---|---|
| | 12.000 | 7.922 | 35.02 | 4795 | 337.50 | 6643.48 | 41.961 | 13.90 |
| | 12.500 | 8.128 | 33.11 | 4994 | 336.10 | 6719.54 | 45.266 | 13.60 |
| | 13.000 | 8.329 | 31.20 | 5203 | 334.43 | 6796.53 | 48.847 | 13.30 |
| | 13.500 | 8.526 | 29.47 | 5422 | 333.86 | 6874.46 | 51.152 | 13.10 |
| | 14.000 | 8.721 | 27.75 | 5422 | 333.18 | 6953.34 | 53.554 | 12.90 |
| | 14.500 | 8.912 | 26.32 | 5808 | 332.11 | 7033.18 | 56.049 | 12.70 |
| | 15.000 | 9.100 | 24.95 | 6087 | 330.88 | 7114.01 | 58.650 | 12.50 |
| | 15.500 | 9.220 | 23.72 | 6384 | 327.38 | 7114.01 | 63.383 | 12.30 |
| | 16.000 | 9.402 | 22.71 | 6384 | 329.27 | 7195.83 | 62.499 | 12.20 |
| | 16.500 | 9.581 | 21.70 | 6698 | 329.85 | 7272.96 | 62.934 | 12.10 |
| | 17.000 | 9.758 | 20.81 | 6698 | 330.34 | 7356.74 | 63.333 | 12.00 |
| | 17.500 | 9.864 | 19.98 | 7032 | 330.48 | 7356.74 | 64.021 | 12.00 |
| | 18.000 | 10.036 | 19.15 | 7386 | 330.99 | 7441.56 | 67.210 | 11.90 |
| | 18.500 | 10.135 | 18.38 | 7386 | 328.89 | 7441.56 | 67.210 | 11.80 |
| | 19.000 | 10.303 | 17.64 | 7762 | 329.34 | 7527.43 | 67.497 | 11.70 |
| | 19.500 | 10.470 | 16.90 | 7762 | 329.84 | 7614.36 | 67.745 | 11.60 |
| | 20.000 | 10.560 | 16.20 | 8162 | 329.96 | 7607.77 | 68.315 | 11.60 |
| | 20.500 | 10.723 | 15.58 | 8162 | 330.40 | 7696.32 | 68.499 | 11.50 |
| | 21.000 | 10.808 | 14.96 | 8587 | 330.48 | 7696.32 | 69.042 | 11.50 |
| | 21.500 | 10.968 | 14.34 | 8587 | 330.97 | 7785.36 | 69.169 | 11.40 |
| | 22.000 | 11.049 | 13.82 | 9038 | 328.48 | 7785.36 | 72.179 | 11.30 |
| | 22.500 | 11.205 | 13.36 | 9038 | 331.39 | 7868.69 | 69.734 | 11.30 |
| | 23.000 | 11.282 | 12.89 | 9518 | 328.84 | 7868.69 | 72.755 | 11.20 |
| | 23.500 | 11.436 | 12.43 | 9518 | 331.82 | 7959.90 | 70.233 | 11.20 |
| | 24.000 | 11.508 | 12.04 | 9518 | 329.18 | 7959.90 | 73.257 | 11.10 |
| | 24.500 | 11.660 | 11.75 | 9518 | 332.09 | 8045.88 | 70.638 | 11.10 |
| | 25.000 | 11.729 | 11.45 | 9518 | 331.97 | 8045.88 | 71.056 | 11.10 |
| | 25.500 | 11.796 | 11.16 | 9518 | 331.86 | 8045.88 | 71.462 | 11.10 |
| | 26.500 | 12.009 | 10.62 | 10029 | 334.65 | 8139.30 | 71.056 | 11.10 |
| | 26.000 | 11.945 | 10.84 | 10029 | 334.80 | 8139.30 | 71.462 | 11.10 |
| | 27.000 | 12.155 | 10.40 | 10573 | 337.48 | 8227.39 | 69.213 | 11.10 |
| | 27.500 | 12.216 | 10.18 | 10573 | 334.90 | 8227.39 | 68.845 | 11.00 |
| *KBr*: $a = 0.64433$ nm; $\lambda_i^0 = 73.70\,\text{S}\,\text{cm}^2\,\text{mol}^{-1}$; $\lambda_i^0 = 78.09\,\text{S}\,\text{cm}^2\,\text{mol}^{-1}$; $q^* = 2.00$; $B = 4.48$ | | | | | | | | |
| | 0.200 | 0.198 | 132.33 | 884 | 19.46 | 143.68 | 0.000 | 75.50 |
| | 0.300 | 0.295 | 129.09 | 881 | 22.70 | 145.60 | 0.000 | 74.20 |
| | 0.400 | 0.395 | 126.31 | 876 | 25.48 | 152.43 | 0.001 | 72.90 |
| | 0.500 | 0.491 | 123.97 | 874 | 27.82 | 154.75 | 0.001 | 71.70 |
| | 0.600 | 0.587 | 122.06 | 871 | 29.73 | 158.12 | 0.001 | 70.50 |
| | 0.700 | 0.683 | 120.39 | 869 | 31.40 | 161.85 | 0.001 | 69.30 |
| | 0.800 | 0.778 | 119.07 | 867 | 32.72 | 165.32 | 0.002 | 68.20 |

(*Continued*)

**TABLE 5.4 (*Continued*)**
**Components of Formula (5.103) to Calculate Electrical Conductivity, S cm²/mol, According to Formula (5.87)**

| Electrolyte | $m$ | $c_k$ | $\Lambda$ | $\eta$ | $\aleph$ | $P_\Lambda$ | $E$ | $\varepsilon_S$ |
|---|---|---|---|---|---|---|---|---|
| | 0.900 | 0.872 | 117.90 | 866 | 33.89 | 168.61 | 0.002 | 67.20 |
| | 1.000 | 0.966 | 117.02 | 864 | 34.77 | 172.11 | 0.002 | 66.10 |
| | 1.250 | 1.194 | 115.33 | 861 | 36.46 | 179.09 | 0.003 | 63.60 |
| | 1.500 | 1.419 | 113.95 | 859 | 37.84 | 186.75 | 0.005 | 61.30 |
| | 1.750 | 1.640 | 112.69 | 857 | 39.09 | 194.38 | 0.006 | 59.10 |
| | 2.000 | 1.859 | 111.44 | 856 | 40.34 | 202.43 | 0.008 | 57.00 |
| | 2.250 | 2.075 | 110.19 | 856 | 41.59 | 210.71 | 0.011 | 55.10 |
| | 2.500 | 2.290 | 108.96 | 856 | 42.82 | 219.25 | 0.014 | 53.30 |
| | 2.750 | 2.503 | 107.79 | 857 | 43.98 | 228.10 | 0.017 | 51.60 |
| | 3.000 | 2.693 | 106.69 | 857 | 45.08 | 232.74 | 0.021 | 50.00 |
| | 3.250 | 2.902 | 105.58 | 859 | 46.18 | 241.93 | 0.025 | 48.50 |
| | 3.500 | 3.086 | 104.35 | 860 | 47.41 | 246.78 | 0.031 | 47.00 |
| | 3.750 | 3.293 | 103.00 | 862 | 48.75 | 256.40 | 0.037 | 45.70 |
| | 4.000 | 3.471 | 101.58 | 864 | 50.17 | 261.47 | 0.044 | 44.40 |
| | 4.250 | 3.676 | 100.14 | 867 | 51.60 | 271.54 | 0.052 | 43.10 |
| | 4.500 | 3.849 | 98.74 | 869 | 52.99 | 276.85 | 0.061 | 42.00 |
| | 4.750 | 4.020 | 97.35 | 871 | 54.37 | 281.92 | 0.070 | 40.90 |
| | 5.000 | 4.188 | 96.03 | 873 | 55.68 | 287.39 | 0.082 | 39.80 |
| | 5.250 | 4.354 | 94.72 | 876 | 56.98 | 292.95 | 0.094 | 38.80 |
| *KCl*: $a = 0.5782$ nm; $\lambda_i^0 = 73.70\,\mathrm{S\,cm^2\,mol^{-1}}$; $\lambda_i^0 = 76.26\,\mathrm{S\,cm^2\,mol^{-1}}$; $q^* = 2.00$; $B = 4.71$ | | | | | | | | |
| | 0.300 | 0.296 | 125.83 | 888 | 24.13 | 706.01 | 0.001 | 74.10 |
| | 0.400 | 0.395 | 124.24 | 887 | 25.72 | 716.21 | 0.001 | 72.90 |
| | 0.500 | 0.493 | 122.66 | 887 | 27.30 | 722.91 | 0.001 | 71.70 |
| | 0.600 | 0.587 | 121.16 | 887 | 28.80 | 722.32 | 0.001 | 70.60 |
| | 0.700 | 0.684 | 119.80 | 886 | 30.16 | 729.66 | 0.002 | 69.60 |
| | 0.800 | 0.781 | 118.44 | 886 | 31.52 | 737.83 | 0.002 | 68.50 |
| | 0.900 | 0.873 | 117.20 | 886 | 32.76 | 738.42 | 0.002 | 67.50 |
| | 1.000 | 0.969 | 116.11 | 886 | 33.85 | 745.93 | 0.003 | 66.30 |
| | 1.250 | 1.206 | 113.54 | 887 | 36.42 | 762.56 | 0.004 | 63.70 |
| | 1.500 | 1.432 | 111.26 | 888 | 38.69 | 771.02 | 0.005 | 61.20 |
| | 1.750 | 1.664 | 109.27 | 890 | 40.68 | 787.50 | 0.007 | 58.70 |
| | 2.000 | 1.883 | 107.44 | 891 | 42.51 | 796.24 | 0.010 | 56.50 |
| | 2.250 | 2.098 | 105.74 | 893 | 44.21 | 805.07 | 0.013 | 54.40 |
| | 2.500 | 2.323 | 104.15 | 897 | 45.79 | 822.38 | 0.016 | 52.50 |
| | 2.750 | 2.532 | 102.61 | 899 | 47.33 | 831.51 | 0.020 | 50.80 |
| | 3.000 | 2.738 | 101.10 | 902 | 48.84 | 840.75 | 0.024 | 49.20 |
| | 3.250 | 2.959 | 99.62 | 907 | 50.31 | 858.76 | 0.030 | 47.60 |
| | 3.500 | 3.160 | 98.17 | 910 | 51.75 | 868.31 | 0.036 | 46.20 |
| | 3.750 | 3.357 | 96.75 | 914 | 53.17 | 877.18 | 0.043 | 44.90 |
| | 4.000 | 3.552 | 95.35 | 917 | 54.56 | 887.02 | 0.051 | 43.60 |
| | 4.250 | 3.745 | 93.95 | 921 | 55.95 | 896.08 | 0.060 | 42.40 |

(*Continued*)

**TABLE 5.4 (*Continued*)**
**Components of Formula (5.103) to Calculate Electrical Conductivity, S cm²/mol, According to Formula (5.87)**

| Electrolyte | $m$ | $c_k$ | $\Lambda$ | $\eta$ | $\aleph$ | $P_\Lambda$ | $E$ | $\varepsilon_S$ |
|---|---|---|---|---|---|---|---|---|
| *KI*: $a = 0.6919$ nm; $\lambda_i^0 = 73.70\,\text{S cm}^2\,\text{mol}^{-1}$; $\lambda_i^0 = 76.84\,\text{S cm}^2\,\text{mol}^{-1}$; $q^* = 2.00$; $B = 3.17$ | | | | | | | | |
| | 0.500 | 0.488 | 118.04 | 859 | 32.50 | 88.51 | 0.001 | 76.60 |
| | 0.600 | 0.581 | 116.69 | 855 | 33.85 | 90.55 | 0.001 | 76.30 |
| | 0.700 | 0.673 | 115.37 | 852 | 35.17 | 92.84 | 0.001 | 75.90 |
| | 0.800 | 0.770 | 114.09 | 847 | 36.45 | 97.53 | 0.001 | 75.60 |
| | 0.900 | 0.860 | 112.88 | 844 | 37.66 | 99.69 | 0.001 | 75.30 |
| | 1.000 | 0.950 | 111.69 | 841 | 38.85 | 102.26 | 0.001 | 74.90 |
| | 1.250 | 1.176 | 108.97 | 835 | 41.57 | 110.01 | 0.002 | 74.10 |
| | 1.500 | 1.399 | 106.45 | 830 | 44.09 | 118.12 | 0.002 | 73.40 |
| | 1.750 | 1.622 | 104.16 | 826 | 46.38 | 126.90 | 0.003 | 69.50 |
| | 2.000 | 1.828 | 102.02 | 824 | 48.52 | 133.01 | 0.004 | 66.10 |
| | 2.250 | 2.031 | 99.98 | 822 | 50.55 | 139.49 | 0.006 | 63.20 |
| | 2.500 | 2.231 | 98.02 | 821 | 52.51 | 146.06 | 0.008 | 60.70 |
| | 2.750 | 2.429 | 96.13 | 821 | 54.40 | 153.17 | 0.010 | 57.70 |
| | 3.000 | 2.624 | 94.25 | 821 | 56.28 | 160.15 | 0.013 | 54.90 |
| *KNO₃*: $a = 0.5980$ nm; $\lambda_i^0 = 73.70\,\text{S cm}^2\,\text{mol}^{-1}$; $\lambda_i^0 = 71.37\,\text{S cm}^2\,\text{mol}^{-1}$; $q^* = 2.00$; $B = 3.93$ | | | | | | | | |
| | 0.010 | 0.010 | 126.37 | 890 | 18.70 | 188.83 | 0.000 | 78.10 |
| | 0.020 | 0.020 | 126.30 | 890 | 18.77 | 189.17 | 0.000 | 77.90 |
| | 0.030 | 0.030 | 126.23 | 890 | 18.84 | 189.77 | 0.000 | 77.60 |
| | 0.040 | 0.040 | 126.17 | 890 | 18.90 | 190.15 | 0.000 | 77.40 |
| | 0.050 | 0.050 | 126.08 | 890 | 18.99 | 190.49 | 0.000 | 77.10 |
| | 0.060 | 0.060 | 125.58 | 890 | 19.49 | 191.10 | 0.000 | 76.90 |
| | 0.070 | 0.070 | 125.07 | 890 | 20.00 | 191.49 | 0.000 | 76.70 |
| | 0.080 | 0.080 | 124.57 | 890 | 20.50 | 191.88 | 0.000 | 76.40 |
| | 0.090 | 0.089 | 124.06 | 890 | 21.01 | 187.51 | 0.000 | 76.20 |
| | 0.100 | 0.099 | 123.56 | 890 | 21.51 | 188.45 | 0.000 | 75.90 |
| | 0.200 | 0.198 | 118.55 | 889 | 26.52 | 192.83 | 0.000 | 73.60 |
| | 0.300 | 0.296 | 115.42 | 889 | 29.65 | 195.53 | 0.000 | 72.80 |
| | 0.400 | 0.393 | 112.33 | 888 | 32.74 | 198.09 | 0.001 | 72.10 |
| | 0.500 | 0.490 | 109.37 | 888 | 35.70 | 201.25 | 0.001 | 71.40 |
| | 0.600 | 0.586 | 106.43 | 888 | 38.64 | 204.05 | 0.001 | 70.80 |
| | 0.700 | 0.681 | 103.79 | 889 | 41.28 | 206.46 | 0.001 | 70.10 |
| | 0.800 | 0.771 | 101.28 | 889 | 43.79 | 206.46 | 0.002 | 69.50 |
| | 0.900 | 0.865 | 98.93 | 889 | 46.14 | 209.50 | 0.002 | 68.90 |
| | 1.000 | 0.958 | 96.83 | 890 | 48.24 | 212.44 | 0.002 | 68.40 |
| | 1.250 | 1.186 | 92.20 | 892 | 52.87 | 218.51 | 0.003 | 67.20 |
| | 1.500 | 1.409 | 88.39 | 894 | 56.68 | 224.33 | 0.003 | 66.30 |
| | 1.750 | 1.629 | 85.14 | 897 | 59.93 | 230.48 | 0.004 | 65.50 |
| | 2.000 | 1.846 | 82.34 | 901 | 62.73 | 236.82 | 0.005 | 64.70 |
| | 2.250 | 2.059 | 79.88 | 906 | 65.18 | 242.99 | 0.006 | 63.90 |

*(Continued)*

**TABLE 5.4 (*Continued*)**
**Components of Formula (5.103) to Calculate Electrical Conductivity, S cm²/mol, According to Formula (5.87)**

| Electrolyte | $m$ | $c_k$ | $\Lambda$ | $\eta$ | ℵ | $P_\Lambda$ | $E$ | $\varepsilon_S$ |
|---|---|---|---|---|---|---|---|---|
| | 2.500 | 2.256 | 77.66 | 908 | 67.40 | 246.31 | 0.007 | 63.30 |
| | 2.750 | 2.463 | 75.69 | 914 | 69.37 | 252.70 | 0.007 | 63.00 |
| | 3.000 | 2.652 | 73.85 | 917 | 71.21 | 256.35 | 0.008 | 62.70 |
| | 3.250 | 2.853 | 72.08 | 924 | 72.98 | 262.73 | 0.009 | 62.40 |
| | 3.500 | 3.034 | 70.36 | 927 | 74.70 | 266.28 | 0.009 | 62.20 |
| *KOH*: $a = 0.52704$ nm; $\lambda_i^0 = 73.70\,\mathrm{S\,cm^2\,mol^{-1}}$; $\lambda_i^0 = 181.99\,\mathrm{S\,cm^2\,mol^{-1}}$; $q^* = 2.00$; $B = 5.85$ | | | | | | | | |
| | 3.750 | 3.613 | 149.28 | 1338 | 106.36 | 911.10 | 0.045 | 42.10 |
| | 4.000 | 3.842 | 144.89 | 1368 | 110.74 | 924.08 | 0.056 | 40.40 |
| | 4.250 | 4.071 | 140.51 | 1399 | 115.11 | 937.25 | 0.067 | 38.90 |
| | 4.500 | 4.298 | 136.17 | 1430 | 119.44 | 950.53 | 0.080 | 37.50 |
| | 4.750 | 4.525 | 132.04 | 1553 | 123.56 | 964.09 | 0.090 | 36.20 |
| | 5.000 | 4.751 | 127.91 | 1608 | 127.67 | 977.85 | 0.106 | 34.90 |
| | 5.250 | 4.975 | 124.40 | 1665 | 131.17 | 990.83 | 0.122 | 33.80 |
| | 5.500 | 5.199 | 120.97 | 1726 | 134.58 | 1004.99 | 0.141 | 32.70 |
| | 5.750 | 5.378 | 117.63 | 1726 | 137.90 | 1004.99 | 0.163 | 31.80 |
| | 6.000 | 5.599 | 114.37 | 1791 | 141.14 | 1018.34 | 0.184 | 30.90 |
| | 6.250 | 5.819 | 111.12 | 1859 | 144.36 | 1031.97 | 0.208 | 30.00 |
| | 6.500 | 6.038 | 108.03 | 1931 | 147.42 | 1045.70 | 0.235 | 29.10 |
| | 6.750 | 6.207 | 104.94 | 1931 | 150.48 | 1045.70 | 0.271 | 28.30 |
| | 7.000 | 6.423 | 101.91 | 2007 | 153.48 | 1059.62 | 0.303 | 27.50 |
| | 7.250 | 6.639 | 99.04 | 2088 | 156.32 | 1073.83 | 0.335 | 26.80 |
| | 7.500 | 6.855 | 96.16 | 2174 | 159.16 | 1088.15 | 0.370 | 26.10 |
| | 7.750 | 7.014 | 93.36 | 2174 | 161.91 | 1088.15 | 0.422 | 25.40 |
| | 8.000 | 7.227 | 90.65 | 2264 | 164.57 | 1101.65 | 0.468 | 24.70 |
| | 8.250 | 7.440 | 87.93 | 2361 | 167.25 | 1116.46 | 0.511 | 24.10 |
| | 8.500 | 7.593 | 85.29 | 2361 | 169.82 | 1116.46 | 0.577 | 23.50 |
| | 8.750 | 7.803 | 82.73 | 2463 | 172.33 | 1130.33 | 0.632 | 22.90 |
| | 9.000 | 8.013 | 80.18 | 2571 | 174.83 | 1144.40 | 0.681 | 22.40 |
| | 9.250 | 8.160 | 77.67 | 2571 | 177.26 | 1144.40 | 0.759 | 21.90 |
| | 9.500 | 8.368 | 75.27 | 2686 | 179.60 | 1159.81 | 0.819 | 21.40 |
| | 9.750 | 8.511 | 72.87 | 2686 | 181.92 | 1159.81 | 0.898 | 21.00 |
| | 10.000 | 8.717 | 70.48 | 2809 | 184.26 | 1174.26 | 0.953 | 20.60 |
| | 10.500 | 9.060 | 66.34 | 2939 | 188.26 | 1187.71 | 1.090 | 19.90 |
| | 11.000 | 9.398 | 62.31 | 3077 | 192.16 | 1202.53 | 1.223 | 19.30 |
| | 11.500 | 9.731 | 58.62 | 3224 | 195.72 | 1217.56 | 1.346 | 18.80 |
| | 12.000 | 10.059 | 54.99 | 3380 | 199.22 | 1232.88 | 1.482 | 18.30 |
| | 12.500 | 10.382 | 51.55 | 3894 | 202.65 | 1247.05 | 1.487 | 17.80 |
| | 13.000 | 10.701 | 48.35 | 4121 | 205.71 | 1261.49 | 1.627 | 17.30 |
| | 13.500 | 11.017 | 46.19 | 4367 | 207.76 | 1277.31 | 1.740 | 16.90 |
| | 14.000 | 11.328 | 44.04 | 4632 | 209.79 | 1292.13 | 1.861 | 16.50 |

(*Continued*)

**TABLE 5.4 (*Continued*)**
**Components of Formula (5.103) to Calculate Electrical Conductivity, S cm²/mol, According to Formula (5.87)**

| Electrolyte | $m$ | $c_k$ | $\Lambda$ | $\eta$ | $\aleph$ | $P_\Lambda$ | $E$ | $\varepsilon_S$ |
|---|---|---|---|---|---|---|---|---|
| | 14.500 | 11.635 | 42.08 | 4918 | 211.72 | 1307.13 | 1.895 | 16.30 |
| | 15.000 | 11.939 | 40.12 | 5226 | 213.59 | 1322.33 | 1.976 | 16.00 |
| | 15.500 | 12.240 | 38.16 | 5560 | 215.52 | 1337.71 | 2.007 | 15.80 |
| $Li_2SO_4$: $a = 0.65149$ nm; $\lambda_i^0 = 38.88\,\mathrm{S\,cm^2\,mol^{-1}}$; $\lambda_i^0 = 51.18\,\mathrm{S\,cm^2\,mol^{-1}}$; $q^* = 3.83$; $B = 30.04$ | | | | | | | | |
| | 0.200 | 0.199 | 76.29 | 994 | 13.77 | 734.80 | 0.005 | 74.50 |
| | 0.300 | 0.297 | 69.65 | 1052 | 20.40 | 754.12 | 0.007 | 72.70 |
| | 0.400 | 0.395 | 63.40 | 1113 | 26.65 | 782.88 | 0.010 | 71.10 |
| | 0.500 | 0.493 | 58.61 | 1179 | 31.44 | 815.23 | 0.013 | 69.40 |
| | 0.600 | 0.591 | 54.09 | 1249 | 35.95 | 847.48 | 0.016 | 68.00 |
| | 0.700 | 0.688 | 50.63 | 1324 | 39.41 | 877.71 | 0.020 | 66.40 |
| | 0.800 | 0.785 | 47.25 | 1405 | 42.79 | 909.72 | 0.023 | 65.00 |
| | 0.900 | 0.882 | 44.54 | 1491 | 45.49 | 942.27 | 0.027 | 63.60 |
| | 1.000 | 0.978 | 41.84 | 1583 | 48.19 | 971.96 | 0.030 | 62.30 |
| | 1.250 | 1.213 | 36.59 | 1790 | 53.43 | 1036.19 | 0.041 | 59.10 |
| | 1.500 | 1.445 | 32.25 | 2027 | 57.76 | 1101.01 | 0.053 | 56.20 |
| | 1.750 | 1.675 | 28.70 | 2302 | 61.29 | 1169.36 | 0.068 | 53.30 |
| | 2.000 | 1.902 | 25.61 | 2621 | 64.37 | 1238.35 | 0.083 | 50.70 |
| | 2.250 | 2.127 | 22.77 | 2990 | 67.19 | 1307.17 | 0.102 | 47.90 |
| | 2.500 | 2.351 | 20.01 | 1596 | 69.79 | 1380.50 | 0.262 | 45.50 |
| | 2.750 | 2.552 | 17.25 | 1658 | 72.48 | 1416.82 | 0.335 | 43.30 |
| | 3.000 | 2.771 | 14.49 | 1794 | 75.15 | 1489.35 | 0.416 | 41.10 |
| $LiCl$: $a = 0.49035$ nm; $\lambda_i^0 = 38.88\,\mathrm{S\,cm^2\,mol^{-1}}$; $\lambda_i^0 = 76.26\,\mathrm{S\,cm^2\,mol^{-1}}$; $q^* = 2.00$; $B = 28.88$ | | | | | | | | |
| | 0.100 | 0.099 | 96.72 | 890 | 18.42 | 310.21 | 0.000 | 76.70 |
| | 0.200 | 0.199 | 93.39 | 922 | 21.75 | 331.17 | 0.000 | 76.20 |
| | 0.300 | 0.297 | 90.07 | 922 | 25.07 | 329.99 | 0.001 | 75.80 |
| | 0.400 | 0.397 | 86.75 | 955 | 28.39 | 343.95 | 0.001 | 75.30 |
| | 0.500 | 0.494 | 83.68 | 955 | 31.46 | 343.95 | 0.001 | 74.90 |
| | 0.600 | 0.590 | 81.72 | 955 | 33.42 | 342.76 | 0.001 | 74.40 |
| | 0.700 | 0.689 | 79.75 | 990 | 35.39 | 353.26 | 0.001 | 73.90 |
| | 0.800 | 0.784 | 77.79 | 990 | 37.35 | 353.26 | 0.001 | 73.50 |
| | 0.900 | 0.884 | 75.83 | 1027 | 39.31 | 367.97 | 0.002 | 73.10 |
| | 1.000 | 0.978 | 74.02 | 1027 | 41.12 | 366.28 | 0.002 | 73.30 |
| | 1.250 | 1.217 | 71.22 | 1066 | 43.92 | 379.93 | 0.003 | 70.60 |
| | 1.500 | 1.454 | 68.41 | 1106 | 46.73 | 392.01 | 0.003 | 68.00 |
| | 1.750 | 1.689 | 66.15 | 1149 | 48.99 | 405.95 | 0.005 | 64.90 |
| | 2.000 | 1.922 | 63.89 | 1194 | 51.24 | 418.30 | 0.006 | 62.50 |
| | 2.250 | 2.153 | 61.69 | 1241 | 53.44 | 430.78 | 0.007 | 60.50 |
| | 2.500 | 2.383 | 59.49 | 1290 | 55.64 | 445.17 | 0.009 | 58.50 |
| | 2.750 | 2.596 | 57.20 | 1290 | 57.93 | 445.17 | 0.011 | 56.80 |

*(Continued)*

**TABLE 5.4 (*Continued*)**
**Components of Formula (5.103) to Calculate Electrical Conductivity, S cm²/mol, According to Formula (5.87)**

| Electrolyte | $m$ | $c_k$ | $\Lambda$ | $\eta$ | $\aleph$ | $P_\Lambda$ | $E$ | $\varepsilon_S$ |
|---|---|---|---|---|---|---|---|---|
| | 3.000 | 2.821 | 54.83 | 1342 | 60.30 | 457.92 | 0.013 | 55.10 |
| | 3.250 | 3.045 | 52.47 | 1397 | 62.66 | 470.81 | 0.015 | 53.70 |
| | 3.500 | 3.267 | 50.25 | 1455 | 64.87 | 485.65 | 0.017 | 52.30 |
| | 3.750 | 3.488 | 48.02 | 1516 | 67.10 | 498.83 | 0.019 | 50.90 |
| | 4.000 | 3.687 | 46.12 | 1516 | 69.00 | 498.83 | 0.023 | 49.50 |
| | 4.250 | 3.904 | 44.41 | 1580 | 70.70 | 512.14 | 0.026 | 48.00 |
| | 4.500 | 4.121 | 42.70 | 1648 | 72.41 | 526.98 | 0.030 | 46.70 |
| | 4.750 | 4.336 | 41.32 | 1719 | 73.79 | 542.48 | 0.034 | 45.30 |
| | 5.000 | 4.524 | 39.94 | 1719 | 75.16 | 540.58 | 0.040 | 44.00 |
| | 5.250 | 4.736 | 38.62 | 1795 | 76.48 | 556.24 | 0.044 | 42.90 |
| | 5.500 | 4.947 | 37.46 | 1874 | 77.63 | 570.15 | 0.049 | 41.70 |
| | 5.750 | 5.157 | 36.29 | 1958 | 78.80 | 586.15 | 0.055 | 40.70 |
| | 6.000 | 5.336 | 35.17 | 1958 | 79.91 | 586.15 | 0.062 | 39.70 |
| | 6.250 | 5.543 | 34.12 | 2046 | 80.95 | 600.37 | 0.069 | 38.60 |
| | 6.500 | 5.750 | 33.07 | 2140 | 81.99 | 616.72 | 0.077 | 37.60 |
| | 6.750 | 5.922 | 32.05 | 2140 | 83.00 | 616.72 | 0.086 | 36.80 |
| | 7.000 | 6.126 | 31.05 | 2239 | 84.00 | 631.26 | 0.094 | 35.90 |
| | 7.250 | 6.329 | 30.06 | 2343 | 84.98 | 645.94 | 0.103 | 35.00 |
| | 7.500 | 6.495 | 29.07 | 2343 | 85.96 | 645.94 | 0.115 | 34.30 |
| | 7.750 | 6.696 | 28.11 | 2454 | 86.91 | 662.83 | 0.124 | 33.50 |
| | 8.000 | 6.857 | 27.15 | 2454 | 87.85 | 662.83 | 0.138 | 32.80 |
| | 8.250 | 7.057 | 26.19 | 2571 | 88.80 | 677.85 | 0.149 | 32.10 |
| | 8.500 | 7.214 | 25.24 | 2571 | 89.74 | 677.85 | 0.164 | 31.50 |
| | 8.750 | 7.411 | 24.30 | 2694 | 90.67 | 693.02 | 0.174 | 30.90 |
| | 9.000 | 7.608 | 23.36 | 2825 | 91.60 | 710.46 | 0.184 | 30.30 |
| | 9.250 | 7.760 | 22.43 | 890 | 92.51 | 710.46 | 0.201 | 29.80 |
| | 9.500 | 7.955 | 21.50 | 922 | 93.43 | 725.98 | 0.210 | 29.30 |
| | 9.750 | 8.103 | 20.58 | 922 | 94.33 | 725.98 | 0.230 | 28.80 |
| | 10.000 | 8.297 | 19.65 | 955 | 95.25 | 741.65 | 0.241 | 28.30 |
| $MgCl_2$: $a = 0.62658$ nm; $\lambda_i^0 = 53.25\,\mathrm{S\,cm^2\,mol^{-1}}$; $\lambda_i^0 = 76.26\,\mathrm{S\,cm^2\,mol^{-1}}$; $q^* = 4.25$; $B = 18.68$ | | | | | | | | |
| | 0.050 | 0.050 | 99.04 | 890 | 30.47 | 2569.83 | 0.001 | 77.70 |
| | 0.060 | 0.060 | 97.83 | 929 | 31.68 | 2574.50 | 0.001 | 77.60 |
| | 0.070 | 0.070 | 96.62 | 929 | 32.89 | 2578.73 | 0.002 | 77.50 |
| | 0.080 | 0.080 | 95.42 | 929 | 34.09 | 2585.99 | 0.002 | 77.40 |
| | 0.090 | 0.090 | 94.21 | 929 | 35.30 | 2590.68 | 0.002 | 77.30 |
| | 0.100 | 0.100 | 93.00 | 929 | 36.51 | 2595.39 | 0.002 | 77.20 |
| | 0.200 | 0.199 | 83.21 | 970 | 46.30 | 2615.97 | 0.005 | 76.30 |
| | 0.300 | 0.298 | 77.36 | 1016 | 52.14 | 2655.87 | 0.007 | 74.80 |
| | 0.400 | 0.397 | 72.14 | 1065 | 57.36 | 2701.64 | 0.010 | 73.40 |
| | 0.500 | 0.496 | 68.86 | 1118 | 60.64 | 2747.66 | 0.013 | 71.90 |

(*Continued*)

**TABLE 5.4 (*Continued*)**
**Components of Formula (5.103) to Calculate Electrical Conductivity, S cm²/mol, According to Formula (5.87)**

| Electrolyte | $m$ | $c_k$ | $\Lambda$ | $\eta$ | $\aleph$ | $P_\Lambda$ | $E$ | $\varepsilon_S$ |
|---|---|---|---|---|---|---|---|---|
| | 0.600 | 0.589 | 66.73 | 1118 | 62.76 | 2740.46 | 0.016 | 70.40 |
| | 0.700 | 0.687 | 64.71 | 1177 | 64.78 | 2787.08 | 0.020 | 69.00 |
| | 0.800 | 0.784 | 62.90 | 1240 | 66.59 | 2829.32 | 0.023 | 67.50 |
| | 0.900 | 0.882 | 61.10 | 1309 | 68.38 | 2877.30 | 0.027 | 66.10 |
| | 1.000 | 0.979 | 59.30 | 1384 | 70.18 | 2920.78 | 0.031 | 64.60 |
| *NaCl*: $a = 0.53835$ nm; $\lambda_i^0 = 52.17\,\mathrm{S\,cm^2\,mol^{-1}}$; $\lambda_i^0 = 76.26\,\mathrm{S\,cm^2\,mol^{-1}}$; $q^* = 2.00$; $B = 6.49$ | | | | | | | | |
| | 0.700 | 0.690 | 95.63 | 953 | 32.80 | 644.25 | 0.002 | 67.80 |
| | 0.800 | 0.784 | 94.39 | 953 | 34.04 | 644.25 | 0.002 | 66.50 |
| | 0.900 | 0.883 | 93.15 | 970 | 35.28 | 651.88 | 0.003 | 65.20 |
| | 1.000 | 0.983 | 91.90 | 988 | 36.53 | 660.84 | 0.003 | 64.00 |
| | 1.250 | 1.220 | 88.86 | 1008 | 39.57 | 669.39 | 0.004 | 61.40 |
| | 1.500 | 1.455 | 85.86 | 1028 | 42.56 | 677.94 | 0.006 | 59.10 |
| | 1.750 | 1.687 | 82.95 | 1049 | 45.47 | 687.34 | 0.008 | 56.90 |
| | 2.000 | 1.916 | 80.09 | 1071 | 48.33 | 696.12 | 0.010 | 54.70 |
| | 2.250 | 2.157 | 77.30 | 1119 | 51.12 | 713.33 | 0.013 | 52.80 |
| | 2.500 | 2.383 | 74.64 | 1145 | 53.77 | 722.43 | 0.016 | 50.80 |
| | 2.750 | 2.606 | 72.04 | 1171 | 56.37 | 731.64 | 0.019 | 49.10 |
| | 3.000 | 2.827 | 69.62 | 1200 | 58.79 | 740.28 | 0.024 | 47.30 |
| | 3.250 | 3.046 | 67.23 | 1229 | 61.17 | 749.71 | 0.029 | 45.70 |
| | 3.500 | 3.263 | 64.99 | 1260 | 63.40 | 759.36 | 0.035 | 44.10 |
| | 3.750 | 3.478 | 62.76 | 1292 | 65.63 | 768.22 | 0.042 | 42.70 |
| | 4.000 | 3.691 | 60.61 | 1326 | 67.77 | 778.10 | 0.049 | 41.30 |
| | 4.250 | 3.902 | 58.46 | 1362 | 69.91 | 787.27 | 0.058 | 39.90 |
| | 4.500 | 4.112 | 56.36 | 1399 | 72.00 | 796.45 | 0.068 | 38.70 |
| | 4.750 | 4.320 | 54.27 | 1438 | 74.08 | 806.69 | 0.079 | 37.50 |
| | 5.000 | 4.526 | 52.17 | 1479 | 76.17 | 816.19 | 0.090 | 36.40 |
| *$NaNO_3$*: $a = 0.57989$ nm; $\lambda_i^0 = 52.17\,\mathrm{S\,cm^2\,mol^{-1}}$; $\lambda_i^0 = 71.37\,\mathrm{S\,cm^2\,mol^{-1}}$; $q^* = 2.00$; $B = 6.13$ | | | | | | | | |
| | 0.200 | 0.199 | 96.52 | 904 | 27.02 | 189.83 | 0.000 | 77.60 |
| | 0.300 | 0.296 | 94.18 | 904 | 29.36 | 189.83 | 0.000 | 77.00 |
| | 0.400 | 0.393 | 91.85 | 912 | 31.69 | 192.11 | 0.001 | 76.00 |
| | 0.500 | 0.491 | 89.54 | 921 | 34.00 | 196.12 | 0.001 | 75.10 |
| | 0.600 | 0.588 | 87.54 | 930 | 36.00 | 199.43 | 0.001 | 74.20 |
| | 0.700 | 0.685 | 85.53 | 940 | 38.01 | 203.08 | 0.001 | 73.30 |
| | 0.800 | 0.777 | 83.67 | 940 | 39.87 | 203.31 | 0.001 | 72.50 |
| | 0.900 | 0.873 | 81.97 | 951 | 41.57 | 207.01 | 0.002 | 71.70 |
| | 1.000 | 0.969 | 80.26 | 962 | 43.28 | 210.99 | 0.002 | 70.90 |
| | 1.250 | 1.204 | 76.17 | 987 | 47.37 | 218.55 | 0.002 | 69.10 |
| | 1.500 | 1.427 | 72.23 | 1001 | 51.31 | 222.44 | 0.003 | 67.80 |
| | 1.750 | 1.657 | 68.43 | 1031 | 55.11 | 230.53 | 0.004 | 66.50 |
| | 2.000 | 1.885 | 65.00 | 1064 | 58.54 | 238.91 | 0.004 | 65.40 |

(*Continued*)

**TABLE 5.4 (*Continued*)**
**Components of Formula (5.103) to Calculate Electrical Conductivity, S cm²/mol, According to Formula (5.87)**

| Electrolyte | $m$ | $c_k$ | $\Lambda$ | $\eta$ | $\aleph$ | $P_\Lambda$ | $E$ | $\varepsilon_S$ |
|---|---|---|---|---|---|---|---|---|
| | 2.250 | 2.097 | 62.19 | 1083 | 61.34 | 243.04 | 0.005 | 64.30 |
| | 2.500 | 2.320 | 60.14 | 1122 | 63.39 | 251.39 | 0.006 | 63.30 |
| | 2.750 | 2.526 | 58.17 | 1143 | 65.36 | 256.04 | 0.007 | 62.40 |
| | 3.000 | 2.727 | 56.26 | 1166 | 67.27 | 259.95 | 0.008 | 60.90 |
| | 3.250 | 2.946 | 54.41 | 1214 | 69.12 | 269.09 | 0.010 | 57.60 |
| | 3.500 | 3.142 | 52.63 | 1240 | 70.90 | 273.60 | 0.013 | 54.80 |
| | 3.750 | 3.335 | 50.90 | 1268 | 72.62 | 278.09 | 0.016 | 52.30 |
| | 4.000 | 3.525 | 49.34 | 1297 | 74.18 | 282.71 | 0.020 | 50.20 |
| | 4.250 | 3.739 | 47.87 | 1360 | 75.65 | 292.45 | 0.023 | 48.30 |
| | 4.500 | 3.925 | 46.49 | 1393 | 77.02 | 297.25 | 0.028 | 46.60 |
| | 4.750 | 4.109 | 45.17 | 1429 | 78.34 | 302.03 | 0.032 | 45.10 |
| | 5.000 | 4.290 | 43.87 | 1466 | 79.63 | 306.96 | 0.037 | 43.70 |
| | 5.250 | 4.469 | 42.63 | 1505 | 80.87 | 311.54 | 0.040 | 43.20 |
| | 5.500 | 4.647 | 41.40 | 1546 | 82.10 | 316.92 | 0.041 | 43.00 |
| | 5.750 | 4.822 | 40.23 | 1590 | 83.27 | 322.03 | 0.042 | 42.80 |
| | 6.000 | 4.996 | 39.07 | 1635 | 84.43 | 327.22 | 0.043 | 42.60 |
| | 6.250 | 5.168 | 38.04 | 1683 | 85.45 | 332.38 | 0.045 | 42.20 |
| | 6.500 | 5.338 | 37.06 | 1734 | 86.43 | 337.26 | 0.047 | 41.80 |
| | 6.750 | 5.468 | 36.15 | 1734 | 87.34 | 337.26 | 0.051 | 41.40 |
| | 7.000 | 5.634 | 35.31 | 1787 | 88.18 | 342.64 | 0.054 | 40.80 |
| | 7.250 | 5.799 | 34.49 | 1843 | 88.99 | 348.08 | 0.056 | 40.30 |
| | 7.500 | 5.963 | 33.71 | 1902 | 89.77 | 353.51 | 0.059 | 39.80 |
| | 7.750 | 6.125 | 32.93 | 1964 | 90.55 | 358.64 | 0.062 | 39.40 |
| | 8.000 | 6.243 | 32.18 | 1964 | 91.29 | 359.09 | 0.065 | 39.00 |
| | 8.250 | 6.402 | 31.45 | 2029 | 92.02 | 364.29 | 0.068 | 38.50 |
| | 8.500 | 6.561 | 30.72 | 2098 | 92.75 | 370.01 | 0.071 | 38.10 |
| | 8.750 | 6.718 | 30.03 | 2170 | 93.44 | 375.81 | 0.073 | 37.80 |
| | 9.000 | 6.827 | 29.35 | 2170 | 94.11 | 375.81 | 0.077 | 37.40 |
| | 9.250 | 6.982 | 28.67 | 2247 | 94.79 | 381.21 | 0.080 | 37.00 |
| | 9.500 | 7.136 | 28.04 | 2327 | 95.42 | 387.06 | 0.081 | 36.70 |
| | 9.750 | 7.238 | 27.40 | 2327 | 96.05 | 387.15 | 0.085 | 36.40 |
| | 10.000 | 7.390 | 26.77 | 2412 | 96.68 | 393.08 | 0.087 | 36.10 |
| *RbCl*: $a = 0.61812$ nm; $\lambda_i^0 = 77.93\,\text{S cm}^2\,\text{mol}^{-1}$; $\lambda_i^0 = 76.26\,\text{S cm}^2\,\text{mol}^{-1}$; $q^* = 2.00$; $B = 4.82$ | | | | | | | | |
| | 0.200 | 0.199 | 129.67 | 886 | 24.52 | 232.25 | 0.000 | 77.80 |
| | 0.300 | 0.301 | 127.64 | 881 | 26.55 | 244.27 | 0.000 | 77.10 |
| | 0.400 | 0.400 | 125.65 | 879 | 28.54 | 250.33 | 0.001 | 76.30 |
| | 0.500 | 0.499 | 123.70 | 877 | 30.49 | 255.96 | 0.001 | 75.70 |
| | 0.600 | 0.597 | 121.81 | 876 | 32.38 | 261.67 | 0.001 | 75.10 |
| | 0.700 | 0.694 | 119.95 | 874 | 34.24 | 267.16 | 0.001 | 74.40 |
| | 0.800 | 0.791 | 118.14 | 872 | 36.05 | 272.49 | 0.001 | 73.90 |

(*Continued*)

**TABLE 5.4 (*Continued*)**
**Components of Formula (5.103) to Calculate Electrical Conductivity, S cm²/mol, According to Formula (5.87)**

| Electrolyte | $m$ | $c_k$ | $\Lambda$ | $\eta$ | $\aleph$ | $P_\Lambda$ | $E$ | $\varepsilon_S$ |
|---|---|---|---|---|---|---|---|---|
| | 0.900 | 0.887 | 116.36 | 871 | 37.83 | 277.57 | 0.001 | 73.30 |
| | 1.000 | 0.983 | 115.20 | 869 | 38.99 | 282.80 | 0.002 | 72.60 |
| | 1.200 | 1.171 | 113.53 | 866 | 40.66 | 292.80 | 0.002 | 71.40 |
| | 1.400 | 1.347 | 112.49 | 865 | 41.70 | 297.29 | 0.003 | 70.20 |
| | 1.600 | 1.522 | 111.50 | 863 | 42.69 | 303.95 | 0.003 | 69.00 |
| | 1.800 | 1.707 | 110.55 | 861 | 43.64 | 316.15 | 0.004 | 67.90 |
| | 2.000 | 1.875 | 109.63 | 860 | 44.56 | 322.12 | 0.004 | 66.90 |
| | 2.500 | 2.312 | 107.34 | 858 | 46.84 | 347.96 | 0.008 | 61.70 |
| | 3.000 | 2.744 | 105.18 | 857 | 49.00 | 375.03 | 0.012 | 57.30 |
| | 3.500 | 3.145 | 102.86 | 857 | 51.31 | 396.63 | 0.018 | 54.00 |
| | 4.000 | 3.538 | 100.54 | 858 | 53.62 | 418.72 | 0.025 | 51.30 |

*Notes:* $m$, mol/kg $H_2O$; $c_k$, mol/dm³; $\Lambda$, S cm²/mol; $\eta$, Pa s.

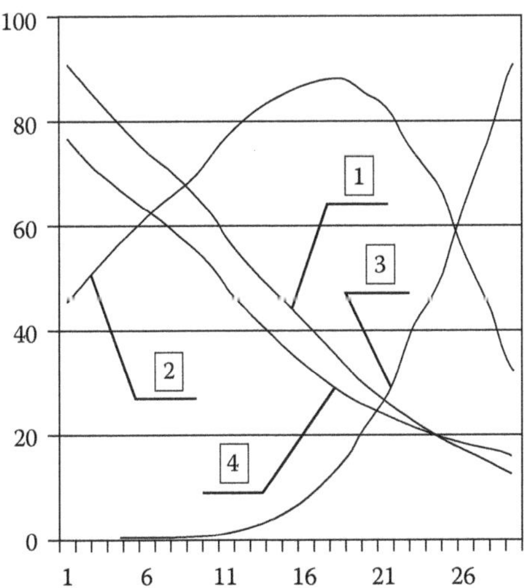


**FIGURE 5.2** Nature of behavior of interparticle forces for electrical conductivity of $CaCl_2$: 1—electrical conductivity, S cm²/mol; 2—relaxation force; 3—electrophoretic retardation; and 4—solution dielectric permittivity.

Figure 5.2 for $CaCl_2$

| Point Number | 1 | 6 | 11 | 16 | 21 | 26 |
|---|---|---|---|---|---|---|
| Concentration, $m$ | 0.20 | 0.70 | 1.50 | 2.75 | 4.00 | 5.25 |

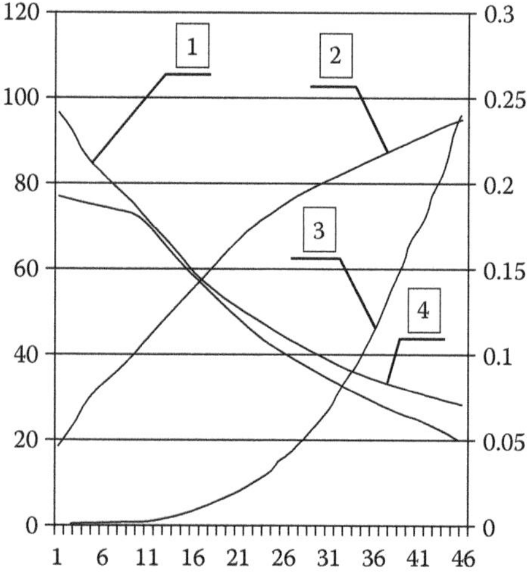


**FIGURE 5.3** Nature of behavior of interparticle forces for electrical conductivity of LiCl: 1—electrical conductivity, S cm²/mol; 2—relaxation force; 3—electrophoretic retardation; and 4—solution dielectric permittivity.

Figure 5.3 for LiCl

| **Point Number** | **1** | **6** | **11** | **16** | **21** | **26** | **31** | **36** | **41** | **46** |
|---|---|---|---|---|---|---|---|---|---|---|
| Concentration, *m* | 0.10 | 0.60 | 1.50 | 1.25 | 2.50 | 5.00 | 6.25 | 7.50 | 8.75 | 10.00 |

With regard to the data from Table 5.4, the following were chosen: 2–1 electrolyte—$H_2SO_4$ and 1–1 electrolyte—LiCl, and in Figures 5.2 and 5.3, the nature of behavior of electrical conductivity, relaxation force, electrophoretic retardation, and dielectric permittivity of the solution are shown.

The axis of abscissas has plotted numbers of concentration points, *m*, mol/kg $H_2O$ from Table 5.4, because, for example, for solution $CaCl_2$, the concentrations were taken from a minimum value of 0.2 to a maximum value of 5.25*m*, for LiCl, accordingly, 0.1–10*m*. Every number of the concentration point attributed its own concentration from Table 5.4:

The values of curve 3 (electrophoretic retardation) and 4 (solution dielectric permittivity) are plotted on the additional axis ordinates on the right in Figure 5.2. The additional axis is meant for curve 3 (electrophoretic retardation) in Figure 5.3. As it is seen from Figure 5.2, the relaxation force (curve 2) increases till decrease of values $\varepsilon_S \approx 16$. It is the third area of concentrations (see Figure 1.1) of polymer aqueous melts of crystalline hydrates or salts, hydroxides, and acids. In this area, the network of H–water bonds is completely disrupted, and steric and many other factors play a defining role in the processes of new preorganization of solution structure formation. The zone of coordination chemistry regularities appears, where the relaxation force at application of weak electric field tries to restore supramolecular formations partially destroyed at their motion toward one another. Electrophoretic retardation (curve 3) smoothly increases, and at dielectric permittivity values of $\varepsilon_S \approx 12$ and lower, its growth slows down and the *saturation* effect appears like that of curve 2. Increase of electrolyte concentration does not already result in its growth. Electrical conductivity (curve 1) in this area of low values of dielectric permittivity decreases considerably.

If we make an analysis of Figure 5.3 for 1–1 electrolyte, it can be seen that relaxation force (curve 2) and electrophoretic retardation (curve 3) approximately uniformly increase in the entire concentration area without any peculiarities. If we compare values of these forces, it can be seen that in 1–1 electrolytes, they are rather lower (Figure 5.2) than in 1–2 or 2–1 electrolytes. Nevertheless, electrical conductivity (curve 1) in the area of low values of dielectric permittivity decreases considerably (curve 2). This points to the fact that heavy electrostatic field of cation or anion 1–2 or 2–1 electrolyte somewhat differently forms structure formation in the solution than 1–1 electrolytes. This problem still needs additional investigation.

## REFERENCES


1. Sivukhin D.V. 2004. *Common Course of Physics*. In: 5. *Electricity*, Vol. 3. FIZMATLIT, Prod. in MFTI, Moscow, Russia, p. 656.
2. Salem R. 2004. *Physical Chemistry: Thermodynamics*. FIZMATLIT, Moscow, Russia, p. 352.
3. Salem R. 2010. *Physical Chemistry: Beginnings Theoretical Electrochemistry*. 2nd edn. KomKniga, Moscow, Russia, p. 320.
4. Harned G., Owen B. 1952. *Physical Chemistry of Solutions of Electrolytes*. Publishing House of Foreign Literature, Moscow, Russia, p. 628.
5. Izmailov N.A. 1976. *Electrochemistry of Solutions*. 3rd edn. Chemistry, Moscow, Russia, p. 488.
6. Robinson R.A., Stokes R.H. 1963. *Electrolyte Solutions*, Vols. 1, 2. Publishing House of Foreign Literature, Moscow, Russia.
7. Mirtskhulava I.A. 1951. Theory concentrated solutions of strong electrolytes. *J. Phys. Chem.* 25 (11): 1347–1354.
8. Mirtskhulava I.A. 1952. Theory concentrated solutions of strong electrolytes. *J. Phys. Chem.* 26 (6): 596–601.
9. Ebeling W., Feistel R., Geisler D. 1976. Zur Theorie der elektrolytischen Leitfyhigkeit bei hiheren Konzentrationen. *Z. Phys. Chem.* (Leipzig) 2: 337–353.
10. Erdey-Gruz T. 1976. *Transport Phenomena in Aqueous Solutions*. Mir, Moscow, Russia, p. 592.
11. Smedly S.I. 1980. *The Interpretation of Ionic Conductivity in Liquids*. Plenum Press, New York, p. 396.
12. Fuoss R.M. 1978. Review of the theory of electrolytic conductance. *J. Solut. Chem.* 17 (10): 771–782.
13. Baldanov M.M., Tanganov B.B., Mokhosoyev M.V. 1990. Electrical conductivity of solutions and Boltzmann's kinetic equation. *J. Phys. Chem.* 62 (1): 88–94.
14. Fuoss R.M. 1978. Conductance—Concentration function for the paired ion model. *J. Phys. Chem.* 82 (22): 2427–2440.
15. Della Monica M., Ceglia A., Agostiano A. 1984. A conductivity equation for concentrated aqueous solutions. *Electrochim. Acta* 29 (7): 933–937.
16. Fuoss R.M. 1959. Conductance of dilute solutions of 1–1 electrolytes. *J. Am. Chem. Soc.* 81 (1): 2659–2662.
17. Fuoss R.M., Onsager L. 1963. The conductance of symmetrical electrolytes. III. Electrophoresis. *J. Phys. Chem.* 67: 628–632.

18. Pitts E., Daly J., Tabor B.E. 1969. Concentration dependence of electrolyte conductance. Part 1. Comparison of the Fuoss-Onsager and Pitt's treatments. *Trans. Faraday Soc.* 65 (3): 849–862.
19. Sidebottom D.P., Spiro M. 1973. Variation of transference numbers with concentration. Test of the Fuoss-Onsager and the Pitts equations. *J. Chem. Soc. Faraday Trans. I* 69 (1): 1287–1312.
20. Fuoss R.M. 1980. Conductimetric determination of thermodynamic pairing constants for symmetrical electrolytes. *Proc. Nat. Acad. Sci. U.S.A.* 77 (1): 33–38.
21. Evans D.F., Tominaga F., Hubbard J.B., Wolines G. 1979. Ionic mobility—Theory meets experiment. *J. Phys. Chem.* 83 (20): 2669–2677.
22. Ohtsuki M. 1984. Transport properties of non-dilute solutions of strong electrolytes. I. Electric conductivity. *Chem. Phys.* 90 (1–2): 11–20.
23. Turq K., Lantelme F. 1979. Transport properties and the time evolution of electrolyte solutions in the Brownian dynamics approximation. *Mol. Phys.* 37 (1): 223–236.
24. Elkind K.M. 1983. Computational method of a direct-current conductivity of aqueous solutions of the strong electrolytes. *J. Phys. Chem.* 57 (9): 2322–2324.
25. Aseyev G.G. 1998. *Electrolytes: Transport Phenomena. Calculation of Multicomponent Systems and Experimental Data on Electrical Conductivity*. Begell House Inc., New York, p. 612.
26. Ermakov V.I., Shcherbakov V.V. 1979. High physico-chemical analysis in solutions. *Vestn. Hark. Univ*. 192: 62–66.
27. Jean-Claude J. 1983. *Treatise Elektrochem*, Vol. 5. New York, pp. 223–237.
28. Das B. 1981. Thermodynamics of electrolytes in dioxane–water mixtures from conductance data. *Thermochim. Acta.* 47 (1): 109–111.
29. Das N.C., Das B. 1980. Thermodynamics of NaCl, NaBr and NaNO in dioxane–water mixtures from conductance measurements. *Thermochim. Acta.* 41 (2): 247–252.
30. Das B. 1980. Thermodynamics of ternary salts in dioxane–water mixtures from conductance measurements. *Thermochim. Acta.* 41 (3): 371–373.
31. Fialkov Yu.A., Zhytomyrskiy A.N. 1987. *J. Phys. Chem.* 61 (2): 390–401.
32. Kuznetsova E.M. 1987. The concentration dependence of the transport numbers of strong electrolytes in aqueous solutions. *J. Phys. Chem.* 61 (10): 2794–2796.
33. Krushnyak E.G. 1980. *Electrochemistry* 16 (11): 1742–1744.
34. Ravdel A.A., Ponomareva A.M. (eds.). 2003. *Quick Reference Physico-Chemical Variables*. Chemistry, St. Petersburg, Russia, p. 232.
35. Vorobyov A.F., Scherbakov V.V., Ksenofontova N.A. 1980. Nature conductivity and ion association in electrolyte solutions. *Thermodynamic Properties of Solutions*, Vol. 1. MFTI, Interhigher Education Institution, Moscow, Russia, pp. 21–34.
36. Horvath A.L. 1989. *Handbook of Aqueous Electrolite Solutions. Physical Properties, Estimation and Correlation Methods*. New York, p. 632.
37. Enderby J.E., Neilson G.W. 1981. The structure of solutions. *Rep. Prag. Phys.* 44: 54.
38. Antropov L.I. 1984. *Theoretical Electrochemistry*. 4-e Izd. The High. Sch., Moscow, Russia, p. 519.
39. Under the editorship of Maksimova I.N. 1987. *Properties of Electrolytes. Metallurgy*, Moscow, Russia, p. 128.
40. Pak-Jon Su., Maksimova S.N. 1985. *J. Appl. Chem.* 58 (3): 491–498.
41. Pak-Jon Su., Maksimova S.N. 1984. The conductivity of solutions of alkali metal nitrates. *Ukr. Chem. J.* 50: 579–586.
42. Karelson M.M., Palm V.A. 1978. The necessity of revision of the basic principles of the theory of solutions of strong electrolytes. *Specificity and Sensitivity of the Methods to Study the Solutions and the Opportunity to Compare their Results: V. Mendeleevskaya Discussion: Mes. of Reports*. Nauka, Leningrad, Russia, pp. 77–81.

43. Krogh-Moc J. 1956. On the proton conductivity in water. *Acta. Chem. Scand.* 10: 331–332.
44. Eigen M., Maeyer L.D. 1958. Self-dissociation and protonic charge transport in water and ice. *Proc. Roy. Soc. Lond.* 24 (12): 505–533.
45. Eigen M., Maeyer L.D. 1959. *Structure of Electrolitic Solution.* New York, p. 320.
46. Chomutov H.E. 1960. On the state of protons in aqueous solutions. *J. Phys. Chem.* 34 (2): 380–386.
47. Shakhparonov M.I. 1976. *Introduction to the Modern Theory of Solutions* (*Intermolecular Interactions. Structure. Simple Liquid*). Higher School, Moscow, Russia, p. 296.
48. Shakhparonov M.I. 1964. Substances with intermolecular hydrogen bonds as proton conductors. *J. Phys. Chem.* 38 (2): 125–132.
49. Shcherbakov V.V., Ermakov V.I., Hubetsov S.B. 1973. Dielectric relaxation and electrical conductivity electrolyte solution. *Physical Chemistry Electrochemistry*, Vol. 75. tr. MKhTI them. D.I. Mendeleev, Moscow, Russia, pp. 87–95.
50. Ermakov V.I., Shcherbakov V.V. 1975. The electrical conductivity of solutions and electrical relaxation in electrolyte solutions. *Electrochemistry* 11 (2): 272–273.
51. Poltoratsky G.M. 1984. *Thermodynamic Characteristics of the Non-Aqueous Electrolyte Solutions.* Chemistry, Leningrad, Russia, p. 302.
52. Ebeling W., Grigo M. 1982. Mean spherical approximation-mass action law theory of equilibrium and conductance in ionic solutions. *J. Solut. Chem.* 11 (3): 151–167.
53. Lozar J., Schuffenecker L., Molinier J. 1992. Determination des proprietes electrochimiques et thermodynamiques de $Pb^{2+}$, $PbCl^{+}$ et $PbCl_2$ a partir de measures de conductivite de solutions aqueisws de chlorure de plomb A 25°C. *Electrochim. Acta.* 37 (13): 2519–2522.
54. Shedlovsky T. 1938. Computation of ionization constants and limiting conductance values from conductivity measurements. *J. Franklin Inst.* 255: 739–743.
55. Fuoss R.M. 1977. Boundary conditions for integration of the equation of continuity. *J. Phys. Chem.* 81 (15): 1529–1530.
56. Fuoss R.M. 1978. Paired ions: Dipolar pairs as subset of diffusion pairs. *Proc. Natl. Acad. Sci. U.S.A.* 75 (1): 16–20.
57. Fuoss R.M., Hsia K.L. 1967. Association of 1–1 salts in water. *Proc. Natl. Acad. Sci. U.S.A.* 57 (6): 1550–1557.
58. Fuoss R.M., Onsager L. 1957. Conductance of an associated electrolytes. *J. Phys. Chem.* 61 (5): 668–682.
59. Fuoss R.M., Onsager L., Skinner B. 1965. The conductance of symmetrical electrolytes. V. The conductance equation. *J. Phys. Chem.* 69 (8): 2581–2594.
60. Valyashko V.M., Ivanov A.A. 1979. On the maximum conductivity isotherms in systems electrolyte salt. *J. Inorg. Chem.* 24 (10): 2752–2760.
61. Lyashchenko A.K. 1994. Structure and structure-sensitive properties of aqueous solutions of electrolytes and non-electrolytes. *Adv. Chem. Phys. Ser.* LXXXVII: 379–426.
62. Lyaschenko A.K., Palitskaya T.A., Lileev A.C. et al. 1995. The concentration zone and the properties of aqueous solutions based on the salt compositions formates Y, Ba, Cu synthesis HTS. *J. Inorg. Chem.* 40 (7): 1209–1217.
63. Lyaschenko A.K., Lileev A.S. 1993. Concentration Zone, the interparticle interaction, and the properties of bi-and multicomponent aqueous solutions of salts of yttrium, barium and copper. *J. Inorg. Chem.* 38 (1): 144–152.
64. Enberby J.E., Nelson G.W. 1980. Structural properties of ionic liquids. *Adv. Phys.* 29 (2): 323–365.
65. Ansell S., Neilson G.W. 2000. Anion–anion pairing in concentrated aqueous lithium chloride solution. *J. Chem. Phys.* 112 (9): 3942–3944.
66. Lyaschenko A.K. 2002. Structural and molecular-kinetic properties of concentrated solutions and phase equilibria of water-salt systems. In: Kutepov A.M. (ed.). *Concentrated and Saturated Solutions.* Nauka, Moscow, Russia, pp. 93–118.

67. Lyaschenko A.K., Zasetskii A.J. 1998. Changes in the structural state, the dynamics of water molecules and properties of solutions of the transition to the electrolyte-aqueous solvent. *J. Struct. Chem.* 39 (5): 851–863.
68. Fedotova M.V., Trostin V.N., Kuznetsov V.V. 2002. In: Kutepov A.M. (ed.). *Concentrated Saturated Solutions*. Nauka, Moscow, Russia, pp. 52–70.
69. Fedotova M.V., Gribkov A.A., Trostin V.N. 2003. Structure of concentrated aqueous solutions of KBr in a wide temperature range. *J. Inorg. Chem.* 48 (10): 1668–1675.
70. Chanda A., Bagchi B. 2000. Beyond the classical transport laws of electrochemistry: New microscopic approach to ionic conductance and viscosity. *J. Phys. Chem.* 104 (39): 9067–9080.
71. Baldanov M.M., Mokhosoev M.V. 1985. The state of the ions in the electrolyte solutions in the approximation of ion plasma. *Dokl. USSR Acad. Sci.* 284 (6): 1384–1387.
72. Lee W.H., Wheaton R.J. 1978. Conductance of symmetrical, unsymmetrical and mixed electrolytes. Part 2. Hydrodynamic terms and complete conductance equation. *Chem. Soc. Faraday Trans.* 74: 1456–1482.
73. Lee E.H., Wheaton R.J. 1978. Conductance of symmetrical, unsymmetrical and mixed electrolytes. *Chem. Soc. Faraday Trans.* 1 (74): 743–766.
74. Kalugin O.H., Panchenko V.G. 2003. Interpretation of the concentration dependence of the conductivity in solutions with low dielectric considering the formation of ion pairs and triplets. *J. Phys. Chem.* 77 (8): 1463–1467.
75. Perelygin I.S., Kilitis L.L., Siskin V.I. et al. 1995. *Experimental Methods of Solution Chemistry: Spectroscopy and Calorimetry*. Nauka, Moscow, Russia, p. 198.
76. Leaist D.G. 1986. Mass transport in aqueous zinc chloride-potassium chloride electrolytes. *Ber. Bunsenges. Phys. Chem.* 90: 797–802.
77. Baldanov M.M. 1986. Approximation of ion plasmas and the theory of solutions of electrolytes. *Izv. Univ. Chem. Chem. Technol.* 29 (8): 38–44.
78. Baldanov M.M., Baldanova D.M., Zhigzhitova S.B., Tanganov B.B. 2006. Plasma-hydrodynamic theory of electrolyte solutions and electrical conductivity. *Dokl. Russ. Acad. Sci. VS* 1 (6): 25–33.
79. Baldanova D.M., Zhigzhitova S.B., Baldanov M.M., Tanganov B.B. 2004. Unified formalism charge conductivities systems: Gas plasma, plasma solids and electrolyte solutions. *Vestn. VSGTU*. 4: 5–10.
80. Baldanova D.M., Zhigzhitova S.B., Baldanov M.M., Tanganov B.B. 2004. The equivalent electrical conductivity of solutions of electrolytes in the approximation of the plasma-hydrodynamic model. *Vestn. VSGTU*. 3: 14–21.
81. Aseyev G.G. 1991. *Theory of Irreversible Processes Occurring in Solutions of Electrolytes*, Vol. 2 (304). Ser. Industry-Wide Issues. NIITEKHIM, Moscow, Russia, p. 72.
82. Aseyev G.G. 2010. Thermodynamics of interactions of electrostatic forces in concentrated electrolyte solutions. *J. Gen. Chem.* 80 (11): 1767–1773.
83. Semenchenko V.K. 1941. *Physical Theory of Solutions*. Gostekhteorizdat, Moscow, Russia, p. 368.
84. Semenchenko V.K. 1966. *Elected Heads of Theoretical Physics*. Education, Moscow, Russia, p. 396.
85. Aseyev G.G. 1998. *Electrolytes: Transport Phenomena. Methods for Calculation of Multicomponent Solutions, and Experimental data on Viscosities and Diffusion Coefficients*. Begell House Inc., New York, p. 548.
86. Zaytsev I.D., Aseyev G.G. 1992. *Properties of Aqueous Solutions of Electrolytes*. CRC Press, Boca Raton, FL, p. 1774.
87. Kon'kova A.V. 2010. *Electrical Conductivity of Solutions of Electrolytes*. STI NI YaU MIFI Publishing House, Seversk, Russia, p. 150.

# 6 Viscosity

## 6.1 SEMIEMPIRICAL APPROACHES TO VISCOSITY DETERMINATION

By now, no theoretical basis has been found for viscosity of concentrated electrolyte solutions [1], though numerous works on viscosity of solutions have been accumulated based on processing of proper experimental data and data from the world scientific literature systematized in [2]. In general terms, prevailing theories of liquid viscosity can be somewhat figuratively divided into those based on *gas* liquid and those based on *crystalline* liquid. The first of them considers liquid as having short- and long-range disorders. Theories of the second type assume that liquid has a regular structure, and momentum transfer occurs from molecules, which oscillate within lattice structure or which shift into closely spaced *holes*, or as a result of both of these phenomena. The selected crystal lattices have most diverse shapes—from cubic ones to those resembling parallel tunnels. These theories relate mainly to liquids, in which molecules do not dissociate into ions.

Problems of the theory of viscosity, its mechanism and dependence on external factors, and internal characteristics of a system are still controversial. Viscosity is a structure-sensitive property characterizing both individual and complex-composition systems. Viscosity is a measure of liquid internal friction forces, which tend to counteract any dynamic change in liquid motion [3].

Modern theories of viscosity are often of empirical and semiempirical nature, in which viscosity is considered a function of a number of parameters, such as temperature, molar or specific volume, and interaction energy [1,4]. As a rule, such models do not claim to be universal and contain a significant number of empirical parameters. Consequently, the applicability of models is limited by a set of systems and conditions.

Many researchers are interested in the thermodynamic theory of rates of chemical reactions, which is developed by Eyring [5]. Flow mechanism is understood as a sequence of equilibrium states, in relation to a certain position of molecules, which changes with flow, and liquid structure is described from the perspective of a certain amount of order. Presentation of this theory is given in many monographs, which address transfer phenomena in liquids [3,6–8].

The Eyring theory was used by researchers to process data on viscosity. For example, Good [9] studied dilute chloride solutions at 20°C. Nightingale and Benck [10] studied some M-solutions at 20°C, such as $NaNO_3$, NaOH, and $Na_2SO_4$. Miller and Doran [11] studied solutions such as NaI, LiCl, and $NaClO_4$ at 25°C and with various concentrations. The results of applying the Eyring formula to calculate electrolyte solutions viscosity were negative.

As it is shown in work [11], concentration dependence of electrolytic solutions viscosity is often explained from the perspective of the Jones–Dole empirical equation:

$$\left(\frac{\eta}{\eta_0}-1\right)\sqrt{c}=1+A+Bc, \tag{6.1}$$

where

$\eta$ and $\eta_0$ are the viscosity of a solution and a solvent, respectively
$c$ is the electrolyte concentration, mol/dm$^3$

Coefficient $A$ is associated with electrostatic interaction of dissolved ions against each other, which gains primary importance only in very dilute solutions. Its value is usually small (for nonelectrolytes it tends to zero), for which reason it is often neglected. Constant $B$ is determined by interaction of dissolved ions with the solvent, that is, according to some researchers, it characterizes ion–dipole interactions, which play a main role in most cases [12].

On the other hand, it is believed that coefficient $B$ is highly specific in relation to the nature of a solvent and the temperature [13], and is associated with entropy of ion dissolution. To discuss ion–solvent interaction, coefficient $B$ is divided into ionic components. Ions having positive values of this coefficient are called *structure-forming* ones, while ions with negative coefficient are called *structure-destroying* ones. As it is emphasized in work [14], there are several factors that influence value $B$ in various solvents. Firstly, structural failure will play a more important role in water and other *three-dimensional* solvents than in less structured solvents and, secondly, for solvents that have comparable solvation ability. This article also points out to a symbate increase in $B$, with increase in size of alkyl of tetraalkyl ammonium ion, and to the tendency of increase in $B$ with reverse dielectric permittivity in aprotic solvents medium, though direct correlation between them is not observed. There are no theoretical formulas to calculate coefficient $B$, but the coefficient value and type of the dependence on temperature and nature of an electrolyte are obtained experimentally for many substances in Kaminsky's works [15,16]. It is hard to understand such a careful attention of researchers to an ordinary empirical coefficient.

These studies relate to dilute solutions. Equation 6.1 is consistent with experiments up to concentration of approximately 0.8 mol/dm$^3$. Introducing a term into it, which contains $c^2$, it is possible to expand its range of applicability toward somewhat larger concentrations.

In physical–chemical works, modified Einstein's equation [13] is used to process experimental data on electrolyte solutions viscosity:

$$\frac{\eta}{\eta_0}=1+2.5V_{ef}c_k+kV_{ef}^2c_k^2, \tag{6.2}$$

where

$V_{ef}$ is the *effective incompressible molar volume*
$k$ is the empirical coefficient

Einstein's equation is valid for dilute suspensions, in which flow lines around adjacent particles do not interact against each other. However, in more concentrated suspensions, interaction of flow lines around adjacent particles becomes significant, that is, turbulence appears. Under these conditions, empirical Vand equation [17] is used:

$$\ln\frac{\eta}{\eta_0}=1+\frac{2.5V_{ef}}{1-QV_{ef}}, \tag{6.3}$$

where $Q$ is the interaction factor, which can be calculated only approximately, not ab initio.

Eagland and Pilling showed that Vand equation is qualitatively valid for concentrated solutions of tetraalkyl ammonium bromides [18] as well. It is clear that this approach is not acceptable for electrolyte solutions.

Angell et al. in their work [19] approached the question of viscosity of aqueous electrolyte solutions somewhat unusually. The authors of this work do not distinguish between solvent and solute; thus, their approach is applicable also to concentrated solutions, in which it is already impossible to distinguish sufficiently large areas that contain pure solvent. Based on the ideas used while describing relaxation processes of vacancy formation in liquid, they generalized the results obtained while measuring viscosity and electrical conductivity of solutions of $Ca(NO_3)_2$ in concentrations of 0–26 mol%, with temperature range from −60°C to 80°C. To describe the dependence of viscosity on concentration at this temperature throughout the range from most dilute solutions to saturated ones, the following semiempirical equation is proposed:

$$\frac{1}{\eta}=A\exp\left(-\frac{B}{x_0-x}\right), \tag{6.4}$$

where

$x$ is the concentration, mol%

$A$ and $B$ are the constants

In accordance with this work, in order to explain the properties of electrolytes of this type, the whole concentration range shall be divided into three areas (approximately, as it is construed in the first part of this work): (1) the segment of low concentrations (dilute solutions), where, in accordance with the general notions, ions surrounded by hydration spheres are dispersed in water, which generally maintains its structure; (2) the section of intermediate concentrations, where solutions have very complicated dependences of properties on concentrations; and (3) the section of relatively high concentrations, where the solution has no areas containing pure water and where all particles in the liquid strongly interact against each other during their motion. It seems that microscopic nonhomogeneities and areas that contain vacancy chains appear in the structure of these solutions. Associations of water molecules and associations of hydrated ions are continuously formed and decomposed in such solutions, and the size of these associations increases with temperature decrease.

### 6.1.1 Multicomponent Solutions

In work [19], the Berecz equation is used to describe the dynamic viscosity of three-component water–acid–salt electrolytic systems (with common anion):

$$\eta = \eta_0 + (1 + Dn + An^2), \tag{6.5}$$

where *D* is the coefficient that, just like in Jones–Dole equation, is additively composed of ionic components, which, in turn, characterize the effect of this ion on solvent structure, that is, coefficient *D* is similar to coefficient *B* in Jones–Dole equation. Coefficient *A* of the Berecz equation characterizes ion–ion interaction of electrolytes and reflects the electrostatic interaction between solvated ions in electrolytic solutions. Electrolytic component concentration, in contrast to Jones–Dole equation, is expressed in mole fractions *n*, which does not depend on temperature. It is impossible to establish exact boundaries of the applicability of the Jones–Dole equation. Concentration areas, which are dominated by one term or another, can be specified roughly. For different electrolytes, they are defined by different adjustable values of coefficients *A* and *B*, which are only qualitatively dependent on properties of the ions, which make up an electrolyte.

In order to describe temperature dependence of viscosity of an electrolyte solution and some other liquids in the range from −10°C to 150°C, Eicher and Zvolensky give the expression containing three empirical parameters [6]:

$$\ln \frac{\eta_0}{\eta} = \frac{A(t-20) + B(t-20)^2}{C+t}, \tag{6.6}$$

where *A*, *B*, and *C* are the constants characterizing this solution.

These and similar studies relate to very dilute solutions. For a wider range of concentrations, there are only empirical formulas for solutions of particular electrolytes, which are valid in an area of certain concentration at one or several temperatures [19,21].

Justified molar additivity of viscosity function in double liquid systems laid the foundation for the analysis of geometry of viscosity isotherms in binary solvents [22–24]. The author of these works classified viscosity isotherms of binary systems depending on the degree of interaction between components. He showed on model systems the dependence of geometry of viscosity isotherms of excess logarithm on the depth of interaction of the components, which make up a binary mixture. For systems characterized by the absence of chemical interaction, viscosity is described by the product of exponential functions of the following form:

$$\eta_{sys} = \prod_i \eta^{x_i}. \tag{6.7}$$

Other equations are also proposed to calculate viscosity [25–27], which are usually less general and are applicable to a particular range of systems with similar physical and chemical properties.

Formula (5.106) performed well for practical calculations (see description to it).

## 6.2 VISCOSITY THEORY

To construct a theory of viscous flow, it is quite substantial from what ideas about the nature of liquid an author of a theory is proceeding, because the times of purely phenomenological interpretation of the issue have already passed, and every author of a theory faces the question of what the elementary process of flowing is and what major characteristics of liquid state are to be set, in order to create a theory that can provide a quantitative expression for viscosity change with temperature, pressure, etc., and it will be well in line with experiments. Description of theories of viscous flow of non-associated liquids for the first half of the last century is found in excellent overviews by Bak [28] and Leont'eva [29]. But we are interested in electrolyte solutions.

Quite a complicated dependence of electrolyte solution viscosity on temperature and pressure is presented in Panchenkov's theory [30]. The derivation of the dependence is based on the mechanism of momentum transfer during joint simultaneous movements of two molecules from adjacent layers of liquid. And the idea is justified that the transfer is performed only when total kinetic energy is less than the bonding energy $\varepsilon$ between them. Using Newton's laws of mechanics, as well as Maxwell–Boltzmann statistics for distribution of particles according to energies, the author obtained the following expression for viscosity:

$$\eta = A\rho^{4/3}T^{1/2}e^{-(\varepsilon/kT)}\left(1-e^{-(\varepsilon/kT)}\right), \tag{6.8}$$

where

$A=\left(12\sqrt{R}\right)/\left(\sqrt{\pi}\sqrt[3]{N_A}\right)\omega^{2/3}M^{-(5/6)}e^{S/R}$

$\rho$ is the density

$T$ is the temperature

$R$ is the gas constant

$\omega$ is the occupied volume of molecules per mole

$M$ is the molecular weight of liquid

$S$ is the bond formation entropy

$N_A$ is the Avogadro number

It is obvious that (6.8) has a significant drawback, associated with the fact that electrolyte concentration is absent here and is taken into account implicitly through density.

On the basis of experimental data processing according to Panchenkov's theory, it is also possible to draw some conclusions about interparticle interactions and structural changes in solutions, depending on temperature, concentration, and nature of a dissolved electrolyte, because Panchenkov's final equation (6.8) includes the bonding energy $\varepsilon$ and the bonding formation entropy $S$. However, viscosity data processing according to Panchenkov runs into difficulties, because Equation 6.8 includes the occupied volume of molecules $\omega$, which changes with concentration, and it is not easy to take it into account.

The theory of medium viscosity change under the influence of electrostatic forces acting between ions for cases of binary electrolytes was successfully developed by Falkenhagen for the first time, and then it was improved by Onsager and Fuoss for their mixtures [31]. They obtained the following equation:

$$\frac{\eta^*}{\eta} = \frac{\kappa R}{80} = \frac{\eta - \eta_0}{\eta}, \tag{6.9}$$

where

$\eta^*$ is the electrostatic component of viscosity
$\kappa$ is the inverse radius of ion atmosphere
$R$ is the ion radius

In the first approximation in (6.9), relative reduction of viscosity is proportional to the ratio of ion radius to radius of its atmosphere.

Later, Falkenhagen obtained theoretical equation [32] to describe the concentration dependence of viscosity of aqueous electrolyte solutions:

$$\frac{\eta}{\eta_0} = 1 + A\sqrt{c}, \tag{6.10}$$

where

$c$ is the concentration, mol/dm$^3$
$A$ is the constant, which is a function of properties of a solvent, mobilities and charges of ions, and temperature

The drawbacks of Equations 6.9 and 6.10 are the same as in the Debye–Hückel theory, underestimation of other interparticle interactions, except electrostatic ones.

Various theories of viscosity are based on the assumption that formation of vacancies in liquid creates conditions for viscous flow. These assumptions are consistent with temperature dependence of viscosity, which is observed in experiments. At constant pressure, viscosity decreases noticeably when temperature increases; on the contrary, at constant volume, temperature coefficient of viscosity is very low compared to the coefficient at constant pressure. If molecule motion requires the presence of vacancies, then it can be assumed that liquid viscosity will be proportional to the number of vacancies in a unit of volume. However, on the basis of some other experiments, it can be concluded that the volume of molecules is almost unchanged with compression of liquid or temperature increase, while the volume and number of vacancies decrease. Thus, in case of volume constancy, the number of vacancies is almost unchanged with temperature increase. This explains very low temperature coefficient of viscosity at constant volume.

The validity of vacancies theory, as applied to nonassociated liquids, is also proved by the formula of Bachinsky, who found the connection of viscosity of liquid with its specific volume, which had a great role in the development of liquid viscosity theory [33]:

$$\eta = \frac{B}{V - b}, \tag{6.11}$$

where

$B$ and $b$ are constants
$V$ is the specific volume

The author assumes that viscosity should depend on the specific volume of a liquid. Consequently, with temperature increase, the distance between liquid molecules will increase, adhesion forces will be lower, and thus viscosity will be lower. Bachinsky formula (6.11) is an approximate one and is not applicable to associated liquids with complicated molecular structure, but, in fact, it gives true grounds for relation of viscosity with the ability of molecules to move freely in liquid volume.

Investigation of viscosity of a large number of electrolyte solutions by Lyaschenko and his colleagues [34] shows that its concentration changes have common characteristic features and are similar to those observed in typical polymerization processes. In electrolyte solutions, values $\eta/\eta_0 = \eta_{H_2O}/c_k$, which characterize viscosity change under the effect of a unit charge in a solution, pass through a minimum and are explained by the occurrence of polymeric ionic groups with high viscosity. For the third area of concentration range, these assumptions are out of doubt.

Baldanov and his colleagues [35] proposed a model to evaluate the viscosity of binary electrolyte solutions within the plasma hydrodynamic model. So, the following equation is proposed for viscosity:

$$\eta = \frac{1.11 \cdot 10^{-12} N_A e^2 \exp(-(\hbar\omega/2kT))}{6\pi\Lambda R_S(1 + (R_S/r_D))}. \tag{6.12}$$

The same remarks remain to Equation 6.12, as to dependence (5.8). Errors of calculation by (6.12) for concentrated solutions reach 35%–40%.

From the analysis of works [36–38] on kinetic properties of electrolyte solutions, it can be concluded that in order to study the concentration dependence of properties of a system of charges, it is necessary to introduce a number of assumptions, adjustable parameters, and allowances for changes in properties of a solute and a solvent, in particular for viscosity change.

From the previous discussion, it can be concluded that the solution theory pays much attention to viscosity study, but one can meet conflicting data on values $\eta$ defined for the same systems. It is still important to develop a quantitative theory based on interparticle interactions potential, obtained in the first part of our presentation, to study viscosity of electrolyte solutions.

## 6.3 GENERAL THEORY OF VISCOSITY

Within the plasma hydrodynamic model [39–41], Baldanov and his colleagues first obtained an equation to calculate electrical conductivity (5.8), which includes solution viscosity, and then derived viscosity from it and obtained dependence (6.12) [31]. Surprisingly, the errors of calculation of electrical conductivity according to (5.8) are quite satisfactory for average concentrations area, and the errors of viscosity calculation according to (6.12) are quite big. We may also go the same way and express viscosity from Equations 5.87, 5.103 through 5.105, and we obtain the following:

$$\eta = \frac{E_\eta}{\Lambda^0 - \aleph - \Lambda}, \tag{6.13}$$

where

$$E_\eta = \frac{397\beta^2 \nu_k c_k \beta_R}{\varepsilon_S}\left(\left|z_i\right| + \left|z_j\right|\right). \tag{6.14}$$

Formula (6.13) calculates solution viscosity absolutely accurately. It was the viscosity value set in formula (5.105) that was obtained. However, it is difficult to explain why viscosity directly depends on the relaxation force of electrical conductivity. Therefore, we proceed to the viscosity theory consideration.

To describe the viscosity theory, Onsager, Falkenhagen, and Fuoss were considering a solution located between two parallel plates *A* and *B*. The plates are spaced apart at distance *h*, as shown in Figure 6.1 [33]. Plate *B* moves at constant speed *V′* in *x*-axis direction, capturing solution particulates. Equilibrium state of ion atmosphere is disrupted by liquid flow, due to which ions in various areas of the ion atmosphere obtain various speeds that depend on the speed of the surrounding solution. In the flowing solution, all the ions located below the central ion move slower than it does and go beyond the limits corresponding to the immobile ion atmosphere. Due to the fact that plate *B* is moving and plate *A* isn't, the ions located above the central ion

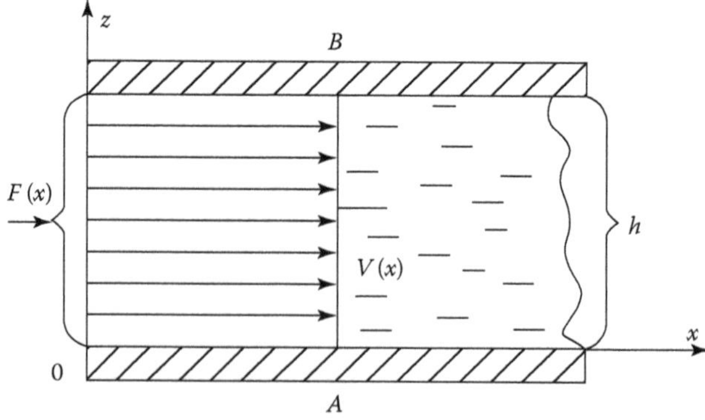


**FIGURE 6.1** Solution shear flow.

move faster and run ahead of it and also go beyond the immobile ion atmosphere. The mathematical interpretation of this situation is very complicated, only the electrostatic theory is involved, and interparticle interactions are completely ignored.

The solution flow model can be presented in a much simpler way. We will consider an electrolyte solution as located between two plates $A$ and $B$. The solution moves along the $x$-axis at constant speed $V'$ under the effect of constant force $F_{yz}$ in plane $yoz$. The distance between the plates is $h$ (see Figure 6.1). The mathematical representation of such model of liquid flow is greatly simplified.

In the $x$-axis direction, constant speed gradient is

$$x\frac{\partial V_y}{\partial x}=\frac{V'}{h}, \quad V_y=0, \quad V_z=0. \tag{6.15}$$

We recall that viscosity coefficient $\eta$ is the voltage related to a unit of area, which is transferred from each layer of liquid to an adjacent layer with the speed gradient equal to one; in other words, it is the force related to a unit of surface, with which the part of the medium located under plane $B$ is acting on the part of the medium located above plane $A$. In the case of laminar motion,

$$F_{yz}=\eta\frac{\partial V_y}{\partial x}. \tag{6.16}$$

In an undisturbed electrolyte, a structure is observed, which is conditioned by long-range and short-range interactions (see Section I). Speed gradient in liquid disrupts this structure, which is restored in some time of relaxation $\tau$. If we consider the general motion of liquid, then speed $V$ at the point located on vector $\vec{R}$, which starts from the origin of coordinates, has projections of components $V_y$ and $V_z$, and the voltage component onto specific surface through the plane, which is perpendicular to the $x$-axis, is

$$F_{yz}=\eta\left(\frac{\partial V_z}{\partial y}+\frac{\partial V_x}{\partial z}\right)=F_{zy}. \tag{6.17}$$

If $F_{yz}^0$ is the general element of voltage matrix for movement in a pure solvent, then

$$F_{yz}-F_{yz}^0=F_{yz}^*$$

is the average combined voltage due to interparticle forces action, and, respectively,

$$\eta-\eta_0=\eta^* \tag{6.18}$$

will increase (decrease) the viscosity of the solution compared with the solvent, which is to be found.

We will consider the general continuity equation (5.26). This equation contains all basic physical prerequisites necessary to resolve the problem of viscosity. While studying the problem of viscosity, external forces $\partial f_{ij}/\partial t$, $\partial f_{ji}/\partial t$ and $\vec{F}(R_i)$, $\vec{F}(R_j)$ are equal to zero, and the continuity equation acquires the following form:

$$\vec{V}(R_i)\nabla_i f_{ij}^0\left(\vec{R}\right)+\vec{V}(R_j)\nabla_j f_{ji}^0\left(\vec{R}\right)$$
$$-\omega_i|z_i|e\left\{\nabla_i\left[f_{ij}^0\left(\vec{R}\right)\nabla_i\psi'_{ij}(R_{ij})\right]\right\}-\omega_j|z_j|e\left\{\nabla_j\left[f_{ji}^0\left(\vec{R}\right)\nabla_j\psi'_{ji}(R_{ji})\right]\right\}$$
$$-\omega_i kT\nabla_i\cdot\nabla_i f'_{ij}(R_{ij})-\omega_j kT\nabla_j\cdot\nabla_j f'_{ji}(R_{ji})=0. \tag{6.19}$$

Further, we take into account that functions $f'$ are symmetric, that is, $f'_{ij}(R_{ij})=f'_{ji}(R_{ji})$ [42]. Then, we obtain

$$\left[\vec{V}(R_i)+\vec{V}(R_j)\right]\nabla f^0-(\omega_i z_i e+\omega_j z_j e)\nabla(f^0\nabla\psi')-(\omega_i+\omega_j)kT\nabla\cdot\nabla f'=0. \tag{6.20}$$

Then, we recall the structure of the solution involved in shear flow in Figure 5.2. In the electrical conductivity theory, complexes of ions $i$ and $j$ move toward each other under the action of an electric field. In the viscosity theory, complexes of ions $i$ and $j$ form a single supermolecule, in which the force acting along the $x$-axis generates distortion and transformation of, one can say, a sphere-like supermolecule into some ellipsoidal or any other one. This additional force aims to reduce the relative speed of supermolecules located in adjacent layers of the solution and causes increase in viscosity, which is a quantitative measure of mobility of solution layers relative to each other.

Based on the previous discussion, we will consider not an independent speed of complexes of $i$ and $j$ ions $\vec{V}(R_i)+\vec{V}(R_j)$ but the speed of a supermolecule

$$\vec{V}(R_i)+\vec{V}(R_j)=\vec{V}_{ij}(R_{ij})=\vec{V}_{ji}(R_{ji}).$$

We assume that radius vector $R_{ij}$, which defines motion of a supermolecule, is directed from the origin of coordinates to its center. At a distance of $R(x,y,z)$ point, the speed vector is equal to

$$\vec{V}(R_{ij})=\vec{V}_{x,ij}+\vec{V}_{y,ij}+\vec{V}_{z,ij}=x\left(\frac{\partial V_{y,ij}}{\partial x}\right)+y\left(\frac{\partial V_{z,ij}}{\partial y}\right)+z\left(\frac{\partial V_{x,ij}}{\partial z}\right)\equiv q_\lambda a_\lambda, \tag{6.21}$$

where $\lambda$ denotes summation and

$$q_1=x,\quad q_2=y,\quad q_3=z,$$

and, omitting subscripts $ij$,

$$a_x = \frac{\partial V_y}{\partial x}, \quad a_y = \frac{\partial V_z}{\partial y}, \quad a_z = \frac{\partial V_x}{\partial z}. \tag{6.22}$$

For further presentation, it is more convenient to introduce a new function:

$$\nabla f^0(R) = \frac{\partial \theta(R)}{R^2 \partial R}. \tag{6.23}$$

With regard to (6.21) and (6.23), the square bracket of Equation 6.20 looks like

$$\vec{V}(R_{ij}) \nabla f^0 = q_\lambda a_\lambda \frac{\partial \theta(R)}{R^2 \partial R}. \tag{6.24}$$

We return to Equation 5.14 and present it in the form of

$$f' = -\varepsilon_0 \varepsilon_S \sum_{i,j} \frac{n_j}{|z_i| e} \nabla \cdot \nabla \psi', \quad (i \neq j). \tag{6.25}$$

We take into account (6.24) and (6.25) in Equation 6.20 and obtain

$$q_\lambda a_\lambda \frac{\partial \theta(R)}{R^2 \partial R} - \sum_{i,j} \left(\omega_i z_i e + \omega_j z_j e\right) \nabla \left(f^0 \nabla \Psi'\right)$$

$$+ \varepsilon_0 \varepsilon_S \sum_{i,j} \frac{n_j}{|z_i| e} \left(\omega_i + \omega_j\right) kT \left(\nabla \cdot \nabla\right)^2 \Psi' = 0, \quad (i \neq j). \tag{6.26}$$

We expand function $\Psi'$ into series and confine ourselves to the first term of the series, and we assume that the extra potential in the theory of electrical conductivity $\Psi'$ depends on the extra potential in the theory of viscosity $\xi$ as follows:

$$\Psi' = q_\lambda a_\lambda \frac{\partial \xi(R)}{R^2 \partial R} + \cdots \tag{6.27}$$

We substitute (6.27) into (6.26), and after some transformations, we have

$$q_\lambda a_\lambda \frac{\partial}{R^2 \partial R} \left[ \begin{array}{l} \left(\nabla \cdot \nabla\right)^2 \xi - \dfrac{1}{\varepsilon_0 \varepsilon_S kT} \displaystyle\sum_i \dfrac{\left(\omega_i z_i e + \omega_j z_j e\right) |z_i| e}{\left(\omega_i + \omega_j\right) n_j} \nabla \left(f^0 \nabla \xi\right) \\ + \dfrac{1}{\varepsilon_0 \varepsilon_S kT} \displaystyle\sum_i \dfrac{\theta |z_i| e}{\left(\omega_i + \omega_j\right) n_j} \end{array} \right] = 0. \tag{6.28}$$

In (6.28), it is necessary to find value θ, which is determined according to dependence (6.23): we write it down in the following form:

$$\frac{\partial \theta^0}{\partial R} = R^2 \nabla f^0. \tag{6.29}$$

We integrate left and right sides of (6.29), and we calculate the integral of the right side in parts. Then, we obtain

$$\theta = R^2 f^0 - 2\int R f^0 \mathrm{d}R. \tag{6.30}$$

We take into account in (6.29) that value $f^0$ is equal to (5.32) and $\varphi^0$ is determined according to (5.52). Then, the product of $R^2 f^0$ is equal to

$$R^2 f^0 = -2n_i n_j \left[\left(\frac{R^3}{\beta}\right) - \frac{5}{3}\beta R - \frac{16}{45}\beta^2\left(\frac{\beta}{R}\right) - \frac{17}{360}\beta^2\left(\frac{\beta}{R}\right)^3\right], \tag{6.31}$$

and integral expression $-2\int R f^0 \mathrm{d}R$ looks as follows:

$$\begin{aligned} -2\int R f^0 \mathrm{d}R &= 4n_i n_j \int \left[\left(\frac{R^2}{\beta}\right) - \frac{5}{3}\beta - \frac{16}{45}\beta\left(\frac{\beta}{R}\right)^2 - \frac{17}{360}\beta\left(\frac{\beta}{R}\right)^4\right] \mathrm{d}R \\ &= 4n_i n_j \left[\left(\frac{R^3}{3\beta}\right) - \frac{5}{3}\beta R - \frac{16}{135}\left(\frac{\beta}{R}\right)^3 - \frac{17}{1800}\left(\frac{\beta}{R}\right)^5 + C_1\right]. \end{aligned} \tag{6.32}$$

With expressions (6.31) and (6.32), value θ is equal to

$$\begin{aligned} \theta = &-2n_i n_j \left[\left(\frac{R^3}{\beta}\right) - \frac{5}{3}\beta R - \frac{16}{45}\beta^2\left(\frac{\beta}{R}\right) - \frac{17}{360}\beta^2\left(\frac{\beta}{R}\right)^3\right] \\ &+ 4n_i n_j \left[\left(\frac{R^3}{3\beta}\right) - \frac{5}{3}\beta R - \frac{16}{135}\left(\frac{\beta}{R}\right)^3 - \frac{17}{1800}\left(\frac{\beta}{R}\right)^5 + C_1\right]. \end{aligned} \tag{6.33}$$

Integration constant $C_1$ is chosen based on condition $\theta \to 0$ with $R \to \infty$. Then, $C_1 = 2n_i n_j(R^3/\beta - 5/3\beta R) - 4n_i n_j(R^3/3\beta - 5/3\beta R)$, and expression (6.33), after simple transformations, acquires the following form:

$$\theta = -2n_i n_j \left[-\frac{16}{45}\beta^2\left(\frac{\beta}{R}\right) - \frac{17}{360}\beta^2\left(\frac{\beta}{R}\right)^3 + \frac{32}{135}\left(\frac{\beta}{R}\right)^3 + \frac{17}{900}\left(\frac{\beta}{R}\right)^5\right]. \tag{6.34}$$

We estimate the values of summands in (6.34). The second and the third summands with equal powers differ from each other by value $\beta^2$, and we recall that $\beta \approx 10^{-9}$, so the second summand can be ignored without serious consequences. By the same reasoning, we ignore the first summand as well. Then, finally,

$$\theta = -\frac{2}{45} n_i n_j \left[ \frac{32}{3} \left( \frac{\beta}{R} \right)^3 + \frac{17}{20} \left( \frac{\beta}{R} \right)^5 \right]. \tag{6.35}$$

We return to Equation 6.28. With regard to (6.35), it looks like

$$q_\lambda a_\lambda \frac{\partial}{R^2 \partial R} \left[ \begin{array}{l} (\nabla \cdot \nabla)^2 \xi - \dfrac{1}{\varepsilon_0 \varepsilon_S kT} \displaystyle\sum_i \dfrac{(\omega_i z_i e + \omega_j z_j e)|z_i| e}{(\omega_i + \omega_j) n_j} \nabla (f^0 \nabla \xi) \\ - \dfrac{2}{45 \varepsilon_0 \varepsilon_S kT} \left[ \dfrac{32}{3} \left( \dfrac{\beta}{R} \right)^3 + \dfrac{17}{20} \left( \dfrac{\beta}{R} \right)^5 \right] \displaystyle\sum_i \dfrac{n_i |z_i| e}{(\omega_i + \omega_j)} \end{array} \right] = 0. \tag{6.36}$$

All quantities in the square bracket of Equation 6.36 are functions of $R$, so we can write down

$$q_\lambda a_\lambda \frac{\partial F(R)}{R^2 \partial R} = 0. \tag{6.37}$$

Then,

$$F(R) = 0,$$

or

$$(\nabla \cdot \nabla)^2 \xi - \frac{1}{\varepsilon_0 \varepsilon_S kT} \sum_i \frac{(\omega_i z_i e + \omega_j z_j e)|z_i| e}{(\omega_i + \omega_j) n_j} \nabla (f^0 \nabla \xi)$$

$$= \frac{2}{45 \varepsilon_0 \varepsilon_S kT} \left[ \frac{32}{3} \left( \frac{\beta}{R} \right)^3 + \frac{17}{20} \left( \frac{\beta}{R} \right)^5 \right] \sum_i \frac{n_i |z_i| e}{(\omega_i + \omega_j)}. \tag{6.38}$$

As Harned and Owen [31] of remarked in chapter 5, according to Onsager, the structure of the left side of the continuity equation for electrical conductivity determination is identical to the structure of the left side of the extra potential determination in the viscosity theory. In our case as well, the structures on the left sides of the continuity equation for electrical conductivity determination (5.31) and (6.38) coincide, though they are obtained by an entirely different method than in the Onsager and Fuoss theory.

## 6.4 SOLUTION VISCOSITY COEFFICIENT (GENERAL SOLVING)

To calculate the component of voltage $BF_{yz}^{*}$ conditioned by interaction of supermolecular formations, their potentials should be known, which, in case of irreversible processes, can be determined only by means of the general continuity equation. Previously, it is necessary to calculate the force caused by interaction of supermolecular formations in two adjacent layers. The force of interaction between two complexes in a dilute solution is equal to

$$\frac{1}{4\pi\varepsilon_0\varepsilon_S}\left|z_i z_j\right|e^2\frac{\vec{R}_{ij}}{R^3},$$

where $\vec{R}_{ij}$ means the vector directed from the first complex to the second one. Later, we will extend the force of interaction between complexes to supermolecular formations as well.

The force of interaction between the $i$th supermolecule, located in the element of volume $\mathrm{d}V_i$, with supermolecules, located in the element of volume $\mathrm{d}V_j$, has the following form:

$$\frac{\left|z_i\right|e}{4\pi\varepsilon_0\varepsilon_S}\sum_j\left|z_j\right|en_{j\,sr}\left(\vec{R}_{ji}\right)\mathrm{d}V_j\frac{\vec{R}_{ji}}{R^3},$$

and the total value of interaction force for all pairs of supermolecules in volumes $\mathrm{d}V_i$ and $\mathrm{d}V_j$ is

$$\frac{1}{4\pi\varepsilon_0\varepsilon_S}\sum_i\left|z_i\right|en_i\mathrm{d}V_i\sum_j\left|z_j\right|en_{j\,sr}\left(\vec{R}_{ji}\right)\mathrm{d}V_j\frac{\vec{R}_{ji}}{R^3}$$

$$=\frac{1}{4\pi\varepsilon_0\varepsilon_S}\sum_i\sum_j\left|z_i z_j\right|e^2 n_i n_{j\,sr}\left(\vec{R}_{ji}\right)\mathrm{d}V_i\mathrm{d}V_j\frac{\vec{R}_{ji}}{R^3}$$

$$=\frac{1}{4\pi\varepsilon_0\varepsilon_S}\sum_{i,j}\left|z_i z_j\right|e^2 f_{i,j}\left(\vec{R}_{ij}\right)\mathrm{d}V_i\mathrm{d}V_j\frac{\vec{R}_{ji}}{R^3}$$

$$=-\frac{1}{4\pi\varepsilon_0\varepsilon_S}\sum_{j,i}\left|z_i z_j\right|e^2 f_{j,i}\left(\vec{R}_{ji}\right)\mathrm{d}V_j\mathrm{d}V_i\frac{\vec{R}_{ji}}{R^3}. \tag{6.39}$$

Thus, according to Equation 2.19, distribution function $f_{j,i}(\vec{R}_{ji})$ appeared in the last summand of Equation 6.39. By integrating in terms of all the elements of volume, which satisfy the condition $x_1 < x_2$, it is possible to calculate the value of total voltage $BF_{yz}^*$ between supermolecules in the $x$-axis direction:

$$-BF_{yz}^* = \frac{1}{4\pi\varepsilon_0\varepsilon_S} \iiint\limits_{x_1<} \iiint\limits_{x_2} \sum_{j,i} |z_i z_j| e^2 f_{ji}\left(\vec{R}_{ji}\right) \frac{x_{ji}}{R^3} \mathrm{d}V_2 \mathrm{d}V_1. \tag{6.40}$$

Thus, if $B = \iint_{x_1<x_2} \mathrm{d}y_1 \mathrm{d}z_1$, then

$$-F_{yz}^* \iint\limits_{x_1<x_2} \mathrm{d}y_1 \mathrm{d}z_1 = \frac{1}{4\pi\varepsilon_0\varepsilon_S} \iiint\limits_{x_1<} \iiint\limits_{x_2} \sum_{j,i} |z_i z_j| e^2 f_{ji}\left(\vec{R}_{ji}\right) \frac{x_{ji}}{R^3} \mathrm{d}V_2 \mathrm{d}x_1 \mathrm{d}y_1 \mathrm{d}z_1, \tag{6.41}$$

or, canceling out certain integrals, we obtain

$$-F_{yz}^* = \frac{1}{4\pi\varepsilon_0\varepsilon_S} \iint\limits_{x_1<} \iint\limits_{x_2} \sum_{j,i} |z_i z_j| e^2 f_{ji}\left(\vec{R}_{ji}\right) \frac{x_{ji}}{R^3} \mathrm{d}V_2 \mathrm{d}x_1. \tag{6.42}$$

Integrating in terms of $x$ and replacing $(x_2 - x_1)$ through $x_{ji}$, we find

$$-F_{yz}^* = \frac{1}{4\pi\varepsilon_0\varepsilon_S} \iiint\limits_{x_1<x_2} \sum_{j,i} |z_i z_j| e^2 f_{ji}\left(\vec{R}_{ji}\right) \frac{x_{ji}^2}{R^3} \mathrm{d}V_{ji}, \tag{6.43}$$

Taking the average from both parts of (6.43), we have

$$F_{yz}^* = -\frac{1}{8\pi\varepsilon_0\varepsilon_S} \iiint\limits_{x_1<x_2} \frac{x^2}{R^3} \sum_{j,i} |z_i z_j| e^2 f_{ji}\left(\vec{R}\right) \mathrm{d}V. \tag{6.44}$$

In Equation 6.44, we need to find dependence $f_{ji}(\vec{R})$ on $\xi(\vec{R})$. First, we write down the dependence of distribution function on potential through Poisson equation in the following form:

$$f_{ji} = -4\pi\varepsilon_0\varepsilon_S \nabla \cdot \nabla\Psi \sum_i \frac{n_j}{|z_i| e}, \tag{6.45}$$

then, we take into account (6.27), and together with (6.45), we substitute it into (6.44):

$$F_{yz}^* = \frac{1}{2} \iiint\limits_{x_1<x_2} \frac{x^2}{R^3} \sum_j n_j |z_j| e q_\lambda a_\lambda \frac{\partial \nabla \cdot \nabla \xi}{R^2 \partial R} \mathrm{d}V. \tag{6.46}$$

We introduce spherical coordinates. We choose integration limits in the first coordinate plane, based on the fact that the solution moves along the $x$-axis. Then, taking into account (6.21), we obtain

$$F^*_{yz} = \frac{xa_x}{2}\sum_j n_j |z_j| e \int\int_0^{\pi/2}\int_0^{\pi/2} \frac{R^5}{R^3}\sin^4\theta\cos\theta\sin\varphi\cos^3\varphi\frac{\partial\nabla\cdot\nabla\xi}{R^2\partial R}\mathrm{d}R\mathrm{d}\theta\mathrm{d}\varphi$$

$$= \frac{xa_x}{40}\sum_j n_j |z_j| e \int \frac{\partial\nabla\cdot\nabla\xi}{\partial R}\mathrm{d}R = \frac{xa_x}{40}\sum_j n_j |z_j| e\nabla\cdot\nabla\xi + C_2, \tag{6.47}$$

where $xa_x$ is the change in the speed of the $i$th supermolecule in relation to the $j$th one along the $x$-axis through plane $yz$. We will determine integration constant $C_2$ later. Consequently, the electrostatic component of viscosity $\eta^*$ is equal to

$$\eta^* = \frac{1}{40}\sum_j n_j |z_j| e\nabla\cdot\nabla\xi + C_2. \tag{6.48}$$

As discussed earlier, the calculation of the extra potential of viscosity $\nabla\cdot\nabla\xi$ can be performed only by means of the general continuity equation (6.38).

## 6.5 SOLUTION VISCOSITY COEFFICIENT (FINAL SOLVING)

We return to Equation 6.38. We introduce Hamiltonian operator under the bracket and take into account that distribution function $f^0$ is determined by (5.32), and after simple transformations, we obtain

$$\nabla Y = \frac{2}{45\varepsilon_0\varepsilon_S kT}\left[\frac{32}{3}\left(\frac{\beta}{R}\right)^3 + \frac{17}{20}\left(\frac{\beta}{R}\right)^5\right]\sum_i \frac{n_i |z_i| e}{\omega_i + \omega_j}, \tag{6.49}$$

where

$$Y = \nabla^2\cdot\nabla\xi - \frac{2}{\varepsilon_0\varepsilon_S kT}\sum_i \frac{(\omega_i z_i e + \omega_j z_j e) n_i |z_i| e}{\omega_i + \omega_j}\varphi^0\nabla\xi. \tag{6.50}$$

We introduce the designations into (6.49) and (6.50):

$$A = \frac{1}{\varepsilon_0\varepsilon_S kT}\sum_i \frac{n_i |z_i| e}{\omega_i + \omega_j}, \quad p = \omega_i z_i e + \omega_j z_j e. \tag{6.51}$$

Expressions (6.49) and (6.50), with regard to designations of (6.51), look like

$$\nabla Y = \frac{2}{45} A \left[ \frac{32}{3} \left( \frac{\beta}{R} \right)^3 + \frac{17}{20} \left( \frac{\beta}{R} \right)^5 \right], \tag{6.52}$$

$$Y = \nabla^2 \cdot \nabla \xi - 2Ap\varphi^0 \nabla \xi. \tag{6.53}$$

We take the integral from the left and the right sides of (6.52):

$$Y = -\frac{2}{45} A \left[ \frac{16}{3} \beta \left( \frac{\beta}{R} \right)^2 + \frac{17}{80} \beta \left( \frac{\beta}{R} \right)^4 + C_3 \right], \tag{6.54}$$

where we do not use integration constant $C_3$ yet, and we will define it later. We denote $\nabla \xi = G$ in (6.53), and with regard to the introduced designation, Equations 6.53 and 6.54 have the following form:

$$\Delta G - 2Ap\varphi^0 G = -\frac{2}{45} A \left[ \frac{16}{3} \beta \left( \frac{\beta}{R} \right)^2 + \frac{17}{80} \beta \left( \frac{\beta}{R} \right)^4 + C_3 \right]. \tag{6.55}$$

We substitute Laplacian operator (5.45) and expression (5.52) into Equation 6.55 and obtain

$$\frac{\partial^2 G}{\partial R^2} + \frac{2}{R} \frac{\partial G}{\partial R} + 4ApG \left[ \left( \frac{R}{\beta} \right) - \frac{5}{3} \left( \frac{\beta}{R} \right) - \frac{16}{45} \left( \frac{\beta}{R} \right)^3 - \frac{17}{360} \left( \frac{\beta}{R} \right)^5 \right]$$

$$= -\frac{2}{45} A \left[ \frac{16}{3} \beta \left( \frac{\beta}{R} \right)^2 + \frac{17}{80} \beta \left( \frac{\beta}{R} \right)^4 + C_3 \right]. \tag{6.56}$$

We find the integral of differential equation (6.56) in the form of power series:

$$G = b_0 + b_1 \frac{\beta}{R} + b_2 \left( \frac{\beta}{R} \right)^2 + b_3 \left( \frac{\beta}{R} \right)^3 + b_4 \left( \frac{\beta}{R} \right)^4 + \cdots, \tag{6.57}$$

$$\frac{\partial G}{\partial R} = -b_1 \left( \frac{\beta}{R} \right) \frac{1}{R} - 2b_2 \left( \frac{\beta}{R} \right)^2 \frac{1}{R} - 3b_3 \left( \frac{\beta}{R} \right)^3 \frac{1}{R} - 4b_4 \left( \frac{\beta}{R} \right)^4 \frac{1}{R} - \cdots, \tag{6.58}$$

$$\frac{\partial^2 G}{\partial R^2} = 2b_1 \left( \frac{\beta}{R} \right) \frac{1}{R^2} + 6b_2 \left( \frac{\beta}{R} \right)^2 \frac{1}{R^2} + 12b_3 \left( \frac{\beta}{R} \right)^3 \frac{1}{R^2} + 20b_4 \left( \frac{\beta}{R} \right)^4 \frac{1}{R^2} + \cdots, \tag{6.59}$$

where $b_i$ means desired coefficients.

As it is seen from the structure of Equation 6.55, it is necessary to calculate the product of $\varphi^0 G$. From (6.57) and the square bracket of the left side of (6.56), after simple transformations, we have

$$\varphi^0 G = \left[\left(\frac{R}{\beta}\right) - \frac{5}{3}\left(\frac{\beta}{R}\right) - \frac{16}{45}\left(\frac{\beta}{R}\right)^3 - \frac{17}{360}\left(\frac{\beta}{R}\right)^5\right]$$

$$\times\left[b_0 + b_1\frac{\beta}{R} + b_2\left(\frac{\beta}{R}\right)^2 + b_3\left(\frac{\beta}{R}\right)^3 + b_4\left(\frac{\beta}{R}\right)^4\right] = b_0\frac{R}{\beta} + b_1 + \left(b_2 - \frac{5}{3}b_0\right)\frac{\beta}{R}$$

$$+\left(b_3 - \frac{5}{3}b_1\right)\left(\frac{\beta}{R}\right)^2 + \left(b_4 - \frac{5}{3}b_2 - \frac{16}{45}b_0\right)\left(\frac{\beta}{R}\right)^3 + \left(b_5 - \frac{5}{3}b_3 - \frac{16}{45}b_1\right)\left(\frac{\beta}{R}\right)^4. \tag{6.60}$$

And in expression (6.60), during multiplication of $\varphi^0 G$, we ignore indices of power $>R^{-4}$, because expansion into series $G$ in (6.57) and the right side of (6.56) do not stipulate for them.

We substitute (6.58) through (6.60) into Equation 6.56, and after simple transformations, we obtain

$$2b_2\left(\frac{\beta}{R}\right)^2\frac{1}{R^2} + 4Ap\left[\begin{array}{l} b_0\dfrac{R}{\beta} + b_1 + \left(b_2 - \dfrac{5}{3}b_0\right)\dfrac{\beta}{R} + \left(b_3 - \dfrac{5}{3}b_1\right)\left(\dfrac{\beta}{R}\right)^2 \\ +\left(b_4 - \dfrac{5}{3}b_2 - \dfrac{16}{45}b_0\right)\left(\dfrac{\beta}{R}\right)^3 + \left(b_5 - \dfrac{5}{3}b_3 - \dfrac{16}{45}b_1\right)\left(\dfrac{\beta}{R}\right)^4 \end{array}\right]$$

$$= -\frac{2}{45}A\left[\frac{16}{3}\beta\left(\frac{\beta}{R}\right)^2 + \frac{17}{80}\beta\left(\frac{\beta}{R}\right)^4 + C_3\right]. \tag{6.61}$$

These transformations, though somewhat cumbersome, are much simpler than in the Onsager, Falkenhagen, and Fuoss theory.

We equate the coefficients of identical powers $R^{-i}$ on the left and the right sides of expression (6.61), and after simple transformations, we obtain a system of equations with coefficients of unknown quantities $b_i$:

| No. | | $b_0$ | $b_1$ | $b_2$ | $b_3$ | $b_4$ | Free Term |
|---|---|---|---|---|---|---|---|
| 1 | $R^{-4}$ | 0 | $-\frac{64}{45}Ap\beta^2$ | 2 | $-\frac{20}{3}Ap\beta^2$ | 0 | $-\frac{17}{1800}A\beta^3$ |
| 2 | $R^{-3}$ | $-\frac{16}{45}$ | 0 | $-\frac{5}{3}$ | 0 | 1 | 0 |
| 3 | $R^{-2}$ | 0 | $-\frac{5}{3}$ | 0 | $Ap$ | 0 | $-\frac{8}{135}A\beta$ |
| 4 | $R^{-1}$ | $-\frac{5}{3}$ | 0 | 1 | 0 | 0 | 0 |
| 5 | $R^{0}$ | 0 | $2p$ | 0 | 0 | 0 | $-\frac{C_3}{45}$ |

To solve this system of equations, we apply the determinant theory. The determinant of our system is easily expanded and has the value $D = 6.67Ap^2$, that is, $D \neq 0$, and the system has an unambiguous solution. Expanding the determinants of unknown quantities, we obtain

$$b_0 = -4.5 \cdot 10^{-3} A\beta^2 C_3, \quad b_1 = \frac{-0.01 C_3}{p}, \quad b_2 = 0,$$
$$b_3 = \frac{-0.06\beta}{p}, \quad b_4 = 1.5 \cdot 10^{-3} A\beta^2 C_3. \tag{6.62}$$

While calculating the determinants of unknown quantities, we ignore the terms of low significance based on the fact that $A \approx 10^{20}$, $p \approx 10^{-3}$, $\beta \approx 10^{-9}$. Proceeding from these data, we can also ignore $b_3$ in (6.62).

We return to (6.57), and after simple transformations, we determine $G$:

$$G = -4.5 \cdot 10^{-3} A\beta^2 C_3 - \frac{0.01 C_3}{p}\left(\frac{\beta}{R}\right) + 1.5 \cdot 10^{-3} A\beta^2 C_3 \left(\frac{\beta}{R}\right)^4. \tag{6.63}$$

We take into account that $\nabla\xi = G$, and we need value $\nabla \cdot \nabla\xi$ in (6.48). We calculate the gradient from both parts of (6.63) and obtain

$$\nabla \cdot \nabla\xi = -\frac{0.01\beta C_3}{p}\ln R - 5 \cdot 10^{-4} A\beta^3 C_3 \left(\frac{\beta}{R}\right)^3. \tag{6.64}$$

We substitute Equation 6.64 into 6.48:

$$\eta^* = -\left[\frac{0.01\beta C_3}{p}\ln R + 1.25 \cdot 10^{-5} A\beta^3 C_3 \left(\frac{\beta}{R}\right)^3\right]\sum_j n_j |z_j| e + C_2. \tag{6.65}$$

We select integration constant $C_2$ from the second boundary condition of (2.60), with $R \rightarrow a$, $\nabla \cdot \nabla\xi \rightarrow C_3(E_{sp} + U_x)$, where $E_{sp}$ is determined according to formula (4.60) and $U_x = \langle E_H \rangle + U_{ot}$ according to formula (4.83), in which averaged value $B$ is taken from Table 5.4. Then, $C_2$ in (6.65) has the form of

$$C_2 = C_3\left\{\left[\frac{0.01\beta}{p}\ln a + 1.25 \cdot 10^{-5} A\beta^3 \left(\frac{\beta}{a}\right)^3\right]\sum_j n_j |z_j| e + \left(E_{sp} + U_x\right)\right\}, \tag{6.66}$$

where $C_3$ is the theoretical constant, individual for every electrolyte.

We substitute (6.66) into (6.65) and obtain

$$\eta^* = C_3 \left\langle \begin{aligned} &\left\{ \frac{0.01\beta}{p} \ln\left(\frac{a}{R}\right) + 1.25 \cdot 10^{-5} A\beta^3 \left[ \left(\frac{\beta}{a}\right)^3 - \left(\frac{\beta}{R}\right)^3 \right] \right\} \sum_j n_j |z_j| e \\ &+ (E_{sp} + U_x) \end{aligned} \right\rangle. \quad (6.67)$$

In Equation 6.67, we can ignore the square bracket for the low significance (because $\beta \approx 10^{-9}$) and obtain

$$\eta^* = C_3 \left\{ \left[ \frac{0.01\beta}{p} \ln\left(\frac{a}{R}\right) \right] \sum_j n_j |z_j| e + (E_{sp} + U_x) \right\}. \quad (6.68)$$

We determine dependence $p$ in (6.68). Proceeding to (6.51), we take into account that supermolecule mobility is determined as follows:

$$\omega_i = \frac{\lambda_i^0}{F|z_i|e},$$

where $F$ the is Faraday constant.

Then, from (6.51), we have

$$p = \frac{\Lambda^0}{F}. \quad (6.69)$$

We consider a binary solution. In (6.68), we write the sum and use ratios (2.48), (3.11), and (3.12), and with regard to (6.69), Equation 6.68 looks like

$$\eta^* = C_3 \left[ \frac{1}{4\pi\varepsilon_0\varepsilon_S} \frac{0.01|z_i z_j| F e^3 N_A \nu_k c_k}{kT\Lambda^0} \ln\left(\frac{a}{R}\right) + (E_{sp} + U_x) \right]. \quad (6.70)$$

We substitute physical constants into Equation 6.70 and obtain

$$\eta^* = C_3 \left[ \Psi_\eta^* + (E_{sp} + U_x) \right], \quad (6.71)$$

where $\Psi_\eta^*$ is the electrostatic component of viscosity and is equal to

$$\Psi_\eta^* = \frac{1554.8|z_i z_j| \nu_k c_k}{\varepsilon_S T\Lambda^0} \ln\left(\frac{a}{R}\right).$$

We present integration constant $C_3$ as $C_3 = C_4 + C_5(E_{sp} + U_x) / \Psi_\eta^*$. We substitute it into (6.71) and obtain

$$\eta^* = \left[C_4 + \frac{C_5\left(E_{sp} + U_x\right)}{\Psi_\eta^*}\right]\left[\Psi_\eta^* + \left(E_{sp} + U_x\right)\right]$$

$$= C_4\Psi_\eta^* + \left(C_5 + 1\right)\left(E_{sp} + U_x\right).$$

Without prejudice to the logic of presentation, we assume that $C_5 + 1 = C_6$ and denote $C_4 = 10^5 B_\eta$ and $C_6 = B_\eta$, where $B_\eta$ is the theoretical constant, individual for every electrolyte. Then, the increase in viscosity caused by the addition of dissolved electrolyte according to formula (6.71) acquires the final form:

$$\eta^* = B_\eta\left[\Psi_\eta + 10^5(E_{sp} + U_x)\right], \tag{6.72}$$

where $\Psi_\eta$ is the electrostatic component of viscosity:

$$\Psi_\eta = \frac{1.55 \cdot 10^8 \left|z_i z_j\right| \nu_k c_k}{\varepsilon_S T \Lambda^0} \ln\left(\frac{a}{R}\right). \tag{6.73}$$

Table 6.1 provides the thermodynamic components of formula (6.72) for a number of electrolytes. In the first column, under the name of electrolyte, auxiliary data for calculation are provided: $a$ is in nm; $\Lambda^0$ is in S cm$^2$/mol, and the averaged value $B$ from Table 5.4. The second column, from left to right, contains electrolyte concentration $m$, mol/kg $H_2O$; the third one contains $c_k$, mol/dm$^3$; the fourth one contains solution viscosity $\eta$, Pa s; the fifth one contains values of the electrostatic component of viscosity $-\Psi_\eta$; the sixth column contains values of supramolecular interactions $-E_{sp}$; the seventh one contains total values of weak chemical and repulsive interactions $U_x = \langle E_H \rangle + U_{ot}$; the eighth one contains values of theoretical constant $B_\eta$, individual for every electrolyte; and the last one contains values of increase (decrease) in viscosity caused by addition of solute $\eta^*$, Pa s. The experimental value of viscosity equal to 893.7 Pa s was taken for 25°C. In Table 6.1, the values of solution viscosity, depending on concentration, were interpolated from the data presented in [2].

From Table 6.1, we chose two 1–2 electrolytes and one 1–1 electrolyte, and supermolecular interactions existing in a solution from very dilute to concentrated ones are shown in Figure 6.2 for $CaCl_2$, Figure 6.3 for $MgCl_2$, and Figure 6.4 for LiCl.

Numbers of points of concentrations $m$, mol/kg $H_2O$, are plotted on $x$-axis; values of thermodynamic variables of Table 6.1 are on $y$-axis: solution viscosity $\eta$, Pa s; electrostatic interactions $-\Psi_\eta$; supramolecular interactions $-E_{sp}$; weak chemical and repulsive interactions $U_x = \langle E_H \rangle + U_{ot}$; values of increase (decrease) in viscosity caused by addition of a solute $\eta^*$, Pa s.

Figure 6.2 through 6.4 show that the behavior of supermolecular interactions in electrolyte solutions is of similar nature, regardless of electrolyte type (1–2 or 1–1).

## TABLE 6.1
## Values of Thermodynamic Quantities Included in Formula (6.72) for Calculation of Viscosity, Pa s

| Electrolyte | $m$ | $c_k$ | $\eta$ | $-\Psi_\eta$ | $-E_{sp}$ | $U_x$ | $B_\eta$ | $\eta^*$ |
|---|---|---|---|---|---|---|---|---|
| *$CaCl_2$: a = 0.66198 nm; B = 19.87; $\Lambda^0$ = 136.06 S cm²/mol* | | | | | | | | |
| | 0.20 | 0.2 | 959.1 | 3.64 | 3.41 | 1.54 | 11.867 | 65.4 |
| | 0.30 | 0.3 | 984.1 | 5.37 | 5.56 | 2.43 | 10.634 | 90.4 |
| | 0.40 | 0.40 | 1009.2 | 7.09 | 7.96 | 3.39 | –9.900 | 115.5 |
| | 0.50 | 0.49 | 1034.2 | 8.70 | 10.67 | 4.42 | –9.401 | 140.5 |
| | 0.60 | 0.59 | 1060.4 | 10.25 | 13.62 | 5.52 | –9.080 | 166.7 |
| | 0.70 | 0.69 | 1090.2 | 11.67 | 16.91 | 6.69 | –8.974 | 196.5 |
| | 0.80 | 0.78 | 1119.9 | 12.99 | 20.52 | 7.92 | –8.843 | 226.2 |
| | 0.90 | 0.88 | 1149.8 | 14.13 | 24.44 | 9.21 | –8.726 | 256.1 |
| | 1.00 | 0.98 | 1179.7 | 15.08 | 28.80 | 10.60 | –8.592 | 286.0 |
| | 1.25 | 1.21 | 1263.2 | 17.49 | 41.24 | 14.46 | –8.345 | 369.5 |
| | 1.50 | 1.44 | 1360.5 | 19.09 | 56.01 | 18.77 | –8.288 | 466.8 |
| | 1.75 | 1.66 | 1471.4 | 19.51 | 74.23 | 23.75 | –8.254 | 577.7 |
| | 2.00 | 1.89 | 1593.8 | 18.51 | 96.38 | 29.44 | –8.192 | 700.1 |
| | 2.25 | 2.11 | 1726.7 | 16.02 | 121.80 | 35.59 | –8.148 | 833.0 |
| | 2.50 | 2.32 | 1888.2 | 11.83 | 154.59 | 43.21 | –8.072 | 994.5 |
| | 2.75 | 2.52 | 2078.4 | 9.34 | 185.97 | 50.82 | –8.199 | 1184.7 |
| | 3.00 | 2.73 | 2241.3 | 0.99 | 228.65 | 59.79 | –7.934 | 1347.6 |
| | 3.25 | 2.94 | 2392.1 | 0.00 | 278.25 | 69.74 | –7.186 | 1498.4 |
| | 3.50 | 3.13 | 2558.4 | 0.00 | 325.00 | 79.77 | –6.788 | 1664.7 |
| | 3.75 | 3.31 | 2770.2 | 0.00 | 375.44 | 90.18 | –6.578 | 1876.5 |
| | 4.00 | 3.51 | 3049.4 | 0.00 | 443.34 | 102.24 | –6.320 | 2155.7 |
| | 4.25 | 3.68 | 3382.0 | 0.00 | 508.45 | 114.65 | –6.319 | 2488.3 |
| | 4.50 | 3.85 | 3708.9 | 0.00 | 575.03 | 127.29 | –6.288 | 2815.2 |
| | 4.75 | 4.05 | 4053.6 | 0.00 | 668.02 | 141.72 | –6.004 | 3159.9 |
| | 5.00 | 4.21 | 4421.6 | 0.00 | 744.59 | 155.05 | –5.984 | 3527.9 |
| | 5.25 | 4.37 | 4795.8 | 0.00 | 831.38 | 169.69 | –5.897 | 3902.1 |
| | 5.50 | 4.53 | 5172.3 | 0.00 | 918.14 | 183.67 | –5.825 | 4278.6 |
| | 5.75 | 4.69 | 5563.6 | 0.00 | 1014.73 | 198.82 | –5.724 | 4669.9 |
| | 6.00 | 4.84 | 5967.9 | 0.00 | 1111.85 | 213.80 | –5.650 | 5074.2 |
| *$Cs_2SO_4$: a = 0.67100 nm; B = 59.39; $\Lambda^0$ = 128.73 S cm²/mol* | | | | | | | | |
| | 0.10 | 0.10 | 927.8 | 7.43 | 1.19 | 2.55 | –5.612 | 34.1 |
| | 0.20 | 0.20 | 939.7 | 14.55 | 2.63 | 5.25 | –3.860 | 46.0 |
| | 0.40 | 0.40 | 945.6 | 28.38 | 5.98 | 10.92 | –2.214 | 51.9 |
| | 0.60 | 0.58 | 966.1 | 41.64 | 9.87 | 16.84 | –2.088 | 72.4 |
| | 0.80 | 0.76 | 1001.4 | 54.53 | 13.99 | 22.82 | –2.357 | 107.7 |
| | 1.00 | 0.95 | 1046.0 | 66.29 | 19.60 | 29.51 | –2.701 | 152.3 |
| | 2.00 | 1.77 | 1283.8 | 119.52 | 54.07 | 64.53 | –3.576 | 390.1 |
| | 3.00 | 2.45 | 1555.2 | 164.94 | 101.32 | 102.97 | –4.051 | 661.5 |
| | 4.00 | 3.04 | 1907.1 | 204.02 | 161.97 | 144.70 | –4.580 | 1013.4 |
| | 5.00 | 3.55 | 2192.9 | 233.58 | 234.60 | 186.98 | –4.620 | 1299.2 |

*(Continued)*

**TABLE 6.1 (*Continued*)**
**Values of Thermodynamic Quantities Included in Formula (6.72) for Calculation of Viscosity, Pa s**

| Electrolyte | $m$ | $c_k$ | $\eta$ | $-\Psi_\eta$ | $-E_{sp}$ | $U_x$ | $B_\eta$ | $\eta^*$ |
|---|---|---|---|---|---|---|---|---|
| *CsCl: a = 0.66865 nm; B = 5.08; $\Lambda^0$ = 153.81 S cm²/mol* | | | | | | | | |
| | 0.10 | 0.10 | 884.5 | 1.67 | 0.20 | 0.14 | 5.328 | −9.2 |
| | 0.20 | 0.20 | 882.1 | 3.33 | 0.40 | 0.28 | 3.353 | −11.6 |
| | 0.30 | 0.30 | 879.4 | 4.93 | 0.66 | 0.43 | 2.773 | −14.3 |
| | 0.40 | 0.39 | 876.4 | 6.54 | 0.90 | 0.59 | 2.524 | −17.3 |
| | 0.50 | 0.49 | 873.5 | 8.07 | 1.22 | 0.75 | 2.369 | −20.2 |
| | 0.60 | 0.58 | 871.3 | 9.60 | 1.53 | 0.92 | 2.199 | −22.4 |
| | 0.70 | 0.68 | 868.9 | 11.04 | 1.91 | 1.10 | 2.089 | −24.8 |
| | 0.80 | 0.77 | 866.3 | 12.53 | 2.27 | 1.27 | 2.029 | −27.4 |
| | 0.90 | 0.86 | 863.7 | 13.97 | 2.65 | 1.45 | 1.977 | −30.0 |
| | 1.00 | 0.95 | 861.2 | 15.38 | 3.07 | 1.64 | 1.933 | −32.5 |
| | 1.25 | 1.18 | 855.9 | 18.68 | 4.34 | 2.14 | 1.808 | −37.8 |
| | 1.50 | 1.40 | 851.8 | 21.73 | 5.84 | 2.68 | 1.684 | −41.9 |
| | 1.75 | 1.63 | 848.3 | 24.30 | 7.69 | 3.25 | 1.581 | −45.4 |
| | 2.00 | 1.83 | 846.4 | 26.92 | 9.52 | 3.81 | 1.449 | −47.3 |
| | 2.25 | 2.04 | 845.2 | 29.13 | 11.64 | 4.40 | 1.333 | −48.5 |
| *$CsNO_3$: a = 0.630 nm; B = 3.54; $\Lambda^0$ = 148.92 S cm²/mol* | | | | | | | | |
| | 0.20 | 0.20 | 877.6 | 5.31 | 0.26 | 0.17 | 2.971 | −16.1 |
| | 0.40 | 0.39 | 869.9 | 10.41 | 0.57 | 0.34 | 2.232 | −23.8 |
| | 0.60 | 0.58 | 861.8 | 15.27 | 0.93 | 0.51 | 2.030 | −31.9 |
| | 0.80 | 0.76 | 854.3 | 19.90 | 1.37 | 0.70 | 1.916 | −39.4 |
| | 1.00 | 0.94 | 846.9 | 24.29 | 1.88 | 0.90 | 1.851 | −46.8 |
| *$Cu(NO_3)_2$: a = 0.73575 nm; B = 17.69; $\Lambda^0$ = 126.93 S cm²/mol* | | | | | | | | |
| | 0.20 | 0.20 | 898.8 | 7.3 | 1.51 | 1.20 | 5.1 | 5.1 |
| | 0.30 | 0.30 | 922.4 | 10.8 | 2.37 | 1.83 | 28.7 | 28.7 |
| | 0.40 | 0.39 | 946.0 | 14.0 | 3.42 | 2.52 | 52.3 | 52.3 |
| | 0.50 | 0.49 | 972.6 | 16.9 | 4.66 | 3.26 | 78.9 | 78.9 |
| | 0.60 | 0.59 | 1005.2 | 20.01 | 5.87 | 4.00 | 111.5 | 111.5 |
| | 0.70 | 0.69 | 1046.8 | 22.43 | 7.50 | 4.85 | 153.1 | 153.1 |
| | 0.80 | 0.78 | 1092.7 | 25.26 | 8.97 | 5.66 | 199.0 | 199.0 |
| | 0.90 | 0.87 | 1141.7 | 27.95 | 10.59 | 6.51 | 248.0 | 248.0 |
| | 1.00 | 0.97 | 1193.9 | 29.57 | 12.90 | 7.51 | 300.2 | 300.2 |
| | 1.25 | 1.20 | 1332.1 | 34.24 | 18.76 | 10.07 | 438.4 | 438.4 |
| | 1.50 | 1.42 | 1471.9 | 37.52 | 26.26 | 12.97 | 578.2 | 578.2 |
| | 1.75 | 1.65 | 1618.1 | 39.05 | 36.05 | 16.38 | 724.4 | 724.4 |
| | 2.00 | 1.86 | 1771.3 | 40.89 | 46.08 | 19.82 | 877.6 | 877.6 |
| | 2.25 | 2.08 | 1946.7 | 38.23 | 60.80 | 24.05 | 1053.0 | 1053.0 |
| | 2.50 | 2.28 | 2158.2 | 36.25 | 76.10 | 28.41 | 1264.5 | 1264.5 |
| | 2.75 | 2.48 | 2402.5 | 32.18 | 94.19 | 33.21 | 1508.8 | 1508.8 |
| | 3.00 | 2.68 | 2663.2 | 25.85 | 115.24 | 38.40 | 1769.5 | 1769.5 |

(*Continued*)

**TABLE 6.1 (*Continued*)**
**Values of Thermodynamic Quantities Included in Formula (6.72) for Calculation of Viscosity, Pa s**

| Electrolyte | $m$ | $c_k$ | $\eta$ | $-\Psi_\eta$ | $-E_{sp}$ | $U_x$ | $B_\eta$ | $\eta^*$ |
|---|---|---|---|---|---|---|---|---|
| | 3.25 | 2.87 | 2965.5 | 16.70 | 140.76 | 44.27 | 2071.8 | 2071.8 |
| | 3.50 | 3.07 | 3308.6 | 4.97 | 169.21 | 50.35 | 2414.9 | 2414.9 |
| | 3.75 | 3.23 | 3706.3 | 0.00 | 194.80 | 56.30 | 2812.6 | 2812.6 |
| | 4.00 | 3.42 | 4143.3 | 0.00 | 232.18 | 63.32 | 3249.6 | 3249.6 |
| | 4.25 | 3.57 | 4611.6 | 0.00 | 262.63 | 69.73 | 3717.9 | 3717.9 |
| | 4.50 | 3.76 | 5114.9 | 0.00 | 310.66 | 77.79 | 4221.2 | 4221.2 |
| | 4.75 | 3.91 | 5637.7 | 0.00 | 349.55 | 85.01 | 4744.0 | 4744.0 |
| | 5.00 | 4.05 | 6163.2 | 0.00 | 389.28 | 92.13 | 5269.5 | 5269.5 |
| | 5.25 | 4.24 | 6686.2 | 0.00 | 455.27 | 101.54 | 5792.5 | 5792.5 |
| | 5.50 | 4.38 | 6829.9 | 0.00 | 504.74 | 109.53 | 5936.2 | 5936.2 |
| *$H_2SO_4$: a = 0.63879 nm; B = 17.06; $\Lambda^0$ = 414.48 S cm²/mol* | | | | | | | | |
| | 0.20 | 0.20 | 941.2 | 1.19 | 5.61 | 1.59 | −9.124 | 47.5 |
| | 0.30 | 0.30 | 967.7 | 1.74 | 8.73 | 2.44 | −9.199 | 74.0 |
| | 0.40 | 0.39 | 987.2 | 2.25 | 12.20 | 3.32 | −8.401 | 93.5 |
| | 0.50 | 0.49 | 1006.6 | 2.77 | 15.67 | 4.22 | −7.940 | 112.9 |
| | 0.60 | 0.59 | 1026.1 | 3.21 | 19.66 | 5.17 | −7.479 | 132.4 |
| | 0.70 | 0.68 | 1044.8 | 3.80 | 23.02 | 6.07 | −7.280 | 151.1 |
| | 0.80 | 0.77 | 1062.6 | 4.20 | 27.38 | 7.07 | −6.889 | 168.9 |
| | 0.90 | 0.87 | 1080.6 | 4.55 | 32.12 | 8.12 | −6.545 | 186.9 |
| | 1.00 | 0.96 | 1099.6 | 4.90 | 36.87 | 9.16 | −6.315 | 205.9 |
| | 1.50 | 1.42 | 1200.4 | 6.44 | 67.94 | 15.60 | −5.216 | 306.7 |
| | 2.00 | 1.85 | 1299.3 | 7.56 | 107.12 | 23.17 | −4.433 | 405.6 |
| | 2.50 | 2.28 | 1399.1 | 6.72 | 160.80 | 32.09 | −3.732 | 505.4 |
| | 3.00 | 2.69 | 1512.0 | 5.43 | 232.15 | 43.58 | −3.187 | 618.3 |
| | 3.50 | 3.08 | 1647.4 | 2.43 | 319.51 | 56.51 | −2.839 | 753.7 |
| | 4.00 | 3.44 | 1797.1 | 0.00 | 411.41 | 69.81 | −2.645 | 903.4 |
| | 4.50 | 3.82 | 1938.6 | 0.00 | 536.68 | 85.57 | −2.316 | 1044.9 |
| | 5.00 | 4.16 | 2070.9 | 0.00 | 678.69 | 103.76 | −2.047 | 1177.2 |
| | 5.50 | 4.49 | 2200.6 | 0.00 | 837.12 | 122.87 | −1.830 | 1306.9 |
| | 6.00 | 4.82 | 2331.6 | 0.00 | 1012.88 | 142.40 | −1.652 | 1437.9 |
| | 6.50 | 5.14 | 2467.1 | 0.00 | 1209.00 | 163.22 | −1.504 | 1573.4 |
| | 7.00 | 5.45 | 2608.5 | 0.00 | 1431.82 | 185.21 | −1.376 | 1714.8 |
| | 7.50 | 5.71 | 2755.2 | 0.00 | 1630.10 | 206.46 | −1.308 | 1861.5 |
| | 8.00 | 6.01 | 2901.7 | 0.00 | 1894.72 | 230.19 | −1.206 | 2008.0 |
| | 8.50 | 6.25 | 3043.4 | 0.00 | 2117.29 | 251.99 | −1.152 | 2149.7 |
| | 9.00 | 6.54 | 3185.6 | 0.00 | 2434.06 | 277.39 | −1.063 | 2291.9 |
| | 9.50 | 6.78 | 3327.9 | 0.00 | 2683.77 | 299.58 | −1.021 | 2434.2 |
| | 10.00 | 7.05 | 3469.5 | 0.00 | 3054.80 | 327.10 | −0.944 | 2575.8 |
| *KBr: a = 0.64433 nm; B = 4.48; $\Lambda^0$ = 151.79 S cm²/mol* | | | | | | | | |
| | 0.10 | 0.10 | 923.6 | 1.02 | 0.16 | 0.09 | −27.491 | 29.9 |
| | 0.20 | 0.20 | 915.8 | 2.00 | 0.35 | 0.19 | −10.222 | 22.1 |

(*Continued*)

**TABLE 6.1 (*Continued*)**
**Values of Thermodynamic Quantities Included in Formula (6.72) for Calculation of Viscosity, Pa s**

| Electrolyte | $m$ | $c_k$ | $\eta$ | $-\Psi_\eta$ | $-E_{sp}$ | $U_x$ | $B_\eta$ | $\eta^*$ |
|---|---|---|---|---|---|---|---|---|
| | 0.30 | 0.29 | 907.9 | 2.99 | 0.54 | 0.29 | –4.393 | 14.2 |
| | 0.40 | 0.40 | 899.4 | 3.88 | 0.78 | 0.40 | –1.351 | 5.7 |
| | 0.50 | 0.49 | 890.3 | 4.82 | 1.00 | 0.51 | 0.636 | –3.4 |
| | 0.60 | 0.59 | 881.8 | 5.72 | 1.25 | 0.63 | 1.882 | –11.9 |
| | 0.70 | 0.68 | 873.5 | 6.58 | 1.53 | 0.75 | 2.740 | –20.2 |
| | 0.80 | 0.78 | 866.6 | 7.41 | 1.82 | 0.87 | 3.244 | –27.1 |
| | 0.90 | 0.87 | 860.2 | 8.21 | 2.13 | 0.99 | 3.587 | –33.5 |
| | 1.00 | 0.97 | 855.5 | 8.99 | 2.47 | 1.12 | 3.695 | –38.2 |
| | 1.25 | 1.19 | 848.6 | 10.91 | 3.35 | 1.46 | 3.523 | –45.1 |
| | 1.50 | 1.42 | 847.6 | 12.61 | 4.37 | 1.81 | 3.042 | –46.1 |
| | 1.75 | 1.64 | 849.2 | 14.12 | 5.54 | 2.20 | 2.546 | –44.5 |
| | 2.00 | 1.86 | 851.2 | 15.40 | 6.91 | 2.61 | 2.157 | –42.5 |
| | 2.25 | 2.08 | 853.1 | 16.38 | 8.47 | 3.05 | 1.861 | –40.6 |
| | 2.50 | 2.29 | 854.0 | 17.09 | 10.26 | 3.51 | 1.665 | –39.7 |
| | 2.75 | 2.50 | 854.0 | 17.48 | 12.29 | 4.01 | 1.541 | –39.7 |
| | 3.00 | 2.69 | 854.2 | 18.37 | 14.07 | 4.48 | 1.415 | –39.5 |
| | 3.25 | 2.90 | 854.9 | 18.14 | 16.61 | 5.03 | 1.304 | –38.8 |
| | 3.50 | 3.09 | 856.4 | 18.63 | 18.79 | 5.55 | 1.171 | –37.3 |
| | 3.75 | 3.29 | 858.3 | 17.70 | 21.93 | 6.16 | 1.059 | –35.4 |
| | 4.00 | 3.47 | 860.5 | 17.67 | 24.54 | 6.72 | 0.936 | –33.2 |
| | 4.25 | 3.68 | 863.2 | 16.00 | 28.48 | 7.42 | 0.824 | –30.5 |
| | 4.50 | 3.85 | 866.6 | 15.37 | 31.58 | 8.01 | 0.696 | –27.1 |
| | 4.75 | 4.02 | 870.4 | 14.64 | 34.93 | 8.64 | 0.570 | –23.3 |
| | 5.00 | 4.19 | 874.5 | 13.57 | 38.59 | 9.31 | 0.448 | –19.2 |
| | 5.25 | 4.35 | 878.6 | 12.23 | 42.48 | 9.98 | 0.337 | –15.1 |
| *KCl: a = 0.57820 nm; B = 4.71;* $\Lambda^0$ *= 149.96 S cm²/mol* | | | | | | | | |
| | 0.20 | 0.20 | 883.1 | 1.21 | 0.66 | 0.21 | 6.375 | –10.6 |
| | 0.30 | 0.30 | 878.9 | 1.81 | 1.02 | 0.32 | 5.908 | –14.8 |
| | 0.40 | 0.40 | 874.7 | 2.35 | 1.43 | 0.44 | 5.699 | –19.0 |
| | 0.50 | 0.49 | 870.5 | 2.89 | 1.85 | 0.56 | 5.553 | –23.2 |
| | 0.60 | 0.59 | 866.3 | 3.50 | 2.23 | 0.68 | 5.405 | –27.4 |
| | 0.70 | 0.68 | 862.3 | 4.01 | 2.70 | 0.80 | 5.317 | –31.4 |
| | 0.80 | 0.78 | 858.3 | 4.48 | 3.20 | 0.94 | 5.252 | –35.4 |
| | 0.90 | 0.87 | 854.6 | 5.07 | 3.64 | 1.06 | 5.115 | –39.1 |
| | 1.00 | 0.97 | 851.4 | 5.53 | 4.21 | 1.20 | 4.964 | –42.3 |
| | 1.25 | 1.21 | 843.7 | 6.60 | 5.71 | 1.57 | 4.662 | –50.0 |
| | 1.50 | 1.43 | 836.8 | 7.80 | 7.22 | 1.95 | 4.355 | –56.9 |
| | 1.75 | 1.66 | 830.9 | 8.64 | 9.16 | 2.39 | 4.077 | –62.8 |
| | 2.00 | 1.88 | 825.6 | 9.65 | 11.03 | 2.82 | 3.811 | –68.1 |
| | 2.25 | 2.10 | 821.3 | 10.59 | 13.07 | 3.27 | 3.553 | –72.4 |

(*Continued*)

**TABLE 6.1 (*Continued*)**
**Values of Thermodynamic Quantities Included in Formula (6.72) for Calculation of Viscosity, Pa s**

| Electrolyte | $m$ | $c_k$ | $\eta$ | $-\Psi_\eta$ | $-E_{sp}$ | $U_x$ | $B_\eta$ | $\eta^*$ |
|---|---|---|---|---|---|---|---|---|
| | 2.50 | 2.32 | 818.6 | 10.86 | 15.68 | 3.79 | 3.301 | −75.1 |
| | 2.75 | 2.53 | 816.5 | 11.48 | 18.09 | 4.28 | 3.054 | −77.2 |
| | 3.00 | 2.74 | 814.7 | 11.97 | 20.68 | 4.80 | 2.836 | −79.0 |
| | 3.25 | 2.96 | 813.5 | 11.55 | 24.21 | 5.41 | 2.643 | −80.2 |
| | 3.50 | 3.16 | 813.0 | 11.66 | 27.28 | 5.98 | 2.448 | −80.7 |
| | 3.75 | 3.36 | 813.0 | 11.69 | 30.52 | 6.57 | 2.264 | −80.7 |
| | 4.00 | 3.55 | 813.5 | 11.49 | 34.01 | 7.18 | 2.094 | −80.2 |
| | 4.25 | 3.74 | 814.0 | 11.20 | 37.74 | 7.82 | 1.937 | −79.7 |
| *KI: a = 0.69190 nm; B = 3.17;* $\Lambda^0$ *= 150.54 S cm²/mol* | | | | | | | | |
| | 0.10 | 0.10 | 875.6 | 1.35 | 0.10 | 0.06 | 13.054 | −18.1 |
| | 0.20 | 0.20 | 860.0 | 2.61 | 0.22 | 0.13 | 12.492 | −33.7 |
| | 0.30 | 0.29 | 844.3 | 3.83 | 0.34 | 0.20 | 12.446 | −49.4 |
| | 0.40 | 0.39 | 828.7 | 5.01 | 0.46 | 0.26 | 12.502 | −65.0 |
| | 0.50 | 0.49 | 813.6 | 6.05 | 0.61 | 0.33 | 12.671 | −80.1 |
| | 0.60 | 0.58 | 799.5 | 7.10 | 0.75 | 0.40 | 12.646 | −94.2 |
| | 0.70 | 0.67 | 786.2 | 8.10 | 0.90 | 0.47 | 12.604 | −107.5 |
| | 0.80 | 0.77 | 773.7 | 8.93 | 1.09 | 0.54 | 12.663 | −120.0 |
| | 0.90 | 0.86 | 762.1 | 9.82 | 1.26 | 0.61 | 12.572 | −131.6 |
| | 1.00 | 0.95 | 750.9 | 10.66 | 1.44 | 0.69 | 12.510 | −142.8 |
| | 1.25 | 1.18 | 725.0 | 12.41 | 1.98 | 0.87 | 12.477 | −168.7 |
| | 1.50 | 1.40 | 702.5 | 13.79 | 2.61 | 1.07 | 12.470 | −191.2 |
| | 1.75 | 1.62 | 684.5 | 15.46 | 3.51 | 1.33 | 11.852 | −209.2 |
| | 2.00 | 1.83 | 667.2 | 17.18 | 4.45 | 1.59 | 11.303 | −226.5 |
| | 2.25 | 2.03 | 650.7 | 18.61 | 5.53 | 1.87 | 10.913 | −243.0 |
| | 2.50 | 2.23 | 636.4 | 19.75 | 6.75 | 2.17 | 10.574 | −257.3 |
| | 2.75 | 2.43 | 624.0 | 20.77 | 8.28 | 2.51 | 10.162 | −269.7 |
| | 3.00 | 2.62 | 613.8 | 21.57 | 10.06 | 2.89 | 9.739 | −279.9 |
| | 3.25 | 2.82 | 604.9 | 21.94 | 12.13 | 3.29 | 9.386 | −288.8 |
| | 3.50 | 3.01 | 596.7 | 21.75 | 14.49 | 3.71 | 9.133 | −297.0 |
| | 3.75 | 3.17 | 589.6 | 22.44 | 16.42 | 4.09 | 8.749 | −304.1 |
| | 4.00 | 3.37 | 583.4 | 21.36 | 19.32 | 4.55 | 8.589 | −310.3 |
| | 4.25 | 3.52 | 579.6 | 21.38 | 21.68 | 4.95 | 8.243 | −314.1 |
| *$KNO_3$: a = 0.59800 nm; B = 3.93;* $\Lambda^0$ *= 145.07 S cm²/mol* | | | | | | | | |
| | 0.20 | 0.20 | 876.1 | 2.56 | 0.35 | 0.15 | 6.358 | −17.6 |
| | 0.30 | 0.30 | 868.1 | 3.82 | 0.54 | 0.23 | 6.218 | −25.6 |
| | 0.40 | 0.39 | 860.0 | 5.04 | 0.74 | 0.31 | 6.170 | −33.7 |
| | 0.50 | 0.49 | 853.4 | 6.23 | 0.95 | 0.40 | 5.939 | −40.3 |
| | 0.60 | 0.59 | 847.0 | 7.39 | 1.18 | 0.48 | 5.775 | −46.7 |
| | 0.70 | 0.68 | 842.0 | 8.55 | 1.41 | 0.57 | 5.504 | −51.7 |

(*Continued*)

**TABLE 6.1 (*Continued*)**
**Values of Thermodynamic Quantities Included in Formula (6.72) for Calculation of Viscosity, Pa s**

| Electrolyte | $m$ | $c_k$ | $\eta$ | $-\Psi_\eta$ | $-E_{sp}$ | $U_x$ | $B_\eta$ | $\eta^*$ |
|---|---|---|---|---|---|---|---|---|
| | 0.80 | 0.77 | 837.6 | 9.76 | 1.61 | 0.65 | 5.228 | –56.1 |
| | 0.90 | 0.87 | 833.2 | 10.85 | 1.87 | 0.74 | 5.053 | –60.5 |
| | 1.00 | 0.96 | 828.6 | 11.89 | 2.13 | 0.82 | 4.936 | –65.1 |
| | 1.25 | 1.19 | 819.2 | 14.44 | 2.82 | 1.05 | 4.599 | –74.5 |
| | 1.50 | 1.41 | 812.5 | 16.76 | 3.55 | 1.27 | 4.264 | –81.2 |
| | 1.75 | 1.63 | 809.0 | 18.86 | 4.35 | 1.50 | 3.902 | –84.7 |
| | 2.00 | 1.85 | 808.4 | 20.77 | 5.22 | 1.74 | 3.517 | –85.3 |
| | 2.25 | 2.06 | 809.0 | 22.51 | 6.17 | 1.98 | 3.172 | –84.7 |
| | 2.50 | 2.26 | 809.3 | 24.35 | 6.98 | 2.19 | 2.899 | –84.4 |
| | 2.75 | 2.46 | 811.6 | 25.55 | 8.02 | 2.43 | 2.636 | –82.1 |
| | 3.00 | 2.65 | 813.3 | 26.94 | 8.89 | 2.64 | 2.423 | –80.4 |
| | 3.25 | 2.85 | 814.4 | 27.80 | 10.05 | 2.88 | 2.268 | –79.3 |
| | 3.50 | 3.03 | 815.1 | 28.91 | 10.97 | 3.08 | 2.137 | –78.6 |
| *KOH: a = 0.52704 nm; B = 5.85;* $\Lambda^0$ *= 255.69 S cm²/mol* | | | | | | | | |
| | 0.30 | 0.30 | 918.0 | 0.99 | 1.55 | 0.40 | –11.362 | 24.3 |
| | 0.40 | 0.40 | 924.3 | 1.33 | 2.11 | 0.55 | –10.586 | 30.6 |
| | 0.50 | 0.50 | 930.6 | 1.62 | 2.77 | 0.70 | –9.998 | 36.9 |
| | 0.60 | 0.60 | 937.0 | 1.97 | 3.39 | 0.85 | –9.617 | 43.3 |
| | 0.70 | 0.70 | 943.3 | 2.22 | 4.18 | 1.02 | –9.239 | 49.6 |
| | 0.80 | 0.79 | 949.6 | 2.56 | 4.85 | 1.18 | –8.995 | 55.9 |
| | 0.90 | 0.90 | 956.0 | 2.78 | 5.74 | 1.37 | –8.710 | 62.3 |
| | 1.00 | 0.99 | 962.3 | 3.11 | 6.47 | 1.54 | –8.528 | 68.6 |
| | 1.25 | 1.25 | 977.8 | 3.58 | 9.06 | 2.04 | –7.933 | 84.1 |
| | 1.50 | 1.49 | 993.0 | 4.13 | 11.70 | 2.56 | –7.483 | 99.3 |
| | 1.75 | 1.73 | 1009.9 | 4.65 | 14.55 | 3.10 | –7.223 | 116.2 |
| | 2.00 | 1.97 | 1027.4 | 5.03 | 17.85 | 3.70 | –6.969 | 133.7 |
| | 2.25 | 2.21 | 1045.2 | 5.40 | 21.63 | 4.36 | –6.681 | 151.5 |
| | 2.50 | 2.45 | 1063.3 | 5.69 | 25.90 | 5.09 | –6.400 | 169.6 |
| | 2.75 | 2.68 | 1082.4 | 5.83 | 30.53 | 5.84 | –6.183 | 188.7 |
| | 3.00 | 2.92 | 1101.8 | 5.80 | 35.72 | 6.65 | –5.966 | 208.1 |
| | 3.25 | 3.15 | 1121.5 | 5.71 | 41.30 | 7.49 | –5.764 | 227.8 |
| | 3.50 | 3.38 | 1141.9 | 5.43 | 47.60 | 8.40 | –5.562 | 248.2 |
| | 3.75 | 3.61 | 1163.2 | 5.00 | 54.80 | 9.41 | –5.348 | 269.5 |
| | 4.00 | 3.84 | 1185.4 | 4.45 | 62.61 | 10.49 | –5.157 | 291.7 |
| | 4.25 | 4.07 | 1209.3 | 3.69 | 71.03 | 11.61 | –5.001 | 315.6 |
| | 4.50 | 4.30 | 1233.8 | 2.72 | 80.34 | 12.79 | –4.840 | 340.1 |
| | 4.75 | 4.53 | 1260.9 | 1.51 | 90.36 | 14.02 | –4.717 | 367.2 |
| | 5.00 | 4.75 | 1288.1 | 0.06 | 101.48 | 15.36 | –4.576 | 394.4 |
| | 5.25 | 4.97 | 1317.3 | 0.00 | 113.06 | 16.70 | –4.396 | 423.6 |

(*Continued*)

**TABLE 6.1 (*Continued*)**
**Values of Thermodynamic Quantities Included in Formula (6.72) for Calculation of Viscosity, Pa s**

| Electrolyte | $m$ | $c_k$ | $\eta$ | $-\Psi_\eta$ | $-E_{sp}$ | $U_x$ | $B_\eta$ | $\eta^*$ |
|---|---|---|---|---|---|---|---|---|
| | 5.50 | 5.20 | 1346.8 | 0.00 | 125.97 | 18.14 | −4.202 | 453.1 |
| | 5.75 | 5.38 | 1378.7 | 0.00 | 134.00 | 19.29 | −4.228 | 485.0 |
| | 6.00 | 5.60 | 1413.0 | 0.00 | 147.95 | 20.79 | −4.084 | 519.3 |
| | 6.25 | 5.82 | 1447.3 | 0.00 | 162.96 | 22.37 | −3.938 | 553.6 |
| | 6.50 | 6.04 | 1485.2 | 0.00 | 179.68 | 24.06 | −3.801 | 591.5 |
| | 6.75 | 6.21 | 1523.4 | 0.00 | 189.93 | 25.43 | −3.828 | 629.7 |
| | 7.00 | 6.42 | 1563.0 | 0.00 | 208.49 | 27.23 | −3.693 | 669.3 |
| | 7.25 | 6.64 | 1606.9 | 0.00 | 227.58 | 29.03 | −3.592 | 713.2 |
| | 7.50 | 6.86 | 1650.7 | 0.00 | 248.76 | 30.95 | −3.475 | 757.0 |
| | 7.75 | 7.01 | 1694.6 | 0.00 | 261.55 | 32.54 | −3.497 | 800.9 |
| | 8.00 | 7.23 | 1738.7 | 0.00 | 284.98 | 34.66 | −3.376 | 845.0 |
| | 8.25 | 7.44 | 1782.8 | 0.00 | 309.52 | 36.76 | −3.260 | 889.1 |
| | 8.50 | 7.59 | 1828.5 | 0.00 | 323.95 | 38.47 | −3.275 | 934.8 |
| | 8.75 | 7.80 | 1876.4 | 0.00 | 351.36 | 40.78 | −3.164 | 982.7 |
| | 9.00 | 8.01 | 1924.2 | 0.00 | 379.39 | 43.03 | −3.064 | 1030.5 |
| | 9.25 | 8.16 | 1971.9 | 0.00 | 395.17 | 44.82 | −3.078 | 1078.2 |
| | 9.50 | 8.37 | 2019.3 | 0.00 | 427.00 | 47.29 | −2.964 | 1125.6 |
| | 9.75 | 8.51 | 2066.7 | 0.00 | 442.57 | 49.02 | −2.981 | 1173.0 |
| | 10.00 | 8.72 | 2114.1 | 0.00 | 475.35 | 51.44 | −2.879 | 1220.4 |
| | 10.50 | 9.06 | 2221.4 | 0.00 | 525.62 | 55.62 | −2.825 | 1327.7 |
| | 11.00 | 9.40 | 2331.2 | 0.00 | 578.40 | 59.80 | −2.772 | 1437.5 |
| | 11.50 | 9.73 | 2450.6 | 0.00 | 632.60 | 63.89 | −2.738 | 1556.9 |
| | 12.00 | 10.06 | 2577.1 | 0.00 | 690.06 | 68.18 | −2.707 | 1683.4 |
| | 12.50 | 10.38 | 2725.9 | 0.00 | 752.76 | 72.71 | −2.694 | 1832.2 |
| | 13.00 | 10.70 | 2876.4 | 0.00 | 819.32 | 77.48 | −2.673 | 1982.7 |
| | 13.50 | 11.02 | 3033.5 | 0.00 | 888.70 | 82.08 | −2.653 | 2139.8 |
| | 14.00 | 11.33 | 3190.7 | 0.00 | 960.69 | 86.86 | −2.629 | 2297.0 |
| | 14.50 | 11.64 | 3366.6 | 0.00 | 1025.31 | 90.74 | −2.646 | 2472.9 |
| | 15.00 | 11.94 | 3542.6 | 0.00 | 1100.31 | 95.31 | −2.636 | 2648.9 |
| | 15.50 | 12.24 | 3727.9 | 0.00 | 1172.76 | 99.42 | −2.641 | 2834.2 |
| | 16.00 | 12.45 | 3918.5 | 0.00 | 1207.98 | 102.41 | −2.736 | 3024.8 |
| | 16.50 | 12.74 | 4112.2 | 0.00 | 1288.20 | 106.72 | −2.724 | 3218.5 |
| | 17.00 | 13.03 | 4335.0 | 0.00 | 1369.31 | 111.11 | −2.735 | 3441.3 |
| | 17.50 | 13.32 | 4557.9 | 0.00 | 1456.18 | 115.62 | −2.733 | 3664.2 |
| | 18.00 | 13.51 | 4695.9 | 0.00 | 1494.13 | 118.81 | −2.765 | 3802.2 |
| *$Li_2SO_4$: a = 0.65149 nm; B = 30.04; $\Lambda^0$ = 90.06 S cm²/mol* | | | | | | | | |
| | 0.10 | 0.10 | 926.3 | 3.50 | 1.46 | 1.01 | −8.245 | 32.6 |
| | 0.20 | 0.20 | 971.1 | 6.73 | 3.19 | 2.11 | −9.910 | 77.4 |
| | 0.30 | 0.30 | 1020.2 | 10.02 | 4.98 | 3.24 | −10.750 | 126.5 |

(*Continued*)

**TABLE 6.1 (*Continued*)**
**Values of Thermodynamic Quantities Included in Formula (6.72) for Calculation of Viscosity, Pa s**

| Electrolyte | $m$ | $c_k$ | $\eta$ | $-\Psi_\eta$ | $-E_{sp}$ | $U_x$ | $B_\eta$ | $\eta^*$ |
|---|---|---|---|---|---|---|---|---|
| | 0.40 | 0.40 | 1069.3 | 13.10 | 6.98 | 4.43 | −11.219 | 175.6 |
| | 0.50 | 0.49 | 1118.4 | 16.01 | 9.22 | 5.69 | −11.507 | 224.7 |
| | 0.60 | 0.59 | 1168.5 | 18.67 | 11.67 | 7.01 | −11.783 | 274.8 |
| | 0.70 | 0.69 | 1222.7 | 21.27 | 14.32 | 8.40 | −12.102 | 329.0 |
| | 0.80 | 0.79 | 1277.7 | 23.57 | 17.24 | 9.85 | −12.405 | 384.0 |
| | 0.90 | 0.88 | 1340.5 | 25.66 | 20.44 | 11.38 | −12.870 | 446.8 |
| | 1.00 | 0.98 | 1403.3 | 27.61 | 23.82 | 12.95 | −13.244 | 509.6 |
| | 1.25 | 1.21 | 1570.1 | 32.11 | 33.12 | 17.12 | −14.058 | 676.4 |
| | 1.50 | 1.45 | 1781.6 | 35.35 | 44.03 | 21.68 | −15.387 | 887.9 |
| | 1.75 | 1.67 | 2010.5 | 37.05 | 57.27 | 26.80 | −16.540 | 1116.8 |
| | 2.00 | 1.90 | 2247.5 | 37.03 | 72.61 | 32.34 | −17.516 | 1353.8 |
| | 2.25 | 2.13 | 2511.7 | 35.49 | 91.23 | 38.70 | −18.381 | 1618.0 |
| | 2.50 | 2.35 | 2507.9 | 31.27 | 112.82 | 45.53 | −16.378 | 1614.2 |
| *LiCl: a = 0.49035 nm; B = 28.88; $\Lambda^0$ = 115.14 S cm²/mol* | | | | | | | | |
| | 0.40 | 0.40 | 945.3 | 1.98 | 2.94 | 2.57 | −22.050 | 51.6 |
| | 0.50 | 0.49 | 956.0 | 2.47 | 3.68 | 3.22 | −21.288 | 62.3 |
| | 0.60 | 0.59 | 971.1 | 2.99 | 4.41 | 3.87 | −21.968 | 77.4 |
| | 0.70 | 0.69 | 986.3 | 3.38 | 5.27 | 4.57 | −22.643 | 92.6 |
| | 0.80 | 0.78 | 1001.4 | 3.87 | 6.03 | 5.22 | −23.023 | 107.7 |
| | 0.90 | 0.88 | 1016.5 | 4.15 | 7.00 | 5.95 | −23.607 | 122.8 |
| | 1.00 | 0.98 | 1031.6 | 4.60 | 7.72 | 6.56 | −23.957 | 137.9 |
| | 1.25 | 1.22 | 1069.4 | 5.63 | 10.18 | 8.51 | −24.045 | 175.7 |
| | 1.50 | 1.45 | 1107.1 | 6.64 | 12.88 | 10.59 | −23.894 | 213.4 |
| | 1.75 | 1.69 | 1144.3 | 7.62 | 16.01 | 12.94 | −23.446 | 250.6 |
| | 2.00 | 1.92 | 1181.4 | 8.50 | 19.30 | 15.34 | −23.095 | 287.7 |
| | 2.25 | 2.15 | 1218.3 | 9.27 | 22.78 | 17.82 | −22.819 | 324.6 |
| | 2.50 | 2.38 | 1255.2 | 9.87 | 26.63 | 20.48 | −22.558 | 361.5 |
| | 2.75 | 2.60 | 1294.0 | 11.08 | 29.88 | 22.98 | −22.266 | 400.3 |
| | 3.00 | 2.82 | 1334.7 | 11.57 | 34.15 | 25.83 | −22.172 | 441.0 |
| | 3.25 | 3.05 | 1376.3 | 11.90 | 38.59 | 28.71 | −22.167 | 482.6 |
| | 3.50 | 3.27 | 1423.2 | 11.97 | 43.42 | 31.75 | −22.403 | 529.5 |
| | 3.75 | 3.49 | 1470.1 | 12.01 | 48.60 | 34.96 | −22.473 | 576.4 |
| | 4.00 | 3.69 | 1518.8 | 13.06 | 52.82 | 38.00 | −22.423 | 625.1 |
| | 4.25 | 3.90 | 1568.6 | 12.93 | 58.86 | 41.65 | −22.397 | 674.9 |
| | 4.50 | 4.12 | 1618.5 | 12.42 | 65.33 | 45.38 | −22.389 | 724.8 |
| | 4.75 | 4.34 | 1672.9 | 11.72 | 72.39 | 49.41 | −22.455 | 779.2 |
| | 5.00 | 4.52 | 1727.3 | 12.78 | 77.68 | 53.06 | −22.292 | 833.6 |
| | 5.25 | 4.74 | 1783.0 | 11.70 | 85.21 | 57.20 | −22.398 | 889.3 |
| | 5.50 | 4.95 | 1842.1 | 10.61 | 93.45 | 61.69 | −22.387 | 948.4 |

(*Continued*)

**TABLE 6.1** (*Continued*)
**Values of Thermodynamic Quantities Included in Formula (6.72) for Calculation of Viscosity, Pa s**

| Electrolyte | $m$ | $c_k$ | $\eta$ | $-\Psi_\eta$ | $-E_{sp}$ | $U_x$ | $B_\eta$ | $\eta^*$ |
|---|---|---|---|---|---|---|---|---|
| | 5.75 | 5.16 | 1901.2 | 9.01 | 101.96 | 66.15 | −22.478 | 1007.5 |
| | 6.00 | 5.34 | 1963.7 | 9.55 | 108.16 | 70.17 | −22.506 | 1070.0 |
| | 6.25 | 5.54 | 2031.2 | 7.83 | 117.94 | 75.25 | −22.514 | 1137.5 |
| | 6.50 | 5.75 | 2098.6 | 5.52 | 128.33 | 80.45 | −22.566 | 1204.9 |
| | 6.75 | 5.92 | 2169.7 | 5.80 | 135.04 | 84.66 | −22.708 | 1276.0 |
| | 7.00 | 6.13 | 2246.3 | 3.32 | 146.16 | 90.10 | −22.780 | 1352.6 |
| | 7.25 | 6.33 | 2323.0 | 0.50 | 158.10 | 95.84 | −22.771 | 1429.3 |
| | 7.50 | 6.49 | 2401.8 | 0.53 | 165.56 | 100.36 | −22.944 | 1508.1 |
| | 7.75 | 6.70 | 2489.1 | 0.00 | 178.58 | 106.35 | −22.090 | 1595.4 |
| | 8.00 | 6.86 | 2576.4 | 0.00 | 186.77 | 111.24 | −22.277 | 1682.7 |
| | 8.25 | 7.06 | 2663.7 | 0.00 | 200.51 | 117.42 | −21.303 | 1770.0 |
| | 8.50 | 7.21 | 2759.6 | 0.00 | 208.87 | 122.31 | −21.557 | 1865.9 |
| | 8.75 | 7.41 | 2857.0 | 0.00 | 223.31 | 128.57 | −20.725 | 1963.3 |
| | 9.00 | 7.61 | 2954.4 | 0.00 | 238.92 | 135.14 | −19.857 | 2060.7 |
| | 9.25 | 7.76 | 3053.3 | 0.00 | 247.78 | 140.15 | −20.065 | 2159.6 |
| | 9.50 | 7.95 | 3155.3 | 0.00 | 263.75 | 146.68 | −19.317 | 2261.6 |
| | 9.75 | 8.10 | 3257.3 | 0.00 | 273.33 | 152.00 | −19.482 | 2363.6 |
| | 10.00 | 8.30 | 3359.3 | 0.00 | 290.79 | 158.99 | −18.706 | 2465.6 |
| *$MgCl_2$: $a$ = 0.62658 nm; $B$ = 18.68; $\Lambda^0$ = 145.07 S cm²/mol* | | | | | | | | |
| | 0.20 | 0.20 | 898.7 | 4.07 | 4.24 | 1.46 | −0.732 | 5.0 |
| | 0.30 | 0.30 | 908.2 | 5.96 | 6.66 | 2.24 | −1.394 | 14.5 |
| | 0.40 | 0.40 | 916.8 | 7.72 | 9.32 | 3.05 | −1.649 | 23.1 |
| | 0.50 | 0.50 | 925.4 | 9.37 | 12.26 | 3.92 | −1.787 | 31.7 |
| | 0.60 | 0.59 | 934.0 | 11.46 | 14.79 | 4.75 | −1.872 | 40.3 |
| | 0.70 | 0.69 | 942.3 | 12.94 | 18.17 | 5.68 | −1.911 | 48.6 |
| | 0.80 | 0.78 | 950.1 | 14.37 | 21.82 | 6.66 | −1.911 | 56.4 |
| | 0.90 | 0.88 | 958.0 | 15.58 | 25.88 | 7.70 | −1.904 | 64.3 |
| | 1.00 | 0.98 | 966.9 | 16.74 | 30.25 | 8.79 | −1.915 | 73.2 |
| | 1.25 | 1.22 | 990.8 | 19.49 | 42.30 | 11.71 | −1.939 | 97.1 |
| | 1.50 | 1.44 | 1017.8 | 22.85 | 54.88 | 14.83 | −1.973 | 124.1 |
| | 1.75 | 1.67 | 1046.6 | 24.20 | 72.04 | 18.55 | −1.968 | 152.9 |
| | 2.00 | 1.91 | 1074.9 | 24.57 | 92.79 | 22.77 | −1.916 | 181.2 |
| | 2.25 | 2.14 | 1105.3 | 23.38 | 117.69 | 27.50 | −1.863 | 211.6 |
| | 2.50 | 2.35 | 1136.9 | 23.89 | 141.30 | 32.27 | −1.830 | 243.2 |
| | 2.75 | 2.58 | 1170.0 | 19.70 | 175.58 | 38.12 | −1.758 | 276.3 |
| | 3.00 | 2.78 | 1204.7 | 18.33 | 207.69 | 44.09 | −1.710 | 311.0 |
| | 3.25 | 3.00 | 1242.8 | 10.40 | 252.56 | 51.04 | −1.647 | 349.1 |
| | 3.50 | 3.20 | 1283.4 | 5.74 | 293.65 | 57.98 | −1.614 | 389.7 |
| | 3.75 | 3.40 | 1325.9 | 0.00 | 339.67 | 65.43 | −1.576 | 432.2 |

(*Continued*)

**TABLE 6.1 (*Continued*)**
**Values of Thermodynamic Quantities Included in Formula (6.72) for Calculation of Viscosity, Pa s**

| Electrolyte | $m$ | $c_k$ | $\eta$ | $-\Psi_\eta$ | $-E_{sp}$ | $U_x$ | $B_\eta$ | $\eta^*$ |
|---|---|---|---|---|---|---|---|---|
| | 4.00 | 3.62 | 1371.6 | 0.00 | 404.26 | 74.25 | −1.448 | 477.9 |
| | 4.25 | 3.81 | 1419.2 | 0.00 | 460.19 | 82.56 | −1.391 | 525.5 |
| | 4.50 | 4.00 | 1467.8 | 0.00 | 522.11 | 91.42 | −1.333 | 574.1 |
| | 4.75 | 4.19 | 1518.7 | 0.00 | 591.59 | 101.16 | −1.274 | 625.0 |
| | 5.00 | 4.37 | 1570.1 | 0.00 | 665.92 | 111.12 | −1.219 | 676.4 |
| | 5.25 | 4.55 | 1623.0 | 0.00 | 742.15 | 121.01 | −1.174 | 729.3 |
| | 5.50 | 4.73 | 1675.9 | 0.00 | 827.02 | 131.76 | −1.125 | 782.2 |
| | 5.75 | 4.91 | 1728.7 | 0.00 | 915.89 | 142.27 | −1.079 | 835.0 |
| | 5.84 | 5.00 | 1747.7 | 0.00 | 971.13 | 147.47 | −1.037 | 854.0 |
| *NaCl: a = 0.53835 nm; B = 6.49; $\Lambda^0$ = 128.43 S cm²/mol* | | | | | | | | |
| | 0.30 | 0.30 | 911.8 | 1.81 | 1.48 | 0.46 | −6.384 | 18.1 |
| | 0.40 | 0.40 | 918.9 | 2.46 | 2.00 | 0.62 | −6.564 | 25.2 |
| | 0.50 | 0.50 | 927.9 | 2.98 | 2.62 | 0.79 | −7.109 | 34.2 |
| | 0.60 | 0.59 | 936.9 | 3.63 | 3.17 | 0.96 | −7.391 | 43.2 |
| | 0.70 | 0.69 | 945.8 | 4.09 | 3.87 | 1.14 | −7.637 | 52.1 |
| | 0.80 | 0.78 | 954.8 | 4.74 | 4.49 | 1.32 | −7.726 | 61.1 |
| | 0.90 | 0.88 | 963.8 | 5.21 | 5.28 | 1.53 | −7.816 | 70.1 |
| | 1.00 | 0.98 | 972.7 | 5.60 | 6.16 | 1.74 | −7.889 | 79.0 |
| | 1.25 | 1.22 | 996.6 | 6.87 | 8.16 | 2.26 | −8.053 | 102.9 |
| | 1.50 | 1.46 | 1021.3 | 8.05 | 10.37 | 2.81 | −8.173 | 127.6 |
| | 1.75 | 1.69 | 1047.9 | 9.09 | 12.83 | 3.40 | −8.327 | 154.2 |
| | 2.00 | 1.92 | 1075.7 | 10.08 | 15.56 | 4.04 | −8.427 | 182.0 |
| | 2.25 | 2.16 | 1105.2 | 10.26 | 19.07 | 4.76 | −8.607 | 211.5 |
| | 2.50 | 2.38 | 1136.0 | 10.88 | 22.47 | 5.49 | −8.695 | 242.3 |
| | 2.75 | 2.61 | 1167.5 | 11.30 | 26.10 | 6.24 | −8.788 | 273.8 |
| | 3.00 | 2.83 | 1201.9 | 11.66 | 30.10 | 7.06 | −8.880 | 308.2 |
| | 3.25 | 3.05 | 1236.7 | 11.73 | 34.46 | 7.91 | −8.961 | 343.0 |
| | 3.50 | 3.26 | 1274.4 | 11.59 | 39.21 | 8.82 | −9.066 | 380.7 |
| | 3.75 | 3.48 | 1312.1 | 11.31 | 44.28 | 9.75 | −9.127 | 418.4 |
| | 4.00 | 3.69 | 1353.2 | 10.67 | 49.81 | 10.75 | −9.238 | 459.5 |
| | 4.25 | 3.90 | 1394.4 | 9.93 | 55.83 | 11.81 | −9.281 | 500.7 |
| | 4.50 | 4.11 | 1439.1 | 8.88 | 62.24 | 12.89 | −9.367 | 545.4 |
| | 4.75 | 4.32 | 1484.3 | 7.38 | 69.19 | 14.04 | −9.444 | 590.6 |
| | 5.00 | 4.53 | 1532.0 | 5.72 | 76.51 | 15.22 | −9.526 | 638.3 |
| | 5.25 | 4.70 | 1581.0 | 6.13 | 82.06 | 16.30 | −9.560 | 687.3 |
| | 5.50 | 4.90 | 1609.0 | 3.99 | 90.25 | 17.58 | −9.330 | 715.3 |
| *$NaNO_3$: a = 0.57989 nm; B = 6.13; $\Lambda^0$ = 123.54 S cm²/mol* | | | | | | | | |
| | 0.20 | 0.20 | 896.8 | 2.27 | 0.43 | 0.23 | −1.250 | 3.1 |
| | 0.30 | 0.30 | 905.4 | 3.40 | 0.65 | 0.35 | −3.158 | 11.7 |
| | 0.40 | 0.39 | 913.1 | 4.52 | 0.89 | 0.47 | −3.924 | 19.4 |

(*Continued*)

**TABLE 6.1 (*Continued*)**
**Values of Thermodynamic Quantities Included in Formula (6.72) for Calculation of Viscosity, Pa s**

| Electrolyte | $m$ | $c_k$ | $\eta$ | $-\Psi_\eta$ | $-E_{sp}$ | $U_x$ | $B_\eta$ | $\eta^*$ |
|---|---|---|---|---|---|---|---|---|
| | 0.50 | 0.49 | 920.8 | 5.57 | 1.15 | 0.59 | −4.409 | 27.1 |
| | 0.60 | 0.59 | 928.4 | 6.62 | 1.43 | 0.72 | −4.738 | 34.7 |
| | 0.70 | 0.69 | 936.1 | 7.63 | 1.73 | 0.85 | −4.984 | 42.4 |
| | 0.80 | 0.78 | 943.4 | 8.74 | 1.98 | 0.98 | −5.103 | 49.7 |
| | 0.90 | 0.87 | 950.4 | 9.70 | 2.31 | 1.12 | −5.206 | 56.7 |
| | 1.00 | 0.97 | 957.4 | 10.62 | 2.66 | 1.26 | −5.300 | 63.7 |
| | 1.25 | 1.20 | 977.4 | 12.89 | 3.56 | 1.62 | −5.644 | 83.7 |
| | 1.50 | 1.43 | 1000.2 | 15.17 | 4.41 | 1.97 | −6.046 | 106.5 |
| | 1.75 | 1.66 | 1025.0 | 16.99 | 5.49 | 2.35 | −6.522 | 131.3 |
| | 2.00 | 1.88 | 1050.6 | 18.52 | 6.68 | 2.75 | −6.985 | 156.9 |
| | 2.25 | 2.10 | 1076.0 | 20.34 | 7.75 | 3.12 | −7.302 | 182.3 |
| | 2.50 | 2.32 | 1103.1 | 21.46 | 9.15 | 3.54 | −7.733 | 209.4 |
| | 2.75 | 2.53 | 1131.1 | 22.88 | 10.36 | 3.93 | −8.100 | 237.4 |
| | 3.00 | 2.73 | 1160.2 | 24.51 | 11.75 | 4.37 | −8.353 | 266.5 |
| | 3.25 | 2.95 | 1190.7 | 25.96 | 14.13 | 5.03 | −8.470 | 297.0 |
| | 3.50 | 3.14 | 1223.7 | 28.00 | 16.23 | 5.67 | −8.558 | 330.0 |
| | 3.75 | 3.34 | 1258.1 | 29.90 | 18.53 | 6.33 | −8.656 | 364.4 |
| | 4.00 | 3.53 | 1295.2 | 31.57 | 20.92 | 7.01 | −8.827 | 401.5 |
| | 4.25 | 3.74 | 1334.0 | 31.69 | 24.29 | 7.80 | −9.138 | 440.3 |
| | 4.50 | 3.92 | 1374.7 | 32.83 | 27.09 | 8.52 | −9.358 | 481.0 |
| | 4.75 | 4.11 | 1417.3 | 33.73 | 30.09 | 9.26 | −9.597 | 523.6 |
| | 5.00 | 4.29 | 1460.7 | 34.42 | 33.23 | 10.02 | −9.838 | 567.0 |
| | 5.25 | 4.47 | 1505.8 | 34.39 | 35.88 | 10.61 | −10.261 | 612.1 |
| | 5.50 | 4.65 | 1551.3 | 33.65 | 38.52 | 11.14 | −10.776 | 657.6 |
| | 5.75 | 4.82 | 1598.3 | 32.85 | 41.18 | 11.67 | −11.298 | 704.6 |
| | 6.00 | 5.00 | 1645.5 | 31.88 | 43.95 | 12.20 | −11.815 | 751.8 |
| | 6.25 | 5.17 | 1692.7 | 30.86 | 47.15 | 12.80 | −12.253 | 799.0 |
| | 6.50 | 5.34 | 1739.8 | 29.82 | 50.38 | 13.41 | −12.667 | 846.1 |
| | 6.75 | 5.47 | 1788.3 | 30.84 | 52.10 | 13.87 | −12.950 | 894.6 |
| | 7.00 | 5.63 | 1838.0 | 29.50 | 55.87 | 14.56 | −13.338 | 944.3 |
| | 7.25 | 5.80 | 1888.1 | 27.87 | 59.71 | 15.25 | −13.748 | 994.4 |
| | 7.50 | 5.96 | 1939.9 | 26.02 | 63.88 | 15.95 | −14.149 | 1046.2 |
| | 7.75 | 6.13 | 1991.7 | 24.10 | 67.93 | 16.63 | −14.561 | 1098.0 |
| | 8.00 | 6.24 | 2046.4 | 24.58 | 70.01 | 17.12 | −14.879 | 1152.7 |
| | 8.25 | 6.40 | 2103.0 | 22.43 | 74.54 | 17.87 | −15.287 | 1209.3 |
| | 8.50 | 6.56 | 2159.6 | 19.76 | 79.20 | 18.59 | −15.751 | 1265.9 |
| | 8.75 | 6.72 | 2219.0 | 16.81 | 83.86 | 19.28 | −16.284 | 1325.3 |
| | 9.00 | 6.83 | 2278.7 | 17.26 | 86.13 | 19.80 | −16.569 | 1385.0 |

(*Continued*)

**TABLE 6.1 (*Continued*)**
**Values of Thermodynamic Quantities Included in Formula (6.72) for Calculation of Viscosity, Pa s**

| Electrolyte | $m$ | $c_k$ | $\eta$ | $-\Psi_\eta$ | $-E_{sp}$ | $U_x$ | $B_\eta$ | $\eta^*$ |
|---|---|---|---|---|---|---|---|---|
| | 9.25 | 6.98 | 2338.5 | 14.30 | 91.28 | 20.56 | −16.995 | 1444.8 |
| | 9.50 | 7.14 | 2401.1 | 10.79 | 96.67 | 21.29 | −17.495 | 1507.4 |
| | 9.75 | 7.24 | 2463.8 | 11.03 | 98.69 | 21.77 | −17.851 | 1570.1 |
| | 10.00 | 7.39 | 2526.5 | 7.19 | 104.44 | 22.52 | −18.323 | 1632.8 |
| | 10.25 | 7.54 | 2590.5 | 3.41 | 110.20 | 23.28 | −18.783 | 1696.8 |
| | 10.50 | 7.64 | 2654.6 | 3.49 | 112.56 | 23.78 | −19.085 | 1760.9 |
| | 10.75 | 7.79 | 2718.7 | 0.00 | 118.78 | 24.57 | −19.372 | 1825.0 |
| | 10.83 | 7.82 | 2747.6 | 0.00 | 119.24 | 24.66 | −19.603 | 1853.9 |
| *RbCl: a = 0.61812 nm; B = 4.82;* $\Lambda^0$ *= 154.19 S cm²/mol* | | | | | | | | |
| | 0.10 | 0.10 | 886.6 | 1.39 | 0.22 | 0.12 | 4.766 | −7.1 |
| | 0.20 | 0.20 | 884.3 | 2.71 | 0.47 | 0.23 | 3.195 | −9.4 |
| | 0.30 | 0.30 | 882.0 | 3.97 | 0.76 | 0.36 | 2.674 | −11.7 |
| | 0.40 | 0.40 | 879.8 | 5.21 | 1.06 | 0.48 | 2.403 | −13.9 |
| | 0.50 | 0.50 | 877.6 | 6.42 | 1.38 | 0.61 | 2.239 | −16.1 |
| | 0.60 | 0.60 | 875.5 | 7.58 | 1.72 | 0.74 | 2.128 | −18.2 |
| | 0.70 | 0.69 | 873.4 | 8.72 | 2.08 | 0.88 | 2.046 | −20.3 |
| | 0.80 | 0.79 | 871.8 | 9.81 | 2.46 | 1.01 | 1.947 | −21.9 |
| | 0.90 | 0.89 | 870.3 | 10.88 | 2.86 | 1.15 | 1.860 | −23.4 |
| | 1.00 | 0.98 | 869.2 | 11.94 | 3.29 | 1.29 | 1.756 | −24.5 |
| | 1.20 | 1.17 | 867.3 | 13.93 | 4.20 | 1.58 | 1.593 | −26.4 |
| | 1.40 | 1.35 | 865.5 | 16.01 | 5.03 | 1.86 | 1.468 | −28.2 |
| | 1.60 | 1.52 | 863.7 | 17.94 | 5.99 | 2.15 | 1.378 | −30.0 |
| | 1.80 | 1.71 | 861.2 | 19.49 | 7.26 | 2.48 | 1.340 | −32.5 |
| | 2.00 | 1.88 | 859.5 | 21.20 | 8.35 | 2.78 | 1.276 | −34.2 |
| | 2.50 | 2.31 | 854.8 | 25.46 | 12.64 | 3.80 | 1.134 | −38.9 |
| | 3.00 | 2.74 | 856.2 | 28.85 | 18.31 | 4.97 | 0.888 | −37.5 |
| | 3.50 | 3.14 | 865.1 | 31.68 | 24.45 | 6.14 | 0.573 | −28.6 |
| | 4.00 | 3.54 | 877.9 | 33.51 | 31.85 | 7.40 | 0.272 | −15.8 |
| | 4.50 | 3.89 | 891.7 | 35.49 | 39.05 | 8.62 | 0.030 | −2.0 |
| | 5.00 | 4.27 | 904.4 | 35.10 | 48.91 | 9.99 | −0.144 | 10.7 |
| | 5.50 | 4.61 | 916.7 | 35.13 | 58.08 | 11.26 | −0.280 | 23.0 |
| | 6.00 | 4.93 | 928.3 | 34.74 | 68.84 | 12.67 | −0.380 | 34.6 |
| | 6.50 | 5.25 | 939.8 | 33.41 | 80.70 | 14.10 | −0.461 | 46.1 |
| | 7.00 | 5.57 | 951.4 | 31.17 | 93.66 | 15.55 | −0.528 | 57.7 |
| | 7.50 | 5.88 | 963.0 | 28.13 | 108.60 | 17.10 | −0.579 | 69.3 |
| | 7.83 | 6.06 | 970.7 | 26.51 | 117.22 | 18.01 | −0.612 | 77.0 |

*Notes:* $m$, mol/kg $H_2O$; $c_k$, mol/dm³; $\eta$ and $\eta^*$, Pa s.

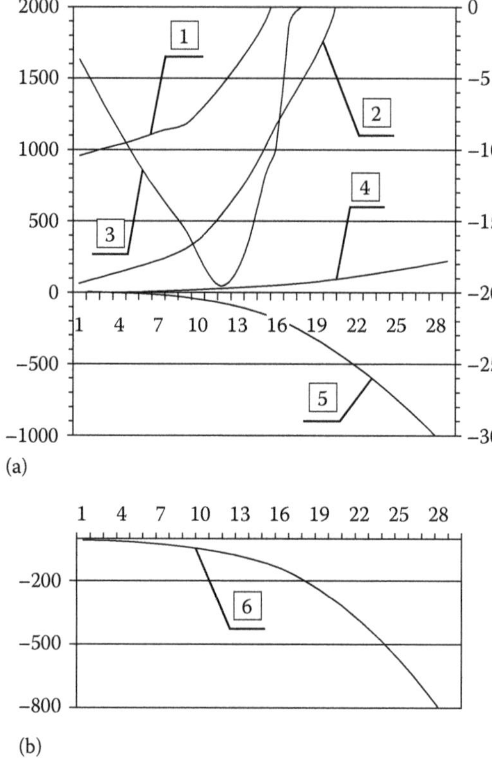


**FIGURE 6.2** Supermolecular interactions (a) (curves 1–5) and extra potential of viscosity (b) (curve 6) for solution of $CaCl_2$: 1—solution viscosity $\eta$, Pa s; 2—increase (decrease) in solution viscosity caused by addition of a solute $\eta^*$, Pa s; 3—electrostatic interactions $-\Psi_\eta$; 4—weak chemical and short-range repulsion forces $U_x = \langle E_H \rangle + U_{oi}$; 5—supramolecular interactions $-E_{sp}$; and 6—extra potential of viscosity $-\nabla \cdot \nabla\xi$. Auxiliary axis is on the right at (a) for curve 3.

Electrostatic forces, curve 3, are gradually increasing in terms of absolute value and also gradually decreasing to zero with electrolyte concentration growth. Analysis of Tables 3.9 and 6.1 shows that for all electrolytes, this value of concentrations is in narrow variance of dielectric permittivities, in which case $\varepsilon_S \approx 30 \div 34$. It is approximately the middle of the second concentration range in Figure 1.1. The peak of curve 3 is blurred and not so clearly manifested, as in Figure 4.13 through 4.15, but the analysis of Table 3.9 and 2.5 allows revealing that the range of electrolyte concentrations is $\varepsilon_S \approx 38 \div 42$, which corresponds to the beginning of the border of the second concentration range in Figure 1.1. Blurring of the peak of curve 3 is explained solely by the fact that the solution moves along the $x$-axis in our case under consideration. It is difficult to say anything concerning the borders of the third concentration range in Figure 1.1, because no practical measurements of viscosity were performed in

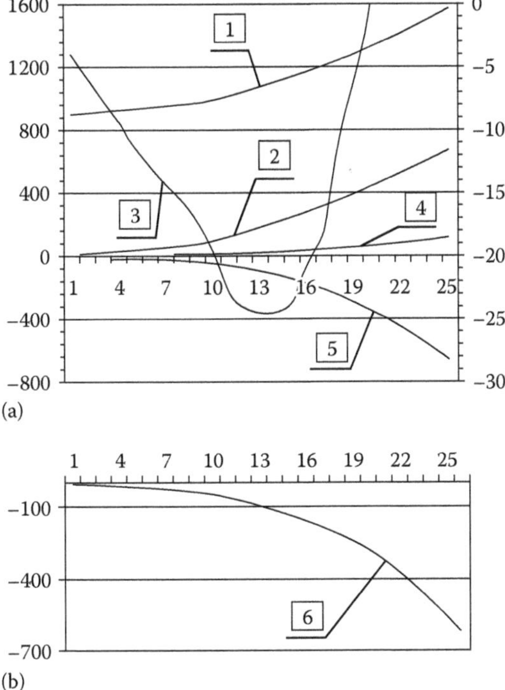


**FIGURE 6.3** Supermolecular interactions (a) (curves 1–5) and extra potential of viscosity (b) (curve 6) for solution of $MgCl_2$: 1—solution viscosity $\eta$, Pa s; 2—increase (decrease) in solution viscosity caused by addition of a solute $\eta^*$, Pa s; 3—electrostatic interactions – $\Psi_\eta$; 4—weak chemical and short-range repulsion forces $U_x = \langle E_H \rangle + U_{ot}$; 5—supramolecular interactions – $E_{sp}$; and 6—extra potential of viscosity – $\nabla \cdot \nabla \xi$. Auxiliary axis is on the right at (a) for curve 3.

this range, in view of the complexity of measurements of viscosity itself. It can be assumed that, as well as for the second concentration range in Figure 1.1, the borders of the third range have not changed, though the solution is moving.

In connection with solution flowing, weak chemical and short-range repulsion forces, curve 4, and supramolecular interactions, curve 5, emerge at the beginning of the appearance of electrolytes in a solution, and increase (decrease) with concentration growth without any peculiarities. In terms of absolute value, these interactions are less than in an immobile solution (Figure 4.13 through 4.15).

Curve 2 of increase (decrease) of solution viscosity, caused by addition of a solute, behaves in proportion to solution viscosity change, curve 1.

The extra potential of viscosity, curve 6, monotonously decreases with the growth of electrolyte concentration in a solution, which indicates that when the solution moves, supermolecules try to occupy a position of minimum potential energy.

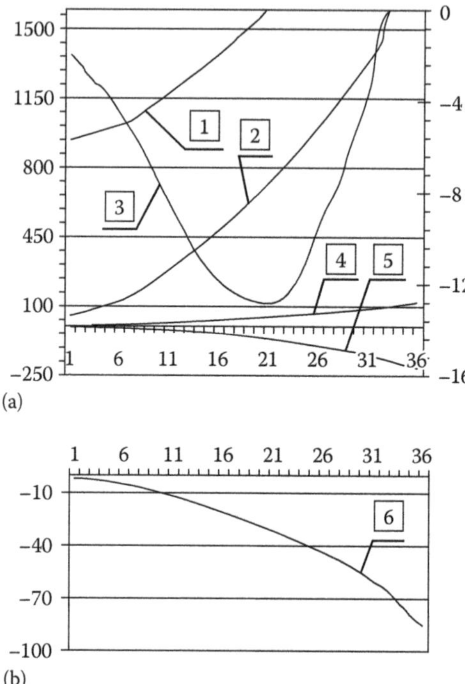


**FIGURE 6.4** Supermolecular interactions (a) (curves 1–5) and extra potential of viscosity (b) (curve 6) for solution of LiCl: 1—solution viscosity $\eta$, Pa s; 2—increase (decrease) in solution viscosity caused by addition of a solute $\eta^*$, Pa s; 3—electrostatic interactions $-\Psi_\eta$; 4—weak chemical and short-range repulsion forces $U_x = \langle E_H \rangle + U_{ot}$; 5—supramolecular interactions $-E_{sp}$; and 6—extra potential of viscosity $-\nabla \cdot \nabla\xi$. Auxiliary axis is on the right at (a) for curve 3.

## REFERENCES

1. Evdokimov I.N., Eliseev N.Y. 2005. Molecular mechanisms of the viscosity of liquids and gases. Part 1. Basic concepts. In: *Oil and Gas*. Russian State University of Oil and Gas, Moscow, Russia, p. 159.
2. Aseyev G.G. 1998. *Electrolytes: Transport Phenomena. Methods for Calculation of Multicomponent Solutions, and Experimental data on Viscosities and Diffusion Coefficients*. Begell House Inc., New York, p. 548.
3. Erdey-Gruz T. 1976. *Transport Phenomena in Aqueous Solutions*. Mir, Moscow, Russia, p. 592.
4. Totchasov E.D., Nikiforov M.Yu., Luk'yanchikova I.A., Al'per G.A. 2001. Comparative analysis of methods for calculating the viscosity of binary mixtures of electrolytes. *J. Appl. Chem.* 74 (5): 797–804.
5. Totchasov E.D., Nikiforov M.Yu., Luk'yanchikova I.A., Al'per G.A. 2002. A semi-empirical method for calculating the dynamic viscosity of binary solvent. *J. Phys. Chem.* 75 (6): 1022–1024.
6. Gleston S., Leydler K., Airing G. 1948. *Theory of Absolute Speeds of Reactions*. Foreign Literature, Moscow, Russia, p. 458.
7. Berd O., St'yuart B., Lightfoot E. 1974. *Transport Phenomena*. Khimiya, Moscow, Russia, p. 288.

8. Hirschfelder Dzh., Certiss Ch., Berd O. 1961. *Molecular Theory of Gases and Liquids.* Foreign Literature, Moscow, Russia, p. 932.
9. Panchenkov G.M. 1974. *The Theory of the Viscosity of Liquids.* Khimiya, Moscow, Russia, p. 156.
10. Good W. 1964. The effect of solute concentration on fluidity and structure in aqueous solution of electrolites. Alkali-metal and ammonium halides. *Electrochem. Acta.* 9: 203–217.
11. Nightingale E.R., Benck R.F. 1959. Viscosity of aqueous sodium fluoride and sodium periodate solutions. Ionic energies and entropies of activation for viscous flow. *J. Phys. Chem.* 63 (10): 1777–1780.
12. Jones Gr., Dole M. 1929. The viscosity of aqueous solutions of strong electrolytes with special reference to barium chloride. *J. Am. Chem. Soc.* 51 (10): 2950–2964.
13. Robinson R.A., Stokes R.H. 1963. *Electrolyte Solutions*, Vols. 1, 2. Publishing House of Foreign Literature, Moscow, Russia.
14. Zhao H. 2006. Viscosity B-coefficient and standard partial molar volumes of amino acids, and their roles in interpreting the protein (enzyme) stabilization. *Biophys. Chem.* 122: 157–183.
15. Kaminsky M. 1957. Ion-solvent interaction and the viscosity of strong-electrolyte solutions. *Discuss. Faraday Soc.* 24: 171–180.
16. Kaminsky M. 1957. Experimentelle Untersuchungen ber die Konzentrationsund Temperatura bhngigkeit der Zhigkeit wriger Lsungen starker Elektrolyte. *Zeitschrift. für. Physikalische. Chemie. Neue. Folge.* 12: 206–231.
17. Vand V. 1948. Viscosity of solutions and suspensions: Theory. *J. Phys. Colloid Chem.* 52 (2): 277–299.
18. Eqgland D., Pilling G. 1972. Viscosity of concentrated aqueous solutions of tetraalkylammoniumbromides. *J. Phys. Chem.* 76 (13): 1902–1906.
19. Angell C.A., Sare J.E., Bressel R.D. 1967. Concentrated electrolyte solution transport theory: Directly measured glass temperatures and vitreous ice. *J. Phys. Chem.* 71 (8): 2759–2761.
20. Bretschneider S. 1966. *Properties of Gases and Liquids. Engineering Methods of Calculation.* Khimiya, Moscow, Russia, p. 536.
21. Benoit H., Goldstein M. 1953. Angular distribution of the light scattered by random coils. *J. Chem. Phys.* 21: 947–952.
22. Partington J. 1951. *Physical Chemistry: An Advanced Treatise*, Vol. 2. London, U.K., p. 325.
23. Fialkov Yu.Ya., Quintus A.A., Zhytomyrskiy A.N. 1977. Justification molar additive function of viscosity in binary liquid systems. *Dokl. Ukrainian Acad. Sci.* 10: 924–926.
24. Fialkov Yu.Ya. 1990. *Solvent Control Means as a Chemical Process.* Khimiya, Leningrad, Russia, p. 240.
25. Fialkov Yu.Ya. 1992. *Physico-Chemical Analysis of Liquid Systems and Solutions.* Science Dumka, Kiev, Ukraine, p. 245.
26. Ali A., Sogbra H. 2002. Molecular interaction study in binary mixtures of dimethylsulfoxide with l,2-dichloroetiiane and 1,1,2,2-tetrachloroethate at 303 K. *Indian J. Phys. B* 76 (1): 23–28.
27. Moumouzias G., Ritzoulis G. 2001. An equation of binary liquid mixtures. *Collect. Czech. Chem. Commun.* 66 (9): 1341–1347.
28. Bak B.V. 1935. The present state of the theory of viscosity. *Prog. Phys. Sci.* 15 (8): 1002–1024.
29. Leont'eva A.A. 1940. Modern theories of viscosity. *Prog. Phys. Sci.* 23 (2): 131–161.
30. Panchenkov G.M. 1950. Properties of tetraalkylammonium salts in aprotornyh solvents. *J. Phys. Chem.* 24: 1390–1395.
31. Harned G., Owen B. 1952. Physical Chemistry of Solutions of Electrolytes. Publishing House of Foreign Literature, Moscow, Russia, p. 628.

32. Falkenhagen H. 1971. *Theore der Electrolite*. Hirzel Verlag, Leipzig, Germany, p. 259.
33. Batchinski A.J. 1953. Investigations on the Interned Friction of Liquids. *J. Phys. Chem.* 8: 644–650.
34. Lyaschenko A.K., Palitskaya T.A., Lileev A.S. et al. 1995. The concentration zone and the properties of aqueous solutions based on the salt compositions formates Y, Ba, Cu synthesis VTSP. *J. Inorg. Chem.* 40 (7): 1209–1217.
35. Baldanov M.M., Baldanova D.M., Zhigzhitova S.B. 2006. Plasma-hydrodynamic method for estimating the transport properties of electrolytes. Sat scientific. Tr. Ser: *Chemistry and Biologically Active Natural Compounds*, Vol. 11. Izd VSGTU, Ulan-Ude, Russia, pp. 112–119.
36. Baranov S.P., Esipova I.A., Saenko E.A., Yufit S.S. 1990. Using the additivity principle to describe the electrical conductivity and viscosity of solutions of lithium perchlorate in acetonitrile. *J. Phys. Chem.* 64 (1): 123–129.
37. Mikulin G.I. ed. 1968. *Questions of Physical Chemistry Solutions of Electrolytes*. Khimiya, Leningrad, Russia, p. 237.
38. Esikova I.A. 1987. Viscosity kontsentrirovainyh electrolyte solutions on their composition. New way of an assessment of koordiiatsionny number of a solvation. *J. Phys. Chem.* 61 (9): 2553–2557.
39. Baldanov M.M., Baldanova D.M., Zhigzhitova S.B., Tanganov B.B. 2006. Plasma-hydrodynamic theory of electrolyte solutions and electrical conductivity. *Dokl. Russ. Acad. Sci. VS* 1 (6): 25–33.
40. Baldanova D.M., Zhigzhitova S.B., Baldanov M.M., Tanganov B.B. 2004. Unified formalism charge conductivities systems: Gas plasma, plasma solids and electrolyte solutions. *Vestn. VSGTU*. 4: 5–10.
41. Baldanova D.M., Zhigzhitova S.B., Baldanov M.M., Tanganov B.B. 2004. The equivalent electrical conductivity of solutions of electrolytes in the approximation of the plasma-hydrodynamic model. *Vestn. VSGTU*. 3: 14–21.
42. Semenchenko V.K. 1941. *Physical Theory of Solutions*. Gostekhteorizdat, Moscow, Russia, p. 368.

# 7 Diffusion Coefficient

## 7.1 INTRODUCTION

Diffusion is transport of a substance from one place to another within one phase without mixing (mechanically or by means of convection). The experiment and theory demonstrate that diffusion takes place at pressure gradient (diffusion under pressure), temperature gradient (thermal diffusion), fields of external forces, and concentration gradient (molecular diffusion). This section considers isothermal isobaric molecular diffusion without gradients conditioned by external-force field.

Concentration equalization by means of self-movement of a substance takes place due to diffusion. In solutions containing one solute, diffusion of such substance takes place from the area of higher concentration to the area of lower concentration. From the point of view of molecular kinetics, a separate particle does not possess preferential movement. However, in case of separating two neighboring elementary volumes, it will turn out that some fraction of particles from the first volume flows toward the second volume and some fraction of particles from the second volume goes in the opposite direction. If concentration in the first elementary volume is higher than in the second one, the number of particles leaving such volume is higher than the number of particles going into it. In other words, a nonvanishing flow of the dilute occurs toward lower concentration [1].

## 7.2 ABOUT DIFFUSION THEORIES

The phenomenological diffusion theory and other kinetic processes are explained in works [2–8]. According to the Onsager linear theory, thermodynamic forces $\overline{X}_j$, inducing irreversible process $j$, are related to flow $\overline{J}_i$ of type $i$ by the ratio

$$\overline{J}_i = \sum_{j=1}^{p} L_{ij}\overline{X}_j, \tag{7.1}$$

where

$p$ is a number of irreversible processes

$L_{ij}$ are kinetic proportionality coefficients

For the majority of evolutionary processes, kinetic coefficients comply with reciprocity relations $L_{ij} = L_{ji}$. The work of Kubo [9] demonstrates that $L_{ij} \approx \int_0^\infty \langle \overline{u}_i(t)\overline{u}_j(t-\tau)\rangle \mathrm{d}\tau$, the angle parentheses here mean averaging of the product of particle velocities of different types by the local equilibrium ensemble. Therefore, Onsager coefficients take into account correlation in motion states of particles of different sorts [10].

Neglect of correlations in motion states of atoms (molecules) leads for the isotropic system to matrix diagonality of coefficients

$$L_{ij} = L_i\delta_{ij}, \quad \delta_{ij} = \begin{cases} 1, & i = j, \\ 0, & i \neq j. \end{cases} \tag{7.2}$$

Some authors believe that this ratio is valid for the majority of practically important cases [11]. The explicit expression for coefficients $L_i$ is found by two ways: either using statistical theories or in the framework of model notions about evolutionary kinetics [12–16].

The Onsager theory is tightly connected with the notion of *local* equilibrium (see, e.g., [17–19]). At availability of irregularities in the macroscopic system with increase of its volume, the relaxation time increases into the equilibrium state, too. In this regard, a situation of transfer of macroscopically small areas of the system into the equilibrium state with further equilibration between these areas is possible. All thermodynamic ratios are valid in local equilibrium macroscopically small areas, but thermodynamic functions depend on spatial coordinates $r$ and time $t$. It is explained by the fact that thermodynamic equilibrium must be set throughout the entire system volume. For implementation of the Onsager theory, deviation of the system from the thermodynamic equilibrium position must be a small value, otherwise nonlinear corrections [20,21] should be introduced. The answer for the question "what deviations should be considered as small values" can be found only in the framework of the statistical theory [22–25], which defines boundaries of applicability of the Onsager model.

The driving force of diffusion is equal to

$$\bar{X}_j = -\nabla\left(\frac{\mu_j}{\theta}\right), \tag{7.3}$$

where

$\mu_j$ is the chemical potential of component $j$
$\theta$ is the temperature in energy units

For example, for the ideal gas, the chemical potential of particles $\mu = \mu_{st}(P, T) + \theta\ln(n/n_{st})$. Here, $\mu_{st}(P, T)$ is a standard value of gas at external pressure $P$, temperature $T$, and gas density $n_{st}$; $n$ is gas density. Then, diffusion flow as per the Onsager theory is equal to

$$\bar{J} = -L\nabla\left(\frac{\mu}{\theta}\right) = -L\left(\frac{\nabla n}{n}\right).$$

Comparing the obtained ratio with Fick's first law $\bar{J} = -D\nabla n$ ($D$ is the diffusion coefficient), when the bond of the kinetic coefficient with the diffusion coefficient was established: $L = Dn$ [16].

It should be stated that self-diffusion coefficient and temperature dependence were established experimentally, which takes the form of Arrhenius law:

$$D = -G\exp(-\beta Q), \tag{7.4}$$

where $G$ and $Q$ are constants for every substance. Using different methods, Zener, Vineyard, Rice, Glyde, and Franklin substantiated applicability of formula (7.4) [4].

The values of matter diffusion coefficients in aqueous solutions depend on the determination method and conditions used in this or that experiment. Thus, the determination of values $D$ in the case of electrolyte solutions is made experimentally using diaphragms [26]. In this case, the diffusion transport of ions through macroporous glass pores separating the solution and clear solvent is examined. In the case of mutually soluble liquids, it is possible to calculate $D$ as per known values of self-diffusion coefficients. In the process of diffusion transport of electrolytes, both cations and anions participate equally, which becomes apparent in a possibility of determining the diffusion *generalized* coefficient only. It is certainly a matter of preference to have values $D$ defined experimentally while calculation values can be used only as guide ones.

Diffusion in electrolyte solutions depends on mobility of ions, which is connected with electrical conductivity. The main differences between these two processes are as follows: (1) electrical conductivity is connected with motion of positively and negatively charged ions in different directions at availability of the external field, and during diffusion either of them move in one direction, and (2) during electrical conductivity at infinite dilution, various electrolyte ions move independently from one another, while at diffusion they are forced to move with the same velocities otherwise separation of electric charge in the solution would take place. One can assume that both processes result from small perturbations of molecular movement: in case of electrical conductivity, the external electric field acts as perturbation, and during diffusion the concentration gradient does. In the initial derivation of the ratio between two effects, which was provided by Nernst, osmotic pressure acted as the driving force at diffusion similarly to the electric field at electrical conductivity. Although, according to the contemporary points of view, osmotic pressure cannot be compared with the actual pressure in the solution. The chemical potential gradient is the effective driving force of diffusion possessing the dimension of force per unit of quantity of the solute [1].

With further development of the theory of solutions as well as various research methods, in particular electrochemical improvement of experimental techniques, the substantial progress in experimental researches of diffusion has been achieved for the recent years [27–80].

Based on the vacancy theory of liquids, in the framework of the transition state theory, there is a connection between viscosity and diffusion [81]:

$$D = \frac{\lambda_1 kT}{\lambda_2 \lambda_3 \eta}, \tag{7.5}$$

where

$\lambda_i$ is the distance between neighboring molecules in directions chosen on axes of coordinates

$k$ is the velocity constant of the viscous flotation process

The same work also demonstrates the connection between the hydrodynamic and transition state theories, when describing diffusion processes, which is expressed by the following equation:

$$D = \frac{kT}{\alpha\pi r\eta}, \tag{7.6}$$

where $r$ is the diffusing particle radius.

As it is known, in the framework of the hydrodynamic theory, considering a diffusing molecule in the solvent as the continuous medium, the coefficient $\alpha$ in the Stokes–Einstein equation is equal to 6. This equation allows using the value of product ($D\eta/T$) [1], proportional to the inverse radius of diffusing particle $r$, as a parameter for the physical–chemical analysis.

Proceeding from the simple hydrodynamic theory, one can expect that the more atomic mass an ion has, the less is its mobility; however, according to the data [82], it appears that the diffusion coefficient of ion $Li^+$ with the smallest atomic mass possesses the smallest value, but, for example, in a series of ions of alkaline metals, the diffusion coefficient increases with increase of the atomic mass. It is well known from the results of measuring equivalent conductivity [82–85] and is generally explained by the fact that the less radius the ions of the same charge have, the more they are hydrated. However, the detailed analysis conducted by Vdovenko et al. [86] shows that such sequence of ion positioning under values of diffusion coefficients cannot be explained simply by increase of hydration shell sizes. We will state that the activation energy of diffusion of ions $\Delta H_i$ (calculated from temperature coefficients) does not substantially differ from the self-diffusion activation energy in clear water [82], which cannot be explained if the difference in values of diffusion coefficients is induced only by the size of particles. In order to interpret this phenomenon, one should consider complicated influence of several factors. The most important role is played by differences in change of the water structure conditioned by dielectric permittivity and ion sizes. Besides, one should take into account special conditions, namely, the fact that ions are usually surrounded by the double hydration shell as Frank et al. [87,88] believe. In inner shell $A$ around the ion, the water molecules are strongly bound with the ion by means of ion–dipole interactions and their mobility is lower than in clear water. In neighboring layer $B$, the molecules are more disordered and more mobile than in the areas rather remote from ions in which their presence does not disrupt the water structure. Hence, the water structure in layer $B$ is somewhat distorted [87,88].

According to the data of Vdovenko et al. [86], Samoylov's heat motion and diffusion theory can be spread to the ions with double hydration shells, too. Samoylov believed that positive hydration transforms to negative-type hydration if the ionic radius is equal to about 0.11 nm. This relates to the ions having the intermediate value between sizes of ions $K^+$ and $Na^+$. If $E_A$ and $E_B$ are energy barriers between two neighboring equilibrium states of the water molecule in the inner and outer layers, that is, activation energy of their displacement, then the average life of molecules in such layers

$$\tau_A = \tau_A^0 \exp\left(\frac{E_A}{RT}\right), \quad \tau_B = \tau_B^0 \exp\left(\frac{E_B}{RT}\right), \tag{7.7}$$

where $\tau_A^0$ and $\tau_B^0$ are the constants determined by oscillation frequency near equilibrium states. Ions in the solution travel in two ways. The first one lies in the fact that they travel without hydration shells, that is, water molecules move away from their ions. It requires activation energy $E_A$. The second one lies in the fact that ions move together with their strongly bound layers $A$ of hydration shells.

General transfer of ion hydrate complexes into the neighboring equilibrium state means overcoming of the high energy barrier, so that is why it is more probable that heat motion and diffusion of hydrated ions take place in several elementary stages, including translation and rotational motions. In work [86], it was assumed that in every elementary stage, there is a maximum of one water molecule moving with the ion. At diffusion of the hydrated ion (ion + layer $A$), there are actually two stages to be distinguished: travel (removal) of some water molecules from their equilibrium states located in layer $B$ closer to the inner hydrate complex and travel of hydrated complex taking of vacancy formed in such a way. The activation energy of the second stage is low, and one can assume that the diffusion activation energy of hydrated complex coincides with value $E_B$, which corresponds to molecule exchange inside layer $B$. Both types of ion travel, if these two processes can be considered as independent, contribute to diffusion according to their statistical masses ($C_A$ and $C_B$):

$$D_i = C_A \exp\left(-\frac{E_A}{RT}\right) + C_B \exp\left(-\frac{E_B}{RT}\right). \tag{7.8}$$

On the other hand, water self-diffusion coefficient $D_w$ is equal to

$$D_w = C_0 \exp\left(-\frac{E_w}{RT}\right), \tag{7.9}$$

where $E_w$ is the barrier height between two neighboring equilibrium states in clear water and $C_0$ is the constant.

From the temperature dependence of the ion diffusion coefficient and water self-diffusion coefficient, one can calculate $E_A$ and $E_B$, introducing some approximate representations. According to the analysis of these results, it can be concluded that hydration shells exist around every ion. For water molecules strongly bound in layer $A$, the energy is $E_A > E_W$, while in the second hydration shell $B$, the energy of water molecules weakly bound with the ion is $E_B < E_W$. Even around ions hydrated as strongly as $Li^+$, one can find the second layer of the distorted structure. On the other hand, around large *negatively hydrated* ions, for example, $Cs^+$, $I^-$ [82], mobility of water molecules closest to them is lower than in clear water. Thereunder, around every ion one can find water molecules both with less and with higher mobility than in clear water. If decrease of mobility in layer $A$ is greater than increase of mobility in layer $B$ (i.e., resulting mobility of water molecules in the hydration shell decreases), the positive hydration phenomenon is observed. However, if increase of mobility in layer $B$ prevails, negative hydration is observed. According to calculation of values $E_A$ and $E_B$, ions $Li^+$ and $Na^+$ move with positive hydration, while ions $K^+$, $Cs^+$, $Cl^-$, and $I^-$ show negative hydration. For ions with the radius of 0.113 nm, $E_A$ is approximately equal to $E_B$; they, like ions $NO_3^-$, do not change average mobility of water molecules around themselves. However, ions $ClO_4^-$ move with positive hydration according to these measurements. Energy $E_B$ possesses about the same value for single-charged ions with the radius exceeding the radius of ion $K^+$. This suggests that around large ions, structural changes in layer $B$ are conditioned mainly by the electrostatic field. This field near every single-charged ion equally depends on the distance. Water molecules around small ions in layer $A$ are positioned more densely than in case of large ions. In regard to the last circumstance, mutual repulsion of water molecules in layer $A$ is stronger, which facilitates their going from layer $A$. The research of the order of values of various effects allows making conclusions about the actual role of the hydration shell in this phenomenon.

As it was stated in work [86], conclusions about the structure of electrolyte solutions can be made as well from research of self-diffusion. In this work, structural factors influencing self-diffusion were studied by means of labeled atoms. Quantitative ratios between transfer coefficients of solution components and ion association were set.

Formation of ion pairs or molecules in electrolyte solutions appears at least in two ways: activity of ions partially decreases and the gradient of Gibbs free energy changes, and resistance of ion friction partially decreases due to formation of one large particle from several smaller ones. If the change in Gibbs free energy is taken into account in the activity coefficient and concentration dependence, set in the given solution experimentally [82], the association does not already change this factor. It is possible by virtue of equilibrium in ion pairs and ions (their chemical potential is identical in the equilibrium state). Therefore, in both states, the change in Gibbs free energy is the same; however, decrease of frictional resistance results in increase of the diffusion coefficient.

Formation of ion pairs or molecules changes as well the electrophoretic effect because ion concentration in binary electrolyte solutions decreases from $c_{ca}$ down to $\alpha c_{ca}$, if $\alpha$ is the electrolyte dissociation degree. Assuming that $D_{ca}^0$ is the diffusion coefficient of ion pairs and nondissociated molecules in ideal solutions, the diffusion coefficient of the partially dissociated or symmetrically associated electrolyte according to the Onsager–Fuoss equation is equal to [82]

$$D = \left[\alpha\left(D^0 + \Delta_1 + \Delta_2\right) + 2\left(1-\alpha\right)D_{ca}^0\right]\left(1 + \frac{\partial \ln \gamma_{ca}}{\partial \ln c_{ca}}\right), \tag{7.10}$$

where $\Delta_i$ are electrophoretic additives.

As Harned and Hudson [89] demonstrated, this equation for zinc sulfate solutions is in good agreement with the experiment. According to the experiment, ion pairing in dilute electrolyte solutions 1–1 is negligible, that is why $D_{ca}^0$ is difficult for being estimated. However, as it was demonstrated for ammonium nitrate solutions [90], it becomes strong in concentrated solutions.

The conditions are similar only for poorly dissociated solutions of weak electrolytes. As a rule, the electrophoretic effect in them is negligible and the major part is taken by diffusion of nonassociated molecules. Dissolved molecules can produce substantial influence on the solvent. It was demonstrated during the diffusion experiment with aqueous solutions of citric [91] and acetic [92] acids; the value $D_{ca}^0$ was found for the $0.657 \cdot 10^{-9}$ and $1.201 \cdot 10^{-9}$ $m^2/s$ accordingly, while limiting values of the diffusion coefficient according to Nernst ($RT\lambda^0/F^2$) for monomeric citrate and acetate ions are equal correspondingly to $0.81 \cdot 10^{-9}$ and $1.088 \cdot 10^{-9}$ $m^2/s$. Hence, the diffusion coefficient of citric acid is considerably lower than the diffusion coefficient of the citrate ion, while it is higher for the acetic acid molecule than for the acetate ion. Assuming that mobility of the acetate ion at its sizes is small and further that the activity coefficient of acetates is relatively large, one can conclude that there is strong interaction of the acetate ion with water molecules or its structure. Relatively high mobility of the monomeric citrate ion, however, points to its strong influence destroying the water structure.

When measuring diffusion coefficients in phosphoric acid solutions in the concentration range of 0.1–6.0 mol/dm³, it was established that the measured values are in good agreement with the calculated ones if taking into account partial dissociation and ratio $D = kT/6\pi\eta r$. However, in friction resistance calculations using values $kT/6\pi\eta r$, one could observe substantial deviations. Hence, one can conclude [82] that sliding is absent on the *surface* of diffusing molecules of phosphoric acid. Therefore, diffusing particles are surrounded by a layer of water molecules moving together with them, that is, they are hydrated.

The more detailed diffusion theory of strong electrolytes in solutions on condition of considerable interaction leading to association is developed by Kiryanov [93].

Based on the Onsager theory for coefficients of conductivity $L_{ij}$ and friction $R_{ij}$, the work of Pikal [94] demonstrated that the term, taking into account formation of ion pairs in the limiting law, certainly results from electrostatic interaction, and introduction of an additional parameter in order to explain mobility of ion pairs is not required.

When writing his book, Erdey-Gruz made a comment that ideas about diffusion in electrolyte solutions are only the first approximation to describing that complicated phenomenon, in spite of the fact that all most important and relevant interactions were taken into account as far as possible. Further researches are required for more veracious and reliable description of the phenomenon [82].

The bond observed between viscosity and diffusion in the framework of the transition state theory is reflected in analogy of the mathematical tool to describe diffusion processes in binary noninteracting solvents. The simplest one is empirical correlation of the following type:

$$D_{AB} = \left(D_{BA}^{0}\right)^{x_A} \left(D_{AB}^{0}\right)^{x_B}, \tag{7.11}$$

mentioned in [95–97], where $D_{BA}^{0}$ and $D_{AB}^{0}$ are diffusion coefficients for components $B$ in clear $A$, and $A$ in clear $B$ accordingly. As it is emphasized in work [97], this empirical dependence, proposed by Wayne, can be obtained as well at some assumptions based on the absolute reaction rate theory.

Self-diffusion coefficients were found in work [98] based on measuring electrical conductivity of aqueous solutions of selenic acid and sodium selenium of various concentrations within the temperature range of 288–318 K. In order to find values of coefficients of ion diffusion at infinite dilution $D_{\pm}^{0}$, the following equation was used:

$$D_{\pm}^{0} = \frac{RT\lambda_{\pm}^{0}}{|z_{\pm}|F^2}.$$

The concept of plasma-like state of ions in solutions, developed in work [99], allows estimating diffusion of electrolyte solutions. Calculation of the diffusion coefficient is made by the Einstein expression

$$D = bkT,$$

where $b$ is the mobility of electrolyte ions.

Baldanov and colleagues propose an equation for calculation of diffusion in concentrated electrolyte solutions [100] in the framework of the plasma-hydrodynamic model:

$$D = \frac{2.22 \cdot 10^{-12} \Lambda kT}{e^2 N_A \exp(\hbar\omega/kT)}, \tag{7.12}$$

where ω is the plasma frequency. Calculation errors under Equation 7.12 reach 15%.

There are no phenomenological models to date, which would realize continuous transition in terms of any of their parameters from one model to another. The role of effective supramolecular interactions is practically ignored, and the nonlocal nature of interactions and the influence of geometric dimensions of particles and dielectric permittivity of a solution on formation of thermodynamic and kinetic properties of electrolytic solutions are not taken into account. There is no attraction of concepts to structure formation processes in the solution. Attempts to explain diffusion using Samoylov's views on primary and secondary hydration did not result in notable success. The common phenomenological approach toward studying thermodynamically equilibrium states of the system is underdeveloped; there are no concepts that can explain diffusion processes from unified positions. Predictive capabilities of existing models are limited by a narrow field of their applicability. That is why investigation of the diffusion phenomenon from the positions of supramolecular interactions is still an important task.

## 7.3 DIFFUSION FLOWS

We will preliminary provide the determination for diffusion flows and diffusion potentials, that is, driving forces. The diffusion coefficient is a proportionality constant between flow and potential. The diffusion coefficient is determined generally by Fick's law:

$$\vec{I} = n\vec{V} = -D\nabla n, \tag{7.13}$$

where

$\vec{I}$ is the flow

$n$ is the number of molecules of the solute, dm$^3$

$\vec{V}$ is the speed, m$^2$/s

$\nabla$ is the gradient

$D$ is the diffusion coefficient, m$^2$/s

In more exact terms, value $\vec{I}$ is flow of the diffusing substance, that is, its quantity passing per time unit through the cross section unit.

The law expressed by Equation 7.13 can be presented as follows, where the flow value is expressed in mol/dm³:

$$\vec{I} = -\vec{L}_{ij}\nabla\mu, \tag{7.14}$$

where

$\vec{L}_{ij}$ is the matrix characterizing the mutual influence degree of $j$ supermolecule on mobility $i$ or, as it is called by many others, the kinetic coefficient

$\mu$ is the chemical potential, J/mol

In more general form, Equation 7.14 is as follows:

$$\vec{I}_j = -\sum_{i=1}^{k} \vec{L}_{ji}\nabla\mu_i, \tag{7.15}$$

if assuming that there is linear dependence between speed and the gradient of solute potentials. As Onsager demonstrated, Equation 7.15 is observed because there is linear dependence between speed and the gradient of solute potentials in electrolyte solutions. Proceeding from this condition, Onsager made a conclusion that matrix coefficients (7.15) are symmetric, that is,

$$\vec{L}_{ji} = \vec{L}_{ij}. \tag{7.16}$$

This condition was proved both using statistical theories and in the framework of model notions about evolutionary kinetics [12–16]. This important result means that the flow of molecules of $j$-type under the action of force equal to one, falling within a unit of quantity of supermolecules of $i$-type, is equal to the flow of supermolecules of $i$-type under the action of force equal to one, falling within a unit of quantity of supermolecules of $j$-type.

From (7.13) and (7.14), we obtain

$$D = \vec{L}_{ij}\left(\frac{\partial\mu}{\partial n}\right)_{P,T}, \tag{7.17}$$

where

$P$ is the pressure, Pa

$T$ is the temperature, K

When considering ideal solutions,

$$\frac{\partial\mu}{\partial n} = \frac{RT}{n}, \tag{7.18}$$

where $R$ is the gas constant, J/(mol K).

From (7.17) and (7.18), we have

$$nD = RT\vec{L}_{ij}. \tag{7.19}$$

When considering real solutions, it is necessary to introduce the activity coefficient [101,102]:

$$nD = RT\Omega\left(1 + n\frac{\partial \ln \zeta}{\partial n}\right), \tag{7.20}$$

where

$n \equiv c_k$, mol/dm$^3$

$f_\pm$ is the activity coefficient of the solute, if $n$ is expressed in mol/dm$^3$

The fundamental ratio $R = kN_A$ is used in (7.20).

The first necessary condition for determining the diffusion coefficient as per Equation 7.20 is to establish kinetic proportionality coefficient $\vec{L}_{ij}$, and activity coefficient $f_\pm$ is expressed by formula (4.84).

## 7.4 GENERAL THEORY OF SUPERMOLECULAR FORCES IN DIFFUSION FLOWS

The absence of electric current for the binary electrolyte solution during consideration of diffusion is a condition for availability of the same velocities for supermolecules of $i$-type and supermolecules of $j$-type.

$$\vec{V}_i = \vec{V}_j = \vec{V}; \quad \vec{I}_i = n_i\vec{V}; \quad \vec{I}_j = n_j\vec{V}. \tag{7.21}$$

If neglecting interactions among supermolecules, in such case,

$$\vec{V} = \vec{K}_i\omega_i = \vec{K}_j\omega_j = -\omega_i\nabla\mu_i = -\omega_j\nabla\mu_j, \tag{7.22}$$

where $\vec{K}_i$ and $\vec{K}_j$ are external forces acting on supermolecules of $i$- and $j$-type, respectively.

If a molecule dissociates into $\nu_1$ cations and $\nu_2$ anions, we obtain distribution of supermolecular forces for chemical potential:

$$\vec{K} = \nu_i\vec{K}_i + \nu_j\vec{K}_j = -\nabla\mu. \tag{7.23}$$

The circumstance that both positive and negative supermolecules move with the same velocity during diffusion is of substantial significance and we have the following condition:

$$\vec{K}_i\omega_i = \vec{K}_j\omega_j. \tag{7.24}$$

Condition (7.24) is met when considering continuity equation (5.26). The terms characterizing the perturbed state during diffusion processes are canceled. Due to such simplification, the distribution of electrostatic and supermolecular interactions gains the symmetrical form (7.24).

On the other hand, some influence is produced by electrophoretic effect in which formula (5.81) for diffusion

$$\vec{K} = \frac{\vec{V}}{\omega}, \quad \rho = \frac{1}{\omega}, \tag{7.25}$$

where $\rho$ is the medium resistance coefficient.

In formula (5.81), we will rename: $\nu_j \equiv E_{D_j}$. Then it will look as follows:

$$E_{D_j} = \vec{V}\kappa_{D_j} \sum_j \rho_j + C_3, \tag{7.26}$$

where

$$\kappa_{D_j} = -\frac{1}{\pi\varepsilon_0\varepsilon_S} \cdot \frac{16\beta^2 \nu_k c_k N_A \beta_R}{135\eta}. \tag{7.27}$$

It can be seen in expressions (7.26) and (7.27) that they participate in distribution of external forces acting on supermolecules in the solution. We will assume in Equation 7.26 that integration constant $C_3$ is equal to

$$C_3 = -\vec{V}\kappa_{D_j} \sum_j \rho_j + \vec{V}\kappa_{D_j} C_3 \sum_j \rho_j. \tag{7.28}$$

We substitute (7.28) into (7.26) and we will obtain

$$E_{D_j} = \vec{V}\kappa_{D_j} C_3 \sum_j \rho_j. \tag{7.29}$$

We assume in (7.29) that

$$C_3 \equiv 10^{23} b_D, \tag{7.30}$$

where $b_D$ is the theoretical constant individual for every electrolyte. In (7.27), we substitute value $\beta$ from (2.48) and constant $C_3$ from (7.30). We will obtain

$$\kappa_D = -1.85 \cdot 10^{20} \left( \frac{1}{\pi\varepsilon_0\varepsilon_S} \right)^4 \left( \frac{|z_i z_j| e^2}{kT} \right)^3 \left( \frac{\nu_k c_k N_A b_D}{\eta R} \right). \tag{7.31}$$

It can be seen from formula (7.31) how substantially electrophoretic effect depends on dielectric permittivity of the solution. We will substitute values of physical constants into expression (7.31) and take into account conversion of values $R$ from m to nm and its intermediate view will be as follows:

$$\kappa_D = -1.19 \cdot 10^{51}\left(\frac{1}{\varepsilon_S}\right)^4\left(\frac{|z_i z_j|}{T}\right)^3\left(\frac{\nu_k c_k b_D}{\eta R}\right). \tag{7.32}$$

We will take into account (7.32) in (7.29): Then,

$$E_{D_j} = \vec{V}\kappa_D \sum_j \rho_j. \tag{7.33}$$

Change in speed of supermolecules under the effect of electrophoretic force produces as well influence on forces $\vec{K}_j$ and $\vec{K}_i$. Total force acting on the supermolecule at permanent motion is equal to the product of its speed by resistance constant $\rho_j$. That is why

$$\vec{K}_j = \rho_j\left(\vec{V} - E_{D_j}\right). \tag{7.34}$$

On the other hand, it is equal by definition to chemical potential gradient μ. Having substituted (7.34) into (7.23), we have

$$\vec{K}_j = \nu_j\rho_j\left(\vec{V} - E_{D_j}\right) + \nu_i\rho_i\left(\vec{V} - E_{D_i}\right) = -\nabla\mu. \tag{7.35}$$

We substitute (7.33) into (7.35) and we will obtain

$$-\nabla\mu = \vec{V}\left[\nu_i\rho_i + \nu_j\rho_j - \kappa_D\left(\nu_i\rho_i^2 + \nu_j\rho_j^2\right)\right]. \tag{7.36}$$

We will take into account in (7.36) that $\rho = 1/\omega$, $\omega = \lambda^0/F|z|e$; we solve the equation concerning $\vec{V}$ and multiply by $n$, and we will obtain $\vec{I}$ according to (7.13) and (7.14):

$$\vec{I} = n\vec{V} = -\vec{L}_{ij}\nabla\mu, \tag{7.37}$$

where kinetic coefficient $\vec{L}_{ij}$ has the intermediate view that will be further reduced:

$$\vec{L}_{ij} = \frac{n\left(\lambda_i^0\lambda_j^0\right)^2}{Fe\left\{\left(\nu_i|z_i|\lambda_j^0 + \nu_j|z_j|\lambda_i^0\right) - Fe\kappa_D\left[\nu_i\left(z_i\lambda_j^0\right)^2 + \nu_j\left(z_j\lambda_i^0\right)^2\right]\right\}}. \tag{7.38}$$

It can be seen from the expression for kinetic coefficient (7.38) that condition (7.16) is observed for it.

We will substitute (7.38) into (7.20) and obtain an intermediate formula for the diffusion coefficient:

$$D = kN_A T\vec{L}_{ij}\left(1 + c_k \frac{\partial \ln f_\pm}{\partial c_k}\right), \tag{7.39}$$

where $c_k$ is concentration, mol/dm³, and the kinetic coefficient will be as follows:

$$\vec{L}_{ij} = \frac{\left(\lambda_i^0 \lambda_j^0\right)^2}{Fe\left\{\left(\nu_i |z_i| \lambda_j^0 + \nu_j |z_j| \lambda_i^0\right) - Fe\kappa_D \left[\nu_i \left(z_i \lambda_j^0\right)^2 + \nu_j \left(z_j \lambda_i^0\right)^2\right]\right\}}. \tag{7.40}$$

Let us proceed to the final derivation of the diffusion coefficient calculation formula taking into account all super- and supramolecular interactions.

## 7.5 DIFFUSION COEFFICIENT (FINAL SOLUTION)

Let us study formula (7.45). We have to differentiate activity coefficient $\partial \ln f_\pm / \partial c_k$. We will use dependencies (4.90) through (4.95) and (4.99). After differentiation, we have

$$\frac{\partial \ln f_\pm}{\partial c_k} = \tau_x - \frac{\tau_{1sp}}{R^4} - \tau_{2sp} - \frac{\Omega_g}{\chi \varepsilon_S T} \Psi_{el}^0. \tag{7.41}$$

Then, in (7.39), taking into account (7.41), we have

$$c_k \frac{\partial \ln f_\pm}{\partial c_k} = \ln f_\pm, \tag{7.42}$$

and the diffusion coefficient (7.39) is found by the formula

$$D = kN_A T\vec{L}_{ij}\left(1 + \ln f_\pm\right). \tag{7.43}$$

We will substitute physical constants into Equations 7.40 and 7.43 and we will take into account correlation of measuring units in diffusion coefficients (m²/s) and electrical conductivity coefficients (S cm²/mol) and transfer constants from the denominator of (7.40) into (7.43). We will obtain from (7.40)

$$\vec{L}_{ij} = \frac{\left(\lambda_i^0 \lambda_j^0\right)^2}{\left(\nu_i |z_i| \lambda_j^0 + \nu_j |z_j| \lambda_i^0\right) - 1.55 \cdot 10^{-14} \kappa_D \left[\nu_i \left(z_i \lambda_j^0\right)^2 + \nu_j \left(z_j \lambda_i^0\right)^2\right]}, \tag{7.44}$$

and from (7.43)

$$D = 1.28 \cdot 10^8 T\vec{L}_{ij}\left(1 + \ln f_{\pm}\right). \tag{7.45}$$

Let us estimate values of summands in the denominator of (7.44) for solution $CaCl_2$ at 25°C and $c_k = 1.886$ mol/dm³. For it, $\nu_i = 1$, $\nu_j = 2$, $\nu_k = 3$, $z_i = 2$, $z_j = 1$. From Table 5.4, we choose $\lambda_i^0 = 59.8$ S cm²/mol, $\lambda_j^0 = 76.26$ S cm²/mol, $\eta = 1579$ P. From Table 3.9, $R = 0.6738$ nm, $\varepsilon_S = 41.6$. We substitute values into the denominator of (7.44) and formula (7.32). The Faraday constant is equal to 96,500 CL/mol, electron charge is $1.6021 \cdot 10^{-19}$ C:

$$\begin{aligned}&\left(1 \cdot 2 \cdot 76.26 + 2 \cdot 1 \cdot 59.8\right) + 96{,}500 \cdot 1.6021 \cdot 10^{-19}\\ &\times\left[1.19 \cdot 10^{51}\left(\frac{1}{41.6}\right)^4\left(\frac{2 \cdot 1}{298}\right)^3\left(\frac{3 \cdot 1.886 \cdot b_D}{1{,}579 \cdot 0.6738}\right)\right]\\ &\times\left[1 \cdot \left(2 \cdot 76.26\right)^2 + 2 \cdot \left(1 \cdot 59.8\right)^2\right] = 272.12 + 3.00 \cdot 10^{26}.\end{aligned}$$

It can be seen from the given example that the first round bracket in expression (7.38) can be neglected. The kinetic coefficient of (7.44) takes the form of

$$\vec{L}_{ij} = \frac{\left(\lambda_i^0 \lambda_j^0\right)^2}{-1.55 \cdot 10^{-14} \kappa_D \Xi}, \tag{7.46}$$

where

$$\Xi = \nu_i\left(z_i \lambda_j^0\right)^2 + \nu_j\left(z_j \lambda_i^0\right)^2. \tag{7.47}$$

We return to Equation 7.32. We multiply coefficients from (7.32) and (7.46) and we will obtain the expression for electrophoretic force:

$$\kappa_D = 1.85 \cdot 10^{37}\left(\frac{1}{\varepsilon_S}\right)^4\left(\frac{|z_i z_j|}{T}\right)^3\left(\frac{\nu_k c_k b_D}{\eta R}\right). \tag{7.48}$$

We will take into account the constant from (7.48) in (7.45):

$$D = 6.92 \cdot 10^{-30} T\vec{L}_{ij}\left(1 + \ln f_{\pm}\right), \tag{7.49}$$

where the kinetic coefficient has the final form of

$$\vec{L}_{ij} = \frac{\left(\lambda_i^0 \lambda_j^0\right)^2}{\kappa_D \Xi}, \tag{7.50}$$

and electrophoretic effect

$$\kappa_D = \left(\frac{1}{\varepsilon_S}\right)^4 \left(\frac{|z_i z_j|}{T}\right)^3 \left(\frac{\nu_k c_k b_D}{\eta R}\right). \tag{7.51}$$

It can be seen from formulas (7.49) through (7.51) that the estimated part of diffusion depends on such thermodynamic values as electrical conductivity, activity coefficients, and viscosity and to a great extent on dielectric permittivity of the solution. Note that activity coefficients depend on electrostatic forces and various supramolecular and weak chemical interactions. As a thermodynamic value, diffusion coefficients include all best achievements of the mentioned properties.

Table 7.1 provides the thermodynamic components of formula (6.72) for a number of electrolytes. The first column contains electrolyte formulas; the second column concentration $m$, mol/kg $H_2O$; the third column concentration $c_k$, mol/dm$^3$; the fourth column diffusion values $D \cdot 10^9$, m$^2$/s; the fifth column kinematic coefficient $L_{ij} \cdot 10^{-16}$; the sixth column electrophoretic retardation $\kappa_D \cdot 10^{12}$; the seventh column activity coefficient $f_\pm$; and the eighth column calculated values of coefficient $b_D$, individual for every electrolyte. Calculation of thermodynamic components was made as per formulas (7.47) and (7.49) through (7.51). Diffusion values in Table 7.1 were interpolated from the data presented in [103], viscosity values were taken from Table 6.1, values $\lambda_i^0$ and $\lambda_j^0$ were taken from Table 5.4, and activity coefficient $f_\pm$ was calculated as per formula (3.33), data for which were taken from Table 3.9.

From Table 7.1, one 1–2 electrolyte and two 1–1 electrolytes were chosen, and thermodynamic interactions of diffusion existing in the solution from very dilute to the concentrated one are shown in Figure 7.1 for $MgCl_2$, in Figure 7.2 for NaCl, and in Figure 7.3 for KOH. Axis $x$ has plotted numbers of concentration points $m$, mol/kg $H_2O$, and axis $y$ has thermodynamic values from Table 7.1: diffusion $D \cdot 10^9$, m$^2$/s; kinematic coefficient $L_{ij} \cdot 10^{-16}$; electrophoretic retardation $\kappa_D \cdot 10^{12}$; activity coefficient $f_\pm$; and coefficient individual for every electrolyte $b_D$.

The axis of abscissas has plotted numbers of concentration points $m$, mol/kg $H_2O$ from Table 7.1 since, for example, for $MgCl_2$ solution, concentrations were taken from the minimum value of 0.2 to the maximum values of 3.5$m$; for NaCl 0.3–4.5$m$ and for KOH 2.0–9.5$m$. Every number of the concentration point attributed its own concentration from Table 6.1:

The analysis of Figures 7.1 through 7.3 shows that all thermodynamic interactions for diffusion take place smoothly, without any runouts. One can state the regularity for isotherms in electrolyte solutions, taking into account the data from Table 6.1 for viscosity and Table 7.1 for diffusion: increase of viscosity values at increase of electrolyte concentration in the solution provokes proportional decrease of diffusion

**TABLE 7.1**
**Thermodynamic Values Entering the Calculation of Diffusion, $D \cdot 10^{-9}$ (m²/s), Based on Formula (7.49)**

| Electrolyte | $m$ | $c_k$ | $D$ | $\bar{L}_{ij}$ | $\kappa_D$ | $f_\pm$ | $b_D$ |
|---|---|---|---|---|---|---|---|
| $CaCl_2$ | 0.20 | 0.20 | 1.136 | 203.590 | 10.215 | 0.4822 | 43.5204 |
| | 0.30 | 0.30 | 1.127 | 230.093 | 9.038 | 0.4664 | 22.0082 |
| | 0.40 | 0.40 | 1.109 | 243.677 | 8.535 | 0.4587 | 13.5206 |
| | 0.50 | 0.49 | 1.115 | 243.990 | 8.524 | 0.4591 | 9.4510 |
| | 0.60 | 0.59 | 1.123 | 233.288 | 8.915 | 0.4646 | 7.3097 |
| | 0.70 | 0.69 | 1.141 | 217.707 | 9.553 | 0.4743 | 6.0122 |
| | 0.80 | 0.78 | 1.159 | 201.499 | 10.321 | 0.4861 | 5.0989 |
| | 0.90 | 0.88 | 1.176 | 180.654 | 11.512 | 0.5044 | 4.6012 |
| | 1.00 | 0.98 | 1.194 | 153.346 | 13.562 | 0.5365 | 4.4182 |
| | 1.25 | 1.21 | 1.230 | 126.079 | 16.495 | 0.5904 | 3.2802 |
| | 1.50 | 1.44 | 1.257 | 102.907 | 20.209 | 0.6649 | 2.6604 |
| | 1.75 | 1.66 | 1.274 | 85.180 | 24.415 | 0.7596 | 2.1784 |
| | 2.00 | 1.89 | 1.283 | 71.343 | 29.150 | 0.8798 | 1.8049 |
| | 2.25 | 2.11 | 1.289 | 60.765 | 34.225 | 1.0285 | 1.5482 |
| | 2.50 | 2.32 | 1.288 | 52.396 | 39.691 | 1.2112 | 1.2553 |
| | 2.75 | 2.52 | 1.283 | 45.620 | 45.587 | 1.4381 | 1.0774 |
| | 3.00 | 2.73 | 1.275 | 40.145 | 51.804 | 1.7153 | 0.9082 |
| | 3.25 | 2.94 | 1.265 | 35.604 | 58.412 | 2.0574 | 0.7611 |
| | 3.50 | 3.13 | 1.251 | 31.800 | 65.399 | 2.4770 | 0.6497 |
| | 3.75 | 3.31 | 1.237 | 28.610 | 72.689 | 2.9886 | 0.5761 |
| | 4.00 | 3.51 | 1.222 | 25.930 | 80.204 | 3.6108 | 0.5219 |
| CsCl | 0.10 | 0.10 | 2.507 | 173.959 | 20.105 | 0.7397 | 5623.4028 |
| | 0.20 | 0.20 | 2.521 | 188.231 | 18.581 | 0.7040 | 2456.2690 |
| | 0.30 | 0.30 | 2.535 | 204.835 | 17.075 | 0.6702 | 1389.1102 |
| | 0.40 | 0.39 | 2.549 | 221.771 | 15.771 | 0.6421 | 904.9093 |
| | 0.50 | 0.49 | 2.563 | 237.242 | 14.742 | 0.6210 | 627.5582 |
| | 0.60 | 0.58 | 2.577 | 251.213 | 13.922 | 0.6048 | 464.7872 |
| | 0.70 | 0.68 | 2.591 | 263.391 | 13.279 | 0.5926 | 354.2541 |
| | 0.80 | 0.77 | 2.605 | 274.888 | 12.723 | 0.5823 | 278.9362 |
| | 0.90 | 0.86 | 2.619 | 286.423 | 12.211 | 0.5730 | 224.5060 |
| | 1.00 | 0.95 | 2.633 | 300.421 | 11.642 | 0.5626 | 181.5753 |
| | 1.25 | 1.18 | 2.668 | 317.443 | 11.018 | 0.5528 | 118.0763 |
| | 1.50 | 1.40 | 2.700 | 334.682 | 10.450 | 0.5439 | 80.8661 |
| | 1.75 | 1.63 | 2.730 | 350.044 | 9.992 | 0.5368 | 58.5318 |
| | 2.00 | 1.83 | 2.768 | 364.164 | 9.604 | 0.5317 | 44.1681 |
| $H_2SO_4$ | 1.00 | 0.96 | 2.152 | −100.019 | −345.659 | 0.1296 | −6.7666 |
| | 1.50 | 1.42 | 2.015 | −94.441 | −366.077 | 0.1308 | −2.9929 |
| | 2.00 | 1.85 | 1.908 | −92.455 | −373.941 | 0.1353 | −1.4887 |
| | 2.50 | 2.28 | 1.815 | −93.901 | −368.181 | 0.1442 | −0.8125 |
| | 3.00 | 2.69 | 1.731 | −98.773 | −350.019 | 0.1574 | −0.3968 |

(*Continued*)

**TABLE 7.1 (*Continued*)**
**Thermodynamic Values Entering the Calculation of Diffusion, $D \cdot 10^{-9}$ (m²/s), Based on Formula (7.49)**

| Electrolyte | $m$ | $c_k$ | $D$ | $\bar{L}_{ij}$ | $\kappa_D$ | $f_\pm$ | $b_D$ |
|---|---|---|---|---|---|---|---|
| | 3.50 | 3.08 | 1.639 | −106.757 | −323.843 | 0.1748 | −0.2090 |
| | 4.00 | 3.44 | 1.530 | −118.261 | −292.340 | 0.1965 | −0.1189 |
| | 4.50 | 3.82 | 1.418 | −136.374 | −253.513 | 0.2223 | −0.0666 |
| | 5.00 | 4.16 | 1.309 | −167.906 | −205.905 | 0.2521 | −0.0340 |
| | 5.50 | 4.49 | 1.216 | −235.084 | −147.065 | 0.2863 | −0.0175 |
| | 6.00 | 4.82 | 1.146 | −446.287 | −77.467 | 0.3248 | −0.0071 |
| | 7.00 | 5.45 | 1.046 | 414.886 | 83.330 | 0.4157 | 0.0047 |
| | 7.50 | 5.71 | 1.005 | 201.667 | 171.433 | 0.4684 | 0.0081 |
| | 8.00 | 6.01 | 0.967 | 130.895 | 264.125 | 0.5262 | 0.0106 |
| | 8.50 | 6.25 | 0.930 | 95.616 | 361.575 | 0.5896 | 0.0125 |
| | 9.00 | 6.54 | 0.897 | 74.647 | 463.150 | 0.6585 | 0.0137 |
| | 9.50 | 6.78 | 0.864 | 60.744 | 569.151 | 0.7331 | 0.0149 |
| | 10.00 | 7.05 | 0.832 | 50.785 | 680.764 | 0.8137 | 0.0155 |
| KBr | 0.10 | 0.10 | 1.968 | 130.803 | 25.323 | 0.7627 | 6861.4414 |
| | 0.20 | 0.20 | 1.967 | 137.928 | 24.014 | 0.7343 | 2985.8145 |
| | 0.30 | 0.29 | 1.966 | 145.335 | 22.791 | 0.7086 | 1756.2617 |
| | 0.40 | 0.40 | 1.963 | 152.090 | 21.778 | 0.6878 | 1150.7802 |
| | 0.50 | 0.49 | 1.960 | 157.334 | 21.053 | 0.6729 | 827.4562 |
| | 0.60 | 0.59 | 1.955 | 161.313 | 20.533 | 0.6620 | 623.3867 |
| | 0.70 | 0.68 | 1.950 | 164.209 | 20.171 | 0.6541 | 485.3946 |
| | 0.80 | 0.78 | 1.944 | 166.322 | 19.915 | 0.6482 | 390.5182 |
| | 0.90 | 0.87 | 1.937 | 168.052 | 19.710 | 0.6432 | 321.8749 |
| | 1.00 | 0.97 | 1.930 | 170.336 | 19.445 | 0.6372 | 266.1974 |
| | 1.25 | 1.19 | 1.912 | 170.055 | 19.478 | 0.6343 | 182.4887 |
| | 1.50 | 1.42 | 1.890 | 168.512 | 19.656 | 0.6335 | 132.8680 |
| | 1.75 | 1.64 | 1.865 | 165.794 | 19.978 | 0.6346 | 100.6235 |
| | 2.00 | 1.86 | 1.841 | 162.606 | 20.370 | 0.6368 | 78.0811 |
| | 2.25 | 2.08 | 1.818 | 158.631 | 20.880 | 0.6411 | 62.4160 |
| | 2.50 | 2.29 | 1.797 | 154.548 | 21.432 | 0.6464 | 50.6075 |
| | 2.75 | 2.50 | 1.781 | 150.066 | 22.072 | 0.6540 | 41.6581 |
| | 3.00 | 2.69 | 1.769 | 146.503 | 22.609 | 0.6604 | 34.8760 |
| KI | 0.10 | 0.10 | 1.867 | 122.557 | 26.168 | 0.7698 | 8032.4482 |
| | 0.20 | 0.20 | 1.861 | 128.241 | 25.008 | 0.7433 | 3691.8347 |
| | 0.30 | 0.29 | 1.886 | 135.967 | 23.587 | 0.7207 | 2252.3967 |
| | 0.40 | 0.39 | 1.913 | 143.146 | 22.404 | 0.7030 | 1552.0740 |
| | 0.50 | 0.49 | 1.942 | 149.584 | 21.440 | 0.6902 | 1140.9606 |
| | 0.60 | 0.58 | 1.968 | 154.395 | 20.772 | 0.6824 | 896.0144 |
| | 0.70 | 0.67 | 1.993 | 158.503 | 20.234 | 0.6766 | 723.6042 |
| | 0.80 | 0.77 | 2.015 | 161.365 | 19.875 | 0.6739 | 598.5416 |
| | 0.90 | 0.86 | 2.036 | 164.144 | 19.538 | 0.6711 | 509.5100 |

(*Continued*)

**TABLE 7.1 (*Continued*)**
**Thermodynamic Values Entering the Calculation of Diffusion, $D \cdot 10^{-9}$ (m²/s), Based on Formula (7.49)**

| Electrolyte | $m$ | $c_k$ | $D$ | $\bar{L}_{ij}$ | $\kappa_D$ | $f_\pm$ | $b_D$ |
|---|---|---|---|---|---|---|---|
| | 1.00 | 0.95 | 2.056 | 166.927 | 19.212 | 0.6682 | 436.2350 |
| | 1.25 | 1.18 | 2.102 | 170.006 | 18.864 | 0.6699 | 317.3722 |
| | 1.50 | 1.40 | 2.146 | 171.981 | 18.648 | 0.6735 | 243.9512 |
| | 1.75 | 1.62 | 2.187 | 173.262 | 18.510 | 0.6782 | 162.1515 |
| | 2.00 | 1.83 | 2.225 | 172.988 | 18.539 | 0.6861 | 114.2502 |
| | 2.25 | 2.03 | 2.262 | 172.719 | 18.568 | 0.6940 | 83.4331 |
| | 2.50 | 2.23 | 2.299 | 172.443 | 18.598 | 0.7020 | 62.9322 |
| | 2.75 | 2.43 | 2.335 | 171.312 | 18.721 | 0.7122 | 46.2876 |
| | 3.00 | 2.62 | 2.371 | 170.268 | 18.835 | 0.7225 | 34.5440 |
| | 3.25 | 2.82 | 2.408 | 169.332 | 18.940 | 0.7328 | 26.2794 |
| | 3.50 | 3.01 | 2.445 | 168.142 | 19.074 | 0.7444 | 20.6064 |
| | 3.75 | 3.17 | 2.483 | 167.083 | 19.195 | 0.7560 | 16.6188 |
| | 4.00 | 3.37 | 2.522 | 165.796 | 19.344 | 0.7688 | 13.5763 |
| | 4.25 | 3.52 | 2.578 | 166.096 | 19.309 | 0.7807 | 11.2528 |
| KOH | 2.00 | 1.97 | 2.890 | 149.625 | 120.233 | 0.9381 | 116.4634 |
| | 2.25 | 2.21 | 2.918 | 142.374 | 126.357 | 0.9935 | 92.3095 |
| | 2.50 | 2.45 | 2.947 | 136.088 | 132.194 | 1.0508 | 73.6541 |
| | 2.75 | 2.68 | 2.976 | 130.006 | 138.377 | 1.1156 | 60.8526 |
| | 3.00 | 2.92 | 3.004 | 124.511 | 144.485 | 1.1845 | 50.6396 |
| | 3.25 | 3.15 | 3.032 | 119.185 | 150.942 | 1.2623 | 42.9542 |
| | 3.50 | 3.38 | 3.059 | 114.336 | 157.343 | 1.3457 | 36.3676 |
| | 3.75 | 3.61 | 3.086 | 109.723 | 163.958 | 1.4382 | 30.4722 |
| | 4.00 | 3.84 | 3.113 | 105.536 | 170.463 | 1.5366 | 25.6743 |
| | 4.25 | 4.07 | 3.138 | 101.513 | 177.218 | 1.6455 | 22.0268 |
| | 4.50 | 4.30 | 3.162 | 97.891 | 183.775 | 1.7606 | 19.0097 |
| | 4.75 | 4.53 | 3.186 | 94.451 | 190.469 | 1.8867 | 16.5606 |
| | 5.00 | 4.75 | 3.209 | 91.483 | 196.647 | 2.0145 | 14.3309 |
| | 5.25 | 4.97 | 3.232 | 88.417 | 203.468 | 2.1631 | 12.7063 |
| | 5.50 | 5.20 | 3.254 | 85.095 | 211.409 | 2.3472 | 11.2829 |
| | 5.75 | 5.38 | 3.275 | 81.660 | 220.302 | 2.5701 | 10.4064 |
| | 6.00 | 5.60 | 3.296 | 78.734 | 228.491 | 2.7984 | 9.4471 |
| | 6.25 | 5.82 | 3.317 | 76.394 | 235.490 | 3.0176 | 8.5029 |
| | 6.50 | 6.04 | 3.338 | 74.497 | 241.487 | 3.2274 | 7.6136 |
| | 6.75 | 6.21 | 3.359 | 72.790 | 247.149 | 3.4437 | 6.9545 |
| | 7.00 | 6.42 | 3.379 | 71.086 | 253.073 | 3.6840 | 6.2785 |
| | 7.25 | 6.64 | 3.398 | 69.315 | 259.541 | 3.9590 | 5.7612 |
| | 7.50 | 6.86 | 3.417 | 67.725 | 265.633 | 4.2425 | 5.2628 |
| | 7.75 | 7.01 | 3.434 | 66.203 | 271.740 | 4.5438 | 4.8450 |
| | 8.00 | 7.23 | 3.448 | 64.678 | 278.148 | 4.8747 | 4.4050 |
| | 8.25 | 7.44 | 3.463 | 63.153 | 284.864 | 5.2474 | 4.0613 |

(*Continued*)

**TABLE 7.1 (*Continued*)**
**Thermodynamic Values Entering the Calculation of Diffusion, $D \cdot 10^{-9}$ ($m^2/s$), Based on Formula (7.49)**

| Electrolyte | $m$ | $c_k$ | $D$ | $\bar{L}_{ij}$ | $\kappa_D$ | $f_\pm$ | $b_D$ |
|---|---|---|---|---|---|---|---|
| | 8.50 | 7.59 | 3.475 | 61.748 | 291.345 | 5.6286 | 3.7739 |
| | 8.75 | 7.80 | 3.485 | 60.380 | 297.947 | 6.0331 | 3.4663 |
| | 9.00 | 8.01 | 3.494 | 59.057 | 304.618 | 6.4730 | 3.2317 |
| | 9.25 | 8.16 | 3.503 | 57.730 | 311.622 | 6.9650 | 3.0396 |
| | 9.50 | 8.37 | 3.510 | 56.647 | 317.577 | 7.4107 | 2.8126 |
| | 9.75 | 8.51 | 3.516 | 55.455 | 324.407 | 7.9523 | 2.6810 |
| | 10.00 | 8.72 | 3.523 | 53.996 | 333.169 | 8.6936 | 2.5399 |
| | 10.50 | 9.06 | 3.535 | 52.086 | 345.391 | 9.8659 | 2.3127 |
| | 11.00 | 9.40 | 3.545 | 50.251 | 358.000 | 11.2418 | 2.1401 |
| | 11.50 | 9.73 | 3.555 | 48.444 | 371.353 | 12.8917 | 2.0239 |
| | 12.00 | 10.06 | 3.564 | 46.834 | 384.124 | 14.7074 | 1.9072 |
| | 12.50 | 10.38 | 3.572 | 45.302 | 397.114 | 16.8098 | 1.8044 |
| | 13.00 | 10.70 | 3.580 | 43.930 | 409.511 | 19.1066 | 1.6956 |
| | 13.50 | 11.02 | 3.587 | 42.624 | 422.060 | 21.7222 | 1.6260 |
| | 14.00 | 11.33 | 3.593 | 41.456 | 433.949 | 24.5490 | 1.5502 |
| | 14.50 | 11.64 | 3.599 | 40.360 | 445.735 | 27.7128 | 1.5541 |
| | 15.00 | 11.94 | 3.605 | 39.374 | 456.894 | 31.1062 | 1.5130 |
| | 15.50 | 12.24 | 3.610 | 38.449 | 467.888 | 34.8596 | 1.5086 |
| | 16.00 | 12.45 | 3.616 | 37.609 | 478.347 | 38.8679 | 1.5149 |
| | 16.50 | 12.74 | 3.621 | 36.825 | 488.524 | 43.2297 | 1.5028 |
| | 17.00 | 13.03 | 3.627 | 36.111 | 498.180 | 47.8268 | 1.4957 |
| | 17.50 | 13.32 | 3.632 | 35.454 | 507.421 | 52.7130 | 1.4828 |
| LiCl | 0.40 | 0.40 | 1.839 | 124.304 | 7.072 | 0.7537 | 500.3527 |
| | 0.50 | 0.49 | 1.816 | 122.541 | 7.174 | 0.7543 | 403.8096 |
| | 0.60 | 0.59 | 1.792 | 119.664 | 7.347 | 0.7601 | 342.4725 |
| | 0.70 | 0.69 | 1.768 | 116.426 | 7.551 | 0.7679 | 297.5239 |
| | 0.80 | 0.78 | 1.744 | 113.085 | 7.774 | 0.7768 | 267.4492 |
| | 0.90 | 0.88 | 1.720 | 109.102 | 8.058 | 0.7898 | 243.6465 |
| | 1.00 | 0.98 | 1.697 | 104.157 | 8.440 | 0.8102 | 236.7310 |
| | 1.25 | 1.22 | 1.647 | 96.358 | 9.123 | 0.8423 | 183.0977 |
| | 1.50 | 1.45 | 1.597 | 87.900 | 10.001 | 0.8875 | 149.4162 |
| | 1.75 | 1.69 | 1.551 | 79.688 | 11.032 | 0.9451 | 121.4417 |
| | 2.00 | 1.92 | 1.505 | 72.253 | 12.167 | 1.0099 | 104.3294 |
| | 2.25 | 2.15 | 1.461 | 66.128 | 13.294 | 1.0735 | 91.9762 |
| | 2.50 | 2.38 | 1.417 | 60.672 | 14.490 | 1.1413 | 81.4125 |
| | 2.75 | 2.60 | 1.374 | 55.690 | 15.786 | 1.2167 | 74.5956 |
| | 3.00 | 2.82 | 1.332 | 51.163 | 17.183 | 1.2998 | 68.1271 |
| | 3.25 | 3.05 | 1.291 | 46.956 | 18.722 | 1.3941 | 63.8619 |
| | 3.50 | 3.27 | 1.251 | 43.212 | 20.344 | 1.4966 | 60.0575 |
| | 3.75 | 3.49 | 1.211 | 39.720 | 22.133 | 1.6130 | 56.6103 |

*(Continued)*

**TABLE 7.1 (*Continued*)**
**Thermodynamic Values Entering the Calculation of Diffusion, $D\cdot 10^{-9}$ ($m^2/s$), Based on Formula (7.49)**

| Electrolyte | $m$ | $c_k$ | $D$ | $\bar{L}_{ij}$ | $\kappa_D$ | $f_\pm$ | $b_D$ |
|---|---|---|---|---|---|---|---|
| | 4.00 | 3.69 | 1.172 | 36.593 | 24.024 | 1.7379 | 53.7170 |
| | 4.25 | 3.90 | 1.134 | 33.673 | 26.107 | 1.8807 | 50.2530 |
| | 4.50 | 4.12 | 1.095 | 31.037 | 28.325 | 2.0337 | 47.6545 |
| $MgCl_2$ | 0.20 | 0.20 | 2.317 | 351.315 | 4.694 | 0.5064 | 18.3230 |
| | 0.30 | 0.30 | 2.250 | 363.147 | 4.541 | 0.4967 | 11.0208 |
| | 0.40 | 0.40 | 2.183 | 355.581 | 4.638 | 0.4954 | 7.8853 |
| | 0.50 | 0.50 | 2.119 | 330.586 | 4.988 | 0.5019 | 6.2910 |
| | 0.60 | 0.59 | 2.056 | 296.461 | 5.562 | 0.5149 | 5.4827 |
| | 0.70 | 0.69 | 1.993 | 263.859 | 6.250 | 0.5305 | 4.9029 |
| | 0.80 | 0.78 | 1.930 | 232.699 | 7.087 | 0.5499 | 4.4869 |
| | 0.90 | 0.88 | 1.866 | 199.015 | 8.286 | 0.5796 | 4.3116 |
| | 1.00 | 0.98 | 1.805 | 162.191 | 10.167 | 0.6308 | 4.3772 |
| | 1.25 | 1.22 | 1.656 | 120.312 | 13.706 | 0.7169 | 3.8215 |
| | 1.50 | 1.44 | 1.524 | 89.858 | 18.352 | 0.8367 | 3.4668 |
| | 1.75 | 1.67 | 1.410 | 68.623 | 24.030 | 0.9960 | 3.1064 |
| | 2.00 | 1.91 | 1.311 | 53.683 | 30.718 | 1.2018 | 2.7458 |
| | 2.25 | 2.14 | 1.222 | 42.837 | 38.496 | 1.4659 | 2.4349 |
| | 2.50 | 2.35 | 1.137 | 34.662 | 47.575 | 1.8038 | 2.2020 |
| | 2.75 | 2.58 | 1.056 | 28.360 | 58.147 | 2.2357 | 1.9545 |
| | 3.00 | 2.78 | 0.977 | 23.391 | 70.499 | 2.7880 | 1.7443 |
| | 3.25 | 3.00 | 0.901 | 19.388 | 85.057 | 3.4948 | 1.5856 |
| | 3.50 | 3.20 | 0.824 | 16.086 | 102.515 | 4.4002 | 1.4616 |
| NaCl | 0.30 | 0.30 | 1.494 | 106.756 | 14.827 | 0.7247 | 1113.6598 |
| | 0.40 | 0.40 | 1.461 | 108.126 | 14.639 | 0.7081 | 770.5272 |
| | 0.50 | 0.50 | 1.462 | 111.039 | 14.255 | 0.6963 | 563.0390 |
| | 0.60 | 0.59 | 1.463 | 113.103 | 13.995 | 0.6886 | 439.4873 |
| | 0.70 | 0.69 | 1.464 | 114.172 | 13.864 | 0.6848 | 351.2349 |
| | 0.80 | 0.78 | 1.465 | 114.995 | 13.764 | 0.6821 | 286.3900 |
| | 0.90 | 0.88 | 1.466 | 115.557 | 13.697 | 0.6804 | 235.2756 |
| | 1.00 | 0.98 | 1.467 | 115.556 | 13.698 | 0.6807 | 197.5388 |
| | 1.25 | 1.22 | 1.475 | 114.867 | 13.780 | 0.6855 | 138.6754 |
| | 1.50 | 1.46 | 1.487 | 113.630 | 13.930 | 0.6936 | 103.4054 |
| | 1.75 | 1.69 | 1.498 | 111.418 | 14.206 | 0.7059 | 80.1602 |
| | 2.00 | 1.92 | 1.509 | 108.816 | 14.546 | 0.7204 | 63.2440 |
| | 2.25 | 2.16 | 1.519 | 105.880 | 14.949 | 0.7373 | 51.3086 |
| | 2.50 | 2.38 | 1.528 | 103.105 | 15.352 | 0.7543 | 41.9368 |
| | 2.75 | 2.61 | 1.536 | 100.120 | 15.809 | 0.7738 | 35.3848 |
| | 3.00 | 2.83 | 1.544 | 97.034 | 16.312 | 0.7956 | 29.7804 |
| | 3.25 | 3.05 | 1.553 | 94.056 | 16.829 | 0.8188 | 25.5729 |
| | 3.50 | 3.26 | 1.561 | 91.198 | 17.356 | 0.8435 | 21.9948 |

(*Continued*)

**TABLE 7.1 (*Continued*)**
**Thermodynamic Values Entering the Calculation of Diffusion, $D \cdot 10^{-9}$ (m²/s), Based on Formula (7.49)**

| Electrolyte | $m$ | $c_k$ | $D$ | $\bar{L}_{ij}$ | $\kappa_D$ | $f_\pm$ | $b_D$ |
|---|---|---|---|---|---|---|---|
| | 3.75 | 3.48 | 1.570 | 88.321 | 17.921 | 0.8706 | 19.3054 |
| | 4.00 | 3.69 | 1.577 | 85.539 | 18.504 | 0.8993 | 16.9644 |
| | 4.25 | 3.90 | 1.585 | 82.787 | 19.119 | 0.9306 | 14.9126 |
| | 4.50 | 4.11 | 1.592 | 80.133 | 19.753 | 0.9634 | 13.3522 |
| | 4.75 | 4.32 | 1.598 | 77.534 | 20.415 | 0.9990 | 11.9533 |
| | 5.00 | 4.53 | 1.604 | 75.086 | 21.080 | 1.0362 | 10.7886 |
| | 5.25 | 4.70 | 1.610 | 72.784 | 21.747 | 1.0750 | 9.7992 |
| | 5.50 | 4.90 | 1.614 | 70.715 | 22.383 | 1.1119 | 8.8814 |

*Notes:* $m$, mol/kg $H_2O$; $c_k$, mol/dm³; $D$, $D \cdot 10^{-9}$ m²/s; $L_{ij}$, $L_{ij} \cdot 10^{16}$; $\kappa_D$, $\kappa_D \cdot 10^{-12}$.

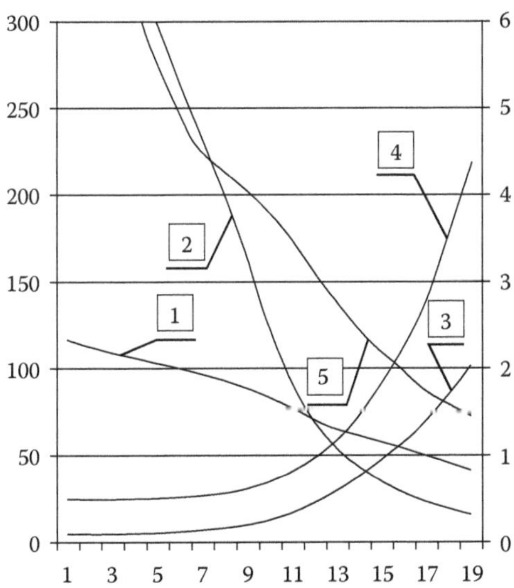


**FIGURE 7.1** Distribution of diffusion thermodynamic interactions for solution $MgCl_2$: 1—diffusion $D \cdot 10^9$, m²/s; 2—kinematic coefficient $L_{ij} \cdot 10^{-16}$; 3—electrophoretic retardation $\kappa_D \cdot 10^{12}$; 4—activity coefficient $f_\pm$; and 5—coefficient $b_D$. The auxiliary axis on the right is for curves 1, 4, and 5.

For $MgCl_2$

| Point Number | 1 | 3 | 5 | 7 | 9 | 11 | 13 | 15 | 17 | 19 |
|---|---|---|---|---|---|---|---|---|---|---|
| Concentration, $m$ | 0.2 | 0.4 | 0.6 | 0.8 | 1.0 | 1.5 | 2.0 | 2.5 | 3.0 | 3.5 |

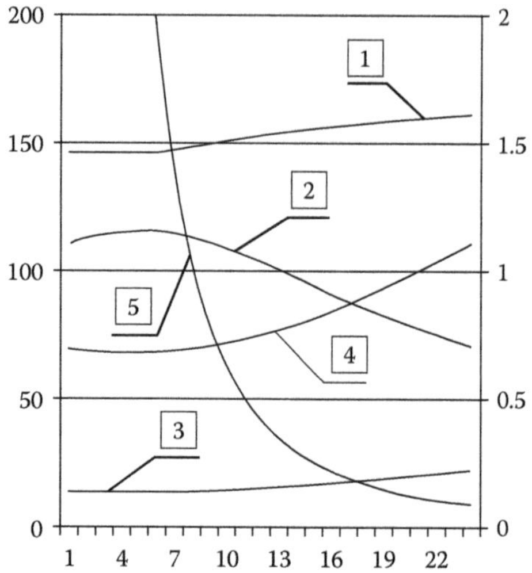


**FIGURE 7.2** Distribution of diffusion thermodynamic interactions for solution NaCl: 1—diffusion $D \cdot 10^9$, m²/s; 2—kinematic coefficient $L_{ij} \cdot 10^{-16}$; 3—electrophoretic retardation $\kappa_D \cdot 10^{12}$; 4—activity coefficient $f_{\pm}$; and 5—coefficient $b_D$. The auxiliary axis on the right is for curves 1 and 4.

For NaCl

| **Point Number** | **1** | **4** | **7** | **10** | **13** | **16** | **19** | **22** |
|---|---|---|---|---|---|---|---|---|
| Concentration, *m* | 0.30 | 0.60 | 0.90 | 1.50 | 2.25 | 3.00 | 3.75 | 4.50 |

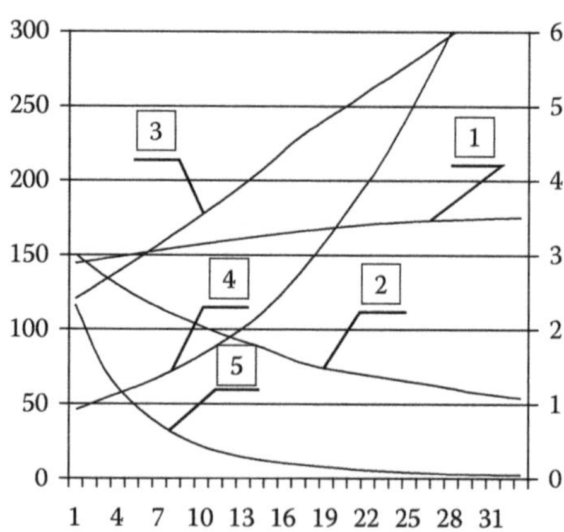


**FIGURE 7.3** Distribution of diffusion thermodynamic interactions for solution KOH: 1—diffusion $D \cdot 10^9$, m²/s; 2—kinematic coefficient $L_{ij} \cdot 10^{-16}$; 3—electrophoretic retardation $\kappa_D \cdot 10^{12}$; 4—activity coefficient $f_{\pm}$; and 5—coefficient $b_D$. The auxiliary axis on the right is for curves 1 and 5.

For KOH

| **Point Number** | **1** | **4** | **7** | **10** | **13** | **16** | **19** | **22** | **25** | **28** | **31** |
|---|---|---|---|---|---|---|---|---|---|---|---|
| Concentration, *m* | 2.00 | 2.75 | 3.50 | 4.25 | 5.00 | 5.75 | 6.50 | 7.25 | 8.00 | 8.75 | 9.50 |

values, and vice versa, which is basically natural. Solutions KOH and NaCl made an exception. In this case, one can actually talk about incorrectness of experimental viscosity or diffusion data.

As it was claimed in the literature, electrophoretic retardation is rather small in diffusion processes and does not produce a decisive influence on it, although it is impossible to refuse from electrophoretic forces taking into account the structure of formula (7.50). Small values of electrophoretic retardation result in a sharp increase of the kinematic coefficient.

## REFERENCES


1. Robinson R.A., Stokes R.H. 1963. *Electrolyte Solutions*, Vols. 1, 2. Publishing House of Foreign Literature, Moscow, Russia.
2. Onsager L. 1931. Reciprocal relations in irreversible processes. *Phys. Rev. I* 37: 405–426 (*Phys. Rev. II* 38: 2265–2279).
3. Smirnov A.A. 1966. *Molecular Kinetic Theory of Metals*. Nauka, Moscow, Russia, p. 488.
4. Shaw D. (ed.). 1975. *Atomic Diffusion in Semiconductors*. Mir, Moscow, Russia, p. 684.
5. Antsiferov V.N., Peshcherenko S.N., Kurilov P.G. 1988. *Mutual Diffusion and Homogenization of the Materials in the Powder*. Metallurgy, Moscow, Russia, p. 152.
6. Lyubov B.Ya. 1969. *Kinetic Theory of Phase Transitions*. Metallurgy, Moscow, Russia, p. 263.
7. Zhuravlev V.A. 1979. *Thermodynamics of Irreversible Processes*. Nauka, Moscow, Russia, p. 135.
8. Voroshnin L.G., Khusid B.M. 1979. *Diffusion Mass Transfer in Multicomponent Systems*. Science and Technology, Minsk, Belarus, p. 256.
9. Kubo S. 1962. *Some Questions Statistical Mechanical Theory of Irreversible Processes. Thermodynamics of Irreversible Processes*. Foreign Literat., Moscow, Russia, pp. 345–421.
10. Svelin A. 1968. *Thermodynamics of Solid State*. Metallurgy, Moscow, Russia, p. 314.
11. Smirnova N.A., Viktorov A.I. 1985. Description of thermodynamic properties of liquids and solutions on the basis of option hole theory. V all-Union school-seminar. In: *Application of Mathematical Methods for the Study and Description of Physical-Chemical Equilibria*. H. 2. Novosibirsk, Russia, pp. 174–178.
12. Berd O., St'yuart B., Lightfoot E. 1974. *Transport Phenomena*. Khimiya, Moscow, Russia, p. 288.
13. Fertsiger Dzh., Caper G. 1976. *Mathematical Theory of Transport Processes in Gases*. Mir, Moscow, Russia, p. 554.
14. Vlasov A.A. 1978. *Nonlocal Statistical Mechanics*. Nauka, Moscow, Russia, p. 264.
15. Kesselman P.M. 1976. *Transport Properties of Real Gases*. Vyshcha Sch., Kiev–Odessa, Ukraine, p. 151.
16. Kogan M.S. 1983. *Introduction to the Kinetic Theory of Stochastic Processes in Gases*. Nauka, Moscow, Russia, p. 272.
17. Chapman S. et al. 1960. *Mathematical Theory of Nonuniform Gases*. Foreign Literat., Moscow, Russia, p. 554.
18. Zubarev D.N. 1969. *Nonequilibrium Statistical Operator Method. Problems of Theoretical Physics*. Nauka, Moscow, Russia, p. 415.
19. Zubarev D.N. 1971. *Nonequilibrium Statistical Thermodynamics*. Nauka, Moscow, Russia, p. 445.
20. Bakhareva I.F., Biryukov A.A. 1970. On the theory of nonlinear nonequilibrium stochastic processes. *Physics* (Izv. Higher Education Institutions) 9: 58–62.

21. Bakhareva I.F., Biryukov A.A. 1974. Stochastic interpretation of equations of nonlinear nonequilibrium thermodynamics. *J. Phys. Chem.* 48: 1959–1964.
22. Mayer Dzh., Geppert-Mayer M. 1980. *Statistical Mechanics*. Mir, Moscow, Russia, p. 544.
23. Huang K. 1966. *Statistical Mechanics*. Mir, Moscow, Russia, p. 520.
24. Bazarov I.P., Nikolaev P.N. 1984. *Theory of Many-Particle Systems*. Moscow State University, Moscow, Russia, p. 312.
25. Smirnova N.A. 1982. *Methods of Statistical Thermodynamics in Physical Chemistry*. Higher School, Moscow, Russia, p. 455.
26. Nepomnyashchii A.B. 2005. *Diffusive Transport of Aqueous Electrolyte Solutions in the Membranes of Porous Glass*. Chemistry, St. Petersburg, Russia, p. 152.
27. Ravdel A.A., Ponomareva A.M. (eds.). 2003. *Quick Reference Physico-Chemical Variables*. Chemistry, St. Petersburg, Russia, p. 232.
28. Aseyev G.G. 1998. *Electrolytes: Transport Phenomena. Methods for Calculation of Multicomponent Solutions, and Experimental data on Viscosities and Diffusion Coefficients*. Begell House Inc., New York, p. 548.
29. Ravdel A.A., Shmuylovich G.A., Samsonov A.N. 1978. Measurement of diffusion coefficients of electrolytes of different valence type and determination of kinetic parameters of diffusive transport. In: *Thermodynamics and Structure Solutions*. Izd IHTI, Ivanovo, Russia, pp. 86–95.
30. Rard J.A., Miller D.G. 1980. Mutual diffusion coefficients of $BaCl_2$-$H_2O$ and KCl-$H_2O$ at 25°C from Rayleigh interferometry. *J. Chem. Eng. Data* 25: 211–215.
31. Han K.N., Kang T.K. 1986. Diffusivity of cobalt and nickel compounds as a function of concentration and temperature. *Met. Trans. B* 17 (September): 425–432.
32. Spallek M., Hertz H.G., Funsch M., Hermann H., Weingartner H. 2010. Ternary diffusion in the aqueous solutions of $MgCl_2^+$ KCl, $CdCl_2^+$ KCl, $ZnCl_2^+$ KCl and Onsager's reciprocity relations. *Ber. Bunsenges Phys. Chem.* 94 (3): 365–376.
33. Patterson R. 1978. Transport in aqueous solutions of group IIB metal salts (298.15). Part 3. Isotopic diffusion coefficients for cadmium ions in aqueous cadmium iodide. *J. Chem. Soc. Faraday Trans.* 74 (1): 93–102.
34. Patterson R., Devine C. 1980. Transport in aqueous solutions of IIB metal salts. Part 7. Measurements and prediction of isotopic diffusion coefficients for iodide in solutions of cadmium iodide. *J. Chem. Soc. Faraday Trans.* 76: 1052–1061.
35. Brun B., Servent M., Salvinien J. 1969. Diffusion study of activity of large ions on water structure. *C. R. Acad. Sci. Ser.* 269: 1–4.
36. Ravdel A.A., Porai Kosice A.B., Sazonov A.M., Shmuylovich G.A. 1973. Concentration and temperature dependence of the diffusion coefficients of salt. *J. Appl. Chem.* 46 (8): 1703–1707.
37. Braun B.M., Weingartner H. 1988. Accurate self-diffusion coefficients of $Li^+$, $Na^+$, and $Cs^+$ Ions in aqueous alkali metal halide solutions from NMR spin-echo experiments. *J. Phys. Chem.* 92: 1342–1346.
38. Woolf L.A., Hoveling A.W. 1970. Mutual diffusion-coefficients of aqueous copper (II) sulfate solutions at 25°C. *J. Phys. Chem.* 74 (11): 2406.
39. Kamakura K. 1983. A method for the determination of mutual diffusion coefficients of electrolytes in water by means of conductivity measurements. *Bull. Chem. Soc. Jpn.* 55 (11): 3353–3355.
40. Noulty R.A., Leaist D.G. 1987. Diffusion in aqueous copper sulfate and copper sulfate-sulfuric acid solutions. *J. Solut. Chem.* 16 (10): 813–825.
41. Zhong E.C., Friedman H.L. 1988. Self-diffusion of ions in solution. *J. Phys. Chem.* 92: 1685–1692.
42. Clunie J., N. Li, Emerson M.T., Baird J.K. 1990. Theory and measurement of the concentration dependence of the differential diffusion coefficient using a diaphragm cell with compartments of unequal volume. *J. Phys. Chem.* 94: 6099–6105.

43. Leaist D.G., Wiens B. 1986. Interdiffusion of acids and bases. HCl and NaOH in aqueous solution. *Can. J. Chem.* 64: 1007–1011.
44. Hui L., Leaist D. 1990. Thermal diffusion of weak electrolytes: Aqueous phosphoric and iodic acids. *Can. J. Chem.* 68: 1317–1322.
45. Nisancioghlu K., Newman J. 1973. Diffusion in aqueous nitric acid solutions. *AICHE J.* 19: 797–801.
46. Eastel A.J., Price W.E., Woolf L.A. 1989. Diaphragm cell for high-temperature diffusion measurements. *J. Chem. Soc. Faraday Trans.* 85 (5): 1091–1097.
47. Leaist D.G. 1987. Diffusion of aqueous carbon dioxide, sulfur dioxide, sulfuric acid, and ammonia at very low concentrations. *J. Phys. Chem.* 91: 4635–4638.
48. Leaist D.G. 1989. The soret effect with chemical reaction in aqueous sulfuric acid solutions. *J. Solut. Chem.* 18 (7): 651–661.
49. Turg N., Lantelme F., Roumegous Y., Chemla M. 1979. Coefficients d'autodiffusion dans les solutions aqueuses de KCl et de LiCl. *J. Chim. Phys.* 68 (3): 527–531.
50. Anderson J., Paterson R. 1975. Application of irreversible thermodynamics to isotopic diffusion. Isotope-isotope coupling coefficients for ions and water in concentrated aqueous solutions of alkali-metal chlorides at 298, 15 K. *J. Chem. Soc. Faraday Trans.* 71: 1335–1351.
51. Hertz H.G., Holz M., Mills R. 1974. Effect of structure on ion self-diffusion in concentrated electrolyte solutions. *J. Chim. Phys.* 71 (10): 1355–1362.
52. Tanaka K., Nomura M. 1987. Measurements of tracer diffusion coefficients of lithium ions, chloride ions and water in aqueous lithium chloride solutions. *J. Chem. Soc. Faraday Trans.* 83 (6): 1779–1782.
53. Rard J.A., Miller D.G. 1990. Ternary mutual diffusion coefficients of $ZnCl_2$-KCl-$H_2O$ at 25°C by Rayleigh interferometry. *J. Solut. Chem.* 19 (2): 129–148.
54. Miller D.G., Ting A.W., Rard J.A. 1988. Mutual diffusion coefficients of various $ZnCl_2$ (0.5 M)-KCl-$H_2O$ mixtures at 298.15K by Rayleigh interferometry. *J. Electrochem. Soc.* 135 (4): 896–904.
55. Noulty R.A., Leaist D.G. 1987. Quaternary diffusion in aqueous KCl-$KH_2PO_4$-$H_3PO_4$ mixtures. *J. Phys. Chem.* 91: 1655–1658.
56. Chang Y.C., Myerson A.S. 1984. Diffusion coefficients in supersaturated solutions. In: Jancic S.J. and de Jong F.J. (eds.). *Proceedings of the Ninth Symposium on Industrial Crystallization*, The Hague, the Netherlands. Elsevier, Amsterdam, the Netherlands, pp. 27–30.
57. Vacek V., Rod V. 1986. Diffusion coefficients of potassium chromate and dichromate in water at 25°C. *Collect. Czech. Chem. Commun.* 51: 1403–1406.
58. Albright J.G., Mathew R., Miller D.G. 1987. Measurements of binary and ternary mutual diffusion coefficients of aqueous sodium and potassium bicarbonate solutions at 25°C. *J. Phys. Chem.* 91: 210–215.
59. Bhatia R.N., Gubbins K.E., Walker R.D. 1968. Mutual diffusion in concentrated aqueous potassium hydroxide solutions. *Trans. Faraday Soc.* 64: 2091–2099.
60. Zagaynov V.M., Sevryughin V.A., Alekseev S.I., Emelyanov M.I. 1986. Effect of lithium ions on the self-diffusion of water molecules in aqueous solutions. In: Nikiforov E.A. (ed.). *Physics of Fluids: Interhigher Education Institution.* Trudy Inst., Kazan, Russia, pp. 110–124.
61. Simonin J.P., Gaillard J.F., Turq K., Soualhia E. 1988. Diffusion coupling in electrolyte solutions. Transient effects on a tracer ion: Sulfate. *J. Phys. Chem.* 92: 1696–1700.
62. Behret H., Schmithals F. 1975. Leitfaehigkeimessungen an Konzentrierten Alkalichlorid-und-nitratlosungen. *Z. Naturforsch.* 30a: 1497–1498.
63. Tanaka K. 1988. Measurements of tracer-diffusion coefficients of sulfate ions in aqueous solutions of ammonium sulfate and sodium sulfate solutions. *J. Chem. Soc. Faraday Trans.* 84 (8): 2895–2897.

64. Harris K.R., Hertz H.G., Mills R. 1978. The effect of structure on self-diffusion in concentrated electrolytes: Relationship between the water and ionic self-diffusion coefficients for structure-forming salts. *J. Chim. Phys.* 75: 391–396.
65. Albright J.G., Mathew R., Miller D.G., Rard J.A. 1989. Isothermal diffusion coefficients for NaCl-$MgCl_2$-$H_2O$ at 25°C. Solute concentration ratio of 3:1. *J. Phys. Chem.* 93: 2176–2180.
66. Mathew R., Albright J.C., Miller D.G., Rard J.A. 1990. Isothermal diffusion coefficients for NaCl-$MgCl_2$-$H_2O$ at 25°C. Solute concentration ratio of 1:3. *J. Phys. Chem.* 94: 6875–6878.
67. Price W.E., Woolf L.A., Harris K.R. 1990. Intradiffusion coefficients for zinc and water shear viscosities in aqueous zinc (II) perchlorate solutions at 25°C. *J. Phys. Chem.* 94 (12): 5109–5114.
68. Bulvin L.A., Ivanitskii P.G., Krotenko V.T., Lyaskovskaya G.N. 1987. Neutron studies of self-diffusion in aqueous electrolyte solutions. *J. Phys. Chem.* 61 (12): 3220–3225.
69. Janz G.J., Oliver B.G., Lakshminarayanan G.R., Mager G.E. 1970. Electrical conductance, diffusion, viscosity, and density of sodium nitrate, sodium perchlorate, and sodium thiocyanate in concentrated aqueous solutions. *J. Phys. Chem.* 74 (6): 1285–1289.
70. Thomas H.C., Ku J.C. 1973. Tracer diffusion-coefficients in aqueous solutions. 1. Method-sodium in sodium chloride. *J. Phys. Chem.* 77 (18): 2333–2335.
71. Rard J.A., Miller D.G. 1987. Ternary mutual diffusion coefficients of 0.5 and 1.0 mol/$dm^3$. *J. Phys. Chem.* 91: 4614–4620.
72. Hawlicka E. 1988. Acetonitrile-water solutions of sodium halides: Viscosity and self-diffusion of $CH_3CN$and $H_2O$. *Z. Naturforsch.* 43a: 769–773.
73. Hawlicka E. 1986. Self-diffusion of sodium, chloride and iodide ions in methanol-water mixture. *Z. Naturforsch.* 41a: 939–943.
74. Hawlicka E. 1987. Self-diffusion of sodium, chloride and iodide ions in acetonitrile-water mixtures. *Z. Naturforsch.* 42a: 1014–1016.
75. Stokes R.H., Phang S., Mills R. 1979. Density, conductance, transference numbers and diffusion measurements in concentrated solutions of nickel chloride at 25°C. *J. Solut. Chem.* 8: 489.
76. Rard J.A., Miller D.G., Lee C.M. 1989. Mutual diffusion coefficient of $NiCl_2$-$H_2O$ at 298.15 K from Rayleigh interferometry. *J. Chem. Soc. Faraday Trans.* 85 (10): 3343–3352.
77. Tanigaki M., Harada M., Eguchi W. 1986. Multicomponent diffusion coefficients in aqueous electrolyte systems. *World Congress III of Chemical Engineering*, Vol. II, Tokyo, Japan, pp. 112–115.
78. Agnew A., Paterson R. 1978. Transport in aqueous solutions of group-IIb metal-salts at 298,15K. Irreversible thermodynamic parameters for zinc-chloride and verification of Onzager reciprocal relationship. *J. Chem. Soc. Faraday Trans.* 74: 2896–2906.
79. Eastel A.J., Giaquinta V., March N.H., Tosi M.P. 1983. Chemical effects in diffusion and structure of zinc chloride in aqueous. *Chem. Phys.* 76: 125–128.
80. Albright J.G., Miller D.G. 1975. Mutual diffusion-coefficients of aqueous $ZnSO_4$ at 25 degrees C. *J. Solut. Chem.* 4 (9): 809–816.
81. Gleston S., Leydler K., Eyring H. 1948. *The Theory ofAbsolute Reaction Rates*. Foreign Literat., Moscow, Russia, p. 458.
82. Erdey-Gruz T. 1976. *Transport Phenomena in Aqueous Solutions*. Mir, Moscow, Russia, p. 592.
83. Smedly S.I. 1980. *The Interpretation of Ionic Conductivity in Liquids*. Plenum Press, New York, p. 396.
84. Elkind K.M. 1983. Computational method of a direct-current conductivity of aqueous solutions of the strong electrolytes. *J. Phys. Chem.* 57 (9): 2322–2324.
85. Sukhotin A.M. (ed.). 1981. *Reference Book on an Electrochemistry*. Chemistry, Leningrad, Russia, p. 436.

86. Vdovenko V.M., Gurikov Yu.V., Legin E.K. 1966. *Research on the Application Docstructure Model to Study the State of Water in Aqueous Solutions.* Publishing house of the Leningrad State University, Leningrad, Russia, p. 124.
87. Frank H.S., Evans M.W. 1945. Free volume and entropy in condensed systems III. Entropy and binary liquid mixtures: Partial molar entropy in dilute solutions: Structure and thermodynamics in aqueous electrolytes. *J. Chem. Phys.* 13: 507–515.
88. Frank H.S., Wen-Yang W. 1957. Ion-solvent interaction. Structural aspects of ion-solvent interaction in aqueous solutions: A suggested picture of water structure. *Discuss. Faraday Soc.* 24: 133–140.
89. Harned H.S., Hudson R.M. 1951. The diffusion coefficient of zinc sulfate in dilute aqueous solutions at 25. *J. Am. Chem. Soc.* 73: 3781–3788.
90. Harutyunyan S., Grigoryan V.V., Kazarian G.A. 1988. Electrical conductivity and viscosity of aqueous solutions of formamide and dimethylformamide. *Arm. Chem. J.* 41 (6): 323–327.
91. Miller G.T.A., Stokes R.H. 1957. Temperature dependence of the mutual diffusion coefficients in aqueous solutions of alkali metal chlorides. *Trans. Faraday Soc.* 53: 642–650.
92. Vitagliano V., Lyons A. 1956. Diffusion in aqueous acetic acid solutions. *J. Am. Chem. Soc.* 78 (18): 4538–4542.
93. Kir'yanov V.A. 1961. Statistical theory of strong electrolytes. *J. Phys. Chem.* 35: 2389–2396.
94. Pikal M.J. 1971. Theory of the Onsager transport coefficients Lij and Rij for electrolytes. *J. Phys. Chem.* 75 (20): 3124–3134.
95. Krestov G.A. (ed.). 1989. *Solutions of Non-Electrolytes in Liquids.* Nauka, Moscow, Russia, p. 263.
96. Reed P., Prausnits Dzh., Sherwood T. 1982. *Properties of Gases and Liquids.* Chemistry, Leningrad, Russia, p. 592.
97. Vlaev L.T., Nikolova M., Gospodinov G.G. 2005. Electrotransport properties of ions in aqueous solutions $H_2SeO_4$ and $Na_2ScO_4$. *J. Struct. Chem.* 46 (4): 655–662.
98. Bhattacharyn S.K., Ulkil U., Kund K.K. 1996. Ion conductance and association studies of KI and some reference electrolytes $KBPh_4$, $Ph_4AsI$ and $n$-$Bu_4NI$ in acetonitrile, N, N-dimethylformamide and their isodielectric binary mixtures at 298,15 K. *Indian J. Chem. A* 35 (12): 1038–1046.
99. Baldanov M.M., Tanganov B.B., Ivanov S.V. 1993. Plasma-state electrolyte solutions and the problem of diffusion. Tez. of reports. *III Ros. Conference in Chemistry and Application of Non-Aqueous Solutions*, Ivanovo, Russia, p. 44.
100. Baldanova D.M., Baldanov M.M., Zhigzhitova S.B., Tanganov B.B. 2007. Theoretical estimates of the diffusion coefficients of electrolyte solutions under the plasma-hydrodynamic model. Sat scientific. tr. Ser. In: *Chemistry and Biologically Active Natural Compounds*, Vol. 12. Izd VSGTU, Ulan-Ude, Russia, pp. 84–88.
101. Harned G., Owen B. 1952. *Physical Chemistry of Solutions of Electrolytes.* Publishing House of Foreign Literature, Moscow, Russia, p. 628.
102. Semenchenko V.K. 1941. *Physical Theory of Solutions.* Gostekhteorizdat, Moscow, Russia, p. 368.
103. Aseyev G.G. 1998. *Electrolytes: Transport Phenomena. Methods for Calculation of Multicomponent Solutions, and Experimental data on Viscosities and Diffusion Coefficients.* Begell House Inc., New York, p. 548.

# Conclusion

Electrolyte solutions are the most complex from the point of view of covering the entire range of supermolecular interactions starting from dilute to concentrated solutions. Quantitative description of all structure formation processes and other thermodynamic phenomena in the solution volume even in steady state is even more complicated. At the present time, researches of electrolyte solutions are underway on several mutually intersecting directions:

- Development of electrostatic interactions with domination of the Debye–Hückel theory, that is, the notion about ion atmosphere and its thickness as well as attempts to spread its provisions to more concentrated solutions are the fundamental points
- Ion solvation, formation of different complexes and their influence on water structure, as well as studying of various interactions: ion–dipole, dipole–dipole, and so on
- To the major extent experimental researches of structure formation issues in a solution, the issues about dielectric permittivity and their influence on behavior of various thermodynamic functions
- Experimental and theoretical researches of weak hydrogen bonds caused by electron shell overlapping and leading to formation of ion or ion-aqueous complexes (supermolecules) and similar trends frequently of low connection with one another

Let us go back to the fundamental equation of the Debye–Hückel theory $\Delta\Psi_j^0(R) = -(4\pi/\varepsilon)\sum |z_i| e \exp\left[-\Psi_j^0(R)e/kT\right]$. This formula was analyzed by dozens of researchers, but we will introduce our point of view too. In the left part of the equation, there is potential $\Psi_j^0(R)$, that is, potential depends on potential. The left part of the expression is a linear function and the right part is the exponential relating to $\Psi_j^0(R)$. No matter how we expand the exponential into series, it remains exponential with some accuracy degree depending on the number of terms in a series. And linear and exponential dependence coincide only at the initial path segment. Therefore, it implies that the initial path segment of potential is nothing else but dilute solutions, not to mention that the Debye–Hückel theory has dielectric permittivity of a solution (water) and there is no consideration of any interparticle forces among ions or ion atmospheres of various ions in sign.

Debye–Hückel used Maxwell–Boltzmann statistics (2.1) in their theory in order to estimate ion concentrations of such type near a definite ion depending on their heat motion. It may be correct for the electrolyte concentration area where their equation was valid, but there is one comment. It was assumed that the average force acting upon one ion is determined by the potential of another ion, but potential energies $U_{ij}$ and $U_{ji}$ will be equal to $\Psi_i^0(R)\,e$ and $\Psi_j^0(R)\,e$, respectively. This resulted in occurrence of these summands in exponential of their theory. Such ordinary change in notation of the values is not very consistent. As stated in Chapter 2, it is known from electrostatics that the potential energy of interaction of two point charges $|z_i|e$ and $|z_j|e$, being in a solution at a distance $R_{ij}$ from each other, can be calculated by formula (2.43). All values are defined easily in it, and thus, we managed to circumvent the presence of the same variables, that is, potentials, in both parts of the equation. This electrostatics law was known at the time of Debye and Hückel.

Let us go back to quantum statistics. If considering a stagnant concentrated solution, aside from thermal fluctuations of ions, there are other fluctuations and their complexes that cannot be estimated by Maxwell–Boltzmann statistics only (2.1). There are fluctuations of concentrations of complexes; fluctuations of orientational, induction, and dispersion interactions; supramolecular weak chemical bonds that occur with participation of interparticle hydrogen bond and overlapping of electron clouds associated with manifestation of the Pauli principle; short-range repulsion forces; and a few others in which description takes place on the quantum level. Taking into account abundance of the described interactions, Fermi–Dirac (2.3) and Bose–Einstein (2.5) quantum statistics were used aside from Maxwell–Boltzmann statistics. The new statistics for concentrated electrolyte solutions (2.9) is obtained on their basis.

The author used the scheme of concentration structural transition toward high concentrations depending on dielectric permittivity values (every value of dielectric permittivity carries certain concentration of the electrolyte in a solution) throughout the stated material. The scheme, given in Figure 1.1, was essentially based on precision experiments and it may still need experimental refinement, but others are not available in the literature.

The scheme of concentration structural transition toward high concentrations depending on dielectric permittivity values became a component part of the obtained supramolecular host–guest structure distribution function in Chapter 2. Further, having taken quantum statistics for concentrated electrolyte solutions (2.9) as the basis, Maxwell's equation was used for electrostatic fields (2.34), which describes the ratio between electric field intensity and electric charges of ions created by these charges and by charge density. Issues concerning drawbacks and inapplicability of using Coulomb's law instead of Maxwell's equation are presented in Section 2.6. In consequence of solving, the potential of supramolecular interactions was obtained under the absence of external fields (2.79). The potential structure is described in detail in Chapter 2.

The potential of supramolecular interactions obviously contained only electrostatic part of forces: $\kappa\beta^2\Psi_{el}^{0}$ (2.79). The rest of the interactions, $U_{sp}$, $U_{vv}$, and $U_x$, were to be defined. In order to define them, thermodynamics of super- and supramolecular interactions was preliminary developed in Chapter 3. According to the provisions of the supramolecular-thermodynamic approach, Section 2.2, the activity theory was chosen for this purpose, which provides an extremely convenient tool for processing experimental data and finding new regularities as once was pointed by Lewis, Bronsted, and their learners and successors. Results obtained by them demonstrate as well that the activity coefficient is a real physical quantity, where value does not depend on the determination method and represents the temperature and concentration function. This approach was also used by Debye and Hückel. When describing electrolyte solution properties from the supramolecular-thermodynamic point of view, it is absolutely not important in what degree the electrolyte is dissociated because this description does not reveal the mechanism of interparticle (super- and supramolecular) interactions. Taking into account these provisions of Chapter 3, all interactions are included into the supramolecular-thermodynamic description of properties of activity coefficients depending on concentration: electrostatic, $U_{sp}$, $U_{vv}$, and $U_x$. Additionally, other macroscopic thermodynamic ratios in the host solution were obtained: minimum interaction distance of electrostatic forces among ions, distance among complexes, and dielectric permittivity depending on concentration that will be repeatedly used during further narration.

Let us go back to interactions $U_{sp}$, $U_{vv}$, and $U_x$. van der Waals interaction forces are obtained based on using interaction energy in the first approximation of the perturbation method and statistical averaging in Chapter 4. Supermolecular interactions are derived by not changing commonness of the approach. Attraction, provoked by chemical interactions, that is, those that are accompanied by collectivization of electrons, in other words weak hydrogen bond, starts increasing at small distances between water dipoles. The most efficient formulas were studied in Section 4.6. When choosing repulsion potential, attention was paid to those that were obtained by means of quantum-chemical calculations. The model of point center of repulsion was used for electrolyte solutions. In order to obtain real calculation formulas, the provisions of thermodynamics from Chapter 3 were applied to all stated interactions. Numerical comparisons of contributions to the solution were conducted for every interaction and majority of them were selected. Based on experimental data of activity coefficients, adjustment of potential of supramolecular interactions is made, and values of empirical coefficients are obtained, which are required for studying nonequilibrium phenomena of Section II. All derived expressions are investigated based on experimental data and can be found in the corresponding tables in the annex.

All provisions of Section II are based on all provisions from Section I or result from them.

Chapter 6 investigates many theoretical works dedicated to electrical conductivity. The research demonstrated that despite certain successes in the development of theoretical concepts, the contemporary theory of electrolyte solutions is underdeveloped and, to that end, the problem of theoretical consideration of electrical conductivity of electrolyte solutions is still topical. Proceeding from the general continuity equation of Section I for electrolyte solutions in the perturbed state, the general continuity equation (5.26) was obtained for nonstationary fields in its final version. In order to solve those specific problems, the equation can be attributed to a different form concerning not only electrical conductivity but also viscosity and diffusion. The differential equation was derived for electrical conductivity, which takes into account the general theory of supermolecular forces outlined in Section I. The expression for relaxation force and electrophoretic effect was obtained on its basis being a component part of the equation of electrical conductivity of electrolytes at weak electric fields and low frequencies. This equation is investigated based on the experimental data set forth in [1].

The analysis of multiple publications for semiempirical and theoretical approaches is conducted for viscosity. The solutions theory pays much attention to viscosity study, but its consistent imperfection is that one can meet conflicting data on viscosity values defined for the same systems. It tells about insufficiently profound attraction of the issues of structural formation and various super- and supramolecular forces into many theories. The issue of explaining viscosity processes based on the potential of supramolecular interactions, presented in Section I, is still relevant. Chapter 6 provides the model of solution shear flow different from generally accepted approaches. This substantially simplified solving of the differential equation obtained on the basis of the general continuity equation for nonstationary fields in Chapter 5. The derived viscosity equation is investigated based on the experimental data set forth in [2].

Chapter 7 of Section II considers isothermal isobaric molecular diffusion without gradients conditioned by external-force field. The existing publications concerning current phenomenological diffusion models are analyzed. The analysis of publications shows that the common phenomenological approach toward studying thermodynamically equilibrium states of the system is underdeveloped; there are no concepts that can explain diffusion processes from unified positions. Predictive capabilities of existing models are limited by a narrow field of their applicability. That is why investigation of the diffusion phenomenon from the positions of supramolecular interactions is still an important task. The quantitative analysis of diffusion flows is made and it is demonstrated that they depend on activity coefficients. Note that activity coefficients depend on electrostatic forces and various supramolecular and weak chemical interactions. The general theory of supermolecular forces in diffusion flows shows that they depend as well on electrophoretic effect, electrical conductivity, and viscosity and to a great extent on dielectric permittivity of the solution. All these interactions and thermodynamic values are considered in Section I of our publication. This diffusion equation is investigated based on the experimental material set forth in [2].

## REFERENCES


1. Aseyev G.G. 1998. *Electrolytes: Transport Phenomena. Calculation of Multicomponent Systems and Experimental Data on Electrical Conductivity*. Begell House, New York, Inc., p. 612.
2. Aseyev G.G. 1998. *Electrolytes: Transport Phenomena. Methods for Calculation of Multicomponent Solutions, and Experimental Data on Viscosities and Diffusion Coefficients*. Begell House, Inc., New York, p. 548.

# Index

## W